NICOLA SCAFETTA

THE FRONTIER OF CLIMATE SCIENCE

Solar variability, natural cycles and model uncertainty

Centro di Ricerca
Previsione Prevenzione e Controllo
dei Rischi Geologici - Ceri

Sapienza
Università di Roma

International Association for Gondwana Research
IAGR

The Frontier of Climate Science

Solar variability, natural cycles and model uncertainty

Author: Nicola Scafetta

First Edition: 2026

Images: reproduced in grayscale

ISBN: 9791298617605

Updated with minor editorial corrections (05/27/2026).

This publication is endorsed by the *International Association for Gondwana Research* (IAGR, https://iagrglobal.org/) and by the *Centre for Research on Prediction, Prevention, and Mitigation of Geological Risks* (*"Centro di Ricerca Previsione, Prevenzione e Controllo dei Rischi Geologici"*) — CERI, Sapienza University of Rome (https://ceri.web.uniroma1.it/). The official logos of IAGR and CERI are reproduced herein with permission, in recognition of their role as institutional partners in this initiative. The scientific content, grounded in peer-reviewed literature, has been assessed for coherence with the principles of geoethics and with the interdisciplinary standards that characterize contemporary Earth Science research. The interpretations and conclusions presented in this work are solely those of the author and do not necessarily represent the official positions or views of IAGR or CERI.

Published by
Complexity Insight Press
Gaeta (LT), Italy
Email: ComplexityInsightPress@gmail.com

The back-cover figure compares the evolution of global lower-troposphere temperatures based on satellite measurements with the predictions of global climate models (SSP2-4.5 scenario). On average, the data show a warming trend nearly half as large as that simulated by the models.

Contents

Foreword by Prof. M. Santosh

A window into the climate change conundrum

Earth has the unique status in the Solar System as a planet having the combination of land (continents), oceans, and atmosphere. Continents (supplying nutrients and serving as the home for various ecosystems), oceans (functioning as the lungs of the planet, regulating climate, and acting as the largest carbon sink), and atmosphere (providing life breath, maintaining stable temperature, and shielding harmful solar radiation) define the trinity of our "*Habitable Planet*".

In the past 4.6 billion years of its history, Earth's climate was significantly influenced by geodynamic factors including solar luminosity, plate tectonic cycles, volcanic activities, and Milankovitch cycles. Thus, the formation, evolution, and disruption of continental assemblies (termed as supercontinent cycles), complemented by the opening and closing of major oceans (termed as the Wilson cycle), significantly influenced the climatic patterns. There were several alternating warm and cool (glacial and interglacial) periods including the well-known Neoproterozoic "*Snowball Earth*" with a thick blanket of ice covering the globe. It is estimated that there were at least eight cycles of ice ages and warmer episodes during the past 800,000 years.

However, the globe getting warmer in the last few decades, purportedly at a rate faster than through natural processes, is largely attributed to greenhouse gas emissions through the burning of fossil fuels, deforestation and other activities of modern human civilization.

Defined under the broader term of climate change, global warming has emerged as one of the focal themes in scientific and political spheres. Although geological events in the past have caused catastrophic punctuations in the planet's climate patterns, there is growing concern over the imminent challenge to the sustainability of the environment and our civilization, particularly in light of booming population growth as well as the rapidly depleting non-renewable natural resources. Thus emerged the concept of "*Net Zero 2050*", aimed at abolishing fossil fuels and transitioning to green energy on a global scale. The objectives of this target place heavy demand on the limited critical metal resources on Earth that are essential for various applications in green energy technology, compounded by the heterogeneous distribution of these resources, thus triggering some of the current political crises.

In this backdrop, this book—authored by Nicola Scafetta on "*The Frontier of Climate Science – Solar Variability, Natural Cycles and Model Uncertainty*"—offers an excellent window into the deep realms of climatology, complex systems physics, and astronomy in addressing three major aspects: (1) how well do we truly understand Earth's climate? (2) what natural forces remain beyond our grasp? (3) is Net-Zero the only viable path forward?

The book is divided into seven major sections as: (1) Climate Science According to the "Scientific Consensus"; (2) Critical Perspectives on the Anthropogenic Global Warming Theory; (3) New Perspectives for Quantifying Climate Change; (4) The Role of the Sun: Alternative Perspectives on Climate Change Attribution; (5) Simulating Climate Change with Empirical and Semi-Empirical Models; (6) Spectral Coherence Between Planetary Harmonics and Solar–Climate Cycles; (7) Toward a New Paradigm.

Through 42 Chapters of excellent narration and elegant presentation, the author Nicola Scafetta, who is among the *"Top 2% Most Influential Scientists in the World"* (Stanford–Elsevier citation database), provides an insightful evaluation of the prevailing concept of anthropogenic causes of global warming, and reveals the uncertainties that plague the climate models and rhetorical strategies, particularly those pivoted by IPCC. From an authoritative analysis, the author formulates insightful perspectives that demystify the exclusive attribution of global warming in the last century to human activities, and places more importance on the dynamic interplay of terrestrial and cosmic forces. This work is an excellent window into climatology as a dynamic science, and calls for adaptive strategies grounded in economic sustainability and social equity to address climate change issues.

I wholeheartedly recommend this book to all libraries, academic institutions, and also to the general public who are interested in the topic of global climate change. Nicola Scafetta's book serves not only as an important reference for researchers and students, but is also a guide for policy makers and the public media.

M. Santosh
Professor of Earth Sciences
Founder and Secretary General of the
International Association for Gondwana Research (IAGR)
Founding Editor-in-Chief of *Gondwana Research*
Editorial Advisor of *Geoscience Frontiers*
China University of Geosciences (Beijing)
Beijing, China

Foreword by Prof. Alberto Prestininzi

Nicola Scafetta, author of *The Frontier of Climate Science*, has been active for over twenty years in the field of theoretical physics applied to complex systems, with particular focus on climatology, meteorology, and the interactions between terrestrial and solar dynamics. His approach is not that of an experimental scientist engaged in laboratory or fieldwork, but rather that of a theoretical physicist who, in the spirit of Galilean science, analyzes observational data to construct, step by step, a holistic and coherent interpretation of the climate system. It is within this framework that complex systems physics finds its place: a discipline that does not settle for reductionist simplifications, but instead seeks emergent regularities, spectral coherences, and interpretative models capable of integrating multiple temporal scales and forcing mechanisms.

The topic of anthropogenic global warming is addressed through a series of fundamental questions that every scientist should ask before formulating hypotheses and proposing solutions: How well do we truly understand Earth's climate? What natural mechanisms still elude us? Are the *"Net Zero"* policies promoted by many governments consistent with the dynamics that govern the climate system?

Answering these questions is essential to assess whether the elimination of entire industrial sectors—with severe repercussions on employment and the transfer of vast financial resources to powerful financial organizations, thereby undermining social achievements won through decades of struggle—is truly necessary.

In *The Frontier of Climate Science*, Scafetta constructs a theoretical and didactic journey that guides the reader through the multiple dimensions of the climate system. The book is conceived as a critical dialogue in which the processes that govern Earth's climate—many of which remain poorly understood or underestimated—are examined in depth. The goal is to distinguish facts from rhetoric, restoring to science its role as a pluralistic, iterative, and non-dogmatic inquiry.

Within this journey, astronomy and astrophysics emerge as key players, interacting with the endogenous dynamics of our planet. Like all celestial bodies in the solar system, Earth receives its primary energy from the Sun. Yet its uniqueness lies in the simultaneous presence of water, landmasses, and a complex atmosphere—formed through geodynamic and physico-chemical processes—that made life possible.

Throughout geological history, complex processes have alternated over millions of years, shaping the current features of Earth's surface. Studying these features allows us to reconstruct past climatic conditions and understand the variations that have occurred over time. It is in this context that the great milestones of geoscience are situated, such as Alfred Wegener's theory of Plate Tectonics and Milutin Milankovitch's orbital cycles, alongside a vast and growing body of scientific literature, as demonstrated by numerous recent studies.

Scafetta's work fits fully within this long trajectory of scientific inquiry, but with a theoretical and systemic perspective. He draws attention to the precarious position of those, such as the Intergovernmental Panel on Climate Change (IPCC), who attribute global warming solely to human activity, starting from the end of the nineteenth century. This attribution is critically examined and compared with geological, paleoclimatic, and modeling evidence, using empirical and semi-empirical tools and integrating data from meteorology, geophysics, and planetary sciences.

The Frontier of Climate Science is divided into seven parts, each addressing a crucial aspect of the climate debate: the vision of "scientific consensus" and the architecture of the IPCC; the epistemological and modeling limitations of the anthropogenic theory; the revision of observational data and systematic biases; the role of the Sun and alternative solar forcings; empirical and harmonic models as simulation tools; the spectral coherence between planetary, solar, and climatic cycles; and the proposal of a new scientific paradigm that is open and pluralistic.

In recent years, the scientific debate on climate has begun to show signs of a more nuanced and open evaluation of the uncertainties and structural limitations of traditional climate models. The recent report *A Critical Review of Impacts of Greenhouse Gas Emissions on the U.S. Climate* (U.S. Department of Energy, 2025), authored by the DOE Climate Working Group—John R. Christy, Judith A. Curry, Steven E. Koonin, Ross McKitrick, and Roy W. Spencer—reflects this shift. Although developed independently, the report acknowledges many of the issues that Scafetta has long investigated and examines in this volume, and it frequently cites his scientific contributions, particularly regarding the representation of natural variability, the divergence between observations and simulations, the role of solar variability, and the need for a more pluralistic approach to climate modeling. This convergence of perspectives, emerging from distinct analytical paths, underscores the timeliness and relevance of the framework presented in *The Frontier of Climate Science*.

The Frontier of Climate Science is thus a work that invites reflection, verification, and debate. It does not offer absolute certainties, but rather tools to interrogate reality with critical thinking and intellectual openness. It is a valuable contribution for those who believe that science must be a process, not a dogma.

Alberto Prestininzi
Full Professor (retired) of Applied Geology
Founder and former Director of the
"Centro di Ricerca Previsione, Prevenzione e Controllo dei Rischi Geologici" (CERI)
Founding Editor and former Editor-in-Chief of the
Italian Journal of Engineering Geology and Environment
Sapienza University of Rome
Rome, Italy

Foreword by Prof. Judith Curry

Climate change is one of the most polarizing subjects in global public discourse. The challenge arises from the deep complexity of the issue, which has scientific, political, cultural, and ethical dimensions, with public perceptions shaped through the lens of a polarized media. In the zeal to eliminate fossil fuels, certain strands of climate science have drifted from critical inquiry, becoming instruments for legitimizing authority and advancing preferred policies. In short, politics has recast a complex scientific issue as an ideological tool.

The concept of the climate "consensus" has been weaponized as a form of social and political pressure, used to suppress debate and discourage intellectual pluralism. While the so-called "consensus" offers a clear and actionable narrative, there's growing realization that both the climate change problem and its solution have been vastly oversimplified. The entire debate has been stripped of depth, rigor, and nuance. Some of the challenges to the mainstream discourse on climate change have been rightfully characterized as "*climate denial*"—equally devoid of rigor and also driven by a political agenda.

Beyond the polarizing rhetoric of "*climate catastrophism*" and "*climate denial*" stands the more rigorous, methodologically grounded perspective of skeptical scholars. Skeptical scholars regard science as a process, not dogma—they don't advocate overthrowing anthropogenic global warming theory for ideological reasons, but rather work to properly test it. Skeptical scholars focus on concerns about the quality of the data and its analysis, the utility of global climate models, and the ideological framing of climate science and policy. Driven by curiosity rather than conformity, they recognize that science is built not on authority but on evidence. They understand that science works by examining assumptions, questioning hypotheses, and evaluating uncertainties.

Following in the tradition that includes recent books such as Steve Koonin's *Unsettled* and my book *Climate Uncertainty and Risk*, the early chapters of Nicola Scafetta's book *The Frontier of Climate Science* challenge key elements of anthropogenic global warming theory. Scafetta clearly justifies his arguments that the exclusive attribution of recent warming to anthropogenic emissions is scientifically premature and that climate models omit or oversimplify critical climate dynamical processes.

Even in the face of serious criticism, flawed theories can persist when there is no conceptually robust alternative. What has been lacking is an alternative theory of climate change that is supported by observations and goes beyond vague attributions to natural climate variability. The seminal contribution of *The Frontier of Climate Science* is a new scientific paradigm that provides a broader interpretive framework capable of resolving the inconsistencies of the current anthropogenic climate change model.

Arguably, the most egregious oversimplification in climate science has been the marginalization of solar variability. Solar variability and its role in climate change remain among the most profound and unresolved issues in contemporary climate science. Scafetta makes a compelling argument that it is time to bring the Sun back to the center of climate discourse. If, as multiple lines of evidence suggest, solar activity has played a significant role in recent global warming, then the global climate forcing datasets and climate models may be structurally misleading, and the attribution of all recent warming to anthropogenic emissions is incorrect.

Beyond solar variability, Scafetta proposes a dynamic interplay of a broad range of forcings—both natural and anthropogenic—operating across a range of spatial and temporal scales in a dynamic, interconnected web of terrestrial and cosmic forces. This is a framework in which solar variability, planetary harmonics, and natural internal climate variability are integral components of the Earth's climate architecture.

Under Scafetta's vision, climate science reclaims its astronomical dimension, governed by gravitational rhythms, orbital geometries, and resonant harmonics that are only partly understood. Scafetta's harmonic planetary approach is not so much a rival to conventional climate models, but rather a complementary and empirically grounded extension. He presents an empirical climate model and attribution framework, whereby more than half of the modern warming is caused by natural and local processes that are underrepresented or omitted in current global climate models.

A healthy scientific culture embraces pluralism, methodological rigor, and open dialogue. Only through this lens can climate science remain credible, adaptive, and truly informative. For a scientific community genuinely committed to open inquiry, Scafetta's framework offers a valuable opportunity for engagement—whether through the refinement of its premises or their rigorous contestation.

The Frontier of Climate Science is a very important contribution toward grasping the true complexity of the Earth's climate. Anchored in the history and philosophy of science, with at times a poetic feel to its narrative, this book provides a crucial step toward shaping a more realistic and mature approach to climate science.

Scafetta says it best in the closing statement of the book:

> *"We may discover that Earth's climate is not merely a product of internal feedbacks and anthropogenic forcings, but a resonant expression of the solar system's deeper architecture."*

Judith Curry
Co-founder and President of *Climate Forecast Applications Network*
Professor Emerita of Atmospheric Sciences and Climatology and
former Chair of the School of Earth and Atmospheric Sciences
Georgia Institute of Technology
Atlanta, Georgia
United States

Introduction

Over the past several decades, climate change has become one of the most prominent and debated subjects in global public discourse. Institutions such as the *Intergovernmental Panel on Climate Change* (IPCC), together with governments, academic bodies, media outlets, and environmental organizations, have increasingly emphasized the role of human activity in driving global warming since the late 19th century. This framing has contributed to the perception of climate change as a critical challenge for future generations, prompting calls for urgent and far-reaching mitigation policies. It has also shaped collective perceptions of climate risk and supported major policy decisions, including large-scale investments in energy transition strategies and the promotion of environmentally responsible behavior.

In parallel, a minority of scholars and professionals have offered alternative perspectives, questioning the magnitude of future warming projected by certain global climate models (GCMs) and advocating a more nuanced interpretation of climate data. These voices highlight the importance of preserving scientific pluralism and of critically examining the assumptions underlying climate projections and policy responses. A similar re-evaluation is reflected in the recent 2025 report by the U.S. Department of Energy, which acknowledges key uncertainties in current climate modelling and attribution frameworks.[1]

Among these contributors are Judith Curry, former director of the School of Earth and Atmospheric Sciences at the Georgia Institute of Technology, who in her 2023 essay *Climate Uncertainty and Risk: Rethinking Our Response* proposed a reassessment of climate strategies. Steven Koonin, theoretical physicist and former U.S. Undersecretary for Energy, emphasized in *Unsettled* (2024) the distinction between the cautious language of scientific reports and the more emphatic tone often found in public discourse. Richard Lindzen, emeritus professor at MIT, has long discussed the uncertainties surrounding climate sensitivity to CO_2 and the limitations of numerical modelling. William Happer, physicist at Princeton University, has likewise argued for a balanced evaluation of the role of CO_2 in climate dynamics.

In Italy, the scientific community has contributed to this debate through initiatives such as the climate-related hearing given by Nobel laureate Carlo Rubbia before the Italian Senate in 2014,[2] and the 2019 *Petition on Anthropogenic Global Warming* signed by over 180 scholars, including renowned scientists

[1]Christy, J.R., Curry, J.A., Koonin, S.E., McKitrick, R., Spencer, R.W., 2025. A critical review of impacts of greenhouse gas emissions on the U.S. climate. U.S. Department of Energy, Washington, DC. Available at: https://www.energy.gov/topics/climate

[2]Available at: https://www.youtube.com/watch?v=ooBPJeDp6k8

such as Antonino Zichichi, Renato Angelo Ricci, Giuliano Panza, Franco Prodi, and Uberto Crescenti.[3] These efforts advocate a cautious and evidence-based approach to environmental policy. The 2022 publication *Dialogues on Climate: Between Emergency and Knowledge*, edited by Alberto Prestininzi, further explores these themes through contributions from sixteen Italian academics.

Beyond the strictly scientific domain, other commentators have examined the socioeconomic implications of climate policy. Bjorn Lomborg, in *False Alarm* (2023), discusses the potential impact of climate strategies on vulnerable populations and proposes pragmatic alternatives. Michael Shellenberger, in *Apocalypse Never* (2020), explores the role of nuclear energy in sustainable development. The documentary *Planet of the Humans* (2019),[4] directed by Jeff Gibbs and produced by Michael Moore, raises questions about the environmental and economic viability of certain green technologies. More recently, *Climate: the Movie (the cold truth)* (2024) has contributed to the public debate.[5] Patrick Moore, co-founder of Greenpeace, has also expressed concerns about the politicization of environmental discourse.

At the European level, the Clintel Foundation (Climate Intelligence),[6] founded in 2019 by geophysicist Guus Berkhout and journalist Marcel Crok, promotes open scientific dialogue on climate issues. While acknowledging the reality of global warming and its possible anthropogenic influences, Clintel calls for a measured approach to climate policy. Its *World Climate Declaration*, signed by approximately 2,000 scholars—including Nobel laureates Ivar Giaever and John F. Clauser—emphasizes the importance of preserving scientific rigor and pluralism. Clintel's 2023 publication *The Frozen Climate Views of the IPCC*[7] offers a scientific critique of prevailing climate narratives.

This work builds on ideas I first presented at the U.S. Environmental Protection Agency's National Center for Environmental Economics (NCEE) on 26 February 2009, during my invited lecture *Climate Change and Its Causes: A Discussion about Some Key Issues*,[8] delivered while I was a researcher in the Physics Department at Duke University (NC, USA) and collaborated with Richard Willson of the ACRIM team on the study of total solar irradiance within the NASA

[3]Antonino Zichichi, Professor Emeritus of Physics at the University of Bologna, Founder and President of the Ettore Majorana Center for Scientific Culture in Erice; Renato Angelo Ricci, Professor Emeritus of Physics at the University of Padova, former President of the Italian Physical Society and of the European Physical Society; Giuliano Panza, Distinguished Professor of Geophysics at the University of Trieste, Academician of the Lincei and of the Italian National Academy of Sciences, known as the XL; Franco Prodi, Distinguished Professor of Atmospheric Physics at the University of Ferrara, Academician of the Italian National Academy of Sciences, known as the XL; Uberto Crescenti, former President of G. D'Annunzio University, Chieti-Pescara, and former President of the Italian Geological Society. The petition is available at: https://www.scienzanazionale.it/e-nata-astri-per-la-ricerca-italiana/petizione-sul-riscaldamento-globale-antropico/

[4]Gibbs, J. (Dir.), Moore, M. (Exec. Prod.), 2019. Planet of the Humans [Film]. Rumble Media.

[5]Martin Durkin (director), Tom Nelson (producer). Available at: https://climatethemovie.net/

[6]Climate Intelligence (Clintel): https://clintel.org/.

[7]Crok, M., May, A. (Eds.), 2023. The Frozen Climate Views of the IPCC. Clintel Foundation. Available at: https://clintel.org

[8]The presentation (pdf and video) is available at: https://web.archive.org/web/20091220063025/http://yosemite.epa.gov/ee/epa/eed.nsf/vwpsw/360796B06E48EA0485257601005982A1

Earth Sciences ACRIMSAT/ACRIM3 project. The present volume further develops those ideas in light of the scientific research I have carried out in the years since.

Scientific foundations and methodological challenges

Climate science is structured around two core tasks: identifying climatic shifts (*"detection"*) and determining their origins (*"attribution"*).

Detection refers to the process of measuring climate data and statistically identifying significant changes within the system, independently of their underlying causes. It represents the initial step in recognizing that a shift has occurred beyond the bounds of short-timescale natural variability. For example, a rise in global average temperature may be considered "detected" when the probability of its occurring by random chance is sufficiently low, typically below 10%.

Attribution seeks to identify the physical causes behind observed changes by evaluating the relative contributions of anthropogenic factors (such as greenhouse gas emissions) and natural influences (such as solar variability, volcanic activity, and others). Since direct experimentation on the climate system is not feasible, attribution relies on climate models and statistical inference, comparing observed outcomes with hypothetical scenarios that exclude human influence. This process entails varying degrees of uncertainty and often depends on expert judgment, particularly in the case of extreme events. The complex interactions within the climate system, combined with limited observational data, make attribution a persistent and evolving scientific challenge.

Several attribution methods are employed to distinguish between natural and anthropogenic drivers of climate change. Among the most widely used is optimal fingerprinting, which applies linear regression and other statistical techniques to compare observed data with model simulations, with the aim of isolating human influence. Time series analysis examines temporal patterns in order to identify causal relationships between natural variability and anthropogenic forcings. Process-based attribution focuses on the physical mechanisms underlying specific changes, and is particularly useful for regional phenomena such as monsoons. Extreme event attribution evaluates how human activity may have influenced the likelihood or intensity of specific events, using probabilistic and narrative approaches to construct counterfactual scenarios.

Both detection and attribution are subject to inherent uncertainties. Detection requires reliable and representative data capable of revealing significant changes, while attribution demands rigorous interpretation of those data in order to determine causality. As a result, many aspects of climate science remain under active investigation, owing both to the complexity of the phenomena involved and to the methodological limitations in collecting and analyzing the evidence.

While acknowledging the reality of global warming and the role of human influence, some scholars—including those cited earlier—adopt a critical stance focused on epistemological and methodological aspects. They distinguish their position from the more polarized views often found in public discourse, which may simplify or exaggerate scientific findings. These scholars stress the importance of maintaining scientific rigor and transparency, particularly in the communication of uncertainties and probabilistic outcomes.

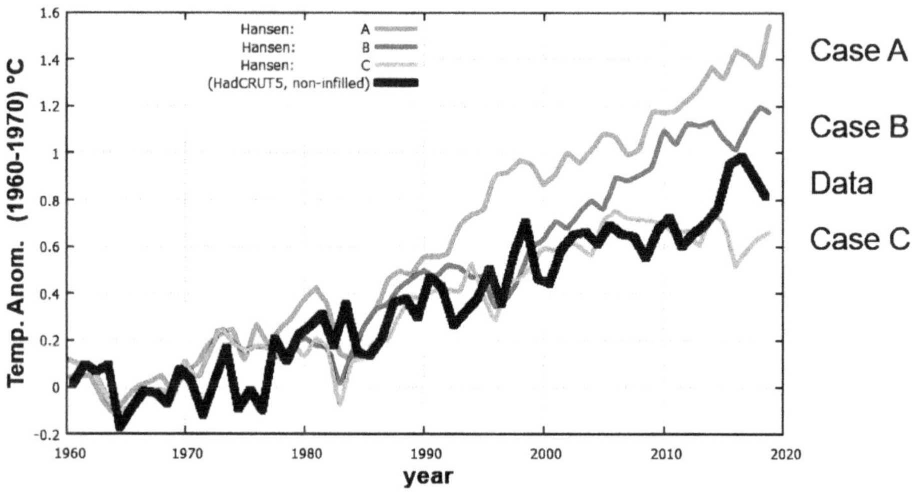

FIGURE I.1. Comparison between the three climate projections presented by James Hansen to the U.S. Congress in 1988 (Scenarios A, B, and C) and a recent reconstruction of global surface temperatures based on the HadCRUT5 dataset Morice et al. (2021). The figure highlights the divergence between Hansen's predictions and the observed data: the black curve, representing actual temperatures, aligns most closely with Scenario C, which assumed a drastic reduction in anthropogenic emissions starting in 1990, whereas real-world emissions have followed a trajectory between Scenario A and Scenario B.

Genuine scientific critiques focus on the reliability of model projections, the selective use of data, and the framing of climate science in media and policy contexts. From this perspective, there is a growing disconnect between the nuanced language of the scientific literature—expressed in terms of probabilities, confidence intervals, and scenarios—and the public narrative, which may translate these into definitive conclusions and urgent alarms.

As with any developing scientific discipline, climate science relies on mathematical models, numerical simulations, indirect estimates, and imperfect measurements, all of which involve margins of uncertainty. Moreover, future climate projections depend on multiple interconnected and difficult-to-predict variables, including technological innovation, policy evolution, demographic trends, feedback mechanisms, and natural variability.

Natural variability itself is influenced by solar and astronomical factors that remain only partially understood and are not comprehensively represented in current climate models. According to some authors, the limited integration of solar and astrophysical knowledge—often a consequence of disciplinary specialization—may constrain the scope of climate analyses.

Despite advances in modeling and the publication of numerous reports, the IPCC has acknowledged that, from 1988 to 2021, it has not been possible to narrow the uncertainty range of the *Equilibrium Climate Sensitivity* (ECS), the long-term increase in global surface temperature expected from a doubling of atmospheric CO_2.

Climate sensitivity remains one of the most debated parameters in climate science, owing to its central role in determining future warming. Surveys and scientific publications have raised questions about the reliability of high-sensitivity models, often referred to as *"hot models,"* suggesting that they may overestimate the rate and magnitude of warming and the associated hazards.

One frequently discussed issue concerns the use of extreme scenarios such as SSP5–8.5, sometimes described as *"business-as-usual."* This scenario projects exponential growth in fossil fuel emissions, tripling current levels by the end of the 21st century. A growing number of experts consider SSP5–8.5 unlikely under current technological and policy trajectories, and suggest that it may be more suitable for stress-testing than for baseline forecasting.

According to official climate reconstructions, global surface temperature has increased by approximately +1.1 °C from the 1850–1900 preindustrial period to the decade 2011–2020. While this represents significant warming, it is lower than some early projections made in the 1990s.

For instance, the projections presented by James Hansen to the U.S. Congress in 1988[9] anticipated a rise in global temperatures by 2020 that exceeded observed values by roughly one third. Hansen's testimony, delivered on June 23, 1988, preceded the formal establishment of the IPCC later that year.

Models, observations, and uncertainties

Figure I.1 compares Hansen's three 1988 projections (Scenarios A, B, and C) with a reconstruction of global surface temperatures based on the HadCRUT5 dataset (Morice et al., 2021).[10] The black curve, representing the observed data, aligns most closely with Scenario C, which assumed a substantial reduction in anthropogenic emissions beginning in 1990 and the achievement of net-zero by 2000.

However, from 1988 to 2020 global CO_2 emissions have steadily increased along a trajectory between Scenarios A and B, with cumulative growth closer to A. The fact that observed temperatures align more closely with Scenario C—despite higher-than-assumed emissions—invites reflection on the climate sensitivity embedded in Hansen's original models and on the complexity of the relationship between anthropogenic emissions and global warming. Natural mechanisms not fully represented in early models, such as oceanic oscillations (e.g., ENSO – *El Niño Southern Oscillation*; AMO – *Atlantic Multidecadal Oscillation*; PDO – *Pacific Decadal Oscillation*) and solar variability, may have contributed to modulating the observed warming.

Cryosphere, sea level, and extreme events

Discrepancies between model projections and observational data continue to motivate scientific inquiry into the robustness of long-term climate forecasts.

[9] James Hansen's testimony before the U.S. Senate Committee on Energy and Natural Resources on June 23, 1988: https://www.sealevel.info/1988_Hansen_Senate_Testimony.html

[10] Morice, C.P., Kennedy, J.J., Rayner, N.A., Winn, J.P., Hogan, E., et al., 2021. An updated assessment of near-surface temperature change from 1850: the HadCRUT5 data set. Journal of Geophysical Research: Atmospheres, 126, e2019JD032361.

NSIDC 12-month averaged records (1979–2025) show a persistent decline in Arctic sea-ice extent, but with a marked slowdown after the mid-2000s, deviating from the more linear decreases anticipated by several climate models (England et al., 2025).[11] The Southern Hemisphere displays even greater complexity: Antarctic sea ice alternates between phases of expansion and contraction, with a modest increase until the mid-2010s followed by a sharp decline. This irregular evolution—already noted in earlier analyses (Scafetta & Mazzarella, 2015a)[12]—suggests that Antarctic sea-ice dynamics are strongly shaped by internal variability, ocean–atmosphere coupling, and regional geophysical processes. These mechanisms remain only partially represented in current climate models, contributing to persistent uncertainties in long-term projections.

Global mean sea level is rising, with satellite altimetry data since 1993 indicating an increase of approximately 3.0–3.5 mm/year, primarily due to ice melt and thermal expansion (Cazenave & Moreira, 2022).[13] However, regional variability challenges the notion of a uniform trend. Geological uplift and subsidence affect local measurements. In geologically stable regions such as the Skagerrak, between Denmark and southern Norway, tide-gauge data indicate a rise of about 1.0–1.5 mm/year, with no significant acceleration (Voortman & De Vos, 2025).[14] When corrected for vertical land motion, similar patterns are observed in other coastal cities. In New York, for instance, sea level has risen at about 3 mm/year since 1900, but with an estimated 1.5 to 2.0 mm/year of this rise attributed to urban subsidence (Boretti, 2025; Parsons et al., 2023).[15]

The complexity of sea level change arises mainly from its dual nature: a steric component, driven by temperature and salinity variations, and a mass component, resulting from the addition of water to the oceans. When temperature increases but salinity decreases—for example, when freshwater from rivers or melting glaciers enters the ocean—the steric effect may be reduced or even neutralized. On a global scale, satellite altimetry estimates suggest that thermal expansion contributes approximately +1.4 mm/year to sea level rise, while mass addition from ice melt and terrestrial water storage accounts for about +1.9 mm/year (Cazenave & Moreira, 2022). In regions where water warming is offset by freshening, the steric contribution may approach zero, causing the local sea level trend to fall below the estimated global average of +3.3 mm/year.

However, satellite altimetry estimates are subject to several methodological limitations that affect their reliability, particularly in coastal and tectonically

[11] England, M.R., Polvani, L.M., Screen, J., Chan, A.C., 2025. Minimal Arctic seaice loss in the last 20 years, consistent with internal climate variability. Geophysical Research Letters, 52, e2025GL116175.

[12] Scafetta, N., Mazzarella, A., 2015. The Arctic and Antarctic sea-ice area index records versus measured and modeled temperature data. Advances in Meteorology, 481834, 1–8.

[13] Cazenave, A., Moreira, L., 2022. Contemporary sea-level changes from global to local scales: a review. Proceedings of the Royal Society A: Mathematical, Physical and Engineering Sciences, 478(2261), 20220049.

[14] Voortman, H. G., De Vos, R., 2025. A global perspective on local sea level changes. Journal of Marine Science and Engineering, 13(9), 1641.

[15] Boretti, A., 2025. Subsidence: critical factor in local sea level rise assessments. Ocean & Coastal Management, 245, 107646.

Parsons, T.E., Wu, P.-C., Wei, M., D'Hondt, S., 2023. The weight of New York City: possible contributions to subsidence from anthropogenic sources. Earth's Future, 11(5), e2022EF003465.

active regions (Birol et al., 2025; Döhne et al., 2024).[16] Instrumental biases between missions, dependence on geoid models, and the need for atmospheric and wet-tropospheric corrections introduce systematic uncertainties. Moreover, satellite altimetry provides spatially averaged measurements over ocean surfaces and lacks direct geodetic anchoring to the solid Earth, making it difficult to disentangle true sea level changes from vertical land motions. Crucially, there is no direct observational benchmark against which satellite-derived global sea level trends can be independently validated. When compared with tide-gauge records—which are locally anchored but sensitive to land uplift or subsidence—discrepancies often emerge, especially in coastal zones. These challenges underscore the importance of integrated approaches that combine satellite altimetry, tide gauges, GPS geodesy, and gravimetric data to obtain robust and physically consistent estimates of sea level trends.

Historically, sea level rose by approximately 120–130 meters during the transition from the Last Glacial Maximum (about 21,000 years ago) to the present interglacial, with rates peaking at 10–15 mm/year during meltwater pulses. In particular, Meltwater Pulse 1A (MWP-1A), which occurred around 14,600 years ago, produced a rapid rise of approximately 12–22 meters over a few centuries, with peak rates exceeding 40 mm/year—one of the fastest natural rises of the late Quaternary, and a benchmark for evaluating modern trends. These values represent the upper bounds of natural sea level rise and provide a critical paleoclimatic reference for assessing contemporary changes, which remain significantly lower in magnitude.

Public discourse often associates climate change with an increase in extreme weather events. However, the scientific literature adopts a more cautious approach. According to the IPCC Sixth Assessment Report (2021, AR6 WG1 Chapter 12 Table 12.12),[17] for many categories of extreme events—including hurricanes, floods, droughts, and wildfires—the evidence for significant long-term changes or future increases remains classified as "limited" or "low," owing to the dominant role of natural variability.

Studies by Roger Pielke Jr. indicate that, when adjusted for population growth and economic development, financial losses from meteorological disasters do not show a consistent upward trend.[18] The perception of increasing severity is often linked to human expansion into vulnerable areas, inflation, and intensified media coverage rather than to demonstrable climatic changes.

The observed increase in recorded global natural disasters over recent decades does not necessarily indicate a rise in the frequency or intensity of natural hazard events. Alimonti et al. (2022)[19] critically examined the major

---

[16] Birol, F., Bignalet-Cazalet, F., Cancet, M., Daguze, J.-A., Fkaier, W., et al., 2025. Understanding uncertainties in the satellite altimeter measurement of coastal sea level: insights from a round-robin analysis. Ocean Science, 21, 133–150.
Döhne, T., Horwath, M., Thomas, M., 2024. Ocean mass change from GRACE/GRACE-FO and satellite altimetry: updated estimates and uncertainty assessment. EGU General Assembly 2024, Abstract EGU24-288.

[17] IPCC, 2021. Climate Change 2021: The Physical Science Basis. Cambridge University Press.

[18] Pielke Jr., R., 2019. Tracking progress on the economic costs of disasters under the indicators of the sustainable development goals. Environmental Hazards, 18(1), 1–6.

[19] Alimonti, G., Mariani, L., Prodi, F., Ricci, R.A., 2022. A critical assessment of extreme events trends in times of global warming. European Physical Journal Plus, 137, Article 112.

international databases and found no robust evidence of a systematic increase in extreme events attributable to anthropogenic climate change. Their analysis underscores the importance of distinguishing between a hazard—a physical phenomenon—and a disaster, which reflects its societal impact. This conceptual distinction is essential for interpreting long-term trends and for designing effective risk reduction strategies. Indeed, the apparent upward trend in disaster records is largely driven by socio-technical factors, including population growth, urban expansion, enhanced detection capabilities, and more comprehensive reporting practices (Alimonti & Mariani, 2023).[20]

Regarding wildfires, global data indicate a decline in the annual extent of burned forested areas over the past two decades (Chen et al., 2023b),[21] attributed to improved land management, agricultural practices, and early warning systems. While dramatic images from regions such as California, Australia, or Greece receive widespread attention, many experts emphasize the importance of local governance, urban planning, and resource allocation in shaping wildfire outcomes.

Emissions, media, and public discourse

Many industrialized nations have begun to reduce greenhouse gas emissions through the adoption of renewable energy sources, improvements in energy efficiency, and the deployment of low-impact technologies. In this context, the European Union stands out, having recorded a steady decline in emissions over the past thirty years (EDGAR).[22] Part of this reduction is also attributable to the relocation of emission-intensive industrial activities to regions such as Asia, where production has increasingly shifted. As a result, the overall progress achieved in many industrialized nations has been offset by the economic and demographic expansion of developing countries, particularly emerging economies such as China and India. This asymmetry raises important questions regarding climate equity, global geopolitical dynamics, and the feasibility of coordinated decarbonization strategies.[23]

The goal of achieving net-zero emissions within a few decades remains ambitious and uncertain unless supported by transformative technological breakthroughs or structural changes in global economic systems. While such transitions are theoretically possible, many observers consider them politically and economically challenging in the short term.

Recent IPCC reports reflect a more cautious and nuanced tone, acknowledging persistent scientific uncertainties. The language has evolved to include clearer distinctions between levels of confidence and probability. Several chapters explicitly recognize knowledge gaps, limitations in the available data, and interpretive divergences among research groups. This shift suggests a growing

[20] Alimonti, G., Mariani, L., 2023. Is the number of global natural disasters increasing? Environmental Hazards, 22(4), 345–368.

[21] Chen, Y., van der Werf, G.R., Randerson, J.T., Rogers, B.M., Mu, M., Li, F., Giglio, L., 2023b. Multidecadal trends and variability in burned area from the fifth version of the Global Fire Emissions Database (GFED5). Earth System Science Data, 15(12), 5227–5247.

[22] EDGARD: Emission Database for Global Atmospheric Research. https://edgar.jrc.ec.europa.eu/

[23] International Energy Agency (IEA): CO_2 Emissions: Global Energy Review 2025. https://www.iea.org/reports/global-energy-review-2025

awareness of the importance of methodological transparency and of the risks associated with consensus-driven approaches.

Despite this evolution, public discourse—particularly in Western media—often employs emotionally charged language, including terms such as "*collapse*," "*imminent crisis*," or "*we only have ten years to save the planet*." While such rhetoric may increase public engagement, it does not always reflect the complexity of the underlying science. Since the 1980s, warnings about limited time to act have been recurrent. For example, a 1989 UN report raised concerns; in 2006, *An Inconvenient Truth* by Al Gore renewed public attention; in 2009, Gore predicted a 75% chance that the Arctic polar cap could be ice-free during summer months within 5–7 years. Yet in 2021, summer Arctic ice still covered approximately $4.72 \cdot 10^6$ km², a value higher than the 2007 minimum of $4.16 \cdot 10^6$ km².

In 2018, the IPCC emphasized the urgency of limiting warming to 1.5 °C within twelve years. In 2020, initiatives such as TED Countdown reiterated the strategic importance of the 2020–2030 decade.[24] Public figures and activists have adopted vivid metaphors—such as "*Our house is on fire*," popularized by the Swedish activist Greta Thunberg et al. (2020)—to convey a sense of imminent danger.[25] However, numerous websites now catalog climate predictions that have not materialized, highlighting the difficulties of long-term forecasting and the importance of distinguishing between advocacy and scientific projection.[26]

This recurring narrative, while effective at mobilizing attention, may oversimplify a complex issue. The proliferation of alarmist or contradictory messages—sometimes described as a "*climate infodemic*"—can distort public perception and contribute to psychological distress, including eco-anxiety. Such dynamics may hinder society's ability to address climate challenges in a rational and proportionate manner.

Polarization in the public debate is increasingly evident. On one side are those advocating immediate and radical measures, viewing climate change as an existential threat. On the other are those who minimize or question the severity of the problem. This polarization complicates the development of balanced and effective solutions, making it difficult to assess cost–benefit ratios and the practical feasibility of proposed policies.

Recognizing climate change as a serious issue is essential. At the same time, addressing it responsibly requires moving beyond the simplistic dichotomy between "*catastrophists*" and "*denialists*." The climate discussion is not a confrontation between opposing fan bases, but a scientific and societal challenge that demands rigor, competence, and intellectual honesty. Interpretations that lack realism or balanced socioeconomic analysis may lead to unintended consequences, since drastic or poorly designed policies risk disproportionately affecting vulnerable populations and exacerbating existing inequalities. What is

[24]TED Countdown: https://countdown.ted.com/

[25]Thunberg, G., Ernman, M., Ernman, B., Thunberg, S., 2020. Our house is on fire: Scenes of a family and a planet in crisis. Penguin Books.

[26]For example: https://extinctionclock.org/;
https://www.survivalworld.com/outdoors/33-climate-predictions-that-never-happened-a-history-of-inaccuracy/;
https://www.aei.org/carpe-diem/50-years-of-failed-doomsday-eco-pocalyptic-predictions-the-so-called-experts-are-0-50/;
https://freebeacon.com/politics/top-10-catastrophic-climate-predictions-that-failed/.

needed is an analytical approach that promotes critical thinking, open dialogue, and evidence-based evaluation of the available strategies.

Toward a balanced scientific approach

In response to this evolving context, alternative visions are gaining attention. These include policies focused on adaptation, risk management, resilience, and sustainable development driven by technological innovation. Some scholars suggest that a moderately warmer climate, if properly managed, could also offer new opportunities.

For instance, warmer winters may reduce cold-related mortality and lower heating energy consumption, particularly in temperate countries such as the United States and Italy, and in colder regions such as Northern Europe, Canada, and Russia. In net terms, the reduction in winter mortality may outweigh the increase in summer mortality due to heat stress, especially in regions with historically harsh winters, although this balance depends on local adaptation and healthcare infrastructure. From an energy perspective, milder winters could reduce the demand for heating fuels, potentially lowering seasonal consumption peaks and the associated emissions. These benefits may be partially offset by increased electricity use for cooling during hotter summers, particularly in regions not traditionally equipped for high temperatures. Additionally, the retreat of Arctic sea ice may facilitate the opening of new maritime trade routes—such as the Northern Sea Route and the Northwest Passage—with potentially significant economic and strategic implications.

This book aims to restore complexity, proportionality, and methodological rigor to the scientific debate on climate change. It does not deny the reality of global warming, nor the anthropogenic contribution to it. Rather, it seeks to examine the fundamental open questions in climate science through the lens of the peer-reviewed literature (about 650 works), which remains largely unknown to the public. Particular attention is paid to uncertainties in detection and attribution, the limitations of predictive models, and the potential side effects of climate policies based primarily on simulations.

The reader is offered an updated overview of emerging research that may contribute to the development of a new paradigm in climate science. The author has contributed to this research over twenty-five years of work conducted between the United States of America and Italy.[27]

[27] Selection of Scafetta's most recent publications on climate change and solar activity:
Scafetta, N., Bianchini, A., 2022. The planetary theory of solar activity variability: a review. *Frontiers in Astronomy and Space Sciences*, 9, 937930.
Scafetta, N., 2023. CMIP6 GCM ensemble members versus global surface temperatures. *Climate Dynamics*, 60, 3091–3120.
Scafetta, N., 2023. Empirical assessment of the role of the Sun in climate change using balanced multi-proxy solar records. *Geoscience Frontiers*, 14(6), 101650.
Scafetta, N., 2024. Impacts and risks of "realistic" global warming projections for the 21st century. *Geoscience Frontiers*, 15(2), 101774.
Scafetta, N., 2026. Detection, attribution, and modeling of climate change: Key open issues. *Gondwana Research*, 152, 92–128.
Scafetta, N., Bianchini, A., 2026. Planetary modulation of solar and climate oscillations. *Habitable Planet*, 2(1), 46–62.

Part 1

Climate Science According to the *"Scientific Consensus"*

CHAPTER 1

Understanding climate change in order to act

1.1. Between science and politics

Global Climate Models (GCMs) are currently among the most sophisticated numerical tools available to the scientific community for simulating and forecasting the evolution of Earth's climate system. These highly complex mathematical models, run on supercomputers, aim to reproduce the physical, chemical, and biological processes that govern the atmosphere, oceans, land surface, and cryosphere. They thereby enable researchers to analyze and simulate the interactions among these components on both global and regional scales.

Within the collaborative framework of the *Coupled Model Intercomparison Project* (CMIP),[1] promoted by the *World Climate Research Programme* (WCRP),[2] leading climate research centers worldwide systematically compare their models, share simulation scenarios, and harmonize computational methodologies. The consortium's objective is to improve understanding of climate projections, identify model uncertainties, and provide a shared scientific foundation for assessing the influence of human activities on the climate.

GCMs enable scientists to test future climate scenarios based on different levels of greenhouse gas emissions and to evaluate potential effects on climate variables such as global and regional temperatures, precipitation patterns, ice extent, sea level rise, and other climatic indicators.

The results of these simulations are extensively used by the *Intergovernmental Panel on Climate Change* (IPCC)[3] in its *Assessment Reports*, which serve as a key reference for the development of environmental, energy, and socioeconomic policies at the international level.

The IPCC is the leading international scientific body responsible for assessing the current state of knowledge on climate change. Established in 1988 by the *World Meteorological Organization* (WMO) and the *United Nations Environment Programme* (UNEP), its mission is to provide policymakers with rigorous, up-to-date, and impartial scientific assessments concerning climate change, its impacts, and strategies for adaptation and mitigation.

The IPCC is structured into three working groups:

- Working Group I examines the scientific basis of climate change;
- Working Group II assesses impacts, vulnerabilities, and adaptation strategies;

[1]Available at: https://wcrp-cmip.org/

[2]WCRP Coupled Model Intercomparison Project (CMIP), 2024. https://wcrp-cmip.org/.

[3]Available at: https://www.ipcc.ch/

- Working Group III evaluates strategies for reducing greenhouse gas emissions.

The IPCC does not conduct original research. Instead, it reviews and synthesizes the findings of thousands of scientific studies published worldwide. Its reports, drafted by selected international experts serving on a voluntary basis, undergo a review process described by the IPCC itself as exceptionally rigorous and transparent.

These reports are considered essential in international climate negotiations and exert a significant influence on global climate policy. In 2007, the IPCC was awarded the Nobel Peace Prize, jointly with Al Gore, for its contribution to raising awareness about climate change and to the strengthening of global agreements. Such agreements are negotiated and ratified within the framework of the *Conference of the Parties* (COP), annual meetings that bring together delegations from nearly every country to address the so-called *"climate crisis."* The COPs began in 1995, following the adoption of the *United Nations Framework Convention on Climate Change* (UNFCCC) at the Rio de Janeiro Earth Summit in 1992. Their purpose is to provide a multilateral negotiation platform aimed at reducing greenhouse gas emissions and protecting the global climate.

A landmark legally binding agreement that emerged from a COP is the Paris Climate Agreement (2016),[4] adopted during COP21 in Paris on December 12, 2015, and entering into force on November 4, 2016. It remains a cornerstone of international climate governance, marking the first time that nearly all countries committed to a collective effort to limit global warming.

1.2. Climate models and the dominant paradigm

The most significant scientific finding from simulations conducted with the CMIP's Global Climate Models (GCMs) is that 100% of the observed warming of Earth's surface between the preindustrial period (conventionally defined as 1850 to 1900) and the decade 2011–2020 is attributed to human activities. Specifically, greenhouse gas emissions from the use of fossil fuels (coal, methane, and petroleum derivatives) are considered the primary drivers of this warming.

This conclusion, widely endorsed by major international scientific institutions, has strongly shaped global political strategies, leading to increasingly ambitious climate mitigation efforts aimed at achieving *"climate neutrality"* or *Net-Zero* by target dates typically set for 2050, 2060, or 2075.

The concept of *Net-Zero* refers to a balance between greenhouse gas emissions released into the atmosphere and those removed through natural processes (e.g., CO_2 absorption by the oceans and vegetation) or by artificial means (e.g., underground CO_2 storage). Achieving this balance requires the deep decarbonization of economic sectors and the offsetting of residual emissions through carbon capture technologies or nature-based solutions such as large-scale reforestation.

GCMs do more than estimate future temperature changes; they enable researchers to explore the effects of different societal development pathways through the so-called *Shared Socioeconomic Pathways* (SSPs). These scenarios

[4] Paris Climate Agreement, 2016. Available at: https://unfccc.int/process-and-meetings/the-paris-agreement

simulate possible futures in terms of emissions, economic growth, energy consumption, and land use, providing a wide range of trajectories. These projections are used to estimate the anthropogenic climate forcing—primarily due to rising atmospheric CO_2 concentrations—which then serves as input for the climate models. The models generate projections of future climate change, helping to assess risks and to guide mitigation and adaptation strategies. These projections are subsequently used to estimate potential risks associated with 21st-century climate changes, including floods, heatwaves, melting ice sheets, sea level rise, and extreme weather events. They thus provide critical information to support the implementation of appropriate global mitigation strategies.

However, behind the apparent scientific rigor of the CMIP's GCM projections driven by the various SSP scenarios lie technically complex challenges that require careful scrutiny. In the scientific literature, the reliability of these models is frequently debated, owing to numerous critical issues that stem from the extraordinary complexity of Earth's climate system and the difficulty of accurately representing all of its dynamics.

1.3. Climate models, empirical alternatives, and the energy transition

One of the central challenges in climate modeling lies in accurately representing natural climate variability, a crucial factor for understanding climate behavior across decadal, centennial, and millennial timescales. Despite continuous technological progress, even the most advanced models struggle to simulate the natural changes that have shaped the climate during the relatively stable interglacial period of the Holocene, which began roughly 10,000 years ago and coincides with the rise of human civilization. Episodes such as the *Holocene Climate Optimum*, the *Roman Warm Period*, and the *Medieval Warm Period* are either poorly represented or entirely absent in the CMIP GCMs. As later sections will show, this limitation also raises questions about the models' ability to distinguish anthropogenic influences from natural forcings over the past century (Scafetta, 2026).[5]

These uncertainties extend beyond natural variability. Another critical issue concerns the reliability of historical temperature reconstructions. The datasets used to trace Earth's climate evolution can be affected by non-climatic factors, such as shifts in the geographic distribution of weather stations, changes in measurement techniques, and urbanization near monitoring sites. Several studies suggest that these distortions may introduce systematic biases, potentially leading to an overestimation of the observed global warming (Scafetta, 2021a; Soon et al., 2023).[6]

A further area of debate concerns the role of solar activity in climate change. Variations in total solar irradiance and in solar magnetic activity remain subjects of ongoing scientific investigation. Recent research (Connolly et al., 2023;

---

[5]Scafetta, N., 2025. Detection, attribution, and modeling of climate change: Key open issues. Gondwana Research, DOI: 10.1016/j.gr.2025.05.001.

[6]Scafetta, N., 2021a. Detection of non-climatic biases in land surface temperature records by comparing climatic data and their model simulations. Climate Dynamics, 56(9–10), 2959–2982.
Soon, W., Connolly, R., Connolly, M., Akasofu, S.-I., Baliunas, S., et al., 2023. The detection and attribution of Northern Hemisphere land surface warming (1850–2018) in terms of human and natural factors: Challenges of inadequate data. Climate, 11, 179.

Scafetta, 2023c)[7] suggests that the Sun's influence on Earth's climate may be significantly greater than that currently accounted for in GCMs, which often treat solar forcing as negligible owing to simplifying assumptions. If the solar contribution has been systematically underestimated, the prevailing attribution of post-1850 warming solely to human activities may require substantial revision.

One of the key parameters in climate modeling is the *Equilibrium Climate Sensitivity* (ECS), defined as the projected increase in global average temperature resulting from a doubling of atmospheric CO_2 concentration once the climate system has reached a new equilibrium. Traditional estimates place the average ECS value at approximately 3 ± 1.5 °C, or slightly higher. However, recent empirical studies suggest that the average ECS may be significantly lower, ranging between one-third and two-thirds of the commonly accepted figure, with several analyses indicating likely values between 1.0 ± 0.5 °C and 2.0 ± 1.0 °C (Lewis, 2023; Scafetta, 2023c,b).[8] If these lower estimates are accurate, they would call for a major reassessment of the more catastrophic climate projections and of the urgency of certain mitigation policies aimed at achieving *Net-Zero*.

These uncertainties raise legitimate concerns about the epistemological robustness of current climate policies, many of which are based on projections that assume a dominant anthropogenic influence and advocate drastic emission reductions in order to meet *Net-Zero* targets. If the models overstate the human impact while underestimating natural cycles, mitigation strategies could prove economically and socially disproportionate to the actual risks.

Alternative models based on empirical approaches that incorporate natural variability, astronomical cycles, and solar effects not yet fully understood suggest that the 21st century may experience only moderate warming, even if global policies fall short of *Net-Zero* objectives. These projections challenge the urgency of aggressive climate actions, such as the rapid phase-out of fossil fuels, the forced decarbonization of industrial sectors, and the widespread electrification of transportation. They raise important questions about the proportionality of such measures relative to the available scientific evidence.

At the same time, growing concerns are emerging about the material sustainability of the energy transition (Brussato, 2021; Pitron, 2020).[9] The technologies required for decarbonization—such as batteries, electric motors, solar panels, and wind turbines—depend on large quantities of rare or critical metals, including cobalt (Co), nickel (Ni), copper (Cu), selenium (Se), silver (Ag), cadmium

[7]Connolly, R., Soon, W., Connolly, M., Baliunas, S., Berglund, J., et al., 2023. Challenges in the detection and attribution of Northern Hemisphere surface temperature trends since 1850. Research in Astronomy and Astrophysics, 23, 105015.
Scafetta, N., 2023a. Empirical assessment of the role of the Sun in climate change using balanced multi-proxy solar records. Geoscience Frontiers, 14(6), 101650.
[8]Lewis, N., 2023. Objectively combining climate sensitivity evidence. Climate Dynamics, 60, 3139–3165.
Scafetta, N., 2023b. CMIP6 GCM ensemble members versus global surface temperatures. Climate Dynamics, 60, 3091–3120.
[9]Brussato, G., 2021. Energia verde? Prepariamoci a scavare. I costi ambientali e sociali delle energie rinnovabili. Montaonda.
Pitron, G., 2020. The Rare Metals War: The dark side of clean energy and digital technologies. Scribe Publications.

(Cd), indium (In), tellurium (Te), and platinum (Pt). According to Groves et al. (2023),[10] the global reserves of these materials may be insufficient to support a full energy transition by 2060, with potential geopolitical and environmental consequences linked to their extraction and trade.

In light of these considerations, a more cautious and pragmatic approach to climate strategy appears advisable. If climate models contain significant margins of uncertainty, then the policies derived from them should reflect this fact, acknowledging the epistemological instability of long-term projections. The adoption of ideologically rigid measures risks producing unintended social, economic, and environmental consequences that may outweigh the very problems they are intended to solve.

1.4. Conclusion

In conclusion, while the GCMs developed within the CMIP framework remain essential tools for studying the climate system, they are not without methodological and computational limitations. These include simplifications in their baseline assumptions and gaps in the input data. Such limitations underscore the need for a more balanced and integrated approach, one that takes into account modeling uncertainties, recognizes the role of natural forcings, values empirical evidence, and carefully weighs the socioeconomic impacts of climate policy.

The remainder of this volume will explore these critical themes in depth—from historical temperature reconstructions to the dynamics of natural variability, and from solar influences to climate sensitivity estimates—with the aim of fostering a broader, more informed, and interdisciplinary reflection that harmonizes environmental protection with economic development and social equity.

[10]Groves, D.I., Santosh, M., Yang, C.-X., 2023. Net zero climate remediations and potential terminal depletion of global critical metal resources. Geosystems and Geoenvironment, 2(1), 100136.

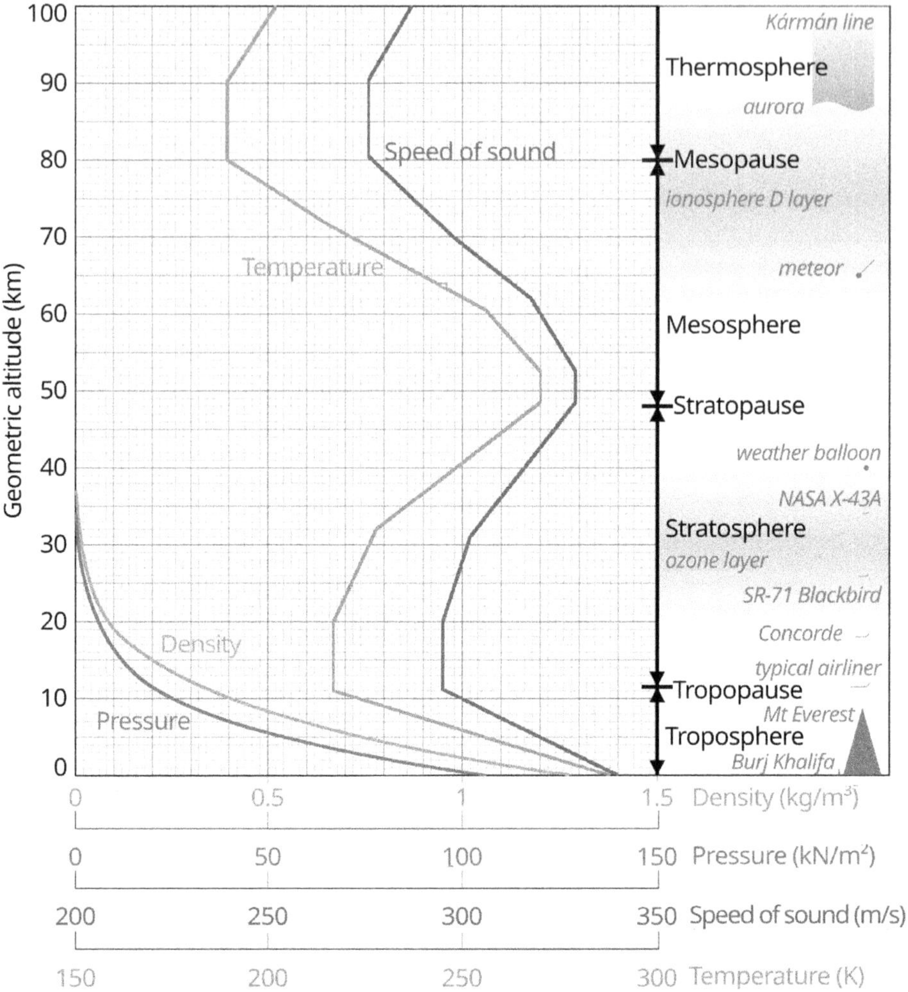

FIGURE 1.1. **Structure and profile of Earth's atmosphere.** Earth's atmosphere is a layered envelope of gases surrounding the planet, composed primarily of nitrogen (78%) and oxygen (21%), with traces of argon, carbon dioxide, and other gases. It is divided into five main layers based on temperature gradients: the troposphere, where weather occurs; the stratosphere, which contains the ozone layer; the mesosphere, where meteors burn up; the thermosphere, characterized by high temperatures and auroras; and the exosphere, the outermost layer, which merges into space. Pressure and density decrease exponentially with altitude, making the upper layers increasingly rarefied. Each layer plays a crucial role in protecting life and in regulating climate and energy exchange.

CHAPTER 2

The climate issue as a global challenge

2.1. Climate and civilization: a longstanding interaction

In contemporary discourse, Earth's climate is often portrayed as a historically stable system, disrupted only by the onset of the industrial era and by the rise in anthropogenic greenhouse gas emissions over the past century. Yet this depiction is an oversimplification and, upon closer scrutiny, a misleading one.

Human history has always been deeply intertwined with a dynamic and ever-changing climate. Environmental fluctuations have repeatedly shaped social, economic, and cultural structures, influencing everything from large-scale migrations to the rise and fall of complex civilizations.

Figure 2.1 presents a schematic reconstruction of Northern Hemisphere temperatures over the past 10,000 years—the Holocene—highlighting quasi-cyclical climate variations of approximately ±1 °C around an average of 15 °C (Dansgaard et al., 1969; Schönwiese, 1995).[1] Notable warm phases include the *Medieval Warm Period* (about 1,000 years ago), the *Roman Warm Period* (about 2,000 years ago), and the *Holocene Climate Optimum* (between 8,000 and 4,000 years ago). Among the cooler intervals, the Little Ice Age stands out, spanning the 14th to 19th centuries, with temperatures up to 2 °C lower than at present.

Public perception and the impact of climate variability. To a layperson, global temperature fluctuations of ±1 °C may seem trivial and barely noticeable on a household thermometer. This misconception often arises from conflating long-term climate variations (which unfold over decades or centuries) with short-term weather changes (daily or seasonal), which involve more dramatic shifts.

In climate science, however, an average global temperature change of ±1 °C is far from negligible. Even modest variations over extended timescales can disrupt climate systems, alter atmospheric and oceanic circulation, shift climate zones, and intensify extreme weather events. For instance, during the *Last Glacial Maximum* (18,000–21,000 years ago), global temperatures were only 4 °C lower than today, yet that was enough to bury much of northern Europe and North America beneath massive ice sheets. The area now occupied by Boston was covered by the Laurentide Ice Sheet, a glacial mass over 1,000 meters thick.

---

[1] Dansgaard, W., Johnsen, S.J., Møller, J., Langway, C.C. Jr., 1969. One thousand centuries of climatic record from Camp Century on the Greenland ice sheet. Science, 166(3903), 377–380.
Schönwiese, C.-D., 1995. Klimaänderungen: Daten, Analysen, Prognosen (2nd ed.). Springer, Berlin.

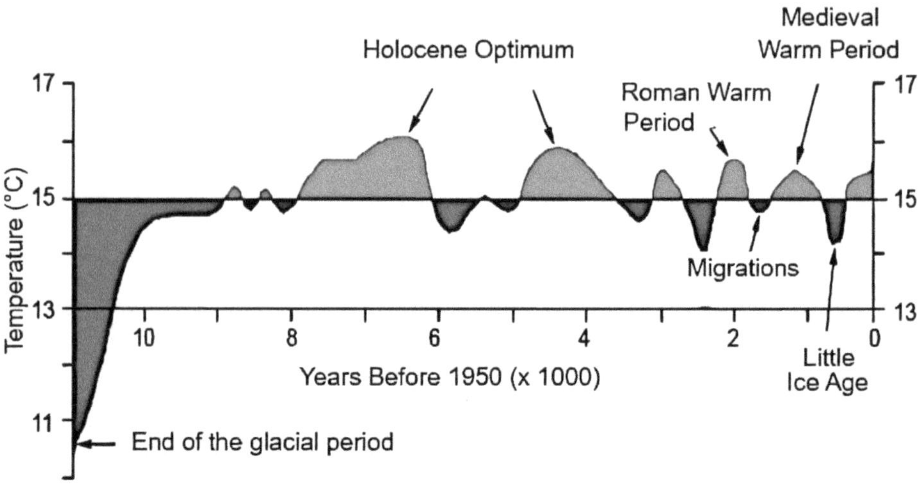

FIGURE 2.1. A Holocene climate reconstruction based primarily on ice core data from Greenland, derived from the Greenland Ice Core Project (GRIP) and the Greenland Ice Sheet Project 2 (GISP2), together with other paleoclimate records (cf.: Dansgaard et al., 1969, 1984; Cuffey & Clow, 1997). (Recreated from: Schönwiese, 1995, chapter 4).

Numerous studies (Guidoboni et al., 2011; Diamond, 2005; Fagan, 2019; White et al., 2018, and many others)[2] have shown that even relatively modest climate changes have had profound effects on human history. In some cases they enabled agricultural expansion and cultural flourishing; in others they triggered economic collapse, famine, and societal upheaval. Shifts in temperature, precipitation, and extreme events have often acted as catalysts for transformative historical developments.

Past climates: A mosaic of opportunity and crisis. Over millennia, Earth has oscillated between warming and cooling phases, each with direct consequences for human societies. For example:

- *Roman Climate Optimum* (1st–4th century AD): agricultural prosperity extended to high latitudes, with vineyards documented even in Britannia;
- *Medieval Warm Period* (9th–13th century): urban revival, demographic growth, and cultural flourishing, including the rise of universities and cathedrals;

[2]Guidoboni, E., Navarra, A., Boschi, E., 2011. The Spiral of Climate: Civilizations of the Mediterranean and Climate Change in History. Bononia University Press.
Diamond, J., 2005. Collapse: How Societies Choose to Fail or Succeed. Penguin.
Fagan, B., 2019. The Little Ice Age: How Climate Made History 1300–1850. Basic Books.
White, S., Pfister, C., Mauelshagen, F. (Eds.), 2018. The Palgrave Handbook of Climate History. Palgrave Macmillan.

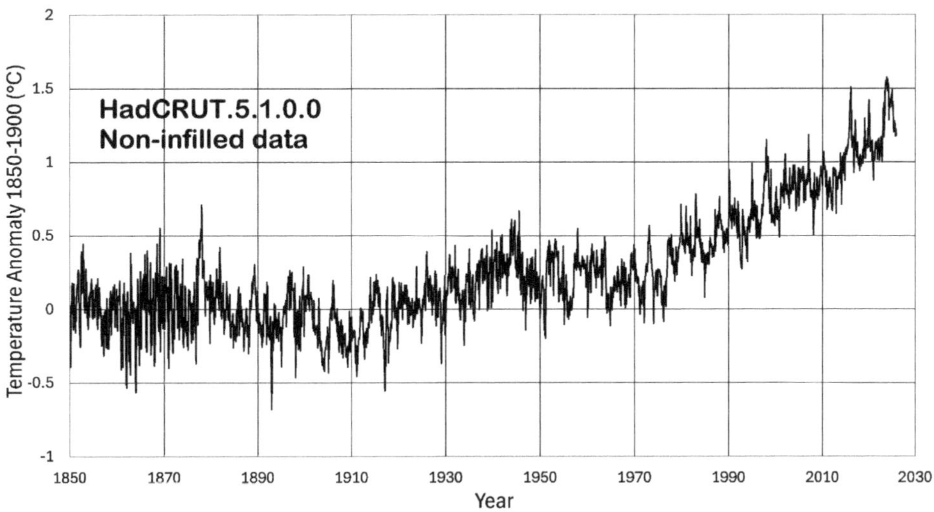

FIGURE 2.2. HadCRUT5 reconstruction of global surface temperature from January 1850 to December 2025 according to the Met Office Hadley center (non-infilled data) (Morice et al., 2021).

- *20th–21st century warming*: although its causes and consequences remain debated, this period has coincided with unprecedented technological, economic, and health advances.

Cold periods:

- *Dark Ages* (5th–8th century): following the fall of the Western Roman Empire, climatic cooling contributed to widespread migrations and societal fragmentation;
- *Little Ice Age* (14th–19th century): marked by crop failures, plagues, and social unrest, as during the *Black Death* (1346–1353), which killed an estimated 25–50% of Europe's population (Horrox, 1994),[3] following the *Great Famine* of 1315–1317, driven by relentless rainfall and adverse weather (Fagan, 2019).

Recent warming and the definition of climate "average". According to the IPCC (2021),[4] global average temperature has risen by approximately +1.1 °C from the preindustrial baseline (1850–1900) to the period 2011–2020. In certain months between 2015 and 2025, surface temperature anomalies have temporarily exceeded the +1.5 °C threshold, as illustrated in Figure 2.2, which is based on the HadCRUT5 dataset from the Met Office Hadley center (Morice et al., 2021).[5]

However, these monthly peaks do not constitute structural climate change. Natural variability—including El Niño events, seasonal cycles, and solar

[3]Horrox, R. (Ed.), 1994. The Black Death. Manchester University Press.
[4]IPCC, 2021. Climate Change 2021: The Physical Science Basis. Cambridge University Press.
[5]Morice, C.P., Kennedy, J.J., Rayner, N.A., Winn, J.P., Hogan, E., et al., 2021. An updated assessment of near-surface temperature change from 1850: the HadCRUT5 data set. Journal of Geophysical Research: Atmospheres, 126, e2019JD032361.

activity—can produce short-lived fluctuations lasting from a few months to about 11 years, the latter corresponding to the Schwabe solar cycle.

For this reason, the World Meteorological Organization (WMO) defines climate change as a statistically significant variation in meteorological averages over periods of at least 30 years. Only on this basis can the +1.5 °C and +2.0 °C thresholds of the Paris Climate Agreement be scientifically validated.

The 30-year average (1995–2024) remains about +1 °C above the preindustrial baseline. A sustained and structural breach of the +1.5 °C threshold has therefore not yet occurred.

Climate crisis or favorable warm phase? Recent global warming is widely attributed to increased greenhouse gas emissions from industrialization and fossil fuel use. This has fueled intense debate around the notion of a *"climate crisis."*

Yet the 20th century also witnessed extraordinary transformations: the Industrial Revolution, medical and agricultural breakthroughs, electrification, and globalization. These developments contributed to a demographic explosion, with the global population growing from roughly 1.5 billion people in 1900 to over 8.2 billion in 2025.[6]

In this context, it is reasonable to argue that 20th-century warming has, on the whole, produced beneficial effects, by enhancing agricultural productivity, reducing famine risks, and supporting socioeconomic progress. Current evidence suggests that, so far, global warming has had net positive impacts on human development, particularly in regions where milder temperatures have expanded arable land and boosted crop yields.

2.2. The new paradigm: imminent climate crisis

Since the 1980s, a widely accepted view has emerged within the international scientific community: that the observed global warming is largely induced by human activity, and in particular by the rise in carbon dioxide (CO_2) emissions resulting from the extensive use of fossil fuels. Over the past few decades, this understanding has contributed to a profound paradigm shift. Climate warming is no longer regarded as a potentially beneficial trend, but rather as one of the most urgent and complex existential threats of the 21st century, with potentially widespread and deeply disruptive consequences for the entire planetary system.

This perspective is now prominently reflected in official reports from the *Intergovernmental Panel on Climate Change* (IPCC) and other leading scientific institutions. According to current projections, the trajectory of warming could lead, within only a few decades, to highly destabilizing climate scenarios, unless effective and timely mitigation measures are implemented to reduce greenhouse gas emissions significantly.

Two critical thresholds were identified in the Paris Climate Agreement (2016)[7] and further emphasized by the IPCC (2018):[8]

- +1.5 °C above preindustrial levels → *orange alert;*

[6]Our World in Data, "How has world population growth changed over time?". Available at: https://ourworldindata.org/population-growth-over-time

[7]Available at: https://unfccc.int/process-and-meetings/the-paris-agreement

[8]IPCC, 2018. Global Warming of 1.5°C. IPCC Special Report.

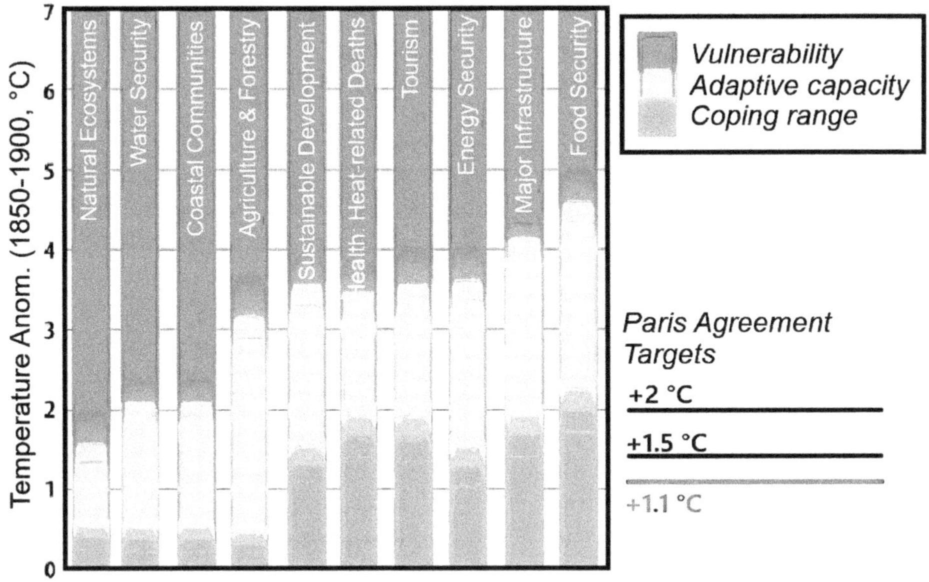

FIGURE 2.3. Impact of climate change on various sensitive systems. The color gradient, from bottom to top, represents a transition from acceptable conditions to significant adaptation needs, and ultimately to critical or potentially dangerous situations (cf. IPCC, 2007, 2013).

- +2.0 °C above preindustrial levels → *red alert*.

Crossing the +2 °C threshold is considered to mark a tipping point beyond which negative impacts could become systemic. These include accelerated biodiversity loss, desertification, ecosystem degradation, threats to food security, damage to infrastructure, and an increased frequency and intensity of extreme weather events (such as floods, hurricanes, droughts, and heatwaves), alongside a progressive and potentially catastrophic rise in sea levels.

Figure 2.3 illustrates the estimated impacts on various sensitive systems, based on IPCC assessments (cf. Kitchen, 2014).[9] The color gradient, from bottom to top, represents a transition from acceptable conditions to significant adaptation needs, and ultimately to critical or potentially dangerous situations. Although in 2025 global warming still remains below the +1.5 °C threshold, the growing concern is that the current balance may be disrupted in the coming decades. According to these projections, the first signs of systemic stress will emerge once the +2 °C threshold is breached, with highly sensitive ecosystems such as coral reefs and the Arctic tundra among the first to undergo changes regarded as irreversible.

Vulnerability and climate justice. Another crucial aspect of the climate crisis is the unequal distribution of risks. While regions such as Western Europe possess the technological, infrastructural, and economic resources to respond more effectively to climate impacts, many communities in the Global South face

[9]Kitchen, D.E., 2014. Global Climate Change: Turning Knowledge into Action. Pearson.

significantly greater vulnerability. These populations are often more exposed to extreme events and have fragile health, agricultural, and social systems that are ill-equipped to cope with the environmental and socioeconomic consequences of global warming.

Climate change is therefore deeply intertwined with issues of social justice and global inequality. Disparities in access to resources, adaptive capacity, and environmental protection risk becoming even more pronounced, potentially transforming an environmental crisis into a historic challenge of equity and planetary solidarity.

2.3. Climate science is still in evolution

Despite the broad scientific consensus on the anthropogenic origin of global warming, it remains essential to maintain a critical and reflective perspective. Earth's climate system is extraordinarily complex, shaped by a dense web of interrelated physical, chemical, and biological variables. In addition to greenhouse gases, other influential factors include atmospheric aerosols, solar variability, oceanic oscillations—such as the *El Niño Southern Oscillation* (ENSO), the *Atlantic Multidecadal Oscillation* (AMO), and the *Pacific Decadal Oscillation* (PDO)—land-use changes, and intricate feedback mechanisms involving cloud formation and the hydrological cycle.

This complexity demands a multidisciplinary scientific approach, open to continuous revision of models and projections. Climate change cannot be reduced to a single narrative; it must incorporate evolving data and account for the dynamic interactions that govern the Earth system. As climatologist Judith Curry (2023)[10] emphasizes, many of these processes remain under active investigation, and the uncertainty surrounding long-term projections remains high.

In contrast, public and media discourse often oversimplifies this complexity, presenting climate change as an incontrovertible fact and frequently accompanying it with catastrophic scenarios portrayed as inevitable. Model simulations, which rely on assumptions and parameters that are rarely disclosed, are sometimes treated as absolute truths. This narrative may produce unintended consequences, including eco-anxiety, a form of psychological distress linked to perceived environmental threats and particularly affecting younger and more sensitive individuals. At the same time, growing polarization in the public debate and an erosion of trust in science can be observed, fueled by communication perceived as dogmatic or one-sided.

Diverging strategies: between radicalism and pragmatism. In response to this evolving landscape, governments and institutions are increasingly adopting polarized strategic approaches, broadly divided into two camps.

(1) *The Radical Transition Model.* This institutional paradigm advocates a rapid shift toward a *Net-Zero* emissions economy by 2050–2060, exemplified by the European Commission's *Green Deal.* It envisions a comprehensive transformation of energy and production systems: phasing out fossil fuels; electrifying transportation; converting industries to low-emission technologies; expanding renewable energy; and

[10]Curry, J., 2023. Climate Uncertainty and Risk: Rethinking Our Response. Anthem Press.

implementing carbon pricing and offset mechanisms. Proponents argue that only a bold paradigm shift can avert catastrophic climate outcomes while promoting social justice, sustainable competitiveness, and technological innovation.

(2) *The Pragmatic Transition Model.* This approach favors a gradual and adaptive transition, tailored to national priorities and socioeconomic realities. Decarbonization should proceed at a pace compatible with economic growth and geopolitical stability, avoiding disruptive imbalances and social tensions. Technological innovation plays a central role, with strategic emphasis on next-generation nuclear reactors, clean hydrogen, advanced bio-fuels, and energy efficiency solutions. This vision seeks to balance climate goals with energy security, economic resilience, and collective well-being, integrating sustainability with equity and pragmatism.

2.4. The "*net-zero*" ambition: between political will and material reality

Based on climate model projections, the IPCC's (2018) *Special Report on Global Warming of 1.5°C*[11] recommended ambitious mitigation strategies aimed at limiting the rise in global average temperature to well below 2 °C above preindustrial levels, and ideally within the 1.5 °C threshold, as outlined in the Paris Climate Agreement (2016). As illustrated in Figure 2.3, the +2 °C threshold is regarded as critical, since exceeding it would significantly increase the risk of severe and potentially irreversible impacts, including ecosystem collapse, the intensification of extreme weather events, sea level rise, and threats to food and water security.

To achieve these goals, the IPCC advocates climate neutrality—commonly referred to as *Net-Zero*—which entails the enforcement of costly mitigation policies to drastically reduce greenhouse gas emissions until a balance is reached between the amount emitted and that removed from the atmosphere by natural or technological means. Proposed strategies include large-scale reforestation, bio-energy with carbon capture and storage (BECCS), and direct air capture using advanced technologies.

A concrete example of policy implementation is the European Union's *Green Deal*,[12] which aims to cut emissions by at least 55% by 2030 (compared with 1990 levels) and to achieve climate neutrality by 2050–2055. Measures include phasing out internal combustion vehicles, transitioning to renewable energy sources (solar, wind, geothermal, and hydroelectric), electrifying industrial processes, and improving energy efficiency in buildings.

Challenges to implementation. Despite its ambition, the realization of *Net-Zero* presents substantial political, technological, and economic challenges. The required investments are estimated to be in the trillions of euros, and concerns about the erosion of European industrial competitiveness are mounting, especially in the absence of comparable commitments from other major emitters such as China, India, and the United States. Additional indirect costs include

[11] IPCC, 2018. Global Warming of 1.5°C. IPCC Special Report.

[12] European Union, 2024. European Green Deal. Available at: https://commission.europa.eu/strategy-and-policy/priorities-2019-2024/european-green-deal_en

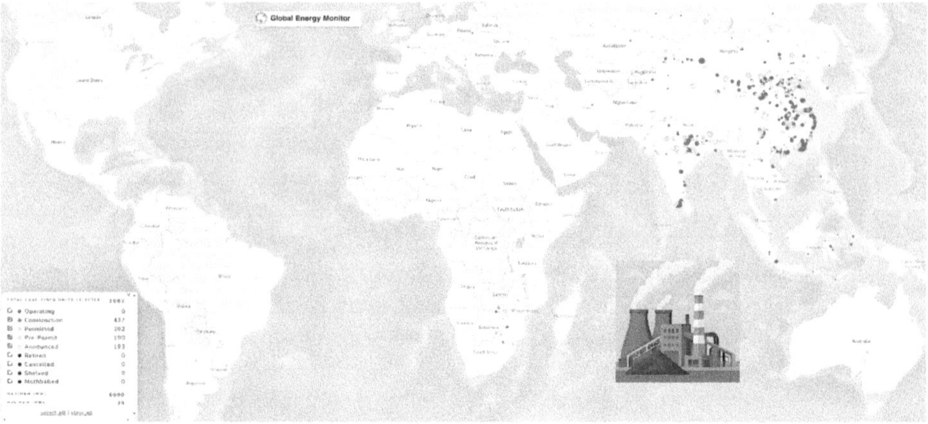

FIGURE 2.4. Over 1,000 new coal-fired power plants are under construction or planned in 2024, with the majority concentrated in Southeast Asia, primarily in China and India (Global Coal Plant Tracker, 2024).

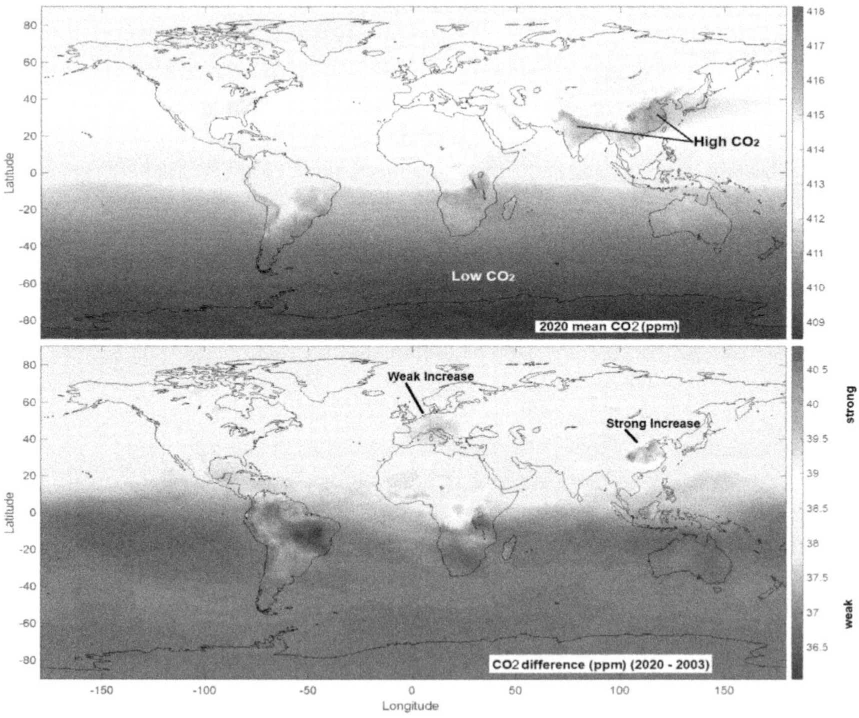

FIGURE 2.5. (Top) Atmospheric CO_2 concentration in 2020 using the CAMS greenhouse gas reanalysis (Agustí-Panareda et al., 2023). (Bottom) Atmospheric CO_2 concentration changes between 2003 and 2020. China and India exhibit the highest atmospheric CO_2 levels and the largest increases. Central Europe and, in particular, Northern Italy show the smallest increases.

unemployment, the forced restructuring of supply chains, and rising energy prices, all of which raise questions about the social sustainability of the transition within a globally competitive economy.

The global context: divergences and constraints. While the EU continues to lead on climate action, global data paint a more sobering picture. In 2024, the 27 EU member states accounted for only about 6% of global greenhouse gas emissions (EDGARD: European Commission, 2025).[13] Meanwhile, global emissions have been rising by approximately 2% per year, driven largely by Asia's rapid industrial expansion. Over 1,000 new coal-fired power plants are under construction or planned, particularly in China and India, as shown in Figure 2.4 (Global Coal Plant Tracker, 2024).[14]

Figure 2.5 further highlights the stark asymmetry between Europe's regulatory ambitions and the energy policies being implemented worldwide. Data from the CAMS global greenhouse gas reanalysis (EGG4) (Agustí-Panareda et al., 2023)[15] or NASA's Orbiting Carbon Observatory (OCO-2)[16] reveal that the regions with the highest concentrations of new coal plants (such as China and India) also exhibit the highest atmospheric CO_2 levels and the largest increases. By contrast, Europe's emissions footprint appears significantly more contained.

Figure 2.6 compares total and per capita emissions across the globe, the EU, China, and Italy. In 2024, European per capita emissions stood at 7.14 tCO_2eq/year, slightly above the global average of 6.56 tCO_2eq. Italy, with 6.3 tCO_2eq/year, was below both averages, while China far exceeded them at 10.8 tCO_2eq/year.

Feasibility of the *"Net-Zero"* goal. This discrepancy raises serious questions about the practical feasibility of *Net-Zero*. According to Hausfather & Peters (2020) and Burgess et al. (2021),[17] current global energy trends are incompatible with the most optimistic decarbonization scenarios, such as the SSP1 pathways, which envisage climate neutrality between 2050 and 2075. Instead, actual trends align more closely with the moderate SSP2 scenarios, which assume that emissions will begin to decline only after 2050, without reaching *Net-Zero* even by 2100.

Figure 2.7 shows that scenarios such as SSP3-7.0 and SSP5-8.5 are considered *"unlikely"* and *"very unlikely,"* respectively, and are generally deemed unsuitable for guiding effective climate policy. Ritchie & Dowlatabadi (2017)[18] critically

---

[13]EDGARD: Emission Database for Global Atmospheric Research. European Commission, 2024. GHG emissions of all world countries. Available at: https://edgar.jrc.ec.europa.eu/report_2024

[14]Global Coal Plant Tracker: 2024. Available at: https://globalenergymonitor.org/projects/global-coal-plant-tracker/tracker/

[15]Agustí-Panareda, A., Barré, J., Balsamo, G., Basart, S., Bennouna, Y., et al., 2023. Technical note: The CAMS greenhouse gas reanalysis from 2003 to 2020. Atmospheric Chemistry and Physics, 23, 3829–3850.
Available at https://ads.atmosphere.copernicus.eu/datasets/cams-global-ghg-reanalysis-egg4-monthly?tab=download.

[16]Available at https://cmr.earthdata.nasa.gov/search/concepts/C2240248790-GES_DISC.html.

[17]Burgess, M.G., Ritchie, J., Shapland, J., Pielke, R., 2021. IPCC baseline scenarios have over-projected CO_2 emissions and economic growth. Environmental Research Letters, 16(1), 014016.

[18]Ritchie, J., Dowlatabadi, H., 2017. Why do climate change scenarios return to coal? Energy, 140, 1276–1291.

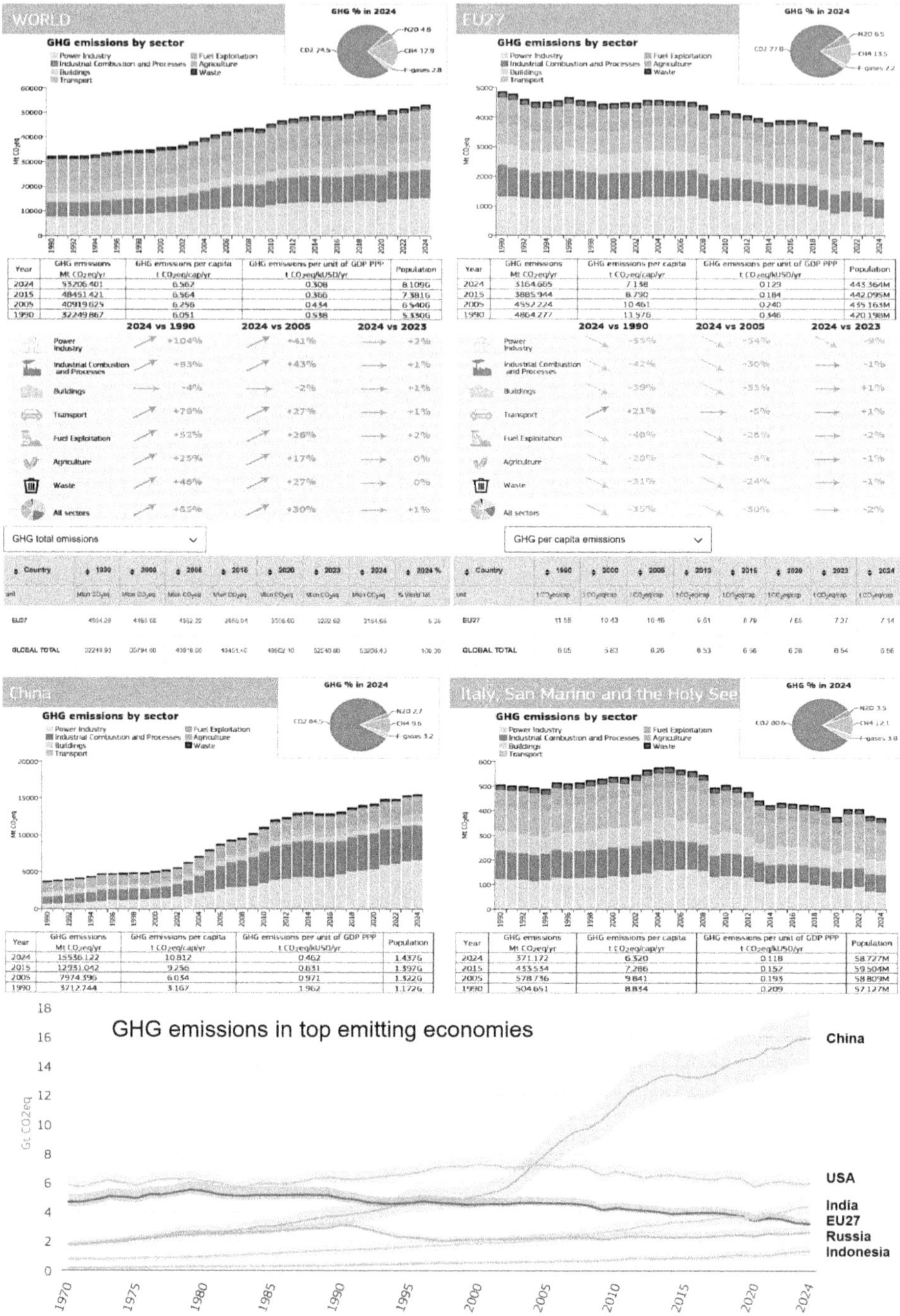

FIGURE 2.6. Top: comparison of total and per capita greenhouse gas emission estimates between the European Union and the global average. Bottom: comparison between China and Italy. (Adapted from EDGARD: European Commission, 2025).

examine why many high-emission climate models, such as RCP8.5, continue to project a resurgence in global coal use despite current decarbonization trends. The authors argue that this outcome stems not from realistic energy forecasts, but from structural assumptions within integrated assessment models that overestimate coal availability and underrepresent technological transitions.

To contextualize the scale of the challenge, and given that the oceans and the biosphere absorb roughly 55%:

- about 17 $GtCO_2$ must be emitted to increase atmospheric CO_2 by 1 ppm;[19]
- raising atmospheric CO_2 from 425 ppm to 1,000 ppm by 2100 would require about 10,000 $GtCO_2$ of emissions;
- current fossil fuel reserves could realistically generate only half the emissions needed to reach such levels;
- the SSP2-4.5 scenario projects a more plausible concentration of about 603 ppm by century's end (Meinshausen et al., 2020).[20]

Italy's position offers a valuable perspective: with per capita emissions below both the global and the European averages, further unilateral reductions could risk undermining national competitiveness without producing any meaningful global impact. A more strategic focus may lie in promoting climate reciprocity and multilateral agreements that ensure fair contributions from major emitters.

Material constraints and economic implications. A rarely discussed but critical constraint is the limited availability of essential raw materials such as copper, nickel, cobalt, tellurium, indium, and the rare earth elements, which are needed for renewable technologies and electric vehicles. Demand for these materials could outstrip supply as early as 2060 (Groves et al., 2023).[21] Without major investments in innovation, recycling, and supply diversification, the energy transition risks creating new geopolitical dependencies and resource-based tensions, potentially escalating into trade or military conflicts.

On the economic front, climate economist Richard Tol (2023)[22] estimates that the global cost of *Net-Zero* policies could exceed their projected benefits. Even under optimistic assumptions, GDP losses of 5.5% are expected by 2100, compared with benefits of only 3%. Danish political scientist Bjorn Lomborg (2023)[23] echoes these concerns, estimating annual costs of $15–$37 trillion, equivalent to 15–37% of current global GDP. He argues that the expected benefits, in terms of reduced climate damage, are modest, and that *Net-Zero* policies may be economically inefficient and potentially detrimental to global development. As

[19]To increase CO_2 by 1 ppm net of absorption, consistent with the concept of *"airborne fraction,"* about 17.3 $GtCO_2$ must be emitted.

[20]Meinshausen, M., Nicholls, Z.R.J., Lewis, J., Gidden, M.J., Vogel, E., et al., 2020. The shared socio-economic pathway (SSP) greenhouse gas concentrations and their extensions to 2500. Geoscientific Model Development, 13(8), 3571–3605.

[21]Groves, D.I., Santosh, M., Yang, C.-X., 2023. Net zero climate remediations and potential terminal depletion of global critical metal resources. Geosystems and Geoenvironment, 2(1), 100136.

[22]Tol, R.S.J., 2023. Costs and benefits of the Paris climate targets. Climate Change Economics, 14(3), 2340003.

[23]Lomborg, B., 2023 (29 November). 'Net Zero' fails the cost-benefit test. The Wall Street Journal. Available at: https://totallyengagedamericans.org/2023/12/net-zero-fails-the-cost-benefit-test/

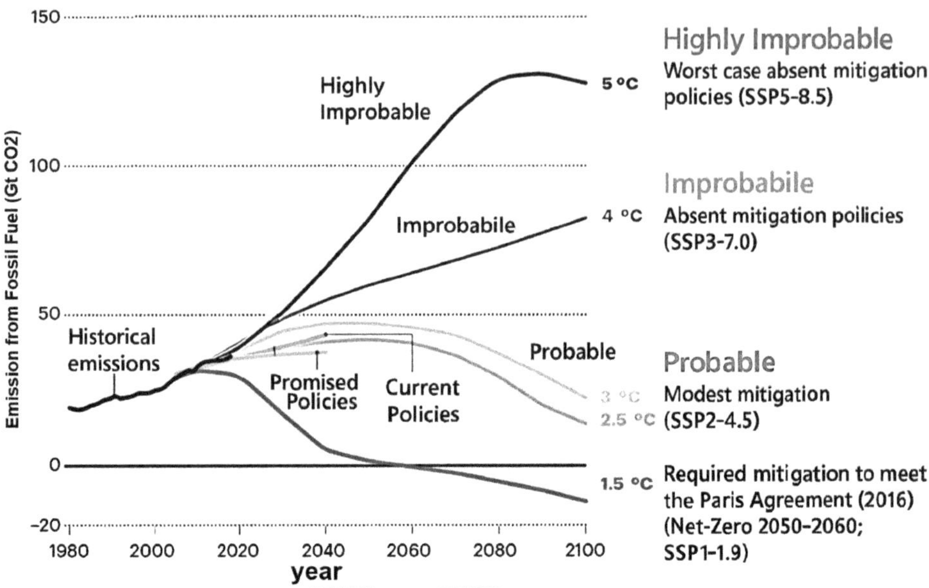

SSP1-1.9		2015	2025	2050	2075	2100	2150	2200	2250	2300	2400	2500
CO_2	ppm	399.9	426.5	437.6	419.7	393.5	349.0	343.4	343.3	342.0	339.2	336.9
CH_4	ppb	1842.0	1875.2	1427.9	1184.3	1036.4	969.8	928.9	881.3	1112.0	870.9	871.4
N_2O	ppb	328.2	335.1	343.6	349.0	354.0	361.4	363.9	362.7	360.6	358.2	357.1
SSP1-2.6		2015	2025	2050	2075	2100	2150	2200	2250	2300	2400	2500
CO_2	ppm	399.9	427.7	469.3	471.0	445.6	411.1	403.2	399.7	396.0	389.5	384.3
CH_4	ppb	1842.0	1865.1	1519.4	1248.4	1056.4	977.4	927.6	875.2	1112.0	863.5	864.0
N_2O	ppb	328.2	334.9	343.5	348.5	353.9	361.2	363.4	362.2	360.0	357.5	356.4
SSP2-4.5		2015	2025	2050	2075	2100	2150	2200	2250	2300	2400	2500
CO_2	ppm	399.9	429.0	506.9	575.5	602.8	626.3	643.1	637.0	621.3	597.8	579.2
CH_4	ppb	1841.9	1960.7	2020.2	1815.7	1683.2	1479.6	1255.4	1038.1	1112.0	999.2	997.3
N_2O	ppb	328.2	336.0	356.2	371.5	377.3	378.3	375.9	371.4	367.0	362.0	359.8
SSP3-7.0		2015	2025	2050	2075	2100	2150	2200	2250	2300	2400	2500
CO_2	ppm	399.9	432.3	540.6	683.0	867.2	1235.3	1456.8	1513.7	1482.8	1423.6	1371.1
CH_4	ppb	1841.9	2006.5	2472.0	2934.1	3372.2	3096.2	2571.6	2107.1	1112.0	1959.1	1938.4
N_2O	ppb	328.2	336.4	361.8	390.7	421.8	471.4	497.2	507.2	510.5	514.3	516.0
SSP4-6.0		2015	2025	2050	2075	2100	2150	2200	2250	2300	2400	2500
CO_2	ppm	399.9	428.3	515.6	606.9	668.4	741.0	783.9	786.2	768.7	739.1	714.0
CH_4	ppb	1841.9	2049.1	2503.7	2688.5	2645.5	2382.6	2111.4	1864.6	1112.0	1791.0	1785.6
N_2O	ppb	328.2	335.9	359.7	383.4	404.7	435.8	451.2	456.1	456.8	457.7	458.1
SSP5-8.5		2015	2025	2050	2075	2100	2150	2200	2250	2300	2400	2500
CO_2	ppm	399.9	431.8	562.8	801.7	1135.2	1737.3	2108.3	2206.4	2161.7	2080.5	2010.0
CH_4	ppb	1841.9	1954.5	2446.5	2672.3	2415.3	1906.9	1515.6	1157.3	1112.0	1038.5	1019.0
N_2O	ppb	328.2	336.3	358.2	377.1	391.8	407.8	413.6	413.0	410.5	407.8	406.6

FIGURE 2.7. Top: Estimated net CO_2 emissions under different *Shared Socioeconomic Pathways* (SSPs). The right-hand axis shows the average global warming projected by GCM climate models relative to the 1850–1900 baseline for each SSP scenario. Negative values indicate cases where CO_2 sequestration exceeds emissions; a value of zero corresponds to Net-Zero. (Recreated from Hausfather & Peters, 2020).
Bottom: Projected greenhouse gas concentrations for each SSP. (Adapted from Meinshausen et al., 2020).

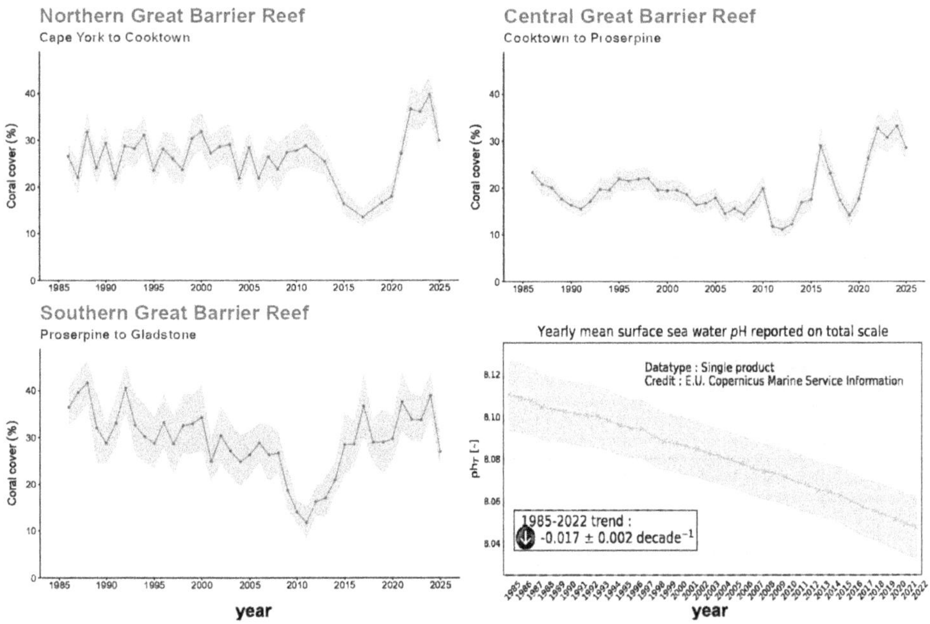

FIGURE 2.8. Comparison between the extent of the Great Barrier Reef in Australia and the yearly mean surface sea water pH. (Adapted from Australian Institute of Marine Science, 2025, 2025; Copernicus Marine Service, 2023).

an alternative, Lomborg advocates greater investment in low-carbon technology, which he views as a more effective and sustainable long-term strategy.

2.5. The ambivalent role of CO_2: life's nutrient or climate threat?

In today's public discourse, and especially in the media, carbon dioxide (CO_2) is frequently labeled a *"pollutant."* Yet from a scientific standpoint, this classification is misleading. While CO_2 is a greenhouse gas with significant implications for the climate system, it is not a chemical pollutant in the strict toxicological sense. Mild physiological effects such as headaches or fatigue occur only at concentrations typically between 3,000 and 5,000 ppm, nearly ten times the present-day average of 425 ppm.

Since the dawn of the industrial era, atmospheric CO_2 has risen by more than one third, from approximately 280–300 ppm to 425 ppm in 2025. Most of the scientific community links this increase to global warming, with associated effects including altered weather patterns, more intense extreme events, and ocean *"acidification"*—although seawater remains alkaline (pH > 7),[24] so that *"reduction in alkalinity"* is a more accurate term. However, the degree of climatic hazard posed by CO_2 depends on the climate system's sensitivity and on the response of the biosphere. If climate sensitivity were zero, even a substantial

[24] The fundamental relation that links pH to the concentration of hydrogen ions in solution is $pH = -\log_{10}[H^{+}]$, where $[H^{+}]$ denotes the molar concentration of hydrogen ions (in moles per liter).

rise in CO_2 would have no effect. Moreover, biospheric responses are typically nonlinear and complex.

For instance, Figure 2.8 compares the extent of Australia's Great Barrier Reef (Australian Institute of Marine Science, 2025, 2025)[25] with the trend in average ocean pH (Adapted from Copernicus Marine Service, 2023).[26] Between 1985 and 2022, the average ocean pH declined from 8.11 to 8.05, likely owing to the rising atmospheric CO_2 concentration. The reef experienced a significant contraction between 1995 and 2015, suggesting a possible link to the declining pH. Yet between 2020 and 2025, coral coverage rebounded despite the continued decline in pH, indicating that other factors, such as ecological resilience, environmental management, or local climate conditions, may have played a major role. On geological timescales, ocean pH has often been lower than at present, ranging between 6.5 and 7.5 during the Archean and Proterozoic eras (Halevy & Bachan, 2017).[27]

At the same time, CO_2 plays a vital biological role. It fuels photosynthesis, the process by which plants produce oxygen and biomass. Elevated CO_2 levels, of up to 1,000–1,500 ppm, generally enhance vegetation growth (Gerhart & Ward, 2010).[28] Any assessment of the risks posed by CO_2 must therefore weigh its uncertain climatic impacts against its well-documented ecological benefits.

Elevated CO_2 enhances photosynthesis and water-use efficiency primarily in C3 plants, which include major cereals such as wheat, rice, and barley; legumes such as soybean, bean, pea, lentil, peanut, and alfalfa; many horticultural crops such as potato, beet, and tomato; oil and fiber crops such as rapeseed, cotton, and tobacco; and nearly all tree species. By contrast, C4 crops such as maize, sorghum, and millet show only modest gains, because their carbon-concentrating mechanism already saturates Rubisco with CO_2. Although this physiological stimulation is clear in controlled environments, its effect in real agricultural systems is limited by nutrient and water constraints, rising temperatures, and more frequent droughts. Elevated CO_2 also reduces protein and micro-nutrient concentrations in plant tissues.

Numerous studies show that rising CO_2 has boosted global agricultural productivity, with estimated gains of up to 40%. Figure 2.9 shows that satellite measurements of the Leaf Area Index (LAI) indicate that 25–50% of vegetated areas have undergone *"global greening,"* including semi-arid and desert regions (cf. Zhang, 2024a).[29] This effect is especially pronounced in China and India, where atmospheric CO_2 concentrations are among the highest (Chen et al.,

[25] Australian Institute of Marine Science, 2025. Annual Summary Report of Coral Reef Condition 2024/25. AIMS.

[26] Copernicus Marine Service, 2023. Global Ocean acidification – mean sea water pH time series and trend from Multi-Observations Reprocessing. LSCE.

[27] Halevy, I., Bachan, A., 2017. The geologic history of seawater pH. Science, 355(6329), 1069–1071.

[28] Gerhart, L.M., Ward, J.K., 2010. Plant responses to low [CO_2] of the past. New Phytologist, 188(3), 674–695.

[29] Zhang, X., Evans, J.P., Burrell, A. L., 2024a. Less than 4% of dryland areas are projected to desertify despite increased aridity under climate change. Communications Earth & Environment, 5(1), 1–9.

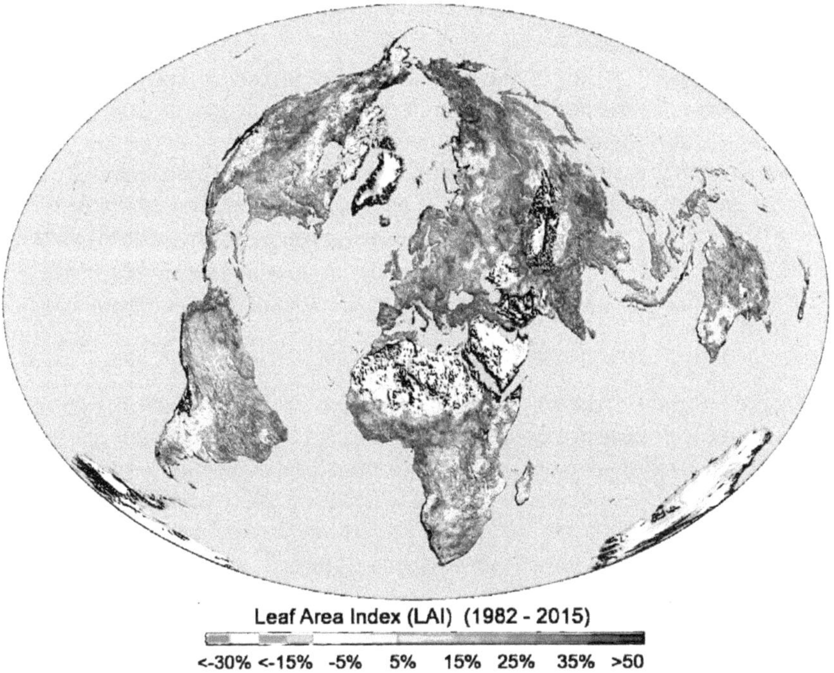

FIGURE 2.9. The "*Greening*" of the Earth due to fertilization from rising atmospheric CO_2 levels. (Adapted from Hille, 2016; Zhu et al., 2016).

2019),[30] as shown in Figure 2.4. According to Zhu et al. (2016)[31] and Piao et al. (2019),[32] CO_2 fertilization accounts for roughly 70% of the greening observed between 1982 and 2009. In Italy, the forested area has nearly tripled, from 4.2 million hectares in 1888 to 11.7 million in 2021, although this is partly due to the rewilding of abandoned farmland (Agnoletti et al., 2022).[33]

Some empirical studies (Zhao et al., 2017; Rezaei et al., 2023)[34] suggest that global warming may already have lowered the yields of major crops, with CO_2 fertilization compensating only partially, and climate projections indicate increasing risks in regions such as the Mediterranean, where drought and heat stress are

[30]Chen, C., Park, T., Wang, X., Piao, S., Xu, B., et al., 2019. China and India lead in greening of the world through land-use management. Nature Sustainability, 2, 122–129.

[31]Zhu, Z., Piao, S., Myneni, R.B., Huang, M., Zeng, Z., et al., 2016. Greening of the Earth and its drivers. Nature Climate Change, 6(8), 791–795.

[32]Piao, S., Wang, X., Park, T., Chen, C., Lian, X., et al., 2019. Characteristics, drivers and feedbacks of global greening. Nature Reviews Earth & Environment, 1(1), 14–27.

[33]Agnoletti, M., Piras, F., Venturi, M., & Santoro, A., 2022. Cultural values and forest dynamics: The Italian forests in the last 150 years. Forest Ecology and Management, 503, 119655.

[34]Zhao, C., Liu, B., Piao, S., Wang, X., Lobell, D.B., et al., 2017. Temperature increase reduces global yields of major crops in four independent estimates. Proceedings of the National Academy of Sciences, 114(35), 9326–9331.

Rezaei, E.E., Webber, H., Asseng, S., Boote, K., Durand, J.L., et al., 2023. Climate change impacts on crop yields. Nature Reviews Earth & Environment, 4: 831–846.

expected to intensify. However, a meta-reanalysis by McKitrick (2025)[35] finds that temperature increases of up to 5 °C should not reduce global agricultural productivity, largely owing to the fertilizing effect of CO_2. The study challenges upward revisions of the Social Cost of Carbon (SCC), arguing that the predicted crop damage is not supported by empirical data.

These insights suggest the need for a pragmatic approach. Among the most promising mitigation strategies is agroforestry, which enhances terrestrial biomass to naturally sequester carbon. Recent research shows that sustainable forest management, particularly in Southern Europe, can play a key role in regulating the carbon cycle (Dalmonech et al., 2022; Marano et al., 2022).[36]

2.6. Conclusion

The growing awareness of climate change carries significant geopolitical consequences. Developing countries face a dual challenge: they must address the impacts of global warming—such as possible prolonged droughts, rising sea levels, and food insecurity—while also pursuing economic growth in order to improve living standards. A fair and sustainable energy transition requires global strategies that respect each country's specific priorities, capacities, and vulnerabilities. Imposing uniform models or centralized mandates risks deepening inequalities and destabilizing fragile societies.

Climate change is not a one-dimensional issue. It is at once natural and cultural, physical and political, scientific and symbolic. This complexity makes it difficult to describe, interpret, and manage. Political decisions must therefore be grounded in a realistic, multidisciplinary understanding, free from ideological distortions and capable of navigating a web of risks and uncertainties.

This volume aims to equip readers with critical tools to engage with climate change—particularly from a scientific perspective—by moving beyond reductive narratives, rhetorical shortcuts, and emotional polarizations that hinder nuanced thinking. The goal is not to downplay the seriousness of the issue, but to confront it with clarity, rigor, and intellectual honesty, avoiding both paralyzing catastrophism and dogmatic denialism. Although seemingly opposed, these extremes share an ideological lens that obstructs constructive dialogue and delays effective, shared solutions.

CO_2 should neither be demonized nor idealized, as some studies on past mass extinctions suggest (Davis, 2023).[37] A balanced perspective must be rooted in scientific evidence and ecological awareness, and must reconcile climate concerns with the biological and economic realities of human development. This calls for a nuanced cost–benefit analysis that takes into account all the relevant dimensions.

---

[35] McKitrick, R., 2025. Extended crop yield meta-analysis data do not support upward SCC revision. Scientific Reports, 15, 5575.

[36] Dalmonech, D., Marano, G., Collalti, A., et al., 2022. Forest management as a strategy for climate change mitigation: insights from southern European plantations. Science of the Total Environment, 838, 159361.

Marano, G., Dalmonech, D., Collalti, A., et al., 2022. Forest management intensity and carbon sequestration potential under climate change scenarios in temperate Europe. Agricultural and Forest Meteorology, 319, 109203.

[37] Davis, W.J., 2023. Mass extinctions and their relationship with atmospheric carbon dioxide concentration: Implications for Earth's future. Earth's Future, 11, e2022EF003336.

CHAPTER 3

The IPCC and the Global Climate Models (GCMs)

3.1. Conceptual origins of the greenhouse effect

The idea that Earth's atmosphere regulates surface temperature dates back to the early 19th century. In 1827, the French mathematician and physicist Joseph Fourier (1768–1830) proposed that the atmosphere acts like a thermal blanket, retaining part of the Sun's energy and warming the planet beyond what would be expected from a simple radiative balance. Although he did not identify the specific gases responsible, Fourier laid the theoretical foundations of what would later become known as the "greenhouse effect":[1]

> *"The temperature of the Earth's globe would be much lower if the atmosphere did not retain the heat transmitted by solar rays." ... "The heat from the Sun passes through the atmosphere, penetrates the surface layers of the ground, and transforms into dark heat [infrared radiation], which tends to dissipate into space. But the atmosphere presents more obstacles to this dark heat than to sunlight."*

In 1856, the American scientist Eunice Newton Foote conducted pioneering experiments on the heat absorption properties of atmospheric gases. Using glass cylinders, thermometers, and an air pump, she demonstrated that carbon dioxide retained more heat than other gases, suggesting that increased CO_2 concentrations could influence Earth's climate:

> *"An atmosphere of that gas would give to our Earth a high temperature; and if, as some suppose, at one period of its history the air had mixed with it a larger proportion than at present, an increased temperature must have resulted."*[2]

Her findings were presented at the American Association for the Advancement of Science (AAAS) by the physicist Joseph Henry, but her contribution was largely overlooked for more than a century, likely because of prevailing gender biases.

In 1861, the Irish physicist John Tyndall extended these early studies with more precise instrumentation. He identified the key greenhouse gases—water vapor (H_2O), carbon dioxide (CO_2), and ozone (O_3)—and demonstrated their ability to absorb the infrared radiation emitted by Earth's surface. His metaphor remains iconic:

---

[1]Fourier, J., 1827. Mémoire sur les températures du globe terrestre et des espaces planétaires. Mémoires de l'Académie Royale des Sciences de l'Institut de France, 6, 569–604.

[2]Foote, E., 1856. Circumstances affecting the heat of the Sun's rays. The American Journal of Science and Arts, 22(66), 382–383.

> *"Water vapor is a blanket more necessary to the vegetable life of England than clothing is to man."*[3]

Tyndall also showed that nitrogen (N_2) and oxygen (O_2), which make up most of Earth's atmosphere, are transparent to infrared radiation, whereas trace gases exert a disproportionate influence on climate dynamics. He even hypothesized that variations in these gases could account for past glacial cycles, foreshadowing the concept of climate sensitivity.

At the end of the 19th century, the Swedish chemist Svante Arrhenius[4] became the first to quantify the relationship between atmospheric CO_2 and global temperature. In his 1896 paper, he estimated that a doubling of CO_2 would result in warming of approximately 5–6 °C, a groundbreaking attempt at climate modeling based on radiative calculations and feedback mechanisms involving water vapor.

Less well known is the fact that Arrhenius later revised his estimates downward. In 1906, he published *Die vermutliche Ursache der Klimaschwankungen*,[5] reducing his estimate to an average of 3.2 °C, and in some scenarios to as low as 1.6 °C, depending on water vapour feedback.

Reflecting the optimism of his era, Arrhenius regarded rising CO_2 levels as potentially beneficial. In *Worlds in the Making* (1908),[6] he wrote:

> *"By the increase of the percentage of carbonic acid in the atmosphere, we may hope to enjoy ages with more equable and better climates."*

The hypothesis was largely set aside owing to the limited empirical data available. It resurfaced in the 1930s, when the English engineer Guy Stewart Callendar (1938) correlated rising temperatures in North America with increased CO_2 from fossil fuel combustion:

> *"The observed rise in global temperature is consistent with the increase in CO_2 caused by the combustion of fossil fuels."*[7]

Despite its significance, Callendar's work was initially dismissed because of the localized nature of the data and skepticism within the scientific community.

The issue resurfaced in the 1950s with the advent of modern spectroscopy and electronic computing. The Canadian physicist Gilbert Plass (1956) confirmed that rising CO_2 levels would affect global temperatures, estimating a warming of 3.6 °C for a doubling of CO_2:

---

[3]Tyndall, J. (1861). The Bakerian Lecture: On the Absorption and Radiation of Heat by Gases and Vapours, and on the Physical Connexion of Radiation, Absorption, and Conduction. Philosophical Transactions of the Royal Society of London, 151, 1–36.

[4]Arrhenius, S., 1896. On the Influence of Carbonic Acid in the Air upon the Temperature of the Ground. Philosophical Magazine and Journal of Science, Serie 5, 41, 237–276.

[5]Arrhenius, S., 1906. "Die vermutliche Ursache der Klimaschwankungen". Meddelanden från K. Vetenskapsakademiens Nobelinstitut Band 1 No 2. English translation: "The Probable Cause of Climate Fluctuations – Svante Arrhenius, A Translation of his 1906 Amended View of "Global Warming", 2014.

[6]Arrhenius, S., 1908. Worlds in the Making: The Evolution of the Universe (H. Borns, Trans.). New York: Harper & Brothers. (original work published in 1906).

[7]Callendar, G.S., 1938. The artificial production of carbon dioxide and its influence on temperature. Quarterly Journal of the Royal Meteorological Society, 64, 223–240.

> *"Global warming will persist for centuries unless CO_2 emissions are reduced."*[8]

In the 1960s and 1970s, systematic atmospheric monitoring began, culminating in the iconic *Keeling Curve* (Keeling, 1960),[9] recorded at the Mauna Loa Observatory in Hawaii. This curve revealed a continuous and accelerating rise in atmospheric CO_2, from 315 ppm in 1958 to approximately 425 ppm in 2024. The seasonal cycle reflects the annual uptake of atmospheric CO_2 by vegetation in the Northern Hemisphere during spring and summer, followed by its release in autumn and winter as photosynthetic activity declines. The seasonal maximum of atmospheric CO_2 is observed in May and the minimum in September–October, since the boreal winter spanning November to April limits photosynthetic uptake.

According to satellite data from NASA's Orbiting Carbon Observatory (OCO-2), shown in Figure 2.4, Hawaii lies in a region where CO_2 concentrations are about 20 g/m^2 (roughly 2 ppm) above the global average, reflecting localized variations in atmospheric composition.

3.2. A turning point in climate science: modeling the greenhouse effect

A pivotal moment in the history of climate science occurred in 1967 with the publication of the groundbreaking study by Syukuro Manabe and Richard Wetherald, entitled *"Thermal Equilibrium of the Atmosphere with a Given Distribution of Relative Humidity."*[10] For the first time, a numerical climate model was developed that integrated radiative transfer with vertical atmospheric convection, quantitatively demonstrating how increased CO_2 concentrations lead to surface warming. Their model estimated that a doubling of CO_2 would result in a global temperature increase of approximately 2 °C, laying the conceptual foundations for what would become the modern Global Climate Models (GCMs).

In 1975, Manabe and Wetherald extended their work in a second study that incorporated atmosphere–ocean interactions.[11] This three-dimensional model yielded a climate sensitivity estimate of 2.9 °C for a doubling of CO_2 and revealed two key insights:

- *polar amplification*: warming is more intense at high latitudes;
- *hydrological intensification*: the water cycle becomes more vigorous under warming conditions.

Manabe's pioneering contributions were formally recognized in 2021, when he was awarded the Nobel Prize in Physics for his role in establishing the physical foundations of climate modeling.

Climate change enters the global agenda. The 1980s marked a surge in international attention to climate change. The 1985 Villach Conference was a watershed moment, placing the issue at the heart of global scientific discourse.

[8] Plass, G.N., 1956. The carbon dioxide theory of climatic change. Tellus, 8(2), 140–154.

[9] Keeling, C.D., 1960. The Concentration and Isotopic Abundances of Carbon Dioxide in the Atmosphere. Tellus, 12(2), 200–203.

[10] Manabe, S., Wetherald, R.T., 1967. Thermal Equilibrium of the Atmosphere with a Given Distribution of Relative Humidity. Journal of the Atmospheric Sciences, 24(3), 241–259.

[11] Manabe, S., Wetherald, R.T., 1975. The effects of doubling the CO_2 concentration on the climate of a general circulation model. Journal of the Atmospheric Sciences, 32, 3–15.

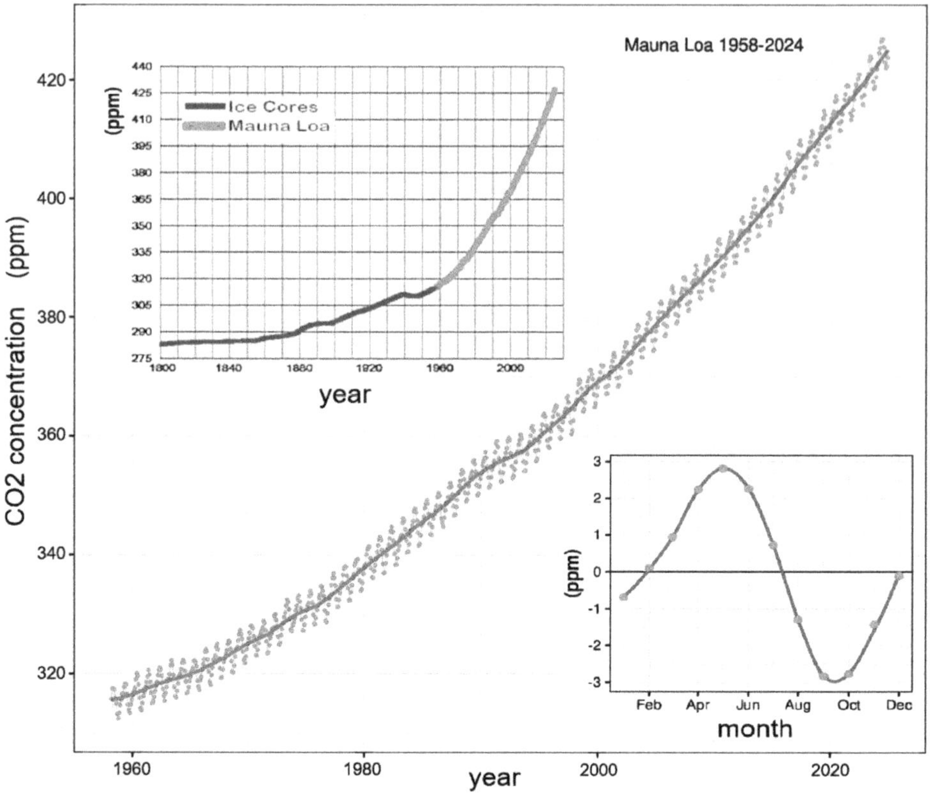

FIGURE 3.1. Keeling Curve: long-term measurements of atmospheric carbon dioxide (CO_2) concentrations recorded at the Mauna Loa Observatory, Hawaii, since 1958 (Keeling, 1960). Insets: (top) extension including the ice.core CO_2 record; (bottom) seasonal cycle in the Keeling Curve.

Just three years later, in 1988, James Hansen's testimony before the U.S. Congress (cf. Figure I.1) galvanized public awareness and political momentum. That same year saw the creation of the Intergovernmental Panel on Climate Change (IPCC),[12] the leading scientific body tasked with assessing the causes, impacts, and possible responses to global warming.

Climate feedback challenges: water vapor and clouds. Despite the growing consensus on the role of CO_2 in climate change, a major source of uncertainty, first acknowledged by Arrhenius, remains unresolved to this day: the response of water vapour and clouds to rising anthropogenic greenhouse gas concentrations.

In the 1960s, water vapour was recognized as a feedback mechanism, yet its behavior remains controversial to the present day (Ma et al., 2025; Simpson et

[12]The Intergovernmental Panel on Climate Change: https://www.ipcc.ch/

al., 2024).[13] In a seminal 1963 study, Fritz Möller (1963)[14] explored the potential amplification of the greenhouse effect due to water vapor. His findings were striking:

- the feedback from water vapor could theoretically produce temperature changes ranging from +1 to +10 °C;
- small changes in absolute humidity or cloud cover could completely offset the radiative impact of increased CO_2.

Möller concluded:

> *"The theory that climate variations are caused by changes in atmospheric CO_2 becomes highly questionable."*

His work remains a powerful reminder of the complexity and uncertainty surrounding climate feedbacks, particularly those involving moisture dynamics and cloud formation. These mechanisms continue to challenge the precision of climate projections and underscore the need for ongoing research.

3.3. Constructing scientific consensus: the IPCC paradigm

Since 1990, the IPCC has played a central role in shaping both the scientific and the political discourse surrounding global warming. Its *Assessment Reports* (ARs)—published in 1990, 1995, 2001, 2007, 2013, and 2021—represent the most systematic and comprehensive effort to collect, synthesize, and interpret scientific knowledge about Earth's climate system and its presumed anthropogenic alteration.[15]

Although the IPCC does not conduct original research, it performs a critical evaluation of the existing scientific literature, which is analyzed and summarized in its official reports. The enormous influence of these documents on international policymaking has elevated the IPCC to the status of a transnational epistemic authority—that is, a body whose assessments shape global climate governance.

Since its founding in 1988, the IPCC's mission has been to determine the extent to which human activity contributes to the observed global warming and to assess its potential future consequences. As stated in Point 2 of the *Principles Governing IPCC Work* (approved at the 14th session in Vienna, 1998):

---

[13] Ma, N., Zhang, Y., Yang, Y., 2025. Recent decline in global ocean evaporation due to wind stilling. Geophysical Research Letters, 52, e2024GL114256.
Simpson, I.R., McKinnon, K.A., Kennedy, D., Lawrence, D.M., Lehner, F., Seager, R., 2024. Observed humidity trends in dry regions contradict climate models. Proceedings of the National Academy of Sciences of the United States of America, 121(1), e2302480120.

[14] Möller, F., 1963. Influence of changes in CO_2 concentration on the radiation balance of Earth's surface and on the climate. Journal of Geophysical Research, 68, 3877–3886.

[15] These are the IPCC's major reports on climate science:
FAR—IPCC, 1990. Climate Change: The IPCC Scientific Assessment. Cambridge University Press.
SAR—IPCC, 1995. Climate Change 1995: The Science of Climate Change. Cambridge University Press.
AR3—IPCC, 2001. Climate Change 2001: The Scientific Basis. Cambridge University Press.
AR4—IPCC, 2007. Climate Change 2007: The Physical Science Basis. Cambridge University Press.
AR5—IPCC, 2013. Climate Change 2013: The Physical Science Basis. Cambridge University Press.
AR6—IPCC, 2021. Climate Change 2021: The Physical Science Basis. Cambridge University Press.

> *"The role of the IPCC is to assess on a comprehensive, objective, open and transparent basis the scientific, technical and socio-economic information relevant to understanding the scientific basis of risk of human-induced climate change, its potential impacts and options for adaptation and mitigation."*

Critiques of the dominant paradigm. Despite its stated commitment to objectivity, some scholars have raised concerns that the IPCC's strong emphasis on anthropogenic attribution may have introduced a form of systemic ideological bias. According to these critics, the selection, interpretation, and presentation of evidence may be disproportionately steered toward reinforcing the dominant hypothesis that human activity is the primary driver of climate change. As a result, studies exploring natural contributions to climate variability may be marginalized, and the uncertainties inherent in climate science may be underrepresented.

The prevailing IPCC paradigm attributes the warming observed since the early 20th century to the increase in anthropogenic greenhouse gas emissions, especially CO_2 from the combustion of coal, oil, and natural gas. The period 1850–1900 is conventionally adopted as the preindustrial baseline, against which subsequent warming is regarded as anomalous and largely human-induced.

The explanatory framework centers on the concept of radiative forcing, defined as a disturbance in Earth's energy balance caused by changes in incoming and outgoing energy fluxes. In this context, greenhouse gases (including human-emitted CO_2) enhance the retention of infrared radiation in the atmosphere, amplifying the natural greenhouse effect, which, however, is primarily driven by water vapor (H_2O). To simulate the climatic consequences of these perturbations—such as changes in global temperature or shifts in weather patterns—scientists rely on Global Climate Models (GCMs), highly sophisticated numerical tools.

The central role and limitations of Global Climate Models (GCMs). GCMs are computational frameworks that solve the fundamental equations of fluid dynamics, thermodynamics, and atmospheric chemistry. The atmosphere and oceans are divided into three-dimensional grids, and the evolution of the climate system is simulated on the basis of emission scenarios and the complex interactions among physical and chemical components (see Figure 3.2).

The results of these simulations form the backbone of the climate projections presented in IPCC reports. They reconstruct past climate trends and generate future scenarios based on key variables such as global average temperature, precipitation, and sea level, using Shared Socioeconomic Pathways (SSPs) to model alternative emission trajectories for the future.

Feedbacks and uncertainty. However, GCMs rely on numerous simplifying assumptions, particularly regarding feedback mechanisms, that is, chain-reaction processes that can either amplify (positive feedback) or dampen (negative feedback) the effects of radiative forcing.

Examples include:

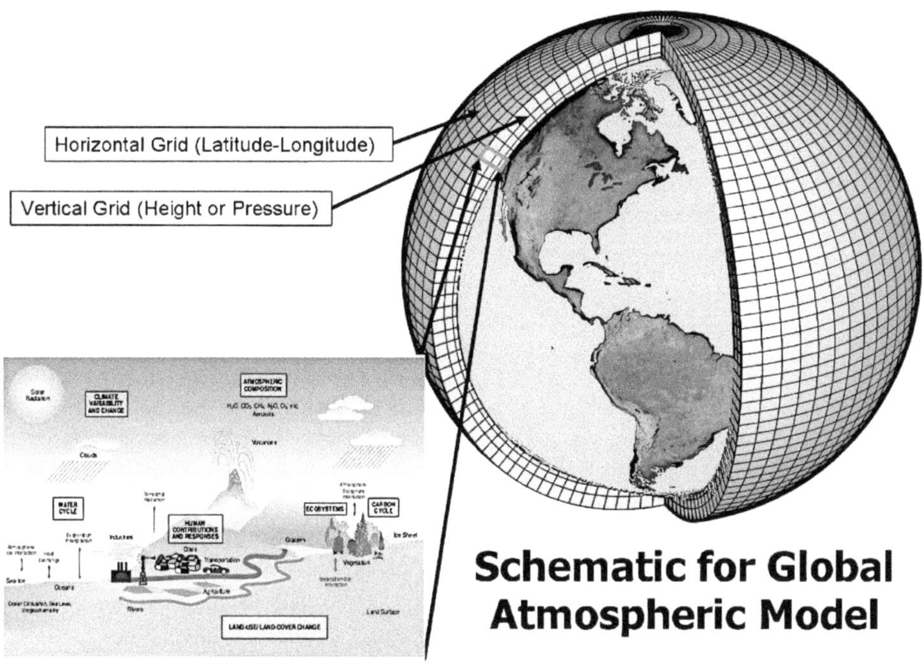

FIGURE 3.2. Schematic representation of Global Climate Models (GCMs) and the climate system.

- *positive feedback*: melting glaciers reduce surface albedo, leading to further warming;
- *negative feedback*: increased evaporation leads to greater cloud cover, which reflects solar radiation and may cool the surface.

Although these processes are incorporated into the models, their representation remains partial and uncertain, especially with respect to:

- water vapour dynamics;
- cloud formation and coverage;
- surface albedo variability;
- ocean circulation patterns.

These uncertainties can lead to significant overestimations or underestimations of climate sensitivity, namely the degree to which the climate responds to changes in radiative forcing such as those derived from changes in greenhouse gas concentrations.

The substantial physical uncertainty surrounding the climate feedbacks makes long-term climate projections inherently fragile, particularly when they are used as the foundation for binding social policies or far-reaching regulatory decisions. This underscores the need for caution, transparency, and continuous refinement in both scientific modeling and policy formulation.

3.4. Scientific validation in contemporary climate science

In the climate change debate, one of the most sensitive and consequential issues concerns the validity of the projections produced by the available GCMs.

Accurately assessing the risks posed by global climate change and evaluating the effectiveness of mitigation strategies require a fundamental prerequisite: verifying the scientific reliability of the tools used to simulate Earth's climate system. This principle of validation is a methodological cornerstone of the IPCC's work.

The validation process involves comparing model simulations with observational data from the recent past, in order to determine whether the models can reliably reproduce historical trends in key climate variables such as temperature, precipitation, and sea level. Statistical tests are employed to assess whether the confidence intervals generated by the models are consistent with the observed variability.

Another essential validation tool is paleoclimate data, that is, the climatic information extracted from ice cores, ocean sediments, tree rings, fossil corals, and other geological archives. These records allow scientists to reconstruct climatic conditions over timescales far longer than those accessible with modern instruments. Integrating insights from complementary disciplines—geology, oceanography, atmospheric physics, dendrochronology—enriches the interpretive framework and deepens our understanding of climate dynamics.

Models, scenarios, and the limits of predictability. Despite methodological advances and increasingly sophisticated modeling techniques, significant margins of uncertainty persist. The climate system is inherently complex, governed by nonlinear interactions among the atmosphere, oceans, biosphere, and human activity. As a result, climate models do not produce definitive predictions, but rather plausible scenarios based on explicit assumptions. This distinction is crucial from a scientific standpoint, yet it is often blurred in public discourse, where projections are sometimes presented dogmatically, as if they were absolute forecasts or incontrovertible climate prophecies.

Persistent uncertainties: an open scientific field. Contrary to the image of a monolithic consensus often conveyed by the media and institutions, climate science remains an epistemologically complex and evolving discipline. The global, dynamic, and nonlinear nature of the phenomena under study makes research in climate science inherently uncertain and subject to continuous revision.

For example, the influential study by Manabe & Wetherald (1975) predicted that a doubling of CO_2 would produce pronounced warming in the upper troposphere and cooling in the stratosphere, a pattern that has been considered a key *"fingerprint"* of anthropogenic climate change. This framework has guided atmospheric interpretation for decades and is routinely reproduced in climate models.

However, recent evidence challenges the persistence of this signal. According to the IPCC's AR6 (2021), the upper troposphere has not warmed as predicted, and the lower stratosphere, after cooling between 1980 and 2000, has since nearly stabilized, with some data even indicating a slight warming. For example, on the basis of radiosonde measurements, Philipona et al. (2018)[16] suggested that the lower stratosphere began warming after 2000, particularly

[16]Philipona, R., Mears, C., Fujiwara, M., Jeannet, P., Thorne, P., et al., 2018. Radiosondes show that after decades of cooling, the lower stratosphere is now warming. Journal of Geophysical Research: Atmospheres, 123(23), 12,509–12,522.

between 15 and 30 km altitude. Santer et al. (2023)[17] confirm that no renewed cooling trend has emerged since then.

This observed behavior—tropospheric warming accompanied by a modest cooling, or even some stratospheric warming—is not consistent with GCM simulations and challenges the validity of the traditional climate fingerprint. While Manabe & Wetherald's work remains foundational, modern observations suggest that the response of the stratosphere to anthropogenic forcing may be more complex and less linear than originally anticipated.

Three critical areas of scientific difficulty. The main challenges in climate attribution and modeling are concentrated in three key areas:

- uncertainties in climate measurements;
- the difficulty of distinguishing natural variability from anthropogenic influence;
- an incomplete understanding of fundamental physical processes.

The robustness of attributing global warming solely to human activity depends heavily on our ability to understand natural forcings and feedback mechanisms. If natural variability is not adequately accounted for, attribution risks becoming not only difficult but potentially misleading.

Attribution is therefore a crucial scientific step: only when the anthropogenic signal clearly emerges above the noise of natural variability can drastic global policies be justified. Where the margins of uncertainty are wide, decisions based solely on model projections may rest on unstable epistemic foundations.

Climate policy under uncertainty: balancing prudence and rationality. Uncertainty is not a justification for denial or paralysis; rather, it calls for a thoughtful and balanced approach to climate policy. When predictive models lack reliability, it is unwise to rely on them alone when making irreversible decisions with far-reaching economic, technological, and social impacts.

In conditions of partial knowledge, precaution must be guided by proportionality. Climate action should reflect the strength of the available evidence, avoiding ideological rigidity. Decarbonization must be environmentally sound, economically viable, and socially equitable. Without this balance, policies risk harming key economic sectors, deepening inequality, and igniting geopolitical tensions, all without delivering meaningful climate benefits.

Effective policymaking requires a pluralistic, multidimensional assessment that integrates:

- scientific findings and their limitations;
- social and economic consequences;
- trade-offs and opportunity costs.

Transparent dialogue among science, institutions, and civil society is essential. Climate models are indispensable tools, but they are not infallible. Governing climate complexity calls for strategic flexibility, innovation, and resilience, not for rigid regulatory uniformity.

---

[17] Santer, B.D., Po-Chedley, S., Zhao, L., Zou, C.-Z., Fu, Q., et al., 2023. Exceptional stratospheric contribution to human fingerprints on atmospheric temperature. Proceedings of the National Academy of Sciences, 120(20), e2300758120.

Toward a more open and dialogical climate science. Scientific credibility is vital for sound climate policy, but it must not be mistaken for absolute certainty. Climate science is a dynamic, self-correcting discipline. It evolves through debate, model refinement, and openness to new evidence. Uncertainty is not a flaw but a core feature of scientific inquiry. It must be openly acknowledged and integrated into decision-making with intellectual honesty. Only a science that embraces its limitations can help shape a sustainable future, through reasoned choices, inclusive dialogue, and adaptive thinking.

3.5. Institutional consolidation of a paradigm

From the late 20th century to the present, the *Anthropogenic Global Warming Theory* (AGWT) has become the cornerstone of international climate policy. Formalized through the *Assessment Reports* of the Intergovernmental Panel on Climate Change (IPCC), AGWT has progressively evolved into the dominant interpretive framework for analyzing global climate dynamics.

Like any complex scientific construct, AGWT did not emerge fully formed, nor is it now immune to revision. Its development reflects a gradual conceptual evolution: from early empirical correlations between rising greenhouse gas concentrations and temperature increases, to the construction of Global Climate Models (GCMs) capable of simulating the dynamic interactions among the atmosphere, oceans, biosphere, and human activity.

Over time, IPCC reports have reinforced the view that the rising atmospheric concentrations of CO_2, methane (CH_4), nitrogen oxides (NO_x), and other greenhouse gases are the primary drivers of the warming observed since the early 20th century. This interpretation is supported by a vast body of observational and model-based literature, yet it is not without blind spots. Questions persist regarding the relative magnitude of the anthropogenic signal compared with the natural climate variability that has shaped Earth's climate over millennia.

Three key areas of uncertainty in the dominant climate paradigm. Although widely accepted in institutional and scientific contexts, AGWT presents at least three critical areas that warrant deeper reflection.

(1) *Reliability of historical global temperature series.* Global average temperatures are reconstructed from heterogeneous datasets: satellite observations, ground-based measurements, and numerical interpolations. These reconstructions form the empirical foundation of AGWT. However, several studies have raised concerns about potential distortions in the official temperature records, due to instrument calibration issues, uneven station distribution, and homogenization procedures. Research by Scafetta & Ouyang (2019),[18] Scafetta (2021a),[19] Katata et

[18]Scafetta, N., Ouyang, S., 2019. Detection of UHI bias in China climate network using Tmin and Tmax surface temperature divergence. Global and Planetary Change, 181, 102989.

[19]Scafetta, N., 2021a. Detection of non-climatic biases in land surface temperature records by comparing climatic data and their model simulations. Climate Dynamics, 56(9–10), 2959–2982.

al. (2023),[20] and Soon et al. (2023)[21] suggests that post-1950 warming may have been partially overestimated owing to urban heat island effects and blending biases. If confirmed, such findings would require a recalibration of global surface warming rates, with significant implications for future projections and policy decisions.

(2) *The role of the Sun, astronomical cycles, and other natural oscillations*. Distinguishing anthropogenic influence from natural forcings such as solar variability, oceanic oscillations, and volcanic activity remains one of the greatest challenges in climate attribution. While models attempt to isolate the *"anthropogenic fingerprint,"* the uncertainties remain substantial. Scholars such as Connolly et al. (2023)[22] and Scafetta (2023c)[23] argue that solar activity—including sunspot cycles, magnetic field variations, and cosmic ray flux—may exert a greater climatic influence than is currently represented in the CMIP6 GCMs. Moreover, Milanković orbital cycles, when combined with internal climate feedbacks, may contribute meaningfully to long-term climate variability. Omitting or under-representing these natural forcings risks compromising the completeness and accuracy of the models and skewing attribution results.

(3) *Climate sensitivity to radiative forcing*. Key mechanisms such as cloud formation, aerosol distribution, regional albedo, ocean dynamics, and solar evolution remain imperfectly understood and are often inadequately represented in GCMs. This uncertainty is reflected in the wide range of equilibrium climate sensitivity (ECS) estimates, defined as the expected temperature increase produced by a doubling of atmospheric CO_2. CMIP6 GCMs estimate ECS values between 1.8 °C and 5.7 °C, indicating substantial structural uncertainty. However, recent studies (including: Lewis, 2023; Scafetta, 2022, 2023b)[24] suggest significantly lower ECS values, ranging from 1 °C to 3 °C. If accurate, these estimates imply that many GCMs may overestimate future warming, necessitating a revision of mitigation strategies and policy targets. Figure 3.3 illustrates that the CO_2 greenhouse-warming framework relies on two linked mechanisms: the direct CO_2 radiative forcing, and the

[20]Katata, G., Connolly, R., O'Neill, P., 2023. Evidence of urban blending in homogenized temperature records in Japan and the United States. Journal of Applied Meteorology and Climatology, 62(8), 1095–1114.

[21]Soon, W., Connolly, R., Connolly, M., Akasofu, S.-I., Baliunas, S., et al., 2023. The detection and attribution of Northern Hemisphere land surface warming (1850–2018) in terms of human and natural factors: Challenges of inadequate data. Climate, 11, 179.

[22]Connolly, R., Soon, W., Connolly, M., Baliunas, S., Berglund, J., et al., 2023. Challenges in the detection and attribution of Northern Hemisphere surface temperature trends since 1850. Research in Astronomy and Astrophysics, 23, 105015.

[23]Scafetta, N., 2023c. Empirical assessment of the role of the Sun in climate change using balanced multi-proxy solar records. Geoscience Frontiers, 14(6), 101650.

[24]Lewis, N., 2023. Objectively combining climate sensitivity evidence. Climate Dynamics, 60, 3139–3165.

Scafetta, N., 2022. Advanced testing of low, medium, and high ECS CMIP6 GCM simulations versus ERA5-T2m. Geophysical Research Letters, 49(6), e2022GL097716.

Scafetta, N., 2023b. CMIP6 GCM ensemble members versus global surface temperatures. Climate Dynamics, 60, 3091–3120.

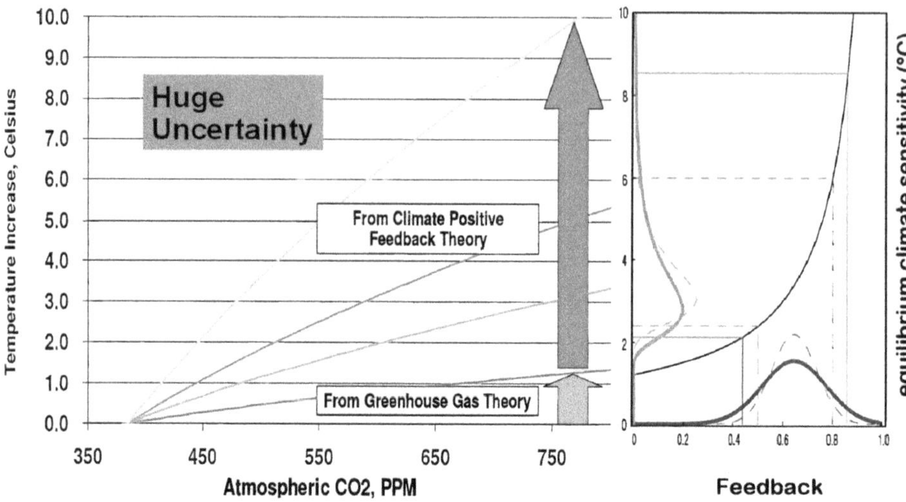

FIGURE 3.3. The CO_2 greenhouse-warming framework relies on two linked mechanisms: (1) the direct radiative warming caused by increasing atmospheric CO_2 (about 1 °C per CO_2 doubling); (2) the additional warming produced by climate feedbacks triggered by higher CO_2 levels, whose magnitude remains highly uncertain. (Recreated from Knutti & Hegerl, 2008).

additional warming produced by climate feedbacks, the magnitude of which remains highly uncertain (Knutti & Hegerl, 2008).[25]

In summary, the reliability of historical temperature records, the representation of natural forcings, and the quantification of climate sensitivity remain open frontiers in climate research. These issues do not invalidate AGWT, but they underscore that the theory is subject to ongoing revision and critical scrutiny.

Properly accounting for these uncertainties is essential to ensure that climate policies are proportionate, evidence-based, and scientifically robust. A more nuanced approach can foster more resilient and adaptive strategies for addressing climate change.

Toward alternative scenarios: moderation or catastrophe? The substantial uncertainties that characterize the current scientific framework open up the possibility that global warming in the 21st century may follow trajectories different from those outlined in the most recent IPCC reports (2021; 2023). In this context, the present volume examines alternative empirical models such as those developed by Connolly et al. (2020)[26] and Scafetta (2023c, 2026, 2024),[27]

[25]Knutti, R., Hegerl, G.C., 2008. The equilibrium sensitivity of the Earths temperature to radiation changes. Nature Geoscience, 1, 735–743.

[26]Connolly, R., Connolly, M., Carter, R.M., Soon, W., 2020. How much human-caused global warming should we expect with business-as-usual climate policies? Energies, 13(6), 1365.

[27]Scafetta, N., 2023c. Empirical assessment of the role of the Sun in climate change using balanced multi-proxy solar records. Geoscience Frontiers, 14(6), 101650.

Scafetta, N., 2024. Impacts and risks of "realistic" global warming projections for the 21st century.

which incorporate elements often underrepresented in mainstream climate simulations, including natural variability, astronomical forcings, and a reduced climate sensitivity.

According to these hypotheses, even under continued greenhouse gas emissions, future warming could remain within moderate bounds, without necessarily triggering the catastrophic impacts frequently anticipated. Although more conservative, such scenarios merit serious consideration. They are grounded in rigorous methodological assumptions and rely on climate data series that, in some cases, differ from those adopted by the IPCC. These methodologies offer an interpretive framework complementary to the prevailing paradigm, raising a pivotal question: *Have the global climate risks been overestimated?*

3.6. Conclusion: climate science as an open process

If the most extreme climate scenarios have indeed been overstated, the implications for social policy could be profound. Many of the most radical mitigation measures, including those embedded in energy transition plans, are predicated on forecasts of imminent and severe climate disruption. Should these projections prove exaggerated or rest on tenuous assumptions, the economic, technological, and social costs of emission reduction strategies may appear disproportionate to the realistically attainable benefits.

Effective climate policy must therefore be grounded in a rational and proportionate assessment of the scientific uncertainties. Prudence—understood not as passivity but as a guiding operational principle—should inform collective decision-making. This means prioritizing:

- infrastructure resilience;
- technological innovation;
- energy diversification;
- a gradual and sustainable reduction of emissions.

Rather than relying on emergency measures driven by worst-case scenarios, policy should reflect strategic foresight and adaptive realism.

A dynamic scientific construct. The theory of anthropogenic global warming is neither a fixed truth nor an untouchable dogma. It is a dynamic scientific construct, subject to continuous refinement. Its credibility does not rest on infallibility, but on its openness to verification, critique, and methodological evolution. Recognizing its margins of uncertainty does not undermine its epistemic foundation; on the contrary, it strengthens its legitimacy when scrutiny confirms its premises through transparent, rigorous, and responsible inquiry. Conversely, if structural flaws are identified, the scientific method demands a reassessment of the theory itself.

The chapters that follow will examine the classical greenhouse effect theory and trace the development of the AGWT across successive IPCC reports from 1990 to the present.

Geoscience Frontiers, 15(2), 101774.

Scafetta, N., 2025. Detection, attribution, and modeling of climate change: Key open issues. Gondwana Research, DOI: 10.1016/j.gr.2025.05.001.

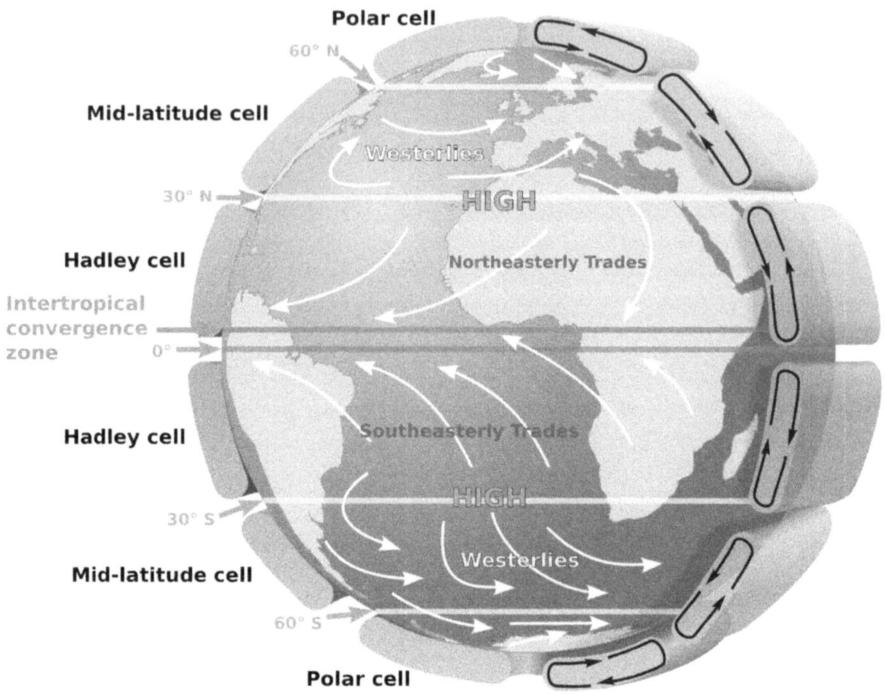

FIGURE 3.4. **Global circulation of the atmosphere.** The three-cell atmospheric model divides each hemisphere into three distinct circulation cells—Hadley, Ferrel, and Polar—that help explain global wind patterns, pressure zones, and heat distribution. The Hadley cell, located between the equator and about 30° latitude, features rising warm air at the equator and descending cooler air in the subtropics, thereby driving the trade winds. The Ferrel cell, between 30° and 60°, operates as a transitional zone in which air flows poleward at the surface and equatorward aloft, generating the prevailing westerlies. The Polar cell, from 60° to the pole, involves cold air sinking at the poles and rising near 60°, producing the polar easterlies. Each cell is associated with characteristic pressure zones: low pressure near the equator and 60° latitudes, due to rising air, and high pressure around 30° and the poles, where the air descends. These pressure patterns influence precipitation: equatorial and subpolar regions tend to be wetter, owing to ascending moist air and frequent cloud formation, whereas subtropical and polar zones are generally drier, owing to descending air and atmospheric stability. This model captures the latitudinal transport of energy and moisture, although it simplifies the complex dynamics of real atmospheric circulation.

CHAPTER 4

Earth's energy balance and the greenhouse effect

4.1. The Sun and the Earth

Understanding Earth's climate dynamics begins with an analysis of the primary energy sources that power the system. At the top of this hierarchy stands the Sun, the thermodynamic engine of life on Earth and the principal driver of atmospheric, oceanic, and biological processes. Although geothermal heat contributes to Earth's energy budget, its impact is negligible on a global scale. The average geothermal flux is approximately 0.087 W/m^2 (Pierrehumbert, 2010),[1] which is about 16,000 times smaller than the solar constant, estimated at roughly 1361 W/m^2.

How the Sun radiates energy. From a physical standpoint, the Sun behaves as a black body, an idealized object that absorbs all incident radiation and re-emits energy in accordance with thermodynamic laws. The Sun's photosphere, its visible surface layer, has a temperature of approximately 5800 K,[2] making it the primary source of electromagnetic radiation reaching Earth's surface.

Two fundamental laws govern black-body radiation.

(1) *Stefan–Boltzmann Law.* The total energy emitted per unit time is proportional to the fourth power of the absolute temperature:

$$P = \varepsilon A \cdot \sigma T^4,$$

where P is the radiated power, A is the surface area, T is the absolute temperature in kelvin, and $\sigma = 5.67 \cdot 10^{-8}$ W/m^2K^4 is the Stefan–Boltzmann constant. The emissivity ε equals 1 for a perfect black body and 0 for a perfectly reflective surface. Earth behaves nearly as a black body, with $\varepsilon = 0.95$–0.98 for its surface (soil, oceans, vegetation) and $\varepsilon \approx 0.99$ for the atmosphere, particularly in the thermal emission band between 8 and 14 μm.

(2) *Wien's Displacement Law.* The wavelength of peak emission is inversely proportional to the absolute temperature:

$$\lambda_{max} \cdot T = b,$$

where $b \approx 2.898 \times 10^{-3}$ m·K. For the Sun, this yields a peak emission near 0.5 μm, at the center of the visible spectrum (0.38–0.75 μm).

[1] Pierrehumbert, R.T., 2010. Principles of Planetary Climate. Cambridge University Press.

[2] To convert from kelvin to Celsius: $T_{(°C)} = T_{(K)} - 273.15$. Thus, 0 °C corresponds to 273.15 K, and 100 °C to 373.15 K.

Both laws are derived from Planck's Law, which describes the spectral distribution of black-body radiation:

$$B(\lambda, T) = \frac{2hc^2}{\lambda^5} \cdot \frac{1}{e^{\frac{hc}{\lambda k_B T}} - 1}$$

where:

- $B(\lambda, T)$ is the energy emitted per unit surface area, per unit time, per unit wavelength, and per unit solid angle (spectral radiance);
- $h = 6.626 \times 10^{-34}$ J·s is Planck's constant;
- $c = 3.00 \times 10^8$ m/s is the speed of light;
- $k_B = 1.381 \times 10^{-23}$ J/K is Boltzmann's constant;
- λ is the wavelength in meters;
- T is the temperature of the black body in kelvin.

Incoming solar irradiance on Earth. The solar irradiance $I(t)$ reaching the top of Earth's atmosphere depends on the instantaneous Sun–Earth distance, $R_{SE}(t)$ (expressed in astronomical units), and on the *Total Solar Irradiance*, $TSI(t)$, whose mean annual value is approximately 1361 W/m², commonly referred to as the "*solar constant*" (Scafetta et al., 2019a; Kopp, 2025):[3]

$$I(t) = TSI(t)/R_{SE}(t).$$

Owing to Earth's albedo, spherical geometry, and rotation, this energy is not absorbed uniformly. Approximately 30% of the incoming solar radiation is reflected back into space, primarily by clouds, ice, oceans, and atmospheric scattering (e.g., Rayleigh scattering, which gives the sky its blue hue). The remaining 70% is absorbed:

- about 20% by the atmosphere;
- about 50% by the surface.

Assuming a uniform distribution across Earth's surface, the average net absorbed solar irradiance is:

$$I_{\text{in,Sun}} = \frac{I_{\text{TSI}} \cdot (1 - a) \cdot \pi R_E^2}{4 \cdot \pi R_E^2} \approx \frac{1361 \cdot 0.7}{4} \approx 238.2 \text{ W/m}^2,$$

where R_E is Earth's radius. This value represents the baseline energy flux that powers Earth's climate system.

Earth's infrared emission and energy balance. Earth itself emits thermal radiation as a black body. With an average surface temperature of 288 K (15 °C), its peak emission occurs at:

$$\lambda_{\text{max}} = \frac{2.898 \times 10^{-3}}{T} \approx \frac{2.898 \times 10^{-3}}{288} \approx 10\ \mu\text{m}.$$

This wavelength lies within the mid-infrared (m-IR) band (4–100 μm).

Energy equilibrium is achieved when the absorbed solar energy equals the emitted infrared radiation:

$$I_{\text{in,Sun}} = I_{\text{out,Earth}}.$$

[3]Scafetta, N., Willson, R.C., Lee, J.N., Wu, D.L., 2019. Modeling quiet solar luminosity variability from TSI satellite measurements and proxy models during 1980–2018. Remote Sensing, 11(21), 2569. Kopp, G., 2025. Solar Irradiance Measurements. Living Reviews in Solar Physics, 22, 1.

Applying the Stefan–Boltzmann law:

$$I_{\text{out,Earth}} = \sigma T_{\text{surf}}^4 = I_{\text{in,Sun}} \Rightarrow$$

$$T_{\text{surf}} = \left(\frac{238.2\,\text{W/m}^2}{5.67 \times 10^{-8}\,\text{W/m}^2\text{K}^4}\right)^{1/4} \approx 254.5\,\text{K} \approx -18.7\,^\circ\text{C}.$$

Yet Earth's actual average surface temperature is about 15 °C, roughly 33–34 °C warmer. This discrepancy is explained by the so-called *"atmospheric greenhouse effect"*.

The atmosphere as a thermal regulator. The infrared radiation emitted by Earth's surface is partially absorbed by greenhouse gases such as water vapor (H_2O), carbon dioxide (CO_2), methane (CH_4), ozone (O_3), and nitrous oxide (N_2O), which re-emit infrared radiation both upward and downward. The major atmospheric gases nitrogen (N_2) and oxygen (O_2), which together account for 99% of dry air, do not absorb infrared radiation.

This process produces a positive radiative feedback that increases the irradiance reaching the surface:

$$I_{\text{total received}} = I_{\text{in,Sun}} + I_{\text{back-IR}},$$

where $I_{\text{back-IR}} = I_{\text{IR}\downarrow}$ is the downward infrared flux emitted by the atmosphere.

Earth's surface therefore receives energy from two sources:

(1) direct solar radiation (visible, UV, near-IR; $\lambda < 4\,\mu\text{m}$);
(2) atmospheric back-radiation (far-IR; $\lambda > 4\,\mu\text{m}$).

This cumulative effect raises the global average surface temperature from −18.7 °C (in the absence of an atmosphere) to +15 °C, making Earth habitable for life as we know it.

4.2. Simplified greenhouse effect calculations

To estimate Earth's surface temperature under the natural greenhouse effect, we refine the basic "no-atmosphere" model. In this simplified approach, the atmosphere is treated as a single radiative layer that:

- absorbs all the infrared (IR) radiation emitted by the surface;
- re-emits the same amount upward and downward.

In this bidirectional model, the surface receives energy from two sources:

- $I_{\text{in, Sun}}$: direct solar radiation;
- $I_{\text{IR}\downarrow}$: infrared radiation re-emitted downward by the atmosphere, equal to the amount emitted toward space, $I_{\text{IR}\uparrow}$.

The surface energy balance then becomes:

$$\sigma T_{\text{surf}}^4 = I_{\text{in,Sun}} + I_{\text{IR}\downarrow}.$$

Since the atmosphere emits an equal infrared flux upward and downward, and the outgoing flux must match the absorbed solar energy:

$$F_{\text{out,Earth}} = I_{\text{IR}\uparrow} = I_{\text{IR}\downarrow} = I_{\text{in,Sun}}.$$

Hence:

$$\sigma T_{\text{surf}}^4 = I_{\text{in,Sun}} + I_{\text{IR}\downarrow} = 2I_{\text{in,Sun}} \Rightarrow T_{\text{surf}} = \left(2\frac{I_{\text{in,Sun}}}{\sigma}\right)^{1/4}.$$

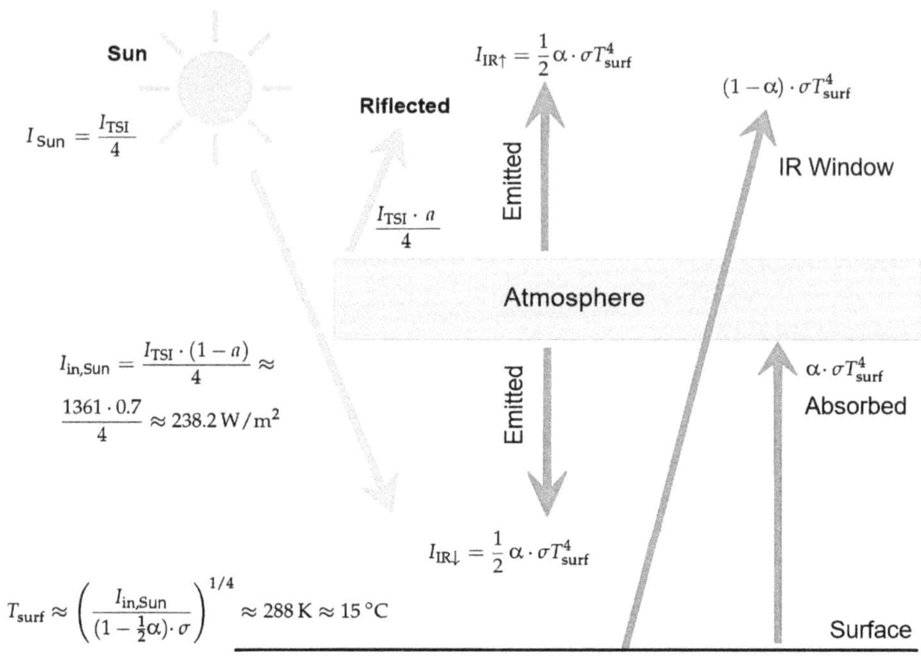

FIGURE 4.1. Educational model illustrating the greenhouse effect of the atmosphere.

Substituting the numerical values yields:

$$T_{surf} = \left(\frac{2 \cdot 238.2}{5.67 \times 10^{-8}}\right)^{1/4} \approx 303.3\,\text{K} \approx 30.1\,°\text{C}.$$

This idealized model, however, overestimates the actual global average temperature (~15 °C), indicating the need for further refinement.

The *Infrared Window Effect*. Not all the infrared radiation emitted by Earth's surface is absorbed by the atmosphere. A significant portion escapes directly to space through the 8–13 μm atmospheric infrared window. The associated transparency varies between 6% and 20%, depending on humidity, greenhouse gas concentrations, and cloud cover.

Energy balance equations. To account for this effect, the radiative balance equations are modified by introducing the transparency factor of the atmospheric window, as shown in Figure 4.1. We define an absorptivity parameter α (ranging between 0 and 1), with the transmittance given by $\tau = 1 - \alpha$.

The three basic energy balance equations of the system are:

(1) *surface balance*: the surface receives the solar radiation plus the infrared radiation re-emitted downward by the atmosphere, and emits:

$$\sigma T_{surf}^4 = I_{in,Sun} + I_{IR\downarrow};$$

(2) *atmospheric balance*: the atmosphere absorbs a fraction α of the radiation emitted by the surface and re-emits it symmetrically upward and

downward:

$$\alpha \cdot \sigma T_{\mathrm{surf}}^4 = I_{\mathrm{IR}\downarrow} + I_{\mathrm{IR}\uparrow} \Rightarrow I_{\mathrm{IR}\downarrow} = I_{\mathrm{IR}\uparrow} = \frac{1}{2}\alpha \cdot \sigma T_{\mathrm{surf}}^4;$$

(3) *top-of-atmosphere balance*: the energy emitted by Earth to space equals the sum of (i) the portion of the surface radiation that passes unimpeded through the atmosphere and (ii) the infrared radiation emitted upward by the atmosphere. This must equal the incoming solar energy:

$$F_{\mathrm{out}} = (1-\alpha) \cdot \sigma T_{\mathrm{surf}}^4 + I_{\mathrm{IR}\uparrow} = I_{\mathrm{in,Sun}}.$$

Combining these equations and assuming, as a first approximation, a single-layer atmosphere, one obtains:

$$(1-\alpha) \cdot \sigma T_{\mathrm{surf}}^4 + \frac{1}{2}\alpha \cdot \sigma T_{\mathrm{surf}}^4 = (1 - \frac{1}{2}\alpha) \cdot \sigma T_{\mathrm{surf}}^4 = I_{\mathrm{in,Sun}}.$$

The model behaves as follows:

- If $\alpha = 0$: fully transparent atmosphere → all infrared radiation escapes → $T_{\mathrm{surf}} = \left(\frac{I_{\mathrm{in,Sun}}}{\sigma}\right)^{1/4} \approx 255\ \mathrm{K}$
- If $\alpha = 1$: fully opaque atmosphere → maximum greenhouse effect → $T_{\mathrm{surf}} = \left(2\frac{I_{\mathrm{in,Sun}}}{\sigma}\right)^{1/4} \approx 303\ \mathrm{K}$
- If $\alpha = 0.78$: partially transparent atmosphere →

$$T_{\mathrm{surf}} \approx \left(\frac{I_{\mathrm{in,Sun}}}{(1-\frac{1}{2}\alpha)\cdot\sigma}\right)^{1/4} = \left(\frac{238.2}{(1-\frac{1}{2}0.78)\cdot 5.67\times 10^{-8}}\right)^{1/4} \approx 288\ \mathrm{K} \approx 15\ °\mathrm{C}.$$

This value of T_{surf} matches the observed global average surface temperature.

Estimating atmospheric temperature. To estimate the atmospheric temperature T_{atm} responsible for the infrared emission to space:

$$\alpha \cdot \sigma T_{\mathrm{atm}}^4 + (1-\alpha) \cdot \sigma T_{\mathrm{surf}}^4 = I_{\mathrm{in,Sun}}.$$

Solving:

$$T_{\mathrm{atm}} = \left(\frac{238.2 - 0.22 \cdot 5.67\times 10^{-8} \cdot 288^4}{0.78 \cdot 5.67 \times 10^{-8}}\right)^{1/4} \approx 242\ \mathrm{K} \approx -31\ °\mathrm{C}.$$

According to the *International Standard Atmosphere* (ISA), this temperature corresponds to an altitude of approximately 7 km, within the upper troposphere, where most of the infrared radiation emitted to space originates.

Observational confirmation from space. Satellite measurements confirm the greenhouse effect. Figure 4.2 shows:

- visible-light images show surface features clearly, since the atmosphere is largely transparent to shortwave radiation;
- infrared images reveal a thermal "fog", indicating that most of the infrared radiation emitted to space originates from clouds, water vapor,

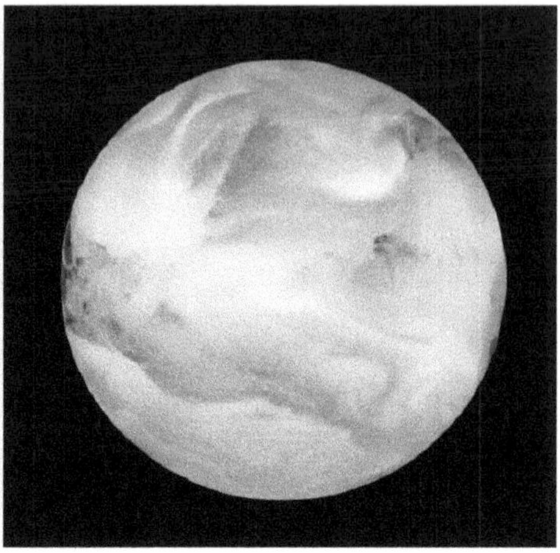

FIGURE 4.2. Satellite image of Earth in visible light (left) and infrared light (right) emitted from Earth to space. The visible image displays solar radiation reflected by the surface and clouds. The infrared image shows radiation predominantly emitted by greenhouse gases (H_2O, CO_2, etc.) and atmospheric clouds, with only a minor contribution from the surface itself.

and other greenhouse gases in the upper troposphere, rather than directly from the surface.

This confirms that the atmosphere acts as a thermal blanket that absorbs and re-emits infrared radiation, thereby modulating Earth's energy release to space.

Robinson & Catling (2014)[4] show that thick planetary atmospheres develop a tropopause near the 0.1-bar pressure level. Their model couples radiative transfer with convection, demonstrating that the transition from a convective lower atmosphere to a radiative upper atmosphere occurs at a characteristic pressure largely determined by infrared opacity. This yields a nearly universal tropopause level across planets endowed with dense atmospheres.

Conclusion: climate equilibrium. In the absence of an atmosphere, Earth's surface temperature would be around –18.7 °C. Owing to the natural greenhouse effect, the actual average value is +15 °C, which is essential for sustaining life.

Earth's climate system is governed by a delicate balance among:

- solar radiation (TSI $\approx$ 1361 W/m^2);
- planetary albedo ($\sim$30% reflection);
- atmospheric absorption and re-emission;
- greenhouse gases (H_2O, CO_2, CH_4, O_3, N_2O).

The climate behaves as a planetary heat engine that redistributes energy across the atmosphere, oceans, biosphere, and lithosphere, and ultimately releases it as infrared radiation. This balance is shaped by forcings, feedbacks, and nonlinear responses operating across timescales ranging from hours to millennia.

[4]Robinson, T.D., Catling, D.C., 2014. Common 0.1 bar tropopause in thick atmospheres set by pressure-dependent infrared transparency. Nature Geoscience, 7(1), 12–15.

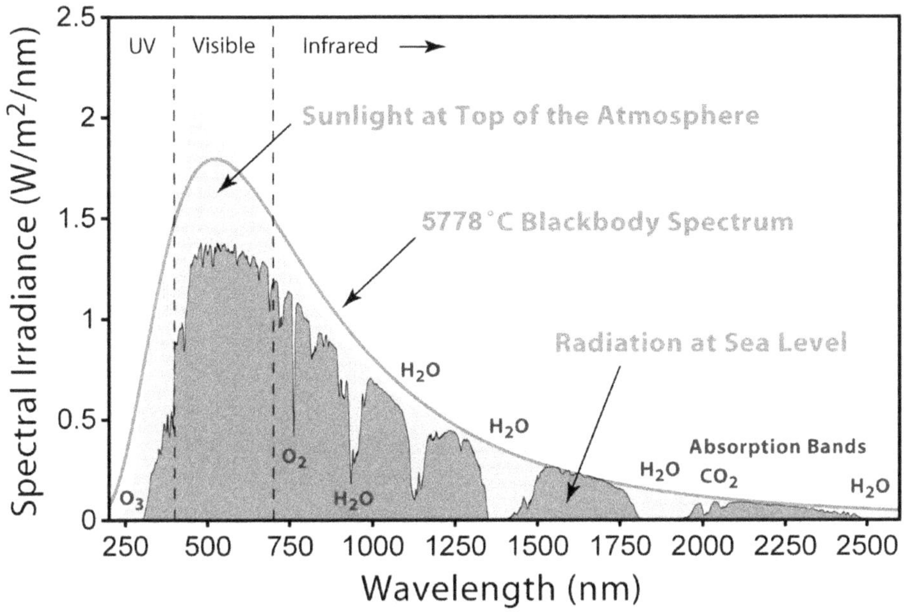

FIGURE 4.3. Spectrum of solar radiation as a function of wavelength (1000 nm = 1 μm), both at the top of the atmosphere and at the surface. The difference between the two curves corresponds to atmospheric absorption. The absorption bands of ozone (O_3), water vapor (H_2O), and carbon dioxide (CO_2) are highlighted.

4.3. Complexity of absorption spectra

Understanding how solar and terrestrial radiation interact with Earth's atmosphere requires examining the spectral behavior of greenhouse gases and the manner in which they absorb and transmit energy across different wavelengths.

Solar radiation and atmospheric filtering. Figure 4.3 compares the solar irradiance spectrum at the top and at the bottom of the atmosphere. The Sun emits radiation with a spectral distribution similar to that of a black body at about 5778 K. As sunlight travels through Earth's atmosphere, part of its energy is absorbed by greenhouse gases within specific spectral bands. Additional losses arise from Rayleigh scattering, which preferentially scatters shorter wavelengths and gives the sky its blue color.

The curves shown are based on ASTM reference spectra, which are widely used in the photovoltaic industry to standardize testing conditions. They highlight the ultraviolet, visible, and near-infrared regions of the solar spectrum.

Selective absorption by greenhouse gases. Figure 4.4 illustrates that greenhouse gases absorb radiant energy selectively rather than uniformly. The key observations include:

- water vapor (H_2O) and carbon dioxide (CO_2) absorb in distinct yet partially overlapping bands;

- CO_2 is especially effective in the 12–15 μm range, where water vapor absorbs only weakly;
- this enables CO_2 to partially close the atmospheric window through which infrared radiation emitted by the surface would otherwise escape directly to space.

Many absorption bands are saturated, meaning that they already absorb nearly all the radiation at those wavelengths. In such regions, further increases in gas concentration yield diminishing returns. However, at the edges of these bands, where absorption is incomplete, even small increases in concentration can broaden the active range, thereby increasing the total energy retained within the atmosphere.

The Beer–Lambert Law and logarithmic absorption. This behavior is governed by the Beer–Lambert Law, which describes how transmittance decreases exponentially with the concentration of an absorbing gas. As a consequence, the absorbance increases logarithmically. This implies that each doubling of the atmospheric CO_2 concentration—from 200 to 400 ppm, 400 to 800 ppm, 800 to 1600 ppm, and so on—produces a roughly equivalent increase in radiative forcing, consistent with the logarithmic nature of infrared absorption.

Estimating contributions to the Greenhouse Effect. Owing to the overlapping absorption bands and to their dependence on meteorological conditions, the individual contribution of each greenhouse gas cannot be determined with absolute precision. Even the IPCC does not provide exact percentages.

General estimates suggest:

- water vapor: about 70% of surface warming;
- CO_2: about 20–25%;
- other gases (CH_4, O_3, N_2O): the remainder.

However, Coe et al. (2021),[5] , using HITRAN spectral data, estimate:

- water vapor: about 29.4 K (about 89%);
- CO_2: about 3.3 K (about 10%);
- others: about 1%;

out of a total of 33 K of warming attributable to the presence of the atmosphere.

The ambivalent role of the greenhouse effect. The natural greenhouse effect is essential for life. Without it, Earth's surface would be too cold to sustain liquid water, and temperature swings between day and night would be extreme. The atmosphere acts as a climate stabilizer, maintaining thermal equilibrium. However, this balance is altered when greenhouse gas concentrations rise as a result of human activities such as fossil fuel combustion, deforestation, and intensive agriculture. Increased CO_2 and CH_4 levels enhance atmospheric opacity, reducing the fraction of infrared radiation that escapes through the atmospheric window and thereby modifying the radiative balance.

[5] Coe, D., Fabinski, W., Wiegleb, G., 2021. The Impact of CO_2, H_2O and Other "Greenhouse Gases" on Equilibrium Earth Temperatures. International Journal of Atmospheric and Oceanic Sciences, 5(2), 29–40.

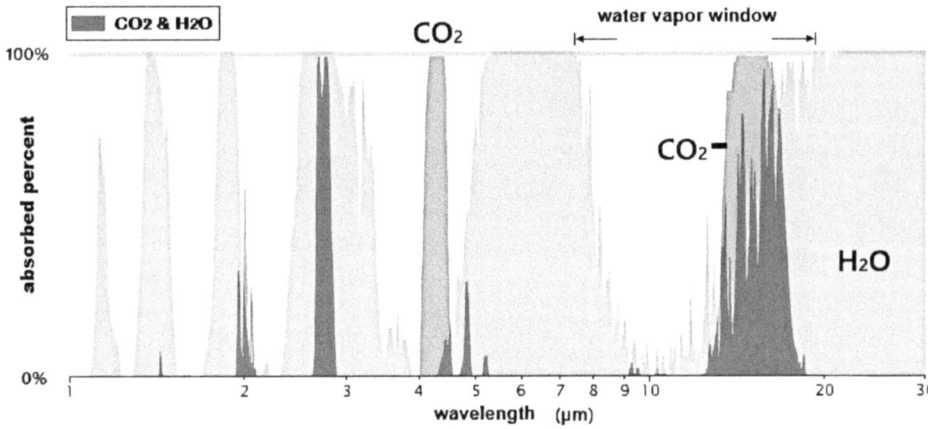

FIGURE 4.4. Absorption bands of water vapor (H_2O) and carbon dioxide (CO_2).

This phenomenon—known as greenhouse effect amplification—is the underlying mechanism identified by the IPCC as the primary driver of modern global warming.

Fragility of equilibrium and predictive complexity. Earth's climate system is highly sensitive to small changes in key parameters:

- albedo shifts (e.g., melting ice);
- cloud cover and ocean circulation;
- natural events (volcanic eruptions, solar cycles);
- anthropogenic forcings.

Because these components are nonlinear and interdependent, the climate may respond with rapid or unexpected transitions. Strong positive feedbacks—such as the methane release from thawing Arctic permafrost—can amplify warming and drive the system toward new, potentially less stable equilibria.

A historical example is the Paleocene–Eocene Thermal Maximum (PETM), which occurred about 55–56 million years ago and was triggered by a massive methane release from oceanic clathrates, volcanic activity, and the oxidation of organic carbon. Global temperatures rose by 5–8 °C, marking one of the most intense warming episodes in Earth's history (Haynes & Hönisch, 2020).[6]

The IPCC interpretation of the energy budget. Figure 4.5 summarizes Earth's average energy budget at equilibrium, based on IPCC (2021, their Figure 7.2).

A detailed breakdown of its main components is given below.

- *Incoming energy (shortwave):*
  - 340 W/m^2 reaches, on average, the top of Earth's atmosphere—one quarter of the Total Solar Irradiance (TSI), as a consequence of Earth's spherical geometry (Figure 4.1);
  - 100 W/m^2 are reflected (albedo $\approx$ 30%);

[6] Haynes, L.L., Hönisch, B., 2020. The seawater carbon inventory at the Paleocene–Eocene Thermal Maximum. Proceedings of the National Academy of Sciences, 117(39), 24088–24095.

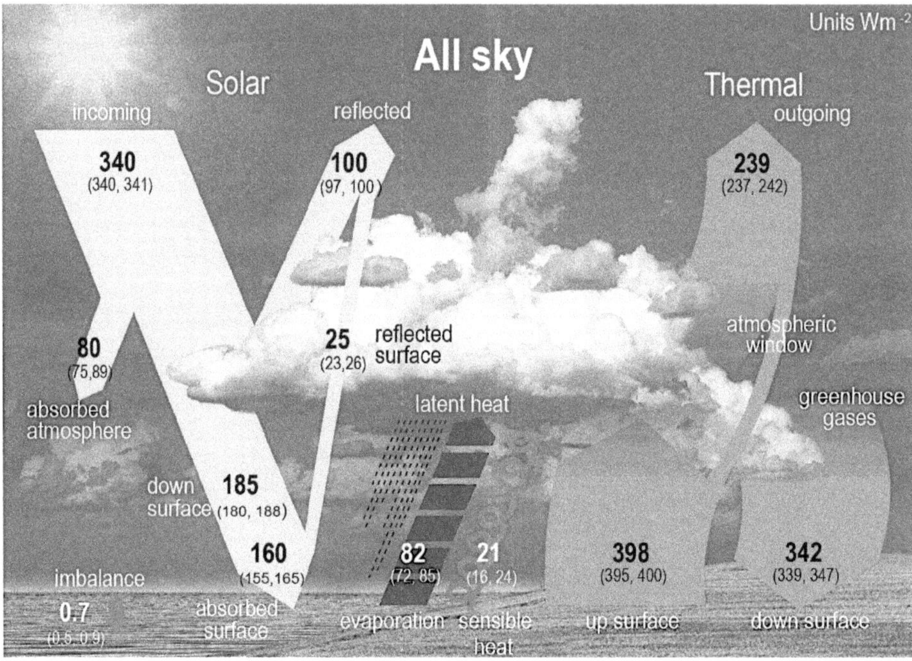

FIGURE 4.5. Earth's energy budget (IPCC, 2021, figure 7.2). The diagram illustrates how the natural greenhouse effect retains part of the outgoing infrared radiation and how greenhouse gases modulate the atmosphere's infrared transparency.

- 240 W/m² are absorbed by the Earth–atmosphere system: 80 W/m² by the atmosphere and 160 W/m² by the surface.

- *Outgoing energy (longwave/infrared):*
  - 398 W/m² are emitted by Earth's surface, but only 40 W/m² escape directly to space; the remainder is absorbed by greenhouse gases and clouds.
- *Atmospheric emission:*
  - 199 W/m² emitted upward to space.
  - 342 W/m² emitted downward to the surface (back-radiation).
- *Non-radiative fluxes:*
  - 82 W/m² via evaporation (latent heat).
  - 21 W/m² via convection (sensible heat).
- *Net balance:*
  - Incoming energy (about 240 W/m²) $\approx$ Outgoing energy;
  - A small residual imbalance (about 0.7–1.0 W/m²) reflects an ongoing planetary warming.

4.4. Toward increasingly complex models

Understanding Earth's energy balance is not a purely theoretical exercise; it amounts to tracing the physical foundations of the climate system. Only through a rigorous knowledge of these mechanisms is it possible to distinguish between unfounded alarmism and genuine risk signals. Climate equilibrium is neither guaranteed nor immutable: it is the product of a long geophysical and geological history, now increasingly influenced by human activity.

The formulas presented earlier are idealized approximations. Although useful for grasping the basic structure of the global energy budget, they omit many critical factors that shape the actual behavior of the climate system.

In reality, Earth is:

- a rotating sphere, tilted relative to the ecliptic;
- covered by a heterogeneous surface (oceans, deserts, forests, ice);
- orbiting the Sun along an elliptical path.

These features introduce anisotropy and spatio-temporal variability, making the climate system far more complex than any static model can capture.

Additional complexity arises from the fact that the efficiency of the greenhouse effect depends on atmospheric pressure. Below approximately 0.1 atm, the atmosphere becomes too tenuous to sustain significant infrared absorption and re-emission (Robinson & Catling, 2014).[7]

Angular distribution of solar irradiance. Rather than assuming a uniform solar irradiation, we must account for the angle of incidence of the solar rays, which varies with time and location. The solar energy received per unit surface area is proportional to the cosine of the solar zenith angle θ_z,[8]:

$$I_{\text{received}}(\varphi, \lambda, t) = \frac{I_{\text{TSI}}}{R(t)^2} \cdot (1 - a(\varphi, \lambda)) \cdot \cos(\theta_z(\varphi, \lambda, t)) \cdot H(t),$$

where:

- φ, λ : latitude and longitude;
- t: time (hour, day, season);
- $H(t)$: 1 during daylight, 0 at night;
- $\theta_z(\varphi, t)$ depends on solar declination, local time, and latitude;
- $a(\varphi, \lambda)$: local albedo;
- $R(t)$ Earth–Sun distance (in AU).

This correction explains why tropical regions receive more energy than the polar ones, and why seasonal and diurnal cycles are fundamental to climate dynamics.

Effects of Earth's rotation. Earth's rotation introduces the day–night cycle, causing irradiance to vary over time. To account for this variability, irradiance must be averaged over the relevant time intervals:

$$I_{\text{average}}(\varphi, \lambda) = \frac{1}{\Delta t} \int_{\Delta t} I_{\text{received}}(\varphi, \lambda, t)\, dt,$$

where Δt may correspond to a day, a month, or a year, depending on the scale of analysis.

Spatial variation of surface properties. Earth's surface is heterogeneous: oceans, ice, vegetation, and urban areas differ in albedo, thermal capacity, and

[7]Robinson, T.D., Catling, D.C., 2014. Common 0.1 bar tropopause in thick atmospheres set by pressure-dependent infrared transparency. Nature Geoscience, 7(1), 12–15.

[8]The zenith angle of a point or object is the angle formed between the observer's local vertical and the line of sight to the object.

emissivity. Consequently, energy absorption and re-emission vary significantly across regions. Climate simulations therefore divide the surface into grid cells and compute the energy balance within each:

$$I_{\text{eff}}(\varphi, \lambda) = I_{\text{average}}(\varphi, \lambda) \cdot (1 - a(\varphi, \lambda)) - L_{\text{IR}}(\varphi, \lambda),$$

where $a(\varphi, \lambda)$ is the local albedo and $L_{\text{IR}}(\varphi, \lambda)$ is the local infrared emission. This approach yields a spatially resolved description of Earth's energy budget, bringing it closer to the framework of full-scale climate models.

How a Global Climate Model (GCM) works. A Global Climate Model (GCM) (see Figure 3.2) simulates the evolution of Earth's climate by dividing the atmosphere, oceans, and surface into a three-dimensional grid of cells. Within each cell, the model solves the governing equations for:

- energy flows (solar input, infrared output);
- mass transport (air and water movement);
- momentum exchange (winds, ocean currents).

Each cell accounts for:

- solar irradiance, corrected for the angle of incidence;
- albedo, absorption, and infrared emission;
- Earth's rotation, including the Coriolis effect;
- surface properties (forests, ice, oceans);
- humidity, cloud cover, and greenhouse gases.

The model integrates these components into a dynamic system that simulates:

- atmospheric circulation;
- cloud formation and precipitation;
- ice formation and melting;
- seasonal cycles and long-term climate trends.

In this way, GCMs generate future climate projections based on different anthropogenic emission scenarios, providing insights into the potential outcomes under various policy choices. The crucial question, however, remains: how reliable are these simulations?

CHAPTER 5

Evolution of the IPCC reports from 1990 to 2023

5.1. The anthropogenic global warming theory and the IPCC

The Anthropogenic Global Warming Theory (AGWT) has become the central framework through which institutional climate science interprets the evolution of Earth's climate. Its most authoritative expression is found in the work of the Intergovernmental Panel on Climate Change (IPCC), established in 1988 under the auspices of the United Nations. The IPCC's mandate is to gather, assess, and synthesize scientific knowledge on climate change, providing policymakers with a structured and evolving reference framework.

AGWT proposes that the rise in global temperatures since the 1850–1900 preindustrial period is primarily the result of human activity. The combustion of fossil fuels, widespread deforestation, and the intensification of agriculture have led to a significant increase in atmospheric greenhouse gases. This alteration of the planet's radiative balance is regarded as the principal driver of modern warming.

Although carbon dioxide remains the most persistent and widely discussed greenhouse gas, methane and nitrous oxide also play important roles, particularly because of emissions from livestock farming and the use of chemical fertilizers. The warming effect is further intensified by the reduction of Earth's albedo—especially through the melting of glaciers and sea ice—and by the increase in atmospheric water vapor. These changes interact with oceanic and atmospheric systems in complex ways, amplifying the overall effect.

Since the 1990s, AGWT has gained increasing scientific support, owing to advances in paleoclimatology, the development of numerical climate models, and a deeper understanding of feedback mechanisms within the climate system. Reconstructions of past climates—derived from ice cores, marine sediments, and tree rings—enable researchers to place current warming in its historical context. At the same time, global climate models have become more sophisticated, allowing simulations that link emissions to specific climatic outcomes.

This body of evidence is reflected in the IPCC's Sixth Assessment Report (AR6), published between 2021 and 2023. The volume entitled *Climate Change 2021: The Physical Science Basis,*[1] produced by Working Group I, attributes the observed warming to human activities with a confidence level exceeding 95%. The report estimates a global temperature increase of 1.09 °C between the preindustrial baseline and the decade 2011–2020, with approximately 1.07 °C directly linked to anthropogenic influence. Notably, warming has been more pronounced over land than over the oceans.

[1]IPCC, 2021. Climate Change 2021: The Physical Science Basis. Cambridge University Press.

Beyond temperature trends, the report highlights a possible increase in the frequency and intensity of extreme events such as heatwaves, heavy rainfall, droughts, and wildfires, which are already affecting ecosystems and human societies. In response, the IPCC advocates a coordinated approach that includes the energy transition, climate adaptation, and international cooperation, particularly in support of vulnerable regions.

Yet AGWT remains a scientific construct rather than an immutable truth. Its strength lies in its capacity to evolve through critique, revision, and the integration of new data. Uncertainties persist, particularly with respect to long-term projections, the response times of geophysical subsystems, and the potential risks of mitigation strategies that may prove economically or socially unbalanced. For this reason, it is essential to maintain a critical and interdisciplinary perspective that aligns scientific rigor with economic realism and social responsibility.

The next section traces the evolution of AGWT across successive IPCC reports, examining how the theory has developed in parallel with our understanding of climate history over the past millennium.

5.2. From plausibility to certainty: AGWT's epistemic path

Tracing the sequence of the six IPCC Assessment Reports reveals the gradual consolidation of AGWT, transforming it from a plausible hypothesis into the dominant explanatory framework of contemporary climate science. This evolution reflects not only the accumulation of empirical evidence but also a shift in epistemological posture—from interpretive caution to growing confidence in attribution.

The First Report (FAR, 1990): initial hypotheses. The *First Assessment Report* (FAR),[2] published in 1990, addressed the question of human-induced climate change with notable restraint. Although it acknowledged that greenhouse gas emissions could contribute to rising global temperatures, the report emphasized that the climate variations observed throughout the 20th century might still be explained, at least in part, by natural variability.

A significant portion of the discussion drew upon the pioneering work of Hubert H. Lamb,[3] whose 1965 study reconstructed past climate episodes such as the Medieval Warm Period (MWP) and the Little Ice Age (LIA). Lamb's reconstruction, based on a wide array of climate proxies ranging from historical documents and botanical records to glaciological and oceanographic data, suggested that climate variations were recurrent and regionally significant. His work, focused mainly on northwestern Europe, was included in FAR without a defined temperature scale, serving more as a qualitative proxy than as a quantitative benchmark.

This paleoclimate reconstruction can be compared, for example, with the historical temperature series of Central England known as the "*Central England*

[2] IPCC, 1990. Climate Change: The IPCC Scientific Assessment. Cambridge University Press.

[3] Lamb, H.H., 1965. The early medieval warm epoch and its sequel. Palaeogeography, Palaeoclimatology, Palaeoecology, 1, 13–37. FAR includes Lamb's paleoclimate reconstruction in their Figure 7.1 without specifying the temperature scale, as it is presented as a proxy for global temperatures.

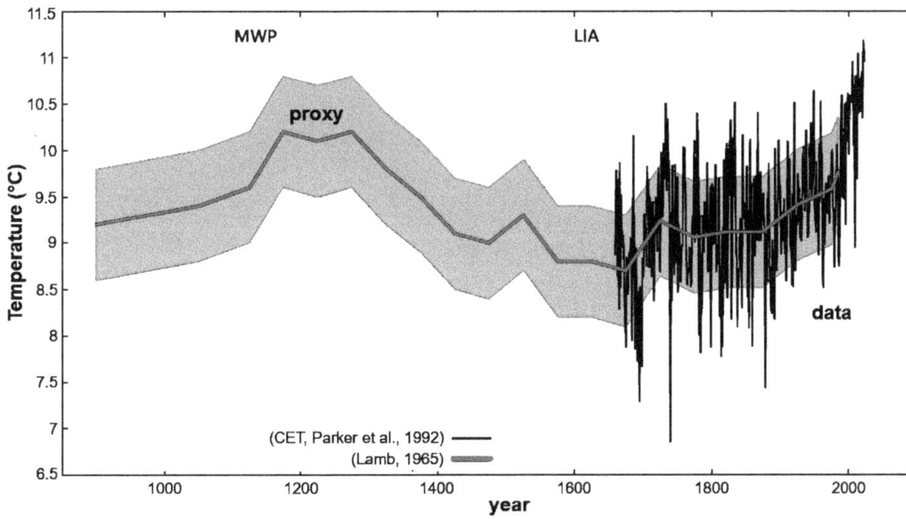

FIGURE 5.1. Black curve: Historical temperature series of Central England (CET) (Hadley Centre Central England Temperature - HadCET dataset; Parker, 1992). Gray curve: Reconstruction of annual mean temperatures in England using a multi-proxy climate model (Lamb, 1965), calibrated and extended to 1985 with the CET.

Temperature" (CET) (Parker, 1992).[4] The comparison is shown in Figure 5.1, which suggests that medieval England may have been as warm as in recent decades.

One of Lamb's most evocative indicators of medieval warmth was the spread of viticulture in England. Drawing on data from the Domesday Book (1086),[5] Lamb noted the presence of 42 vineyards across various counties, including some north of the 53rd parallel. This expansion, fostered by warmer and drier summers, allowed England to compete with continental wine-producing regions. The subsequent disappearance of these vineyards at the onset of the LIA constituted, for Lamb, a clear indication of how human activities could be shaped by natural climate shifts.

In this context, FAR represented a foundational moment in the development of AGWT. It acknowledged the potential influence of human activities while maintaining a balanced view, leaving ample room for natural dynamics in the interpretation of climate change.

[4]Parker, D.E., T.P. Legg, C.K. Folland, 1992. A new daily Central England Temperature Series, 1772–1991. International Journal of Climatology, 12, 317–342.

[5]The Domesday Book, compiled in 1086 by order of William the Conqueror, is one of the oldest agrarian inventories in medieval Europe and a key source for studying the economy and social structure of Norman England. Among its many records—lands, pastures, mills, forests, and tax revenues—it lists 42 vineyards located in various counties of England and Wales. These vineyards were often associated with religious institutions and mostly located in southern England, favored by a milder climate. However, some were situated north of the 53rd parallel, suggesting an unusually mild climate for the time, which is consistent with the onset of the Medieval Warm Period.

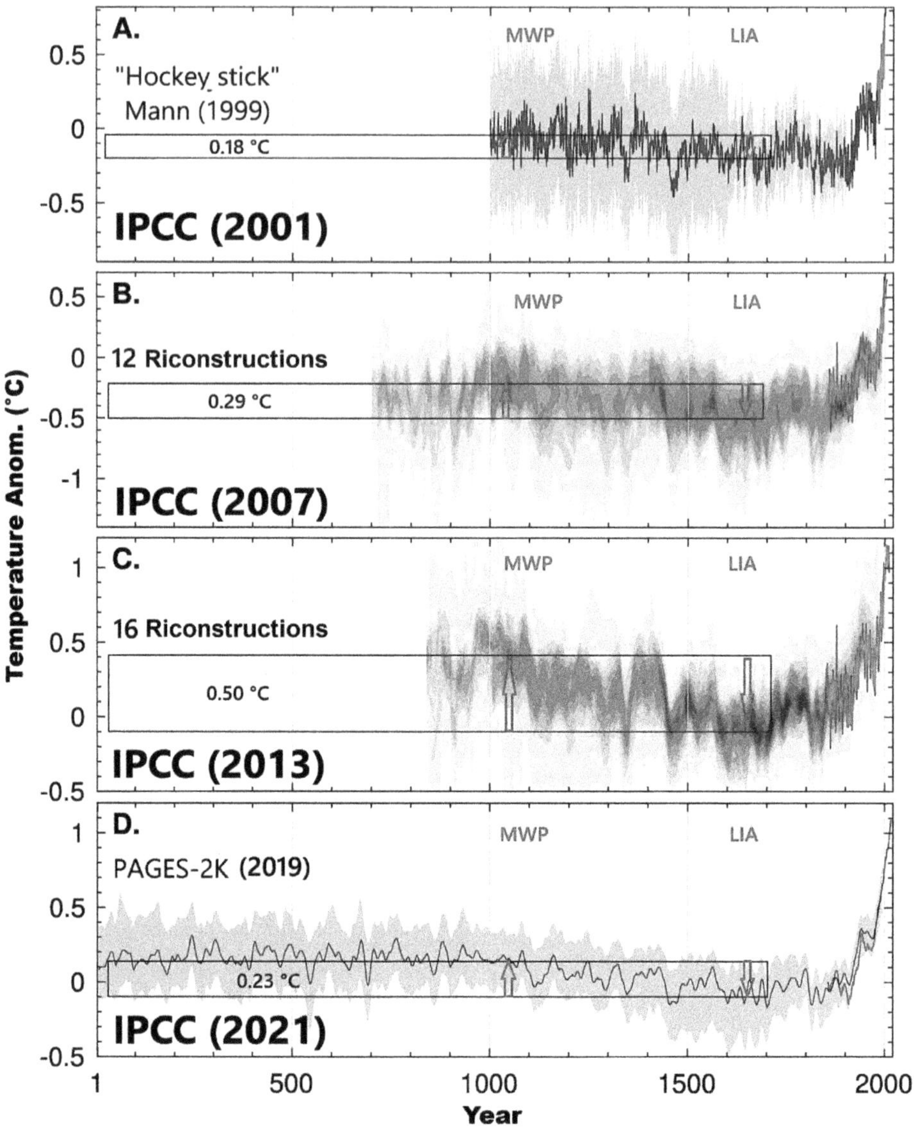

FIGURE 5.2. Comparative sequence of major temperature graphs for the Northern Hemisphere and globally published in IPCC reports from TAR (2001) to AR6 (2021), illustrating the transition from single reconstructions (e.g., Mann et al., 1999) to multiple-curve ensembles (AR4, AR5) and finally to the integrated dataset PAGES-2K Consortium (2019) in AR6. Instrumental temperature data are shown since 1850. Temperatures in the boxes indicate the difference between the Medieval Warm Period (MWP) and the Little Ice Age (LIA). (Adapted from Anchukaitis & Smerdon, 2022).

The Second Report (SAR, 1995): first anthropogenic signals. The *Second Assessment Report* (SAR),[6] released in 1995, marked a significant epistemological shift. For the first time, the IPCC stated that there was *"evidence suggesting a discernible human influence on global climate."* Although still cautious in tone, this formulation was more assertive than the one adopted in FAR.

SAR estimated that roughly half of the warming observed during the 20th century could be attributed to anthropogenic greenhouse gas emissions. It also acknowledged the mid-century cooling phase (1940s–1970s), likely influenced by natural variability and by the increased aerosol emissions associated with human activity. This nuanced interpretation began to position humanity not merely as a possible contributor but as a measurable agent within the climate system.

The Third Report (TAR, 2001): the era of the *"Hockey Stick"*. The *Third Assessment Report* (TAR),[7] published in 2001, introduced one of the most iconic and controversial images in climate science: the *"hockey-stick graph"* developed by Mann, Bradley, and Hughes (1999).[8] Based on tree-ring proxies from the Northern Hemisphere, the graph depicted a millennium of relatively stable temperatures followed by a sharp warming in the 20th century. Being nearly flat for nine centuries and then curving upward after 1900, it resembled a *"hockey stick"* and conveyed a powerful message: recent warming was unprecedented over the last thousand years. This graph is shown in Figure 5.2A.

The image quickly became emblematic of the climate change narrative, especially after its inclusion in Al Gore's documentary *An Inconvenient Truth* (2006),[9] which helped bring the issue to global public attention and contributed to Gore's Nobel Peace Prize in 2007.

However, the graph also sparked an intense scientific debate. Critics such as Soon & Baliunas (2003) and Soon et al. (2003a),[10] McIntyre & McKitrick (2003),[11] and Deming (2006)[12] questioned the statistical methods, the robustness of the proxy network, and the apparent suppression of well-known climate episodes such as the MWP and LIA. They also raised concerns about the reliability of tree-ring proxies as indicators of hemispheric or global temperature.

Despite the controversy, TAR marked a turning point. It reinforced the view that current warming patterns were anomalous relative to historical variability and shifted the scientific discourse toward recognizing anthropogenic forcing as the primary driver of 20th-century climate change. The *"hockey stick"* became

[6]IPCC, 1995. Climate Change 1995: The Science of Climate Change. Cambridge University Press.
[7]IPCC, 2001. Climate Change 2001: The Scientific Basis. Cambridge University Press.
[8]Mann, M.E., Bradley, R.S., Hughes, M.K., 1999. Northern hemisphere temperatures during the past millennium: inferences, uncertainties, and limitations. Geophysical Research Letters, 26, 759–762.
[9]Gore, A., 2006. An Inconvenient Truth [Film documentary]. Paramount Classics.
[10]Soon, W., Baliunas, S., 2003. Proxy climatic and environmental changes of the past 1000 years. Climate Research, 23(2), 89–110.
Soon, W., Baliunas, S., Idso, C., Idso, S., Legates, D.R., 2003a. Reconstructing climatic and environmental changes of the past 1000 years: A reappraisal. Energy and Environment, 14(2–3), 233–296.
[11]McIntyre, S., McKitrick, R., 2003. Corrections to the Mann et al. (1998) proxy database and Northern Hemispheric average temperature series. Energy & Environment, 14(6), 751–771.
[12]Deming, D., 2006. U.S. Senate Hearing on Climate Change and the Media (Dec 6, 2006).

not only a scientific artifact but also a symbol of the epistemological transition from plausibility to growing certainty.

The Fourth Report (AR4, 2007): consolidation of scientific consensus. Published in 2007, the *Fourth Assessment Report* (AR4) [13] marked a decisive moment in the institutional endorsement of AGWT. Despite the lingering controversy surrounding the *"hockey-stick"* graph, the IPCC reaffirmed its validity and more strongly attributed 20th-century warming to human activities. This conclusion was supported by simulations from the CMIP3 project, which employed General Circulation Models (GCMs) capable of simulating coupled atmospheric and oceanic interactions.

AR4 presented a more comprehensive view of climate history, incorporating twelve paleoclimate reconstructions. The climate history of the last millennium according to AR4 is shown in Figure 5.2B. Across these reconstructions, the estimated cooling from the MWP to the LIA averages about 0.29 °C— slightly greater than that of Mann's original *"hockey stick"*, but still much smaller than the warming observed during the 20th century. The report concluded that it was *"very likely"* (more than 90% probability) that more than 90% of post-1950 warming was due to anthropogenic influence, particularly greenhouse gas emissions.

This represented a consolidation of AGWT: no longer a peripheral hypothesis, it became central to contemporary climate science and to the broader scientific and political discourse. Yet debates concerning natural climate variability persisted, especially among critics who questioned the declared level of certainty and the scope of model-based attribution.

The Fifth Report (AR5, 2013): a revision of the paleoclimate framework. The *"Fifth Assessment Report"*, (IPCC, 2013)[14] released in 2013, further strengthened the attribution of climate change to human activity. It stated that it was *"extremely likely"* (95–100% probability) that human influence was the dominant cause of global warming since the mid-20th century, on the basis of simulations from the CMIP5 GCMs.

At the same time, AR5 introduced a more nuanced view of preindustrial climate variability. New reconstructions—such as those by Moberg et al. (2005),[15] Ljungqvist (2010),[16] Mann et al. (2008),[17]—pointed to a greater natural variability than previously acknowledged. Some reconstructions indicated cooling between the MWP and LIA of up to 1 °C, notably in the work of Christiansen & Ljungqvist (2012),[18] which challenged the near-flatness of earlier temperature

[13] IPCC, 2007. Climate Change 2007: The Physical Science Basis. Cambridge University Press.

[14] IPCC, 2013. Climate Change 2013: The Physical Science Basis. Cambridge University Press.

[15] Moberg, A., Sonechkin, D.M., Holmgren, K., Datsenko, N.M., Karlén, W., 2005. Highly variable Northern Hemisphere temperatures reconstructed from proxy data. Nature, 433, 613–617.

[16] Ljungqvist, F.C., 2010. A new reconstruction of temperature variability in the extra-tropical northern hemisphere during the last two millennia. Geografiska Annaler: Series A, Physical Geography, 92, 339–351.

[17] Mann, M., Zhang, Z., Hughes, M., Bradley, R., Miller, S., et al., 2008. Proxy-based reconstructions of hemispheric and global surface temperature variations over the past two millennia. Proceedings of the National Academy of Sciences, 105(36), 13252–13257.

[18] Christiansen, B., Ljungqvist, F.C., 2012. The extra-tropical Northern Hemisphere temperature in the last two millennia: Reconstructions of low-frequency variability. Climate of the Past, 8, 765–786.

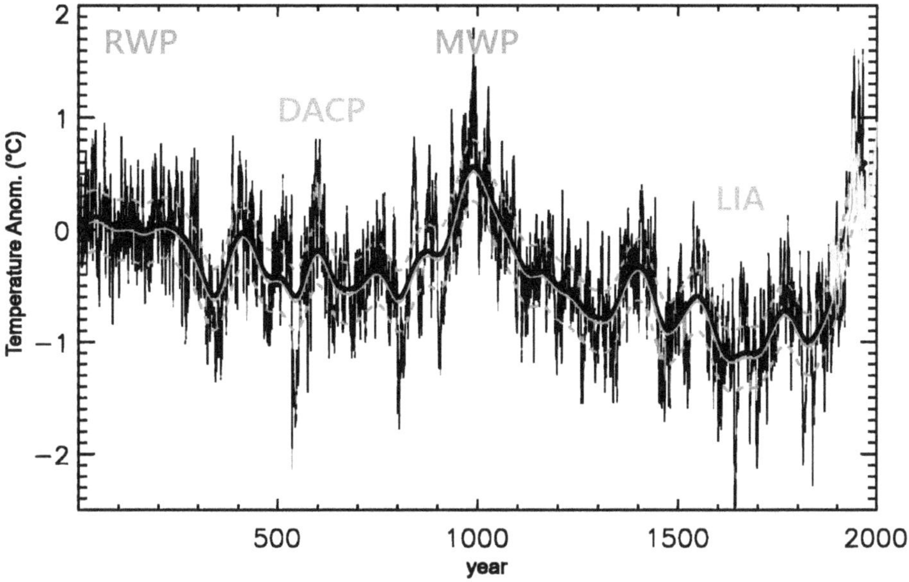

FIGURE 5.3. Reconstruction of extra-tropical Northern Hemisphere temperatures. Calibration period: 1880–1960 AD. (Adapted from Christiansen & Ljungqvist, 2012).

profiles. The climate history of the last millennium according to AR5 is shown in Figure 5.2C, based on a set of 16 paleoclimate reconstructions.

As a result, the original *"hockey-stick"* graph quietly receded from prominence: the revised reconstructions were included without explicitly rejecting their earlier versions. Instead, the IPCC adopted a more diversified portrayal of the preindustrial climate, reflecting a broader range of proxy data and methodological approaches. AGWT, however, remained firmly supported by the CMIP5 GCM simulations.

This period also saw a crisis of public trust. Between 2009 and 2011, the unauthorized release of emails from the Climate Research Unit (CRU) at the University of East Anglia—an event dubbed *"Climategate"*[19]—attracted widespread media attention and prompted allegations of data manipulation. Critics such as Andrew Montford (2010)[20] and Ross McKitrick[21] argued that the emails revealed attempts to control peer review and suppress dissent, raising concerns about transparency and scientific integrity. Although multiple investigations cleared the scientists of misconduct, the episode underscored the importance of openness and methodological rigor in climate science.

[19]Available at: https://en.wikipedia.org/wiki/Climatic_Research_Unit_email_controversy

[20]Montford, A.W., 2010. The hockey stick illusion: Climategate and the corruption of science. London: Stacey International.

[21]Available at: https://www.rossmckitrick.com/uploads/4/8/0/8/4808045/climategate.10yearsafter.pdf

The Sixth Report (AR6, 2021–2023): toward definitive attribution? With the Sixth Assessment Report, published between 2021 and 2023, the IPCC reached a new level of coherence in its attribution framework. It concluded that the entire global warming observed between 1850 and 2020 was attributable, with very high probability, to human activities—particularly to greenhouse gas emissions. The anthropogenic contribution was quantified as a temperature increase between 0.8 and 1.3 °C, a figure now considered robust by most of the scientific community. AGWT was supported by the CMIP6 models.

AR6 also introduced a new paleoclimate reconstruction from the PAGES-2K Consortium (2019),[22] which offered an intermediate synthesis between the earlier *"hockey-stick"* profiles and more conservative reconstructions. This reconstruction is shown in Figure 5.2D: from the MWP to the LIA, the average cooling is only 0.23 °C, very similar to the *"hockey stick"* of Mann et al. (1999).

However, the decision to represent two millennia of climate history by means of a single reconstruction has raised important epistemological concerns. Esper et al. (2024a)[23] questioned the adequacy of this approach, warning against the oversimplification of complex climate dynamics. They argue that reducing the diversity of paleoclimate evidence to a single statistical synthesis risks obscuring the richness of historical variability.

Further methodological critiques, such as those by McKitrick (2019) and Crok & May (2023, p. 37-39)[24], highlight potential issues with proxy selection and data interpretation. These concerns suggest that the IPCC's choice may reflect not only scientific judgment but also a narrative preference that emphasizes anthropogenic influence while minimizing the role of natural variability.

In light of these developments, it is essential to preserve the plurality of paleoclimate reconstructions and to maintain a scientific framework open to debate, revision, and the continuous evolution of evidence. AGWT may be the prevailing theory, but its legitimacy depends on its capacity to withstand scrutiny and to adapt to new insights.

5.3. Ambiguities in the IPCC paleoclimate reconstructions

Figure 5.2 suggests that the reconstruction of climate history remains a central issue in the debate surrounding the Anthropogenic Global Warming Theory (AGWT). As discussed in previous chapters, understanding the magnitude, frequency, and persistence of natural climate variations over past centuries and millennia is essential to assess whether current global warming is truly unprecedented or whether it lies, at least in part, within the bounds of Earth's natural variability.

Over the years, several studies have challenged the paleoclimate reconstruction proposed by Mann et al. (1999), particularly with regard to the Medieval Warm Period (MWP). Among the most notable are those by Moberg et al. (2005) and Ljungqvist (2010), which suggest that the MWP may have been warmer than Mann reported.

---

[22]PAGES-2K Consortium, 2019. Consistent multi-decadal variability in global temperature reconstructions and simulations over the Common Era. Nature Geoscience, 12(8), 643–649.

[23]Esper, J., Smerdon, J.E., Anchukaitis, K.J., Cook, E.R., et al., 2024. The IPCC's reductive Common Era temperature history. Communications Earth & Environment, 5, 222.

[24]Crok, M., May, A. (Eds.), 2023. The Frozen Climate Views of the IPCC. Clintel Foundation.

Moberg et al. (2005) offer a reconstruction of Northern Hemisphere temperatures over the last two millennia by combining proxy data with different temporal resolutions: high-resolution tree rings and low-resolution marine and lake sediments. By means of a wavelet transform approach, the authors integrate these data into a multiscale framework. One key insight is that tree rings are well suited to capturing short-term climate fluctuations, whereas sedimentary proxies are more effective in revealing long-term trends. This methodological synthesis allows for a broader representation of multi-centennial variability, revealing a warmer MWP and a pronounced cooling in the 17th century. The study underscores the importance of integrating heterogeneous proxies in order to better understand the amplitude and dynamics of natural climate variability.

The contributions of Ljungqvist (2010) and Christiansen & Ljungqvist (2012)[25] further deepen this understanding. Ljungqvist (2010) reconstructs temperatures using a wide array of paleoclimate proxies—many of which had not previously been employed at the hemispheric scale—and clearly identifies the major climate phases of the Common Era, from the *Roman Warm Period* to the *Little Ice Age*. This reconstruction shows that the decadal mean temperatures during the MWP and RWP were comparable to, or even exceeded, those of recent decades. The amplitude of centennial-scale variability exceeds 0.6 °C, suggesting a substantial role for natural forcings.

In their 2012 study, Christiansen & Ljungqvist refine the methodology in order to preserve low-frequency variability, which is often dampened by traditional regression approaches. Their reconstruction, shown in Figure 5.3, confirms the presence of the RWP and MWP, with a notably deep LIA reaching −1.0 °C below the average of the last 2,000 years. The method is reported to be effective in reducing distortions and in retrieving coherent climate signals across diverse proxy datasets. A similar trend is observed in an independent reconstruction of annual mean temperatures in northeastern Europe (Klimenko et al., 2014), which reinforces the robustness of these findings.

Taken together, these studies suggest that natural climate variability has been more pronounced than some IPCC-favored reconstructions have indicated. They highlight the need for a more nuanced representation of climate history in order to properly contextualize and attribute modern warming.

Other studies have pointed to episodes such as the Holocene Climatic Optimum, which likely exceeded current thermal levels at regional and hemispheric scales. These findings underscore the importance of a better understanding of natural (solar, volcanic, and orbital) forcings and of their role in shaping long-term climate dynamics.

Variations in graphical representation and methodological choices. Beyond modeling techniques, the comparison among paleoclimate reconstructions also involves epistemological and communicative dimensions. The choice of proxy sources, aggregation methods, and graphical representations all influence how past climates are interpreted and communicated. A clear visual demonstration of this evolution is offered by the sequence of paleoclimate graphs published in the IPCC reports from TAR (IPCC, 2001) to AR6 (IPCC, 2021). The transition

[25]Christiansen, B., Ljungqvist, F.C., 2012. The extra-tropical Northern Hemisphere temperature in the last two millennia: Reconstructions of low-frequency variability. Climate of the Past, 8, 765–786.

is striking: from a single reconstruction in TAR (based on Mann et al., 1999), to multiple-curve ensembles in AR4 and AR5, and then back to a single reconstruction from the PAGES-2K Consortium (2019) in AR6. In this regard, Crok & May (2023, p. 35-36) note that references to the Medieval Warm Period (MWP) and to the Little Ice Age (LIA) quietly disappeared from the final version of the IPCC's Sixth Assessment Report (AR6, 2021), despite having been included in the First Order Draft (FOD) and further emphasized in the Second Order Draft (SOD) following a referee's comment. The significance of preindustrial climate variability was downplayed in the final version of AR6—apparently without referee approval—by asserting that these two episodes were not well defined and varied regionally.

Figure 5.2 shows the record derived from instrumental temperature data since 1850, while the temperature differences between the MWP and LIA are highlighted in shaded rectangles.

Discrepancies and uncertainties in natural variability. The discrepancies among these reconstructions are significant. Mann's original curve suggested a cooling of about 0.18 °C between the MWP and LIA, while AR4 raised this value to 0.29 °C. AR5 reported an average of 0.5 °C, with some reconstructions indicating up to 1 °C. AR6 returned to a curve more similar in shape to that of TAR, although based on a different dataset. These variations reveal substantial uncertainty about the true magnitude of natural climate variability, which directly affects model validation and the estimation of climate sensitivity.

Recent critiques have focused on the representation adopted in AR6. Esper et al. (2024a) argue that the use of a single global annual curve in the Summary for Policymakers, based on the PAGES-2K dataset, constitutes an excessive simplification. They observe that previous reports offered a more pluralistic view, incorporating multiple reconstructions that better reflected the methodological diversity and the inherent uncertainty of paleoclimate science. Esper et al. (2024a) conclude that this reductionist approach risks distorting public understanding of preindustrial climate variability and its relevance for correctly assessing modern warming attributions. They call for a more transparent and nuanced strategy, one that embraces the complexity of the available evidence rather than streamlining it into a single narrative.

Ultimately, the progression of paleoclimate reconstructions in the IPCC reports reflects not only technical refinement but also a transformation in the visual and rhetorical paradigm through which Earth's climate history is communicated. These shifts shape not only scientific understanding but also public perception and policy discourse.

5.4. McIntyre's 2021 critique of PAGES-2K proxy reconstruction

McIntyre & McKitrick (2003)[26] provided the first major critique of the *"hockey stick"* graph by Mann et al. (1999). Nearly two decades later, Stephen McIntyre returned to the debate with a detailed 2021 series of articles published

[26]McIntyre, S., McKitrick, R., 2003. Corrections to the Mann et al. (1998) proxy database and Northern Hemispheric average temperature series. Energy & Environment, 14(6), 751–771.

on his website Climate Audit,[27] offering a meticulous critique of the temperature reconstruction featured in the IPCC's Sixth Assessment Report (AR6), with particular focus on the PAGES-2K Consortium (2019) multiproxy dataset. His analysis identifies systemic issues in proxy selection, data truncation, and methodological transparency that, in his view, compromise the scientific integrity of the new *"hockey stick"* prominently displayed in the AR6 Summary for Policymakers.

In his 2021 article *The Decline and the Stick,* McIntyre revisits the long-standing controversy surrounding the *"decline"* in tree-ring density data—first highlighted in Briffa et al. (1998)[28]—which had been concealed in earlier reconstructions through selective data omission. McIntyre argues that the same pattern persists in AR6, where proxies exhibiting post-1960 divergence are excluded or truncated, thereby artificially reinforcing the narrative of an unprecedented modern warming.

His regional analyses of the PAGES-2K dataset expose further weaknesses. In the 0–30°N and 0–30°S latitudinal bands, McIntyre finds that many proxies are short coral series beginning after 1850, providing no insight into preindustrial climate variability. Only a handful of proxies extend into the first millennium, and those that do often show cooling trends inconsistent with the final reconstruction. In the 30–60°S zone—dominated by oceanic regions—he notes the absence of marine sediment proxies, despite their availability in earlier versions of the dataset. He suggests that their exclusion may be due to "inconvenient" trends, such as declining temperatures in recent centuries.

McIntyre's scrutiny of the Asian tree-ring chronologies reveals that many series included in PAGES-2K were unpublished and had not undergone peer review. Some chronologies, such as *"Asia_207"*, exhibit anomalous end spikes that are not present in the raw data and were likely introduced by undocumented smoothing or signal-processing algorithms. He emphasizes that the lack of archived code and metadata prevents independent replication and undermines the credibility of the reconstruction.

In his concluding article, *The IPCC AR6 Hockey Stick,* McIntyre critiques the graphical representation of global temperature over the past 2000 years. He argues that the reconstruction is not only methodologically opaque but also selectively curated to suppress historical variability—particularly the *Medieval Warm Period* and the *Little Ice Age*. The result, he contends, is a politically convenient narrative rather than a robust empirical synthesis.

Taken together, these articles suggest that the AR6 IPCC paleoclimate reconstruction relies on selective data inclusion, algorithmic opacity, and editorial choices that diminish the visibility of natural climate variability. McIntyre

---

[27]McIntyre, S., 2021:
https://climateaudit.org/2021/11/02/the-decline-and-the-stick/
https://climateaudit.org/2021/09/15/pages-2019-0-30n-proxies/
https://climateaudit.org/2021/09/02/pages19-0-30s/
https://climateaudit.org/2021/08/26/pages2019-30-60s/
https://climateaudit.org/2021/08/15/pages19-asian-tree-ring-chronologies/
https://climateaudit.org/2021/08/11/the-ipcc-ar6-hockeystick/

[28]Briffa, K.R., Schweingruber, F.H., Jones, P.D., Osborn, T.J., Harris, I.C., Shiyatov, S.G., Vaganov, E.A., 1998. Trees tell of past climates: but are they speaking less clearly today? Philosophical Transactions of the Royal Society B: Biological Sciences, 353(1365), 65–73.

calls for greater transparency, reproducibility, and methodological pluralism in climate science, particularly when such reconstructions are used to justify far-reaching policy decisions.

5.5. Conclusion

The trajectory outlined by the IPCC's Assessment Reports represents the evolution of a scientific paradigm, shaped by conceptual exploration, methodological refinement, internal debate, and periodic synthesis. These documents do not merely reflect the current state of climate knowledge; they actively shape its interpretation and steer its institutional framing.

The Anthropogenic Global Warming Theory (AGWT), now widely institutionalized, has emerged from a complex interplay of empirical observations, modeling frameworks, and consensus-building processes. This has yielded a coherent narrative of global climate change—one that has become the foundation for international policy, public discourse, and academic research. Yet coherence, while valuable, is not synonymous with completeness. It is worth asking whether this narrative captures the full scope of scientific inquiry or whether it has come at the expense of interpretive nuance, epistemological openness, and the inclusion of alternative explanatory frameworks. Although some recent tree-ring studies, such as Esper et al. (2024b),[29] suggest that recent warmth may be unparalleled over the past 2,000 years, the worldwide historical evidence supporting, for example, Roman and Medieval warming (Lamb, 1995; Guidoboni et al., 2011),[30] cannot be dismissed. Indeed, paleoclimatic reconstructions should be validated against historical documentation to the greatest extent possible.

Scientific paradigms are not static truths but evolving constructs subject to revision, challenge, and transformation. If climate records and models systematically underrepresent natural variability and solar or planetary influences, the conclusions they yield may prove incomplete or skewed.

[29]Esper, J., Torbenson, M., Büntgen, U., 2024b. 2023 summer warmth unparalleled over the past 2,000 years. Nature, 627, 769–773.

[30]Lamb, H.H., 1995. Climate, History and the Modern World (2nd ed.). London: Routledge. ISBN: 978-0415127341.

Guidoboni, E., Navarra, A., Boschi, E., 2011. The Spiral of Climate: Civilizations of the Mediterranean and Climate Change in History. Bononia University Press.

CHAPTER 6

Results of the global climate models

At the core of contemporary climate science, and of the efforts to attribute global warming, lie mathematical instruments as sophisticated as they are contested: the Global Climate Models (GCMs). Continuously refined over the past decades, these computer programs have served as the backbone of IPCC assessments since the mid-1990s. They provide a dynamic representation of Earth's climate system under varying radiative forcing scenarios and act as a key interpretive tool for understanding recent climate trends and informing environmental policy.

Despite the persistent uncertainties in the reconstruction of past climates—particularly episodes such as the Medieval Warm Period (MWP) and the Little Ice Age (LIA)—the attribution of the warming observed since the 1850–1900 baseline relies heavily on GCM outputs. By contrasting natural and anthropogenic scenarios, these models offer crucial insights into the relative influence of the various climate drivers.

Confidence in GCM projections has grown in parallel with their increasing use. The Fourth, Fifth, and Sixth IPCC Assessment Reports (2007; 2013; 2021)[1] have relied heavily on the assumption that these models simulate Earth's climate system with sufficient accuracy to inform long-term policy decisions. Yet their complexity does not eliminate their limitations. GCMs are built upon specific theoretical assumptions, structural simplifications, and parameterizations that can significantly influence the results. This calls for a critical reflection on their reliability, especially when their outputs underpin decisions with profound social and economic consequences.

6.1. The Coupled Model Intercomparison Project (CMIP)

The development and comparison of GCMs are coordinated through the *Coupled Model Intercomparison Project* (CMIP, https://wcrp-cmip.org/), an international initiative led by the World Climate Research Programme (WCRP). Since 1995, CMIP has organized successive phases—CMIP1 through CMIP6—bringing together hundreds of researchers to evaluate the models produced by institutions worldwide.

Over time, CMIP has expanded significantly, with the early phases including fewer models:

- CMIP1 & CMIP2 (1990s) → 18 models.
- CMIP3 (2000–2005) → 20 models used for AR4 (IPCC, 2007).

[1]IPCC, 2007. Climate Change 2007: The Physical Science Basis. Cambridge University Press. IPCC, 2013. Climate Change 2013: The Physical Science Basis. Cambridge University Press. IPCC, 2021. Climate Change 2021: The Physical Science Basis. Cambridge University Press.

- CMIP5 (2010–2014) → 40 models from over 20 international groups for AR5 (IPCC, 2013).
- CMIP6 (from 2013 to present) → 55 models, basis for AR6 (IPCC, 2021; Eyring et al., 2016).

CMIP has been instrumental in promoting transparency and standardization in climate modeling. However, the results remain sensitive to the initial conditions, the choice of forcings, and the treatment of feedback mechanisms. Many processes—especially those occurring at scales smaller than the model grid—must be approximated through parameterizations, which introduce uncertainties that are difficult to assess.

6.2. Structure and functioning of the global climate models

GCMs are computational frameworks designed to simulate the evolution of Earth's climate by dividing the planet into a dense three-dimensional grid (see Figure 3.2). Within each grid cell, the model solves the fundamental equations governing the conservation of energy, mass, and momentum, in accordance with the laws of fluid dynamics and thermodynamics, including the Navier–Stokes equations.

This structure enables GCMs to simulate a wide range of climate variables, from atmospheric temperature, pressure, and humidity to ocean currents, salinity, and heat distribution. In particular, GCMs represent the exchange of energy between the land surface and the atmosphere, which is essential for understanding the planet's radiative balance.

To these physical components are added external forcings: greenhouse gases such as CO_2, CH_4, and N_2O; aerosols from industrial activity and natural sources; variations in solar irradiance; and volcanic eruptions, which can inject reflective particles into the stratosphere. Land-use changes, which alter surface albedo and affect the carbon cycle, are also incorporated.

Each simulation incorporates these elements in order to explore the response of the climate system over time. Depending on the socioeconomic scenarios adopted—such as the *Shared Socioeconomic Pathways* (SSPs) used in CMIP6—GCMs also project changes in global temperature, precipitation patterns, glacier retreat, and other indicators of climate change throughout the 21st century.

Simulations and comparison with observational data. A GCM is typically assessed by its ability to reproduce past climate behavior. This procedure, known as hindcasting, consists in simulating historical trends using known forcings and comparing the results with the observational data. Although it provides a valuable test of model robustness, it cannot guarantee future accuracy.

One of the most challenging aspects of climate modeling concerns phenomena that occur below the resolution of the model grid, such as cloud formation and convective processes. These are represented through empirical parameterizations which, although grounded in observation and theory, introduce uncertainties that can affect the simulation outcomes.

Moreover, observational data are themselves not free from limitations. Global climate records are based on instrumental measurements that span only the last 175 years and are unevenly distributed: significant gaps remain over oceanic, polar, and southern regions. Paleoclimate reconstructions provide longer-term insights but rely on indirect indicators and complex interpretive frameworks, which carry their own uncertainties.

Given these constraints, it is unrealistic to expect GCMs to replicate the climate history of recent centuries with perfect fidelity. Their output must be interpreted with caution, acknowledging the structural limitations and the uncertainty margins inherent in any simulation.

6.3. Feedback mechanisms and climate sensitivity

A central concept in climate modeling is climate sensitivity, defined as the extent to which the global mean temperature responds to changes in radiative forcing. GCMs typically quantify this response by means of two closely related metrics:

(1) "*Equilibrium Climate Sensitivity*" (ECS): the long-term increase in global mean surface temperature resulting from a sustained doubling of atmospheric CO_2 concentration relative to preindustrial levels (from approximately 280 to 560 ppm), once the climate system has reached a new equilibrium;

(2) "*Transient Climate Response*" (TCR): the temperature increase at the time of CO_2 doubling under a scenario of gradual increase (typically 1% per year), which reflects the pace at which the climate system approaches equilibrium.

Basic radiative physics, neglecting climate feedbacks, suggests that a doubling of atmospheric CO_2 would produce approximately 1 °C of global warming. This estimate is derived from the Stefan–Boltzmann law, by means of the approximation:

$$\Delta T_{2\times CO_2} = \frac{\Delta F_{2\times CO_2}}{4\sigma T^3} \approx 1\ ^\circ\mathrm{C},$$

where $\Delta F_{2\times CO_2} = 3.7\ \mathrm{W/m^2}$ is the radiative forcing associated with CO_2 doubling (from ~280 to ~560 ppm) (cf. Figure 8.4C), and $T \approx 255$ K is the average global emission temperature, without taking into account amplifying feedbacks such as water vapor, clouds, or ice albedo (Rahmstorf, 2008).[2]

However, the climate system is not linear. Feedback mechanisms such as increased water vapor, changes in cloud cover, and shifts in surface albedo can amplify or dampen the initial warming induced by radiative forcing. It is the representation of these feedbacks within the GCMs that determines the effective ECS value.

The CMIP6 generation of GCMs yields ECS values ranging from 1.8 °C to 5.7 °C, with a central tendency around 3 to 4 °C, which reflects the difficulty of accurately modeling the climate feedbacks. This uncertainty leads to divergent future climate projections even when the same forcing functions are assumed. Recent studies have raised concerns that many CMIP6 models may overestimate climate sensitivity, producing projections that are "*hotter*" than historical observations would suggest (Hausfather et al., 2022; Scafetta, 2023b).[3]

---

[2]Rahmstorf, S., 2008. Anthropogenic climate change: revisiting the facts. In: Zedillo, E. (Ed.), Global Warming: Looking Beyond Kyoto. Brookings Institution Press, Ch. 3, pp. 34–53.

[3]Hausfather, Z., Marvel, K., Schmidt, G.A., Nielsen-Gammon, J.W., Zelinka, M., 2022. Climate simulations: recognize the 'hot model' problem. Nature, 605(7908), 26–29.

***"Climate Feedbacks"*: regulators of the system.** The most critical aspect of the GCM design is the representation of the climate feedback processes, that is, the mechanisms that either amplify (positive feedback) or dampen (negative feedback) the system's response to external forcings, such as an increase in atmospheric CO_2. Climate feedbacks are central to understanding climate sensitivity and play a decisive role in shaping future projections.

The main processes include:

- **water vapor feedback**—rising temperatures intensify evaporation, thereby increasing atmospheric H_2O, the most important greenhouse gas (this is one of the most significant positive feedbacks);
- **albedo feedback**—the reduction of snow and ice decreases Earth's surface reflectivity (albedo), leading to a greater absorption of solar radiation and to further warming;
- **cloud feedback**—one of the greatest sources of uncertainty in GCMs, as clouds can exert both cooling effects (by reflecting sunlight) and warming effects (by trapping infrared radiation), depending on their altitude, thickness, and distribution;
- **lapse-rate feedback**—changes in the vertical temperature gradient affect the atmosphere's ability to dissipate heat, modulating the greenhouse effect;
- **carbon-cycle feedback**—warming may promote the release of CO_2 from soils, oceans, and permafrost, further increasing greenhouse gas concentrations;
- **vegetation feedback**—changes in vegetation cover alter both albedo and carbon fluxes, influencing the radiative and biogeochemical balance of the climate system.

Taken together, these processes show that climate sensitivity is not a fixed value but a dynamic synthesis of the interactions between forcings and feedbacks. Accurately quantifying these mechanisms is essential for assessing the magnitude of future climate change and the effectiveness of mitigation strategies.

Mathematical formulation. The concept of climate feedback can be formalized by examining how the system responds to a radiative forcing ΔF. The equilibrium temperature change ΔT is related to this forcing through the climate response parameter λ, expressed as:

$$\Delta T = -\frac{\Delta F}{\lambda} = S \cdot \Delta F.$$

Here, λ (in W/m^2K) represents the change in net outgoing radiative flux per degree of surface warming, and $S = -1/\lambda$ is the *"Equilibrium Climate Sensitivity"* (ECS) per unit of forcing.

This parameter can be decomposed into the Planck component λ_P—the fundamental stabilizing feedback of a black body—and the sum of all other feedbacks λ_i:

$$\lambda = \lambda_P + \sum_i \lambda_i.$$

Scafetta, N., 2023b. CMIP6 GCM ensemble members versus global surface temperatures. Climate Dynamics, 60, 3091–3120.

Parameter	Central Estimate AR6	"Very likely" Interval
Planck	−3.22	−3.4 to −3.0
Water Vapor + Lapse Rate (WV+LR)	+1.30	+1.1 to +1.5
Surface Albedo	+0.35	+0.10 to +0.60
Clouds	+0.42	−0.10 to +0.94
Non−CO_2 Biogeophysical/ Biogeochemical	−0.01	−0.27 to +0.25
Net Feedback (ECS)	−1.16	−1.81 to −0.51

FIGURE 6.1. Central estimates and *"very likely"* ranges of climate feedback processes, λ_i, in $Wm^{-2}K^{-1}$ according to AR6 (IPCC, 2021, Table 7.10). The *"net feedback"* is $\lambda = \lambda_P + \sum_i \lambda_i$.

The Planck response describes how much irradiance is emitted to space per degree of warming. According to the Stefan–Boltzmann law:

$$I_{out} = \epsilon\sigma T^4,$$

where $\sigma = 5.67 \times 10^{-8}\, W\, m^{-2}\, K^{-4}$ is the Stefan–Boltzmann constant. Taking the derivative with respect to temperature yields:

$$\frac{dI_{out}}{dT} = 4\epsilon\sigma T^3.$$

Assuming a global mean emission temperature of 255 K and an emissivity $\epsilon = 1$, this yields approximately

$$\frac{dI_{out}}{dT} = -\lambda_P = 4 \cdot 5.67 \times 10^{-8} \cdot (255)^3 \approx 3.76\, \frac{\mathrm{W}}{\mathrm{m^2K}}.$$

The CMIP6 models, however, calculate a Planck response λ_P between –3.0 and –3.4 W/m²K, primarily owing to stratospheric warming in GCM calculations (IPCC, 2021; Cronin & Dutta, 2023).[4] A more negative value implies stronger radiative cooling and therefore a lower climate sensitivity; a less negative value implies weaker dissipation and a higher sensitivity. The central CMIP6 GCM estimate is:

$$\lambda_P \approx -3.22\, \frac{\mathrm{W}}{\mathrm{m^2K}}.$$

The negative sign reflects the stabilizing nature of the Planck response: as temperature increases, more energy escapes to space.

Quantifying feedbacks and climate gain. The terms λ_i represent the various climate feedbacks. Positive feedbacks (e.g., water vapor and high clouds)

[4]Cronin, T.W., Dutta, I., 2023. How well do we understand the Planck feedback? Journal of Advances in Modeling Earth Systems, 15(7), e2023MS003729.

have $\lambda_i > 0$ and reduce the stabilizing effect of λ_P, whereas negative feedbacks (e.g., low clouds or albedo changes) have $\lambda_i < 0$ and enhance radiative cooling.

To analyze the relative contributions, each feedback is normalized to the Planck response. Thus, each feedback process contributes a term λ_i (positive feedbacks have $\lambda_i > 0$, negative feedbacks have $\lambda_i < 0$), normalized relative to the Planck response:

$$g_i = \frac{\lambda_i}{\lambda_P}.$$

The total feedback factor f is:

$$f = -\sum_i g_i = -\sum_i \frac{\lambda_i}{\lambda_P}.$$

Since $\lambda_P < 0$, a positive feedback ($\lambda_i > 0$) increases f, whereas a negative feedback ($\lambda_i < 0$) decreases it.

The temperature response in the absence of feedbacks is:

$$\Delta T_0 = -\frac{\Delta F}{\lambda_P}.$$

With feedbacks, the amplified temperature becomes:

$$\Delta T = -\frac{\Delta F}{\lambda} = -\frac{\Delta F}{\lambda_P \cdot (1 + \sum_i g_i)} = \frac{\Delta T_0}{1 + \sum_i g_i} = \frac{\Delta T_0}{1 - f}.$$

The total climate gain is:

$$\text{gain} = \frac{\Delta T}{\Delta T_0} = \frac{1}{1 - f}.$$

This formulation shows that as f approaches 1, the denominator tends to zero and the temperature response diverges, indicating a potential point of instability. If $f \to 1$ or $f > 1$, the linear approximation breaks down entirely and the system can no longer reach a new equilibrium—a situation mathematically analogous to a runaway greenhouse regime.

Summary: application to CO_2 doubling. For a doubling of atmospheric CO_2, the canonical forcing is:

$$\Delta F_{2\times CO_2} \approx 3.7 \text{ W/m}^2.$$

In the absence of feedbacks ($f = 0$), and assuming $\lambda_P = 3.22$ Wm^{-2}K^{-1} :

$$\Delta T = \frac{\Delta T_0}{1 - f} = \Delta T_0 = -\frac{\Delta F}{\lambda_P} = \frac{3.7}{3.2}\,°C = 1.16\,°C.$$

With feedbacks, the physical interpretation is as follows:

- if $f < 0$ (net negative feedback), then $0 < \Delta T < \Delta T_0$;
- if $f = 0$, then $\Delta T = \Delta T_0$;
- if $0 < f < 1$, then $\Delta T > \Delta T_0$ (amplified warming);
- if $f \approx 1$, the system approaches instability ($\Delta T \to \infty$), i.e., the initial climate perturbation self-amplifies indefinitely;
- if $f > 1$, then $\Delta T < 0$ for $\Delta F > 0$, and no stable equilibrium exists (runaway greenhouse in the linear model).

According to AR6, and assuming $\lambda_P \approx -3.2$ Wm^{-2}K^{-1} across all models, the value of f in CMIP6 GCMs ranges from approximately $f = 0.35$ to $f = 0.80$, meaning that the total climate "*feedback*" in GCMs falls within the interval:

$$1.12\ \frac{\mathrm{W}}{\mathrm{m^2K}} \leq -f \cdot \lambda_P = \sum_i \lambda_i \leq 2.56\ \frac{\mathrm{W}}{\mathrm{m^2K}} < -\lambda_P.$$

This corresponds to ECS values between about 1.8 °C and 5.8 °C. The CMIP GCMs estimate that the net climate feedback amplifies warming, but that it remains within a range that allows the system to reach a new thermodynamic equilibrium; that is, $0 < f < 1$. Figure 6.1 shows the central estimates and "*very likely*" ranges of the individual feedback values λ_i reported in AR6 (IPCC, 2021, Table 7.10).

6.4. Comparing climate simulations using only natural and all forcings

One of the most compelling arguments in support of the Anthropogenic Global Warming Theory (AGWT) emerges from a simple computational experiment: the comparison between two classes of climate simulations produced by GCMs within the CMIP framework.

(1) Natural-only simulations: these include only solar variability and volcanic forcing.
(2) All-forcings simulations: these include the natural forcings together with anthropogenic contributions (greenhouse gases, aerosols, land-use changes).

The divergence between these two sets of simulations is striking. The natural-only simulations produce a nearly flat temperature trend from the preindustrial period to the present, implying that the climate system would have remained relatively stable in the absence of human influence. By contrast, the simulations that include anthropogenic forcings closely reproduce the observed warming, both globally and regionally.

This contrast is evident in panels (B) and (D) of Figure 6.2, which show that the natural-only simulations diverge significantly from the instrumental record. According to these models, natural forcings cannot account for the magnitude or pace of the warming observed since 1850–1900, not even partially.

The IPCC's Sixth Assessment Report (2021) reiterates this conclusion, stating that nearly 100% of the warming observed since the preindustrial era is attributable to human activity, particularly to the rise in CO_2 and other greenhouse gases. These attribution experiments form a cornerstone of AGWT.

Yet, as the subsequent sections will explore, several methodological questions remain open: the representation of uncertainty, the selection of observational datasets, and the robustness of the underlying model assumptions. The strength of AGWT lies not only in its explanatory power but also in its capacity to withstand critical scrutiny and to evolve in response to new evidence.

6.5. Quantitative attribution of warming

Figure 6.3, adapted from AR6 (IPCC, 2021, fig. SPM.2), quantifies the contributions to the observed global warming between the preindustrial baseline (1850–1900) and the 2010–2019 decade. According to the CMIP6 simulations, natural forcings—such as solar variability and volcanic activity—together with internal climate system variability, contributed negligibly to the net temperature

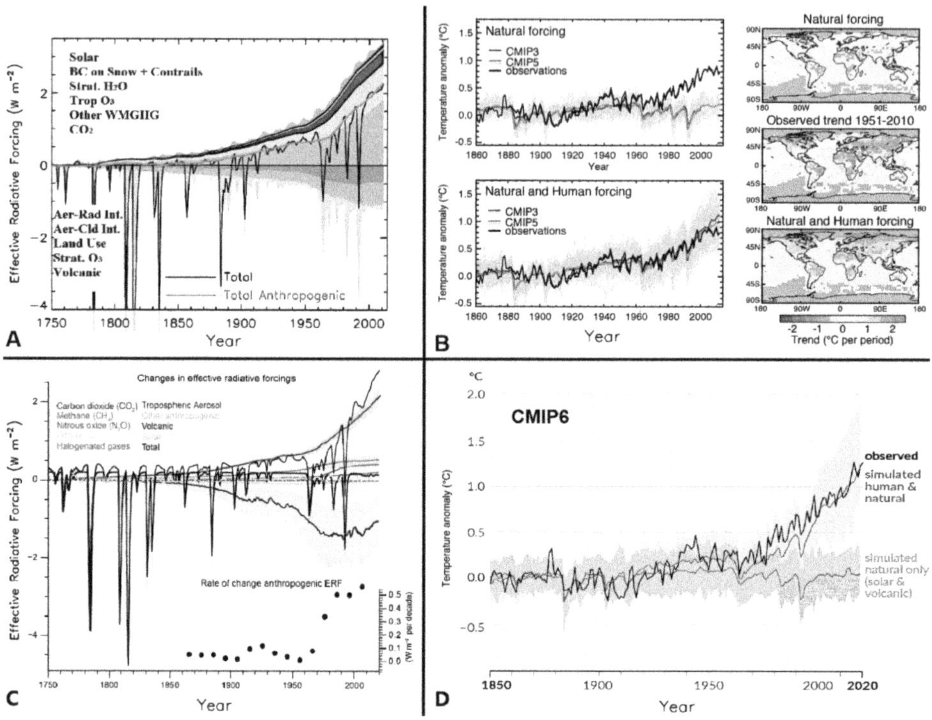

FIGURE 6.2. (A) Radiative forcing functions used in CMIP5 GCMs. (Adapted from IPCC, 2013, fig. 8.18).
(B) Observed global surface temperature variations (black line), compared with CMIP3 and CMIP5 GCM simulations using only natural forcings (volcanoes and solar activity) and adding anthropogenic forcings (greenhouse gases, aerosols, land use). (Adapted from IPCC, 2013, FAQ 10.1, fig. 1).
(C) Radiative forcing functions used in CMIP6. (Adapted from IPCC, 2021, fig. 2.10).
(D) Observed global surface temperature variations (black line), alongside CMIP6 GCM simulations distinguishing natural-only forcings and all-forcings scenarios. (Adapted from IPCC, 2021, fig. SPM.1).

change, with an average impact close to 0 °C. By contrast, anthropogenic forcings produced a net warming of approximately +1.07 °C—essentially the entire observed increase.

The breakdown of the climate contributions according to the CMIP6 GCMs is as follows:

- well-mixed greenhouse gases (CO_2, CH_4, N_2O) of human origin account for roughly +1.5 °C of warming.
- their effect is partially offset by anthropogenic aerosols, such as sulfates and black carbon, which exert a cooling influence of about −0.4 °C.

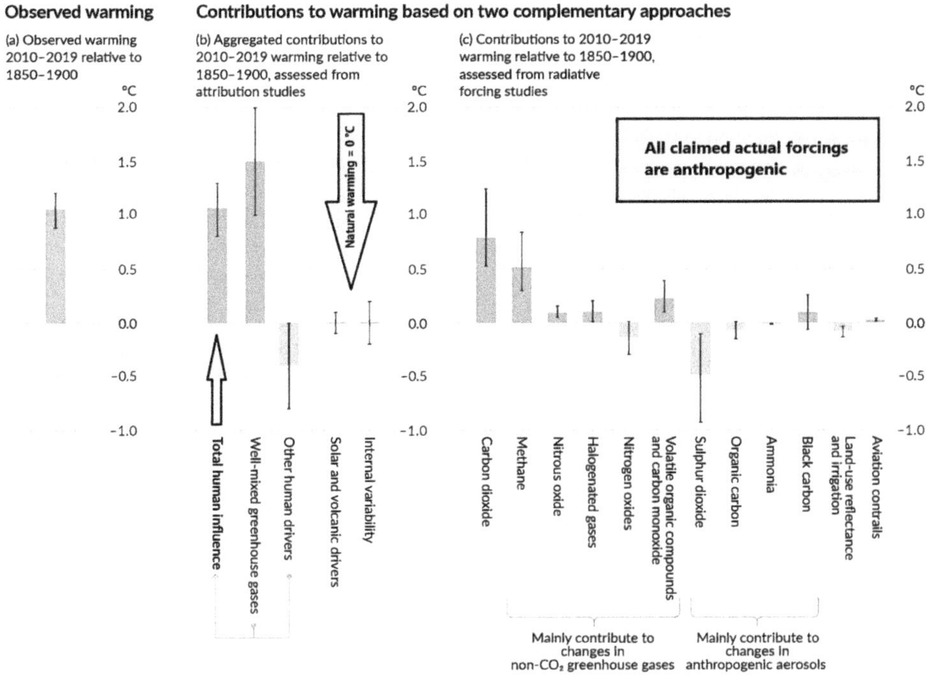

FIGURE 6.3. Attribution of global warming according to the radiative forcing functions used in the CMIP6 GCMs between 1850–1900 and 2010–2019. Natural forcings (solar variability and volcanic activity) and internal variability contribute approximately 0 °C, while anthropogenic forcings account for the full observed warming. (Adapted from IPCC, 2021, *"Summary for Policymakers"*, Figure SPM.2).

- other anthropogenic factors, including land-use changes, contribute modestly but remain relevant to the overall radiative balance;
- solar and volcanic drivers, together with internal variability, account for about 0 °C.

In particular, Figure 6.2C suggests that a significant portion of the global surface warming observed in recent decades may be attributed to the global reduction in aerosol emissions, which has led to an increase in solar radiation reaching the surface. Indeed, Mallet et al. (2024)[5] demonstrated that the decline in biomass burning aerosols (BBA) over tropical Africa significantly altered regional climate dynamics through ocean–atmosphere–aerosol interactions. Dong et al. (2023)[6] identified decadal trends over the North Atlantic sector, linking the enhanced surface solar radiation and cloud cover changes to shifts in aerosol concentrations.

[5]Mallet, M., Voldoire, A., Solmon, F., Nabat, P., Drugé, T., Roehrig, R., 2024. Impact of biomass burning aerosols (BBA) on the tropical African climate in an ocean–atmosphere–aerosol coupled climate model. Atmospheric Chemistry and Physics, 24, 12509–12535.

[6]Dong, B., Sutton, R.T., Wilcox, L.J., 2023. Decadal trends in surface solar radiation and cloud cover over the North Atlantic sector during the last four decades: drivers and physical processes. Climate Dynamics, 60, 2533–2546.

In Europe, Schilliger et al. (2024)[7] used satellite data to investigate the causes of the detected solar brightening, pointing to aerosol reductions as a key factor. Complementary evidence from long-term night sky brightness measurements also suggests that atmospheric conditions and instrumental aging can influence radiative assessments, as shown by Fiorentin et al. (2025).[8] On the global scale, Hodnebrog et al. (2024)[9] estimate that the recent reduction in aerosol emissions between 2001 and 2019 may have contributed to an increase in Earth's energy imbalance of approximately 0.2 ± 0.1 W/m² per decade.

Taken together, these findings support the hypothesis that the recent decline in aerosols has contributed to an increased radiative forcing and to regional warming. This assessment is consistent with the IPCC's conclusion that the warming observed from 1850 to 2020 can be explained only by anthropogenic forcings. No computer simulation based solely on natural drivers can replicate the magnitude or the pace of the observed temperature increase, not even partially. This computer-based evidence forms a central pillar of the Anthropogenic Global Warming Theory (AGWT), providing a quantitative and theoretical foundation for its attribution claims.

6.6. Future scenarios and the Shared Socioeconomic Pathways (SSP)

Once retrospective attribution has been established, GCMs are employed to explore possible future trajectories of the climate system. In this context, the IPCC's Sixth Assessment Report adopted a structured set of scenarios known as "*Shared Socioeconomic Pathways*" (SSPs). These pathways describe plausible futures for global development, incorporating variables such as economic growth, demographic trends, technological innovation, and climate policy.

Each SSP corresponds to a distinct radiative forcing trajectory from 2015 to 2100, expressed in watts per square meter (W/m²), which reflects the emissions associated with each socioeconomic storyline.

SSP1–2.6 envisions a sustainable future with low population growth, a rapid energy transition, and strong climate policies. It targets carbon neutrality by 2075, with radiative forcing reaching 2.6 W/m² by 2100. A more ambitious variant, SSP1–1.9, assumes Net-Zero by 2050–2055 and underpins the European Union's Green Deal.

SSP2–4.5 represents a middle-ground scenario, characterized by moderate progress and partial mitigation. Emissions stabilize by mid-century, but Net-Zero is not achieved. Radiative forcing reaches 4.5 W/m² by 2100.

[7]Schilliger, L., Wild, M., Folini, D., Hakuba, M., 2024. An investigation on causes of the detected surface solar radiation brightening in Europe using satellite data. Journal of Geophysical Research: Atmospheres, 129(6), e2024JD041101.

[8]Fiorentin, P., Cavazzani, S., Bertolo, A., Ortolani, S., Binotto, R., Saviane, I., 2025. SQM Ageing and Atmospheric Conditions: How Do They Affect the Long-Term Trend of Night Sky Brightness Measurements? Sensors, 25(2), 516.

[9]Hodnebrog, Ø., Myhre, G., Samset, B.H., Sager, P.L., Smith, C.J., et al., 2024. Global aerosol emission reductions drive positive trend in Earth's energy imbalance. Nature Communications Earth & Environment, 5, 317.

SSP3–7.0 reflects a fragmented world marked by geopolitical tensions, high population growth, and weak climate governance. Emissions rise significantly, with radiative forcing reaching 7.0 W/m².

SSP5–8.5 is often labeled "business-as-usual," and assumes a continued reliance on fossil fuels and a tripling of CO_2 emissions by 2075. Radiative forcing reaches 8.5 W/m² by the end of the century.

Although some researchers, such as Hausfather & Peters (2020),[10] consider SSP3–7.0 and SSP5–8.5 to be unlikely or even implausible, these scenarios remain valuable reference points for assessing high-impact climate risks.

Climate projections to 2100: a wide range of outcomes. The CMIP6 simulations presented in Figure 6.4 outline a broad spectrum of possible climate futures. Under extreme emissions—as in SSP5–8.5—global warming could even approach 8 °C above preindustrial levels, particularly if equilibrium climate sensitivity (ECS) values above 5 °C prove accurate.

At the other end of the spectrum, SSP1–2.6 offers a technically feasible path to limiting warming below the critical 2 °C threshold, consistent with the goals of the Paris Climate Agreement (2016). Achieving this outcome would require profound transformations in energy systems, aggressive decarbonization, and coordinated global action.

This dual range of projections underscores the decisive role of policy choices. Although the magnitude of future warming is embedded in the models through climate sensitivity parameters, the actual trajectory will depend on the socioeconomic, technological, and geopolitical decisions made in the coming decades.

The IPCC's *"Reasons for Concern"*. To complement the numerical projections, IPCC (2023)[11] identified five categories of global climate risk known as *"Reasons for Concern"* (RFC), which highlight areas in which rising temperatures may generate significant impacts:

RFC1: Unique and vulnerable systems, such as coral reefs and Arctic ecosystems.

RFC2: Extreme events, including heatwaves, floods, and droughts.

RFC3: Unevenly distributed impacts, disproportionately affecting low-income regions.

RFC4: Aggregated global effects on health, infrastructure, and economies.

RFC5: Large-scale and irreversible changes, such as ice sheet destabilization.

The severity and likelihood of these risks increase with the intensity of warming, reinforcing the urgency of mitigation and adaptation strategies.

6.7. Concluding remarks

GCMs supporting the AGWT. The attribution of global warming to human activity rests not only on empirical temperature trends but also, crucially, on the simulations produced by General Circulation Models (GCMs). Initiatives such

[10] Hausfather, Z., Peters, G.P., 2020. Emissions—the 'business as usual' story is misleading. Nature, 577(7792), 618–620.

[11] IPCC, 2023. Climate Change 2023: Synthesis Report. IPCC, Geneva, Switzerland.

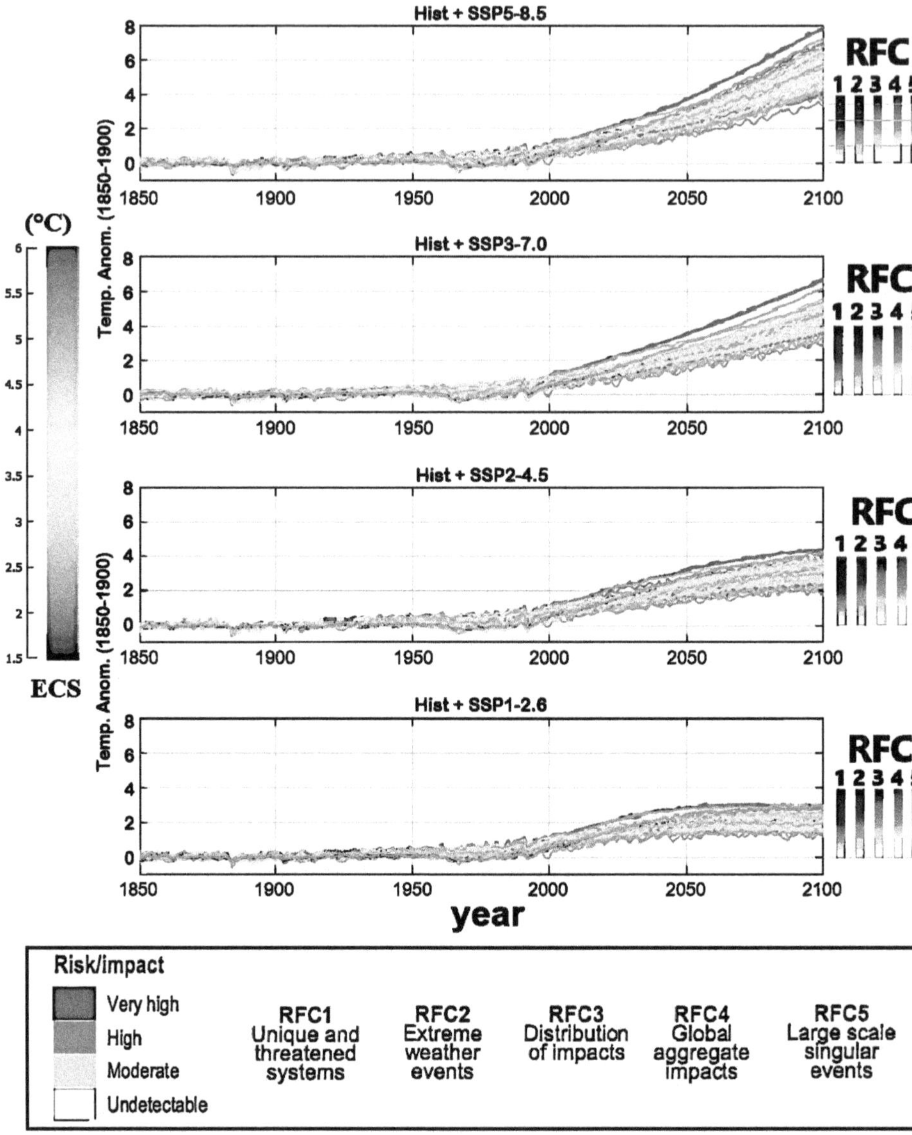

FIGURE 6.4. Mean projections from the CMIP6 GCM ensemble from 1850 to 2100, using historical effective radiative forcing functions for the period 1850–2014 (see 6.2C), and forcings corresponding to the four socioeconomic scenarios SSP1–2.6, SSP2–4.5, SSP3–7.0, and SSP5–8.5 for the subsequent period. Curves are color-coded by the equilibrium climate sensitivity (ECS) of the models considered. Panels on the right show climate risks and impacts associated with different levels of global warming, according to the five *"Reasons for Concern"* (RFC) identified by IPCC (2023). (Recreated from Scafetta, 2024).

as CMIP3, CMIP5, and CMIP6 have provided the theoretical backbone of the Anthropogenic Global Warming Theory (AGWT).

Developed by an international network of research institutions, these models are employed by the IPCC to compare climate scenarios with and without anthropogenic forcings. Only the simulations that include greenhouse gases and aerosols of human origin reproduce the warming observed during the 20th and 21st centuries. The comparative graphs between observations and model outputs featured in the IPCC reports have become some of the most persuasive visual arguments in support of AGWT.

Climate modeling has evolved significantly: from the early 2000s to the present, the representation of atmospheric, oceanic, and terrestrial processes has become more detailed. The interactions among atmosphere, oceans, cryosphere, and biosphere are now simulated with greater fidelity, owing to advances in cloud microphysics, carbon–vegetation dynamics, and land-use–albedo modeling, although many of these processes remain only partially understood.

Despite this progress, uncertainties persist. The chaotic nature of the climate system and the incomplete understanding of its feedbacks continue to limit model precision.

Structural limits and epistemological challenges. GCMs, despite their sophistication, remain inductive tools subject to structural uncertainties and systemic biases. These issues will be examined in greater depth in the following chapters. The IPCC acknowledges several discrepancies between model predictions and observations but considers them insufficient to undermine the broader framework. Nonetheless, they raise important questions about model accuracy in specific vertical and regional domains.

GCMs are regarded as indispensable for understanding climate dynamics and exploring future scenarios, but they are not infallible. Their central role in shaping global policy, especially in the context of radical decarbonization, raises delicate epistemological and ethical concerns. As George Box (1976) famously noted,

> *"all models are wrong, but some are useful."*[12]

The challenge is not to reject modeling, but to assess its assumptions, boundaries, and implications critically. A responsible approach must avoid both dogmatism and technocratic overreach, favoring instead a reflective integration of scientific insight with methodological caution.

Remaining open to alternative hypotheses. The narrative emerging from the GCMs is coherent and compelling, yet inevitably shaped by the modeling paradigm itself. These models are designed to incorporate anthropogenic forcings and have shown skill in reproducing past warming. However, this apparent reliability should not be mistaken for definitive validation.

Confidence in model projections depends on several key factors: the realism of the initial conditions, the treatment of poorly constrained climate feedbacks, and the capacity to simulate natural variability across multidecadal to millennial

[12]Box, G.E.P., 1976. Science and Statistics. Journal of the American Statistical Association, 71(356), 791–799.

scales. These elements are central to distinguishing internal variability from externally forced signals.

GCMs should be regarded as heuristic tools: imperfect yet essential, designed to explore plausible futures rather than to deliver certainties. Their value lies in organizing complex dynamics within a testable, iterative framework, rather than in providing final truths.

Yet, despite these caveats, climate policy continues to rely heavily on model outputs to guide transformative decisions. This dependence calls for a more reflective stance, one that acknowledges the limits of predictive modeling and embraces a multidisciplinary, evolving understanding of Earth's climate.

Such an approach must remain open to alternative hypotheses, empirical anomalies, and the possibility that solar, astronomical, or geophysical forcings may play a more significant role than is currently assumed. Only a pluralistic and intellectually rigorous climate science can ensure that our future is shaped not by dogma, but by thoughtful, evidence-based inquiry.

CHAPTER 7

Concluding remarks on IPCC attribution frameworks

7.1. Evolution of anthropogenic attribution in IPCC reports

Over more than three decades—from the First Assessment Report (1990, FAR,) to the Sixth (2021; 2023, AR6)—the *Intergovernmental Panel on Climate Change* (IPCC) has progressively refined its narrative on the role of human activity in global warming. What began as a cautious hypothesis has evolved into a confident attribution, culminating in the assertion that virtually all observed warming since the preindustrial era is anthropogenic in origin.

From shared causality to exclusive responsibility. In the early 1990s, the IPCC maintained a balanced view: twentieth-century warming was regarded as the result of both natural and human-induced factors, with a tentative 50/50 attribution. This stance reflected the preliminary state of climate science at the time, which was constrained by limited observational data and early-stage modeling.

As empirical evidence accumulated and climate models grew more sophisticated, the IPCC progressively shifted its position. By AR6, the conclusion had become unequivocal:

> *"It is unequivocal that human influence has warmed the atmosphere, ocean and land."* (Summary for Policymakers, SPM A.1).

The report explicitly attributes nearly 100% of the temperature increase observed between 1850–1900 and 2011–2020 to anthropogenic greenhouse gas emissions and other human-induced factors.

Models, radiative forcings, and statistical confidence. The cornerstone of this attribution is the comparison between model simulations and observational data. When GCMs include only natural forcings such as solar variability and volcanic eruptions, they fail to reproduce the observed warming trend. Once anthropogenic forcings are included, however—in particular greenhouse gases and aerosol changes—the agreement with the historical temperature record improves markedly.

This evidence leads AR6 to conclude, with high statistical confidence, that the warming observed over the past century is entirely anthropogenic. Figure 6.3 illustrates this stark contrast: natural forcings and internal variability contribute approximately 0 °C to the warming between 1850–1900 and 2010–2020. This claim carries profound scientific and political implications.

Scientific foundations of AGWT: two crucial assumptions. This level of attribution rests on two foundational assumptions:

(1) the reliability of the global temperature record. Were these datasets compromised by systematic errors (e.g., urban heat island effects, instrument biases, or discontinuities), the empirical basis for attribution would be substantially weakened.
(2) the validity of the CMIP GCMs. Although increasingly sophisticated, climate models remain simplifications of reality. They rely on empirical parameters and schematic representations of complex processes such as cloud formation, ocean circulation, and albedo feedback. Structural deviations in model architecture can therefore affect the robustness of attribution.

While the IPCC advocates the anthropogenic origin of warming, it also acknowledges substantial uncertainties in projecting future impacts, owing to poorly constrained feedback mechanisms and regional variability.

Future scenarios, model uncertainty, and interpretive limits. AR6 presents a range of future projections based on Shared Socioeconomic Pathways (SSPs). Among these, SSP5–8.5 stands out as the most extreme, envisioning high economic growth and continued fossil fuel dependence, with potential warming of between +4 and +8 °C by 2100. Many researchers nevertheless regard this scenario as implausible, even though it remains a benchmark for high-risk assessments.

The IPCC itself notes that confidence in many extreme impacts remains low, owing to limitations in data, models, and the understanding of climate feedbacks. Table 12.12 of AR6 (IPCC, 2021, p. 1856) acknowledges low confidence in attributing increases in the frequency or intensity of events such as floods, droughts, landslides, frosts, storms, and tropical cyclones. Marine phenomena such as coastal erosion and ocean heatwaves, as well as changes in global circulation, also fall within this category (Figure 15.1).

By contrast, moderate to high confidence is reserved for phenomena more directly linked to rising temperatures and CO_2 concentrations: increases in mean temperature, heatwaves, sea-level rise, ice melt, ocean acidification, and shifts in precipitation patterns. Even in these cases, however, regional and temporal details remain uncertain, and the modeling of local impacts continues to pose significant challenges.

7.2. Between institutional certainties and epistemic uncertainties

The evolution of the IPCC's position reflects both the growing body of empirical evidence and the inherent complexity of the Earth's climate system. Although the dominant narrative has coalesced around the anthropogenic hypothesis, this consolidation must not foreclose critical reflection.

Climate is shaped by a dense network of forcings, feedbacks, and interdependent variables. Modeling such a system remains an ambitious and, in many respects, experimental undertaking. The IPCC's role should therefore be understood not as that of a purveyor of absolute truths, but as that of a synthesizing body that informs policy while acknowledging the probabilistic nature of its assessments.

The structural incompleteness of climate knowledge. Despite remarkable scientific progress, key scientific uncertainties persist. Three challenges are particularly pressing:

(1) *Refining climate models*. The CMIP6 models still display several discrepancies in regional projections, divergences from observational data, and substantial variability in climate sensitivity estimates. GCMs are valuable tools, yet they remain deterministic approximations of a chaotic system, built on empirical parameters and simplified representations of poorly understood processes.

(2) *Expanding and improving observational data*. Reliable projections depend on robust, consistent datasets. Integrating satellite measurements, buoy networks, and ground-based observations with paleoclimate proxies requires methodological coherence and constant vigilance against systematic biases.

(3) *Disaggregating climate impacts*. Even if global trends were fully constrained, climate impacts vary widely across geographic, social, and ecological contexts. A more granular approach is needed to distinguish anthropogenic signals from natural variability (e.g., ENSO, NAO, AMO) and to assess local resilience and vulnerability.

The value of knowledge in planning. The purpose of climate research is not to produce alarmist or reassuring narratives, but to reduce uncertainty in support of informed decision-making. As policy comes to rely increasingly on scientific evidence, assessments must be rigorous and transparent, clearly communicating error margins, confidence levels, and interpretive limits.

Effective climate science does not impose dogma. Rather, it equips society with analytical tools to navigate uncertainty, fostering a balance between environmental action and socioeconomic sustainability. In an interconnected world, strategic planning must be interdisciplinary, integrating climate knowledge with economic, technological, and social insight.

The scientific challenge. Climate change represents a scientific, political, and ethical challenge of historic proportions. Its complexity, however, should not transform the debate into ideological orthodoxy or marginalize dissenting perspectives. The role of science—especially in domains with normative implications—is to illuminate uncertainty, not to conceal it behind assertive formulas.

It is within the dialectical space between knowledge and prudence that a balanced relationship between research, governance, and citizenship can emerge. Only by preserving a critical spirit, epistemological transparency, and methodological openness can climate change be addressed as both a scientific problem and a shared civic responsibility.

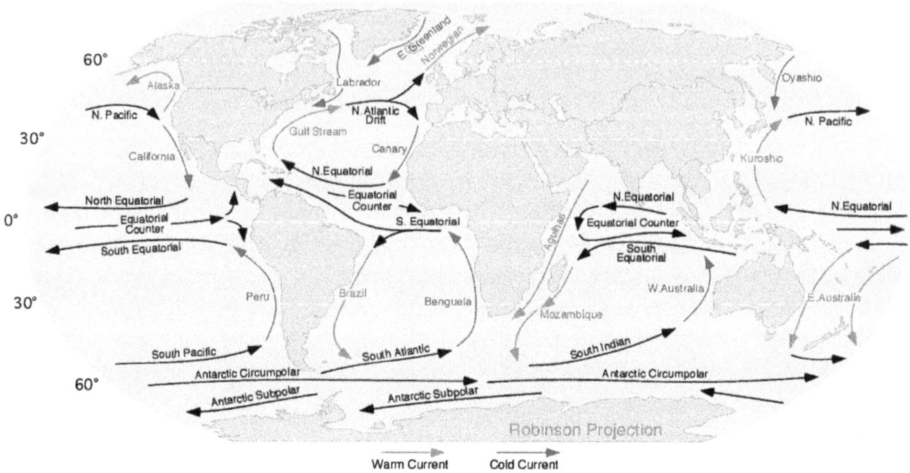

FIGURE 7.1. **Ocean surface currents.** Ocean gyres are vast, circular systems of surface currents that dominate the major ocean basins, shaped by global wind patterns, the Coriolis effect, and continental boundaries. Each subtropical gyre—in the North and South Atlantic, the North and South Pacific, and the Indian Ocean—rotates clockwise in the Northern Hemisphere and counterclockwise in the Southern Hemisphere, forming a closed loop of four main currents: fast, warm western boundary currents (e.g., the Gulf Stream and the Kuroshio); slow, cold eastern boundary currents (e.g., the Canary and California Currents); and two transverse currents that connect them across equatorial and mid-latitude zones. These gyres redistribute heat from the tropics toward the poles, influence regional climates, and shape marine ecosystems by regulating zones of upwelling and downwelling, which in turn control nutrient supply and biological productivity. The centers of gyres tend to be oligotrophic, with low nutrient availability and limited marine life, whereas their margins often sustain rich fisheries. Beyond the subtropical gyres, other major features include the Antarctic Circumpolar Current—the only uninterrupted flow around the globe—and equatorial countercurrents, which move counter to the prevailing trade winds and thereby balance zonal water transport. Together, these surface currents interact with the deep thermohaline circulation, driven by temperature and salinity contrasts, to regulate global climate, transport nutrients and oxygen, and structure oceanic biogeography across both horizontal and vertical dimensions. Their variability on seasonal and decadal timescales also modulates phenomena such as El Niño and the Atlantic Multidecadal Oscillation, with significant impacts on weather extremes and the distribution of marine resources.

Part 2

Critical Perspectives on the *Anthropogenic Global Warming Theory*

CHAPTER 8

Science and pseudo-science: preserving the boundary

8.1. Climate science between consensus and critical inquiry

In recent decades, the *Anthropogenic Global Warming Theory* (AGWT) has become a central framework within both the scientific community and international policymaking. According to this paradigm, human activity—and in particular the emission of greenhouse gases—plays a significant role in the observed planetary warming. As with any mature scientific field, however, widespread acceptance does not imply the closure of debate.

Numerous researchers continue to offer well-founded critiques that do not reject the anthropogenic component but rather seek to refine its quantification, explore its mechanisms, and assess its representation in climate models. These contributions, grounded in alternative methodological approaches and data interpretations, enhance scientific rigor, stimulate innovation, and broaden the scope of inquiry.

In public discourse, however, scientifically grounded critiques are sometimes conflated with oversimplified or non-evidence-based claims. For instance, the assertion that "*climate has always changed, therefore humans cannot be responsible for current warming*" reflects a misunderstanding of attribution science rather than a valid scientific argument. It is therefore essential to distinguish between legitimate critical inquiry and conceptual misconceptions.

Substantive critique typically addresses complex and technically nuanced issues, such as:

- the difficulty of accurately measuring global temperatures over long timeframes;
- the uncertainty surrounding the climate system's sensitivity to increased atmospheric CO_2;
- the role of clouds and water vapor as amplifiers within feedback mechanisms;
- the limitations of climate models in projecting long-term scenarios;
- the influence of natural cycles—solar, oceanic, and astronomical—within the broader warming context.

These perspectives do not advocate abandoning the AGWT framework; rather, they emphasize the importance of subjecting it to continuous empirical testing and theoretical refinement. As with any scientific theory, climate science must remain open to revision, falsification, and the integration of new evidence. This process is essential for preserving its credibility and adaptability as it seeks to elucidate one of the most complex and consequential natural systems.

8.2. When climate debate becomes theater

A striking paradox marks the public discourse on climate change: while the scientific evidence grows increasingly sophisticated and nuanced, media narratives tend to gravitate toward simplification and polarization, oscillating between the rhetoric of "*climate catastrophism*" and that of "*climate denial*." This disjunction between analytical depth and communicative brevity risks fostering misinterpretations and rhetorical distortions that obscure the underlying complexity of the issue.

One such distortion is the "*straw man*" argument, a rhetorical strategy that misrepresents a position in order to refute a simplified or distorted version of it. The mechanism typically unfolds as follows:

(1) a nuanced, evidence-based position is presented;
(2) the position is simplified or altered until it no longer reflects its original meaning;
(3) the distorted version is refuted, often with apparent ease;
(4) the refutation is then portrayed as applicable to the original argument.

While this strategy may appear to resolve disagreement, it generally undermines substantive dialogue and contributes to a public understanding that lacks depth and precision.

In climate discourse, both proponents and critics of the AGWT may, intentionally or inadvertently, fall into this rhetorical trap. In either case, the result is a diminished space for authentic, constructive scientific exchange.

This section examines how such rhetorical strategies can affect the integrity of climate communication. It highlights the importance of distinguishing between ideological posturing and scientifically grounded critique, and calls for a restoration of analytical rigor and openness in public debate.

Straw man arguments in support of AGWT. In efforts to promote awareness of, and action on, climate change, certain public arguments—even when well-intentioned—oversimplify scientific concepts. Although simplification can aid communication, excessive reductionism risks distorting the issue and undermining public trust in science.

Examples of such rhetorical shortcuts include:

- "*Denying AGWT is denying science.*" This conflates methodological skepticism with rejection of the scientific method. Critiquing model assumptions is part of scientific inquiry, not denialism.
- "*If warming continues, Earth will become uninhabitable.*" Apocalyptic framing may attract attention, but it tends to exaggerate projected outcomes. Even under high-emission scenarios, adaptation and resilience strategies remain viable. Scientific communication should prioritize proportionality and evidence.
- "*Critics are funded by the oil industry.*" Ad hominem arguments and sweeping generalizations can suppress legitimate dissent. While conflicts of interest must be scrutinized, many critical perspectives arise from independent research and data-driven analysis.
- "*CO_2 is the only factor that matters.*" Climate systems are multifactorial. While CO_2 can play an important role, other variables—such as

FIGURE 8.1. On the left, an example of a meteorological phenomenon (a lightning bolt during a thunderstorm) studied by meteorology; on the right, an example of climate (desert) studied by climatology.

cloud dynamics, oceanic circulation, and solar variability—are also significant and merit consideration.

- *"All scientists agree."* Scientific consensus reflects a convergence of evidence, not unanimity. Divergent interpretations and alternative hypotheses are essential to the advancement of knowledge. As is often attributed to Galileo, the validity of a scientific argument rests not on the number of its supporters but on the strength of its reasoning: *"In science, the authority of a thousand is not worth the humble reasoning of a single individual."* Empirical rigor outweighs popularity.

Another frequent oversimplification is causal inflation, whereby isolated weather events such as storms, heatwaves, or floods are presented as direct evidence of anthropogenic climate change. This approach overlooks the distinction between:

- *weather*—short-term atmospheric conditions;
- *climate*—long-term statistical patterns of weather over decades.

Without robust holistic analysis, attributing individual events to global warming may lead to:

(1) *misattribution*: not all extreme events are causally linked to anthropogenic emissions; attribution requires rigorous methodology;
(2) *credibility risk*: when simplified narratives are contradicted by observed anomalies (e.g., unusually cold winters), public confidence in climate science may be undermined.

Climate change is a well-documented phenomenon, supported by extensive empirical research. Its communication, however, must reflect the complexity of the evidence and avoid rhetorical shortcuts. Scientific credibility is best preserved through transparency, methodological rigor, and a commitment to nuance.

Straw man arguments in opposition to AGWT. Simplification is not exclusive to proponents of the AGWT; critics, too, may employ rhetorical strategies that distort the scientific debate and divert attention from substantive issues. These arguments, although often presented as rational critiques, frequently lack empirical grounding and methodological rigor.

Rather than contributing to a deeper understanding of climate dynamics, such claims serve to dismiss, trivialize, or delegitimize climate research. Examples include:

- *"One or two degrees more is not a problem because we already live in different climates."* This statement overlooks the significance of the rate and scale of climate change. Even modest increases in the global temperature can trigger systemic effects—including shifts in precipitation patterns, an increased frequency of extreme events, ecosystem disruptions, and economic repercussions.
- *"Climate has always changed, so there is nothing to worry about."* Although historical climate variability is well documented, current changes are claimed to be occurring at an accelerated pace and to be driven, at least in part, by anthropogenic factors. Comparing present-day dynamics with past fluctuations without accounting for causality and rate of change misrepresents the nature of the phenomenon.
- *"Science makes mistakes, so it must be wrong this time too."* Scientific progress is inherently iterative, and errors and revisions are integral to the refinement of theories. Dismissing a theory on the basis of the mere possibility of error undermines the epistemological foundations of scientific inquiry.
- *"It's cold today, so global warming is a hoax."* This reflects a misunderstanding of the distinction between weather and climate. Climate refers to long-term statistical patterns, whereas weather describes short-term atmospheric conditions. Isolated cold events do not contradict broader warming trends.
- *"Climatologists are in it for money and fame."* Ad hominem critiques shift the focus from evidence to personal motives. The scientific process is designed to safeguard against bias through peer review, data transparency, and methodological diversity. Sweeping accusations against the research community detract from the evaluation of empirical findings.

Expressing skepticism toward the AGWT is not inherently problematic; on the contrary, critical inquiry is a cornerstone of scientific advancement. For such critique to be constructive, however, it must be informed by evidence, grounded in methodology, and articulated with intellectual integrity. Simplistic analogies and rhetorical shortcuts risk obscuring the complexity of climate science and exacerbating polarization.

A meaningful dialogue on climate futures requires the ability to distinguish between serious scientific critique and rhetorical distortion. Only through such discernment can public discourse evolve toward greater clarity, allowing science to continue informing policy and societal choices with credibility and precision.

Let us now analyze some common critiques in detail.

8.3. Human adaptability to climate

A frequently cited argument used to question climate risk is that humans have successfully inhabited a wide range of environments—from arid deserts to polar tundra, and from tropical rainforests to temperate zones. On this view, humans could readily adapt to a global mean temperature increase of 1 or 2 °C without significant difficulty.

Figure 8.2 shows the global distribution of climate zones according to the Köppen classification system, which categorizes climates into five primary groups (A to E) on the basis of latitude, precipitation, and seasonal temperature patterns. Figure 8.3 compares average global temperatures on July 1 and January 1 between 1979 and 2000, illustrating the planet's climatic diversity and seasonal variability.

At first glance, the notion that a modest temperature increase poses a serious threat may appear overstated. This perspective, however, oversimplifies the issue by focusing exclusively on human adaptability, without accounting for broader ecological and systemic consequences.

Valid considerations. Despite its limitations, the adaptability argument contains elements worth acknowledging:

- *human resilience*: technological innovation, agricultural diversification, and architectural adaptation have enabled human survival in extreme climates;
- *civilizational diversity*: societies have historically flourished under a wide range of meteorological and climatic conditions;
- *adaptation potential*: were climate change to proceed gradually and predictably, there would be time to adjust infrastructure, production systems, and energy models accordingly.

The scientific limitations of this argument lie in what it omits.

- *Rate of change*: historical adaptations occurred over centuries or millennia, whereas contemporary temperature shifts are unfolding over decades, potentially exceeding the adaptive capacity of both societies and ecosystems, particularly in vulnerable regions.
- *Systemic impacts*: even a modest increase of 1.5–2 °C can trigger cascading effects, including:
  - the melting of glaciers and rising sea levels;
  - disrupted precipitation patterns, leading to droughts and floods;
  - reduced agricultural yields, water scarcity, and food insecurity;
  - an increased frequency and intensity of extreme events (e.g., hurricanes, wildfires, heatwaves).
- *Uneven consequences*: climate change does not affect all regions uniformly. While some areas may benefit from milder winters, others—such as low-lying coastal zones—may face higher risks. Vulnerable populations are likely to be disproportionately affected.
- *Ecological constraints*: unlike humans, ecosystems cannot rapidly adapt. Biodiversity depends on relatively stable conditions, and rapid climatic shifts threaten habitats, accelerate species extinction, and disrupt vital

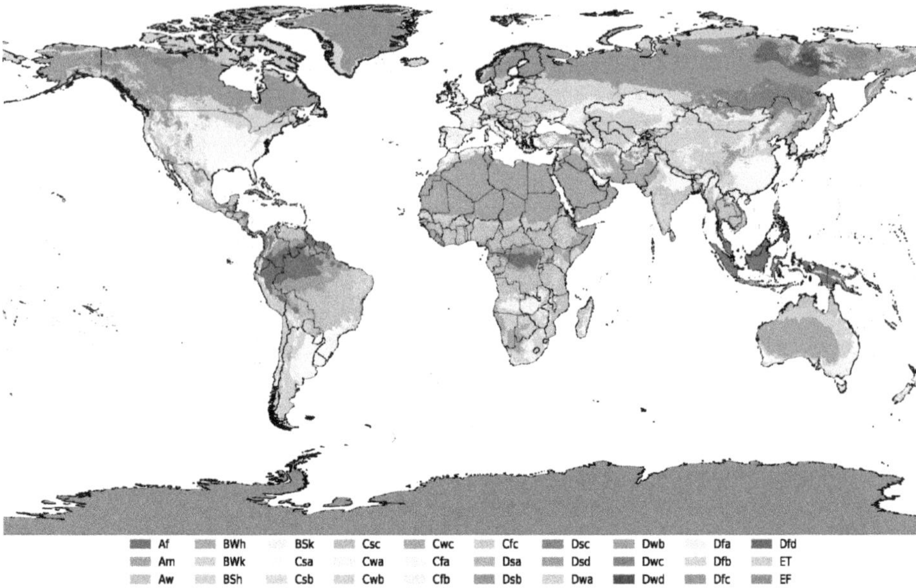

FIGURE 8.2. The Köppen classification system organizes global climates into five principal categories:
A – Humid megathermal climates: intertropical zones with monthly temperatures consistently above +18 °C. Includes equatorial (Af), monsoon (Am), and tropical dry winter/summer climates (Aw/As).
B – Arid climates: regions with low precipitation. Includes hot/cold desert (BWh/BWk) and hot/cold steppe climates (BSh/BSk).
C – Humid mesothermal climates: coldest month between –3 °C and +18 °C. Includes subtropical/oceanic (Cfa/Cfb), Mediterranean (Csa/Csb), and monsoon-influenced climates (Cwa/Cwb).
D – Boreal microthermal climates: found only in the Northern Hemisphere, with cold winters and short summers. Includes humid continental (Dfa/Dfb), monsoon continental (Dwa/Dwb), and subarctic climates (Dfc/Dfd).
E – Polar climates: warmest month below +10 °C. Includes tundra (ET) and ice cap climates (EF), characterized by minimal vegetation and widespread permafrost.

biogeochemical cycles. The disruption of coral reefs is a prominent example of this fragility.

Conclusion: from simplification to oversight. The assertion that *"humans already live in diverse climates, therefore global warming is manageable"* may appear intuitively plausible, but it does not withstand scientific scrutiny. It overlooks the pace of change, the unequal distribution of impacts, and the vulnerability of both ecosystems and socioeconomic systems.

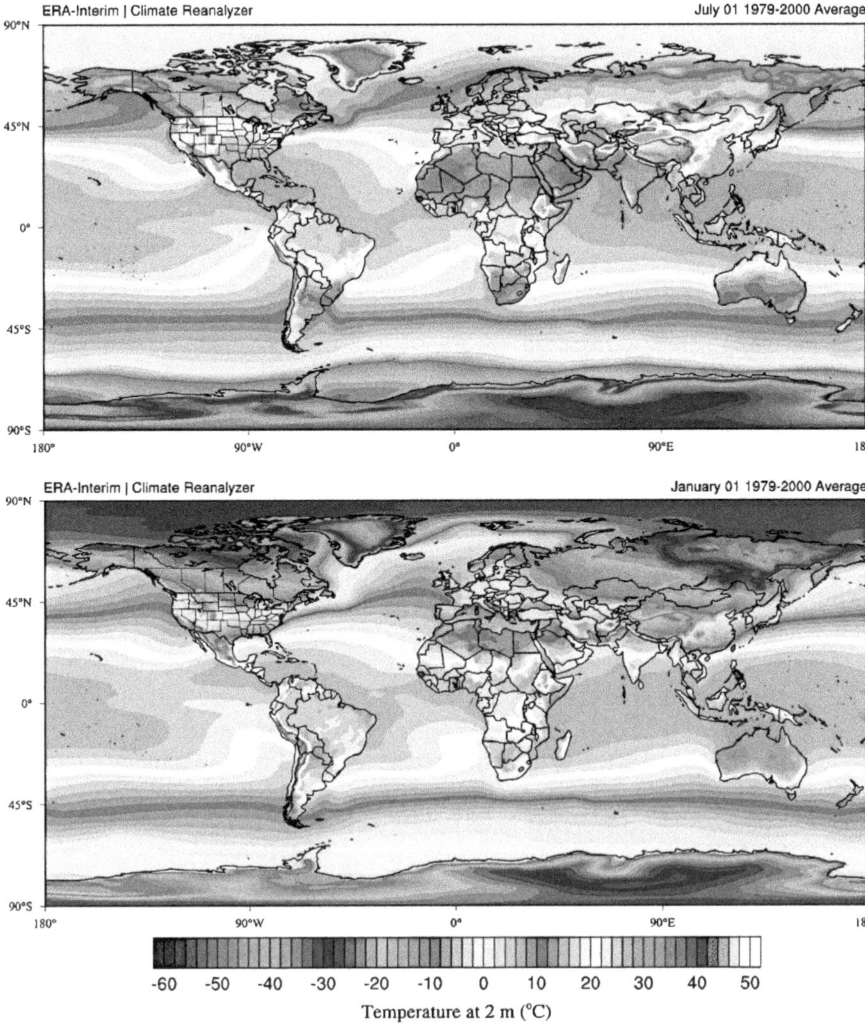

FIGURE 8.3. Distribution of average temperatures between 1979 and 2000 on July 1st and January 1st.

As with other rhetorical simplifications, this argument may offer superficial reassurance while obscuring deeper complexities. In the context of climate science, acknowledging complexity is not optional; it is essential for an accurate understanding of the system and an effective response to its changes.

For perspective: the temperature difference between the Last Glacial Maximum (LGM), approximately 21,000 years ago, and the present is estimated at 4–6 °C, roughly equivalent to the difference between the climates of Rome and London. Yet the LGM was not merely colder; it featured extensive ice sheets, profoundly altered ecosystems, sea levels 120–130 meters lower than today, and atmospheric CO_2 concentrations of around 180–200 ppm, which significantly affected plant growth and biospheric dynamics. In sum, human adaptability is a documented strength, but it is not without limits. It should not be invoked to dismiss the risks associated with rapid climate change.

8.4. Climate is not weather

A common misunderstanding in public discourse on climate change is the conflation of short-term weather variability with long-term climate trends. The fact that a given location may experience seasonal temperature fluctuations exceeding 40 °C does not imply that a global mean increase of 1–2 °C is negligible. On the contrary, even modest shifts in the global temperature reflect structural changes in Earth's energy balance, with potential consequences for atmospheric circulation, oceanic dynamics, precipitation regimes, and biogeochemical cycles.

Historical sensitivity to small climatic shifts. Historical records demonstrate that relatively minor changes in mean temperature can exert a significant influence on human societies. Two well-documented examples illustrate this sensitivity.

- *Medieval Warm Period* (MWP, 9th–14th century)—A possible warming of approximately 1 °C contributed to agricultural expansion in Europe, prolonged droughts in parts of the Americas and Asia (potentially linked to the decline of the Maya civilization), and the establishment of Viking settlements in Greenland—which were later abandoned as regional climates cooled.
- *Little Ice Age* (LIA, 14th–19th century)—A cooling of roughly 1 °C relative to the MWP led to widespread crop failures, glacial advances that damaged infrastructure, severe winters, and sociopolitical instability. The 17th-century cooling, estimated at 1–2 °C below present-day levels, has also been associated with the decline of the Ming Dynasty in China and increased unrest across Europe.

These episodes underscore the capacity of even modest climatic shifts to disrupt agricultural productivity, public health, geopolitical stability, and ecological systems.

Limitations of the "few degrees" argument. The assertion that *"a few degrees more is not a problem"* may appear intuitively plausible, but it oversimplifies the issue. Climate change concerns not so much individual thermal comfort as the rate, distribution, and systemic consequences of environmental transformation. Key considerations include:

- *rate of change*: rapid warming over decades contrasts with historical transitions that unfolded over centuries or millennia;
- *uneven impacts*: geographic regions and population groups differ in exposure and adaptive capacity;
- *systemic stress*: biodiversity, agriculture, water resources, and public health systems face increasing pressure;
- *disruption of planetary cycles*: seasonal rhythms, precipitation patterns, species migrations, and food availability may all be altered.

Global warming should not be equated with isolated temperature increases or seasonal anomalies. It represents a systemic shift in the interaction among energy flows, ecological processes, and human systems. Addressing climate change requires a commitment to analytical rigor, methodological transparency, and an appreciation of complexity. These qualities are central to scientific inquiry.

8.5. CO_2 absorption bands and the persistence of radiative forcing

A recurring argument in critiques of the *Anthropogenic Global Warming Theory* (AGWT) is the claim that increasing atmospheric carbon dioxide (CO_2) no longer contributes meaningfully to global warming, owing to the "*saturation*" of its infrared (IR) absorption bands. According to this interpretation, once a spectral band is fully absorbed, additional CO_2 would have no further radiative effect. This view, however, is inconsistent with established atmospheric physics and radiative transfer theory (Gussmann, 1967),[1] where the concept of "*saturation*" differs fundamentally from the chemical notion of "*maximum capacity.*" In the former case, the effect continues but is progressively attenuated; in the latter, it ceases entirely.

Logarithmic response of the CO_2 radiative forcing. The concept of saturation arises from a simplified understanding of radiative interactions. While the central portion of the CO_2 absorption band near 15 microns is saturated at lower altitudes, the spectral wings—that is, the outer edges of the band—remain active and continue to absorb IR radiation. Because Earth's atmosphere is vertically stratified, with temperature and density decreasing with altitude, radiative processes occur across multiple layers. This structure allows incremental absorption to persist even at elevated CO_2 concentrations.

The cumulative effect follows a logarithmic relationship: each doubling of the CO_2 concentration produces a measurable, though diminishing, increase in radiative forcing. This principle is well established in climate science and underpins many global climate-model projections.

An assessment by skeptical scientists. A recent study by van Wijngaarden & Happer (2023)[2]—researchers known for their critical stance toward alarmist climate scenarios—calculated that doubling atmospheric CO_2 from 400 to 800 ppm reduces the upward IR flux by approximately 3 W/m^2 at the mesopause (approximately 86 km altitude), under cloud-free conditions and a surface temperature of 288.7 K. These findings are consistent with estimates reported by the IPCC and other mainstream climate research, reinforcing the conclusion that CO_2 continues to exert a radiative effect even at elevated concentrations. van Wijngaarden and Happer (2020)[3] also evaluated the climate sensitivity of a simplified atmospheric model that assumes a water-vapor feedback maintaining constant relative humidity. They obtained an equilibrium climate sensitivity (ECS) of approximately 2.3 °C, which lies at the lower end of the "very likely" range reported by the IPCC (AR6, 2021), spanning from 2.0 °C to 5.0 °C, with a central estimate of 3.0 °C.

[1]Gussmann, E.A., 1967. Saturation in Fraunhofer lines. The Observatory, 87, 233–236.

[2]van Wijngaarden, W.A., Happer, W., 2023. Atmosphere and greenhouse gas primer. arXiv, DOI: 10.48550/arXiv.2303.00808.

[3]van Wijngaarden, W. A., Happer, W. H., 2020. Dependence of Earth's thermal radiation on five most abundant greenhouse gases. arXiv:2006.03098.

MODTRAN simulations. The MODTRAN model[4] simulates IR absorption and emission under a range of atmospheric conditions. Figure 8.4 illustrates how increasing CO_2 alters the IR spectrum.

- Panel A shows the broadening of the 15-micron band at 100 km altitude. The curve at 800 ppm diverges slightly from that at 400 ppm, indicating enhanced absorption. The curve at 0 ppm, which represents an atmosphere without CO_2, shows an absorption increase of 3.05 W/m², primarily due to water vapor. This value closely matches the estimate by van Wijngaarden & Happer (2023).
- Panel B demonstrates enhanced downward IR radiation toward the surface, with a 3.61 W/m² increase as CO_2 rises from 400 to 800 ppm.
- Panel C plots the logarithmic relationship between the molar fraction of CO_2, C_{CO_2}, and the radiative forcing, using the formula:

$$\Delta F_{\text{CO}_2} \approx 3.7 \cdot \log_2 \left(\frac{C_{CO_2}}{C_0} \cdot \frac{P_s}{P_o} \right) \quad \text{W/m}^2,$$

  where $C_0 = 280$ ppm is the preindustrial CO_2 baseline and $P_0 = 1$ atm. At $C_{CO_2} = 425$ ppm (the 2025 average) and $P_s \approx P_0$, the forcing is approximately $\Delta \text{F}_{\text{CO}_2} \approx 2.2$ W/m². This formula carries an estimated uncertainty of 10% (Myhre, 1998; Etminan et al., 2016).

Interpretation and climatic significance. The logarithmic nature of the CO_2 radiative forcing reflects the saturation of central absorption bands together with the continued activity of the spectral wings. The vertical layering of the atmosphere enables radiative interactions across altitudes, and models such as MODTRAN confirm that absorption persists even at elevated CO_2 levels.

A superficial reading of the near-overlap between the 400 and 800 ppm curves might suggest saturation. However, the difference—approximately 1.3% of the total IR flux—is climatically significant. CMIP6 global climate models indicate that this increase in radiative forcing could correspond to a global temperature rise of between 1.8 and 5.7 °C, depending on feedback strength. In 2025, with atmospheric CO_2 averaging around 425 ppm, the estimated radiative forcing, $\Delta \text{F}_{\text{CO}_2} \approx 2.2\ \text{W/m}^2$, remains a key driver of the observed and projected warming.

8.6. Atmospheric pressure and surface temperature

A critique of the AGWT advanced in certain skeptical circles is the empirical model proposed by Nikolov & Zeller (2017),[5] which challenges the conventional radiative theory of the greenhouse effect. Rather than attributing surface warming to the selective absorption of infrared radiation by greenhouse gases, the authors argue that planetary surface temperature is primarily governed by atmospheric pressure, regardless of chemical composition. In their view, the

[4] MODTRAN Infrared Light in the Atmosphere (University of Chicago) is available at: https://climatemodels.uchicago.edu/modtran/

[5] Nikolov, N., Zeller, K., 2017. New Insights on the Physical Nature of the Atmospheric Greenhouse Effect Deduced from an Empirical Planetary Temperature Model. Environment Pollution and Climate Change, 1(2), 1000112.

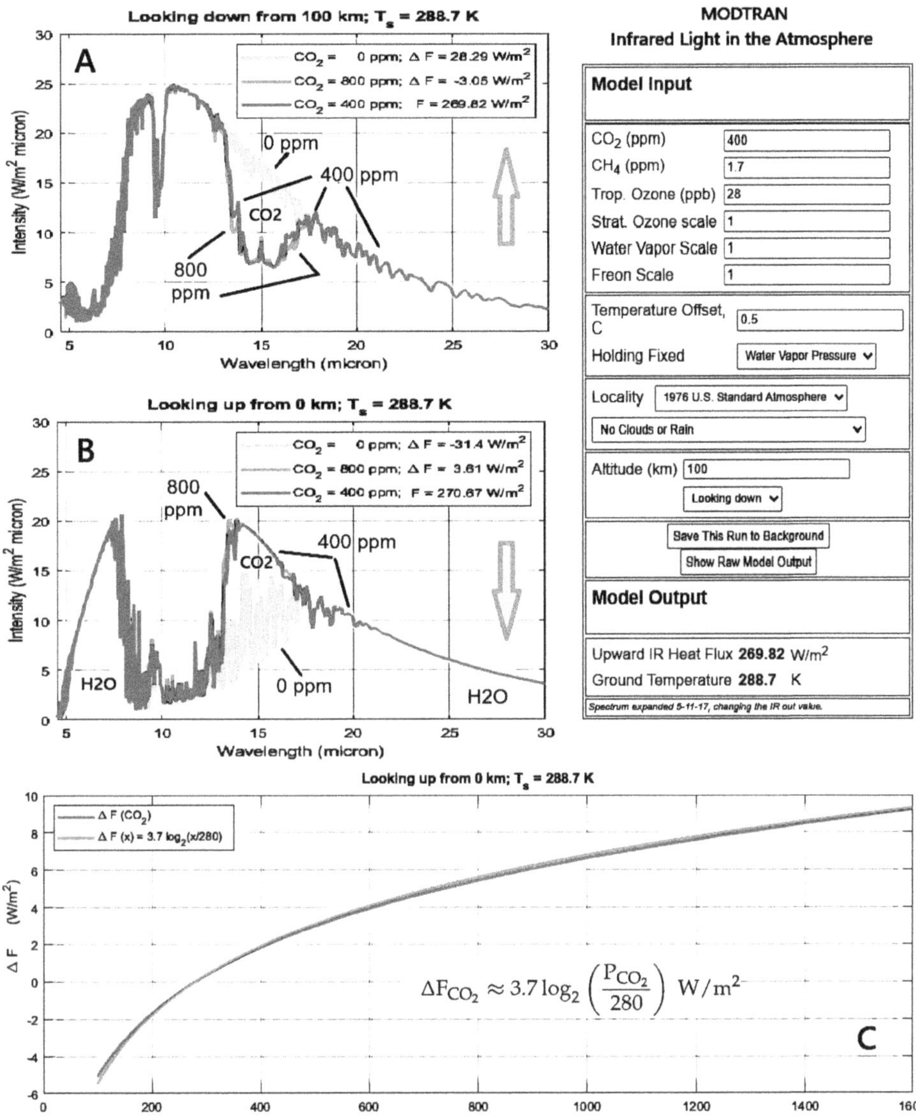

Figure 8.4. Applications of the MODTRAN model for infrared absorption in the atmosphere.
[A] Atmospheric infrared absorption spectrum (at 100 km altitude), calculated under standard cloud-free conditions, with surface temperature of 288.7 K (15.4 °C), for CO_2 concentrations of 0 ppm, 400 ppm, and 800 ppm, in a configuration similar to that hypothesized by van Wijngaarden & Happer (2023).
[B] Under the same conditions as [A], the infrared radiation emitted by the atmosphere and directed toward Earth's surface is calculated.
[C] Logarithmic trend of radiative forcing as a function of increasing atmospheric CO_2 concentration.

warming effect arises from adiabatic compression and the thermal capacity of the atmosphere, not from the radiative trapping of IR radiation.

The model is expressed through the following exponential formula:

$$T_s = T_{na} \cdot \exp\left[0.174205\left(\frac{P_s}{P_r}\right)^{0.150263} + 1.83121\times10^{-5}\left(\frac{P_s}{P_r}\right)^{1.04193}\right].$$

Here, T_s is the surface temperature; T_{na} is the radiative equilibrium temperature in the absence of an atmosphere, determined by the total solar irradiance (TSI) received by the planet and its albedo; P_s is the surface pressure; and P_r is a reference pressure. The authors show that this empirical formula provides a reasonably good fit to observed planetary data, and conclude that a planet's surface temperature is not governed by the greenhouse effect, but rather by atmospheric pressure alone.

While the model offers a satisfactory statistical fit, its physical interpretation is questionable. By attributing surface temperature solely to atmospheric pressure, without accounting for radiative transfer or the spectral behavior of greenhouse gases, it risks conflating correlation with causation. The absence of a mechanistic foundation renders the model epistemologically fragile, particularly in light of alternative, physically grounded explanations.

The foundational work of Manabe & Strickler (1964)[6] provides a physically grounded framework for understanding how atmospheric pressure influences surface temperature. Their radiative–convective model assumes a gray atmosphere and incorporates both radiative equilibrium and convective adjustment. A simplified approximation expresses the surface temperature as:

$$T_s = T_{na}\cdot\left(1+\frac{3}{4}\cdot\tau\right)^{1/4} \quad\Rightarrow\quad T_s \approx T_{na}\cdot\left(1+\frac{3}{4}\cdot c\cdot P_s\right)^{1/4},$$

where τ is the infrared optical depth and the equation is rewritten as a function of the surface pressure P_s. This assumes a homogeneous atmosphere and a constant path length, since at wavelength λ, $\tau(\lambda) = \alpha(\lambda)\cdot L\cdot C_{CO_2}\cdot P$, where $\alpha(\lambda)$ is the specific absorption coefficient of CO_2 at wavelength λ (dependent on temperature and pressure); P is the total pressure (in atm); C_{CO_2} is the molar fraction of CO_2 (e.g., 400 ppm $= 4\times10^{-4}$); and L is the optical path length (in meters). Here, $c = 0.836\,\text{atm}^{-1}$ is an empirically calibrated constant under standard conditions ($T_s = 288\,\text{K}, T_{na} = 255\,\text{K}$), and P is the surface pressure in atmospheres.

Moreover, according to Robinson & Catling (2014) (see Section 4.2), many planetary atmospheres emit thermal radiation to space from a characteristic pressure level near 0.1 bar, largely independent of surface conditions. By combining this insight with the barometric equation, which relates pressure to altitude in a stratified atmosphere, one finds that increasing the surface pressure deepens the atmospheric column below the emission level. Assuming an adiabatic lapse rate with effective specific heat c_p^*, the surface temperature is given by

$$T_s = T_e\left(\frac{P_s}{P_e}\right)^{R/c_p^*},$$

[6]Manabe, S., Strickler, R.F., 1964. Thermal Equilibrium of the Atmosphere with a Convective Adjustment. Journal of the Atmospheric Sciences, 21(4), 361–385.

where $R/c = 0.286$ for dry air and T_e is the effective emission temperature of the planet at pressure P_e. Thus, for fixed P_e, the surface temperature T_s increases nonlinearly with the surface pressure P_s, following a power-law dependence.

In conclusion, several theoretical frameworks already establish an approximate relationship between atmospheric surface pressure and temperature. The Nikolov–Zeller hypothesis that a planet's surface temperature is independent of the chemical composition of its atmosphere appears coincidental rather than causally grounded. In any event, it could only be validated by observing a planet endowed with an atmosphere entirely devoid of greenhouse gases—an empirical condition for which no evidence currently exists.

8.7. Climate sensitivity: the crucial uncertainty

While the radiative impact of carbon dioxide (CO_2) is well established, the key scientific uncertainty lies in the climate system's response—that is, in its sensitivity to radiative forcing. This sensitivity depends on several interrelated factors:

- the nature and magnitude of feedback mechanisms, which may amplify (positive feedback) or dampen (negative feedback) warming;
- the accurate identification and quantification of both natural and anthropogenic forcings;
- the precision and reliability of the observational data used to validate climate models.

Contrasting perspectives on feedbacks. van Wijngaarden & Happer (2023) argue that the climate system tends to resist perturbations, invoking Le Chatelier's principle, according to which a perturbed system seeks a new equilibrium. They suggest that climate feedbacks are predominantly negative and therefore attenuate warming.

The view that net climate feedback may be slightly negative is consistent with the "*Iris*" hypothesis formulated by Lindzen et al. (2001) and Lindzen & Choi (2011),[7] which is based on satellite observations (ERBE and CERES). The hypothesis posits that rising tropical temperatures lead to the formation of high, thin cirrus clouds that enhance the infrared emission to space, thereby cooling the system. Under this scenario, a doubling of CO_2 would produce a temperature increase of only 0.6–0.8 °C, lower than the 1.16 °C expected in the absence of feedback ($f = 0$, see Section 6.3). A missing iris effect has been proposed as a possible cause of erroneous hydrological change and of the high climate sensitivity exhibited by current models (Mauritsen & Stevens, 2015).[8]

By contrast, other studies have found evidence for a net positive feedback (e.g. Sherwood et al., 2020).[9] The IPCC Sixth Assessment Report (AR6, 2021) estimates the equilibrium climate sensitivity (ECS) for CO_2 doubling as follows:

[7]Lindzen, R.S., Chou, M.-D., Hou, A.Y., 2001. Does the Earth have an adaptive infrared iris? Bulletin of the American Meteorological Society, 82, 417–432.
Lindzen, R.S., Choi, Y.-S., 2011. On the observational determination of climate sensitivity and its implications. Asia-Pacific Journal of Atmospheric Sciences, 47, 377–390.

[8]Mauritsen, T., Stevens, B., 2015. Missing iris effect as a possible cause of muted hydrological change and high climate sensitivity in models. Nature Geoscience, 8(5), 346–351.

[9]Sherwood, S., Webb, M., Annan, J., Armour, K., Forster, P., et al., 2020. An assessment of Earth's climate sensitivity using multiple lines of evidence. Reviews of Geophysics, 58, e2019RG000678.

- likely range: 2.5–4.0 °C;
- very likely range: 2.0–5.0 °C;
- best estimate: ∼3.0 °C.

The sign of climate feedback may also evolve over time. Initial negative responses (e.g., cloud adjustments) may be followed by slower positive ones (e.g., ice melting and albedo changes), potentially amplifying the long-term warming.

Le Chatelier's principle: a partial analogy. Applying Le Chatelier's principle to the assessment of climate feedback requires caution. According to the formalism presented in Section 6.3, the relevant condition is:

$$\lambda_P(1 - f) < 0,$$

where $\lambda_P = -3.2\ \mathrm{W/m^2K^{-1}}$ represents the Planck response. This implies that the feedback factor f must remain below 1 to ensure system stability, but not necessarily below zero. In other words, even a positive net feedback is compatible with the stabilization of climate equilibrium, provided that it does not exceed the compensatory effect of the Planck response.

A runaway greenhouse effect would require $f \gtrsim 1$, a condition not predicted by any CMIP6 model. Instead, all global climate models simulate stabilization at a new equilibrium:

- with no feedback ($f = 0$), a doubling of CO_2 yields approximately 1.16 °C of warming;
- with positive or negative feedback, the final temperature adjusts above or below this baseline, respectively.

Thus, the behavior of the climate system as simulated by the general circulation models (GCMs) currently in use remains consistent with Le Chatelier's principle.

The Venus analogy: a non-replicable scenario. The concept of a *runaway greenhouse effect* is often illustrated through the case of Venus. This scenario, however, is not considered plausible for Earth under known geological and atmospheric conditions. During the Cambrian period, atmospheric CO_2 concentrations reached 4,000–5,000 ppm (Royer et al., 2001),[10] yet Earth did not undergo thermal runaway. This suggests the existence of internal regulatory mechanisms capable of redistributing energy even under elevated CO_2 levels.

Venus differs fundamentally from Earth in several respects: its proximity to the Sun, the absence of oceans, and an atmosphere composed of more than 96% CO_2. These factors likely contributed to a self-reinforcing dynamic that led to its present equilibrium, with surface temperatures exceeding 450 °C. Venus is therefore not a suitable analogue for Earth's climate trajectory.

Conclusion. The assertion that the CO_2 absorption bands are fully saturated is supported neither by radiative transfer theory nor by empirical data. Models such as MODTRAN, together with observational studies and theoretical analyses, consistently demonstrate that increasing CO_2 concentrations continue to produce radiative forcing.

[10] Royer, D.L., Berner, R.A., Beerling, D.J., 2001. Phanerozoic atmospheric CO_2 change: evaluating geochemical and paleobiological approaches. Earth-Science Reviews, 54(4), 349–392.

The central uncertainty in climate science thus lies not in the existence or persistence of the greenhouse effect—which remains active and quantifiable—but in the system's sensitivity to that forcing, which depends on the interplay of feedback mechanisms. Clarifying this sensitivity is essential for refining projections and informing policy decisions.

8.8. CO_2 and climate: complexity beyond correlation

A frequently cited argument among those who question the role of carbon dioxide (CO_2) in climate change is the perceived lack of correlation between atmospheric CO_2 concentrations and global temperature trends (Davis, 2017).[11] One of the most commonly cited examples is the period from 1940 to 1970, during which CO_2 emissions increased substantially while global temperatures exhibited a slight cooling trend. From this observation, some infer that CO_2 cannot be a primary driver of warming.

This reasoning, however, overlooks the complexity of Earth's climate system, which is influenced by multiple interacting forcings operating across diverse spatial and temporal scales.

Multifactorial drivers of climate variability. Global temperature does not respond to CO_2 in a direct or linear manner. Several factors contribute to climate variability, including:

- fluctuations in solar activity;
- oceanic oscillations (e.g., ENSO, PDO, AMO);
- volcanic eruptions and stratospheric aerosol loading;
- anthropogenic aerosols from industrial activity;
- other greenhouse gases (e.g., methane, nitrous oxide);
- feedback mechanisms that modulate the initial forcings.

Given this complexity, a perfect correlation between CO_2 and temperature is neither expected nor required to establish causality.

The 1940–1970 cooling episode. Following World War II, global CO_2 emissions rose markedly as a result of industrial expansion and increased fossil fuel consumption (Figure 3.1). Yet between 1940 and 1975, global surface temperatures declined slightly (Figure 2.2). This apparent divergence prompted speculation in the 1970s about a potential cooling trend, exemplified by the Newsweek article "*The Cooling World*" (April 28, 1975) by Peter Gwynne, which synthesized emerging concerns based on limited datasets and early climate models. Climatologists such as Stephen Schneider briefly entertained this possibility, attributing it to increased sulfate aerosols and reduced solar irradiance.

While the article reflected the scientific uncertainty of the time, subsequent research clarified that the mid-century cooling was likely driven by transient factors (Peterson et al., 2008),[12] including increased atmospheric aerosols reflecting solar radiation, the 1963 eruption of Mount Agung, which injected volcanic particles into the stratosphere, and natural variability in oceanic and solar cycles.

[11] Davis, W.J., 2017. The relationship between atmospheric carbon dioxide concentration and global temperature for the last 425 million years. Climate, 5(4), 76.

[12] Peterson, T.C., Connolley, W.M., Fleck, J., 2008. The Myth of the 1970s Global Cooling Scientific Consensus. Bulletin of the American Meteorological Society, 89(9), 1325–1338.

These cooling influences temporarily offset the warming effect of rising CO_2 concentrations. Climate models from the Coupled Model Intercomparison Project (CMIP) confirm that twentieth-century temperature trends can be accurately reproduced only when both natural and anthropogenic forcings are taken into account.

Figures 6.2A and 6.2C illustrate how sulfate aerosols and volcanic activity contributed a negative (cooling) forcing during this period. The CMIP models do not treat CO_2 as the sole driver, but as one component within a multifactorial system.

Correlation, causality, and natural cycles. The assertion that CO_2 does not influence climate on the basis of imperfect correlation is a logical fallacy. Atmospheric CO_2 levels are themselves influenced by temperature through ocean–atmosphere interactions. According to Henry's Law, warmer sea surfaces release more CO_2 into the atmosphere owing to reduced gas solubility. This explains why temperature changes typically precede CO_2 variations in paleoclimate records and in short-term observational studies (Humlum et al., 2013; Koutsoyiannis, 2024; Mazzarella et al., 2013).[13] These findings reflect the existence of a natural CO_2 cycle and its associated feedback processes. They do not, however, refute the role of CO_2 as a radiative forcing agent, nor the fact that anthropogenic emissions have contributed substantially to the increase in atmospheric CO_2 over the past century.

Models, simulations, and observable reality. The IPCC CMIP6 models demonstrate that the warming observed during the 20th century can be reproduced only when all relevant forcings are taken into account. These models do not assume that CO_2 is the sole driver, nor do they treat the climate system as linear. Rather, they simulate the dynamic interactions among the atmosphere, oceans, biosphere, cryosphere, and human activity. Dismissing the role of CO_2 on the basis of imperfect correlation or a delayed temperature response oversimplifies the issue. Just as complex diseases may have multiple contributing factors, climate change results from a combination of forcings. CO_2 remains a major contributor, although not the only one. The cooling observed between 1940 and 1970 does not invalidate the influence of CO_2; rather, it underscores the importance of rigorous, multidimensional analysis in understanding a system as complex as Earth's climate.

8.9. Does a global mean surface temperature exist?

Some scholars argue that the global mean surface temperature (GMST) is not a physically or mathematically meaningful quantity. It has also been claimed that

[13]Humlum, O., Stordahl, K., Solheim, J.-E., 2012. The phase relation between atmospheric carbon dioxide and global temperature. Global and Planetary Change, 100, 51–69.
Koutsoyiannis, D., 2024. Stochastic assessment of temperature–CO_2 causal relationship in climate from the Phanerozoic through modern times. Mathematical Biosciences and Engineering, 21(7), 6560–6602.
Mazzarella, A., Giuliacci, A., Scafetta, N., 2013. Quantifying the Multivariate ENSO Index (MEI) coupling to CO_2 concentration and to the length of day variations. Theoretical and Applied Climatology, 111, 601–607.

different averaging methods applied to heterogeneous, non-equilibrium temperature fields (arithmetic, geometric, harmonic, spatial, temporal, or weighted means) yield incompatible results, with no unique thermodynamic interpretation (e.g.: Cohler, 2025; Essex et al., 2007).[14]

This objection is only partially valid. It is true that the GMST is not a thermodynamic variable in the strict sense, since it does not correspond to a directly measurable physical quantity. Nevertheless, it functions as a climatic index that is specifically defined and derived from a set of explicit physical assumptions, statistical procedures, and modeling frameworks. The use of the arithmetic mean follows from the fact that, for air, temperature is linearly proportional to the molecular energy of the gas, which makes this averaging method the most physically coherent choice. Although the GMST is a model construct, it remains meaningful and useful as a global climate indicator when correctly interpreted. What matters most is acknowledging that the reported GMST values are subject to several sources of uncertainty—temporal, spatial, and methodological—including coverage gaps, homogenization procedures, and potential systematic biases (e.g., uncorrected urban heat contamination). All of these factors must be assessed and taken into account, as discussed in Part 3 of this book.

8.10. Conclusion: the role of scientific skepticism in climate research

The public debate on climate change tends to crystallize into polarized positions and oversimplified narratives. On one side, the *Anthropogenic Global Warming Theory* (AGWT) is sometimes presented as an incontrovertible truth; on the other, it is categorically rejected, often through arguments that lack empirical or theoretical rigor. The core of the scientific discussion, however, lies elsewhere—namely, in the quality of the data, the appropriateness of their interpretation, and the reliability of the models used to simulate future scenarios.

Critiquing the AGWT does not amount to rejecting science. On the contrary, the scientific method is founded on the continuous questioning of hypotheses, the examination of underlying assumptions, and the quantification of uncertainties. The climate system is among the most complex of natural phenomena, shaped by interactions among the atmosphere, oceans, biosphere, and geosphere, and governed by feedbacks and inertia operating across multiple temporal scales. Reducing this complexity to binary assertions—such as *"entirely anthropogenic"* or *"entirely natural"*—oversimplifies the phenomenon by obscuring its multifactorial nature.

Data, models, and methodology: the foundations of scientific inquiry. Climate models are indispensable tools for understanding and projecting climate dynamics. Their credibility, however, depends on their ability to reproduce and accurately predict observed phenomena, manage uncertainty, and incorporate feedback mechanisms. Without continuous validations against empirical evidence, models risk becoming detached from observable reality.

[14] Cohler, J., 2025. The father of lies hijacking climate science: global mean surface temperature does not exist. Journal of American Physicians and Surgeons, 30(4), 122–126.
Essex, C., McKitrick, R., Andresen, B., 2007. Does a global temperature exist? Journal of Non-Equilibrium Thermodynamics, 32(1), 1–27.

The Cooling World

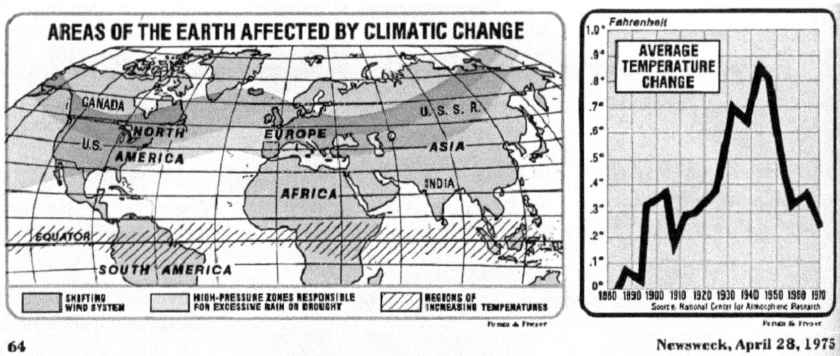

Figure 8.5. *"The Cooling World"*, written by Peter Gwynne, Newsweek, April 28, 1975.

In the April 28, 1975, issue of Newsweek, the science writer Peter Gwynne presented a synthesis of emerging climatological concerns regarding a perceived cooling trend in global temperatures. Drawing on the scientific literature and expert opinion available at the time, the article reported that average global temperatures had declined over the preceding three decades, prompting speculation about potential long-term climatic shifts.

The article attributed this cooling to several factors, including increased atmospheric concentrations of aerosols and industrial pollutants, which were thought to reduce the solar radiation reaching Earth's surface. It cited early climate models suggesting that continued cooling could lead to significant environmental consequences, such as expanded polar ice coverage, altered agricultural zones, and a heightened risk of drought and famine in vulnerable regions. Although the tone of the article was cautious, it conveyed a sense of urgency regarding the need for improved climate monitoring and preparedness. It also mentioned speculative geoengineering proposals—such as the artificial melting of the polar ice caps or the redirecting of solar energy—as hypothetical interventions, although these were not endorsed. In retrospect, the article reflects the scientific uncertainty of the era, characterized by limited global datasets and nascent climate modeling capabilities. Although the cooling trend ultimately proved to be temporary and was later overtaken by a warming trend, the piece remains a historical artifact illustrating the evolving nature of climate science discourse in the mid-twentieth century.

CHAPTER 9

The rhetoric of scientific *"consensus"* in the climate debate

In contemporary climate discourse, appeals to scientific consensus are frequently invoked not only as a form of validation for the *Anthropogenic Global Warming Theory* (AGWT), but also as a rationale for adopting ambitious mitigation strategies, such as the *Net-Zero* targets set out in the 2016 Paris Climate Agreement. While consensus may carry communicative and political weight, it is essential to recognize that science is not governed by majority opinion, but by the capacity of hypotheses to withstand empirical scrutiny.

Richard Feynman, Nobel laureate in Physics (1965), articulated this principle in his lectures and writings:

> *"The experts who are guiding you might be wrong... science teaches the value of doubt and freedom of thought. (...) I believe we live in a non-scientific age in which almost all communications and television—words, books, and so on—are non-scientific. As a result, there is a considerable amount of intellectual tyranny in the name of science."*[1]

Feynman's observation underscores a foundational tenet of scientific inquiry: authority alone does not confer validity. Scientific progress depends on the continuous testing of claims, including those advanced by the so-called "experts." It is therefore essential to distinguish between rhetorical consensus, which may reflect sociopolitical alignment, and demonstrable scientific consensus, which emerges from reproducible evidence and critical evaluation.

9.1. The claim of a *"settled"* climate science

The notion that climate science is *"settled"* has gained prominence in institutional and academic discourse, often serving to reinforce a sense of urgency. Marcia McNutt (2015),[2] then president of the *U.S. National Academy of Sciences*, called for decisive action in response to climate risks. Similarly, Filippelli et al. (2021)[3] describe the planet as being in a state of *"crisis,"* while Atwoli et al.

---

[1]Feynman, R.P.. Science is believing in the ignorance of experts. Pratica Bioenergetica. Retrieved from *https://praticabioenergetica.it/la-scienza-e-credere-nell-ignoranza-degli-esperti/*

[2]McNutt, M., 2015. The beyond-two-degree inferno. Science, 349, 7.

[3]Filippelli, G., Beal, L., Rajaram, H., AghaKouchak, A., Balikhin, M.A., et al., 2021. Geoscientists, who have documented the rapid and accelerating climate crisis for decades, are now pleading for immediate collective action. Geophysical Research Letters, 48, e2021GL096644.

(2021)[4] characterize climate change as the *"greatest global public health threat"* of the 21st century. These positions align with the policy orientation of COP21 and the IPCC, which advocate rapid and comprehensive emission reduction strategies.

While such statements reflect growing concern within the scientific and medical communities, the underlying scientific debate remains active. Frequently cited consensus figures—such as the 97% reported by Cook et al. (2013)[5] or the 99% claimed by Lynas et al. (2021)[6]—have been subject to methodological scrutiny. Critics, including Montford (2013)[7] and Legates et al. (2015),[8] have raised questions about survey design, data selection criteria, and the semantic framing of the key statements.

For example, the phrase *"humans are causing global warming"* admits multiple interpretations. If understood as implying the full attribution of warming to anthropogenic activity, as suggested in IPCC AR6 (2021), it tends to exclude natural variability. Earlier IPCC reports, however, such as the First Assessment Report (FAR, 1990) and the Second Assessment Report (SAR, 1995), adopted more cautious language, acknowledging human activity as one contributing factor among others.

This evolution in attribution language illustrates that consensus figures often reflect communicative simplification rather than uniform scientific agreement. Scientific consensus, when invoked rhetorically, may obscure the nuances of ongoing research, the diversity of methodological approaches, and the legitimate range of interpretations within the climate science community.

9.2. Appeals and declarations: a plurality of scientific positions

Over the past several decades, numerous public appeals have shaped the climate debate, often signed by prominent scientists. In 1997, the Union of Concerned Scientists issued the *World Scientists' Call for Action*, signed by more than 1,500 scientists, warning of a *"discernible human influence"* on climate and defining global warming as *"one of the most serious threats to the planet."*[9] In 2017, Ripple et al.[10] published the *Second Notice*, endorsed by more than 15,000 scientists, calling for rapid decarbonization. The 2015 Mainau Declaration and

[4]Atwoli, L., Baqui, A.H., Benfield, T., Bosurgi, R., Godlee, F., et al., 2021. Call for Emergency Action to Limit Global Temperature Increases, Restore Biodiversity, and Protect Health. The New England Journal of Medicine, 385, 1134–1137.

[5]Cook, J., Nuccitelli, D., Green, S.A., Richardson, M., Winkler, B., et al., 2013. Quantifying the consensus on anthropogenic global warming in the scientific literature. Environmental Research Letters, 8, 024024.

[6]Lynas, M., Houlton, B.Z., Perry, S., 2021. Greater than 99% consensus on human-caused climate change in the peer-reviewed scientific literature. Environmental Research Letters, 16, 114005.

[7]Montford, A., 2013. Consensus? What consensus? GWPF Note 5.

[8]Legates, D.R., Soon, W., Briggs, W.M., Monckton, C., 2015. Climate consensus and 'misinformation': A rejoinder to agnotology, scientific consensus, and the teaching and learning of climate change. Science & Education, 24, 299–318.

[9]Union of Concerned Scientists. (1997). World Scientists' Call for Action. UCS Publications.

[10]Ripple, W. J., Wolf, C., Galetti, M., et al., 2017. World Scientists' Warning to Humanity: A Second Notice. BioScience, 67(12), 1026–1028.

the open letter from 1,800 academics issued ahead of COP21 conveyed a similar sense of urgency.[11]

More recently, the tone has become increasingly direct. Scientist Rebellion has stated: *"First, we were concerned. Then, we were alarmed. Now, we are terrified."*[12] In 2023, 1,447 scientists—including 33 IPCC authors—signed an appeal published in *The Guardian* urging public activism with the words *"We are terrified. We need you."*[13] In Italy, 96 scientists, including the Nobel laureate Giorgio Parisi, called on the media to treat climate change as a national emergency.[14] In Australia, more than 260 researchers linked devastating bushfires to anthropogenic influence.[15]

Alongside these alarmist positions, however, skeptical and critical voices have also been heard. The *Leipzig Declaration* (1995)[16] challenged the prevailing consensus, while the *Oregon Petition* argued that increased CO_2 does not necessarily lead to catastrophic warming.[17] The Climate Intelligence Foundation (CLINTEL), through its *World Climate Declaration*, gathered approximately 2,000 signatories—including the Nobel laureates Ivar Giaever and John F. Clauser—who stated that *"There is no climate emergency."*[18] Even mainstream scientists, such as Hans von Storch, have warned that excessive alarmism may undermine scientific credibility.[19]

Public opinion. In Italy, a 2025 survey by *Termometro Politico* found that 26% of Italians regard climate change as media manipulation, citing the view that *"heatwaves have always existed"* and accusing environmentalism of ideological bias.[20] A further 12.4% acknowledged warming but dismissed it as a threat, maintaining that *"we will adapt as we always have."* Globally, the UNDP's *People's Climate Vote* (2024), based on over 75,000 respondents across 77 countries, found that 80% favored stronger government action, while 53% reported feeling more worried than in the previous year—an indicator of rising *"climate anxiety."*[21] Yet

---

[11]Lindau Nobel Laureate Meetings. (2015). Mainau Declaration 2015 on Climate Change. UOW (University of Wollongong). (2015). Open letter from 1,800 academics ahead of Paris climate conference.

[12]Scientist Rebellion. (2025). Global Scientists' Letter on Climate Emergency. ScientistRebellion.org.

[13]Boyle, L. (2023). More than 1,000 climate scientists urge public to become activists. The Guardian, 4 December 2023.

[14]Phys.org. (2023). Italian scientists urge media to take climate change seriously. Phys.org, 26 July 2023.

[15]Morton, A. (2020). Bushfires: Scientists call for climate action after fires. The Guardian, 17 January 2020.

[16]Singer, S. F., Idso, C., Baliunas, S. (1995). The Leipzig Declaration on Global Climate Change.

[17]Robinson, A. B., Robinson, N. E., Soon, W. (2007). Environmental effects of increased atmospheric carbon dioxide. Oregon Institute of Science and Medicine.

[18]CLINTEL. (2022). 'There is No Climate Emergency' – World Climate Declaration, signatories over 1,100 and growing (updated to over 1,600 by 2023).

[19]von Storch, H. (2006). Statement before the U.S. House of Representatives, Committee on Energy and Commerce, Subcommittee on Oversight and Investigations, 19 July 2006.

[20]Termometro Politico Polls: *"6 out of 10 Italians are concerned about climate change"*. Available at https://www.termometropolitico.it/1626673_sondaggi-politici-elettorali-tp-04-07-2025.html

[21]Il Post: *"8 out of 10 people want their governments to do more against the climate crisis"*. Available at: https://www.ilpost.it/2024/06/20/sondaggio-undp-cambiamento-climatico-italia/

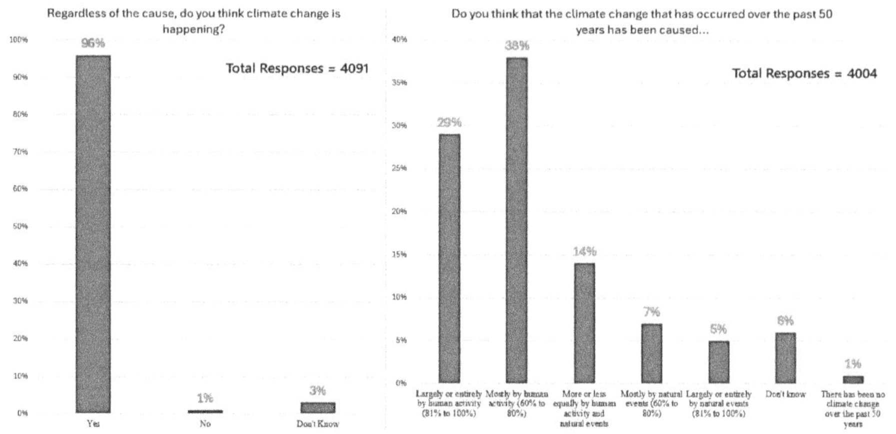

FIGURE 9.1. Distribution of opinions among more than 4,000 members of the "*American Meteorological Society*". The graph on the right shows the level of agreement on the existence of global warming from 1960 to 2010. The one on the left represents attribution estimates of the causes, distinguishing between anthropogenic contributions and natural factors in the observed warming. (Recreated from Maibach et al., 2016).

this concern coexists with growing eco-fatigue and skepticism toward top-down environmental policies.

These data indicate that public opinion is far from monolithic. While environmental awareness is widespread, many citizens question the dominant narrative and call for more transparent and inclusive approaches.

Conclusion. Ultimately, the climate debate is not solely about data; it concerns narratives, framing, and the intersection of science, politics, and public perception. Scientific consensus, while important, must not be conflated with unanimity nor used to silence legitimate critique. A healthy scientific culture embraces pluralism, methodological rigor, and open dialogue. Only through this lens can climate science remain credible, adaptive, and genuinely informative.

9.3. Analysis of a professional survey

More detailed surveys that distinguish between degrees of attribution reveal a broader spectrum of scientific opinion than is commonly portrayed. The 2016 national survey conducted by Maibach et al. (2016),[22] comprising more than 4,000 members of the *American Meteorological Society* (AMS), provides a nuanced snapshot of professional views (Figure 9.1).

While 96% of respondents agreed that climate change is occurring:

- only 29% attributed it predominantly (81–100%) to human activity;
- 38% regarded human influence as dominant but not exclusive (60–80%);

[22] Maibach, E., Perkins, D., Francis, Z., Myers, T., Engblom, A., et al., 2016. A 2016 national survey of American Meteorological Society member views on climate change. George Mason University, Center for Climate Change Communication.

- 14% reported a mixed contribution (40–60%);
- 19% leaned toward natural causes or expressed uncertainty.

These results challenge the notion of a unanimous consensus on the AGWT as the near-exclusive explanation for warming. Moreover, in the same survey, approximately two-thirds of respondents expressed skepticism about the effectiveness of radical mitigation policies, highlighting a disconnect between institutional narratives and professional assessments.

9.4. Persistent scientific uncertainties

Despite decades of research, key uncertainties persist in climate science. A central example is the *Equilibrium Climate Sensitivity* (ECS), namely the projected temperature increase resulting from a doubling of atmospheric CO_2.

In 1979, Charney et al.[23] estimated that ECS lay between 1.5 °C and 4.5 °C. Remarkably, this range has remained largely unchanged across successive IPCC reports, including AR6 (2021) (cf. Sherwood et al., 2020),[24] despite significant advances in climate modeling and data collection (Figure 9.2). Undorf et al. (2022)[25] note that this wide interval—a relative variation of more than 300%—continues to pose a methodological challenge.

Over the past 50 years, climate models have improved in simulating processes such as convection, ocean dynamics, and aerosol–cloud interactions. Yet feedback mechanisms—particularly those involving low tropical clouds—remain complex and chaotic, limiting the ability to narrow ECS estimates. A growing number of recent studies, however, point toward lower ECS values (Gervais, 2016).[26] For example, Lewis (2023)[27] and Scafetta (2022, 2023c,b)[28] have obtained low ECS values on the basis of paleoclimatic data, instrumental climate records, and natural oscillations, as illustrated on the right side of Figure 9.2.

Sherwood & Forest (2024)[29] argue that, although the ECS range has not narrowed, the scientific understanding of its drivers has improved. Lewis (2025)[30]

[23]Charney, J.G., Arakawa, A., Baker, D.J., Bolin, B., Dickinson, R.E., et al., 1979. Carbon dioxide and climate: A scientific assessment. Washington, DC: The National Academies Press.

[24]Sherwood, S., Webb, M., Annan, J., Armour, K., Forster, P., et al., 2020. An assessment of Earth's climate sensitivity using multiple lines of evidence. Reviews of Geophysics, 58, e2019RG000678.

[25]Undorf, S., Pulkkinen, K., Wikman-Svahn, P., Bender, F.A.-M., 2022. How do value-judgments enter model-based assessments of climate sensitivity? Climatic Change, 174, 19.

[26]Gervais, F., 2016. Anthropogenic CO_2 warming challenged by 60-year cycle. Earth-Science Reviews, 155, 129–135.

[27]Lewis, N., 2023. Objectively combining climate sensitivity evidence. Climate Dynamics, 60, 3139–3165.

[28]Scafetta, N., 2022. Advanced testing of low, medium, and high ECS CMIP6 GCM simulations versus ERA5-T2m. Geophysical Research Letters, 49(6), e2022GL097716.

Scafetta, N., 2023c. Empirical assessment of the role of the Sun in climate change using balanced multi-proxy solar records. Geoscience Frontiers, 14(6), 101650.

Scafetta, N., 2023b. CMIP6 GCM ensemble members versus global surface temperatures. Climate Dynamics, 60, 3091–3120.

[29]Sherwood, S.C., Forest, C.E., 2024. Opinion: Can uncertainty in climate sensitivity be narrowed further? Atmospheric Chemistry and Physics, 24, 2679–2688.

[30]Lewis, N., 2025. Comment on "Opinion: Can uncertainty in climate sensitivity be narrowed further?" by Sherwood and Forest (2024). Atmospheric Chemistry and Physics, 25, 8821–8829.

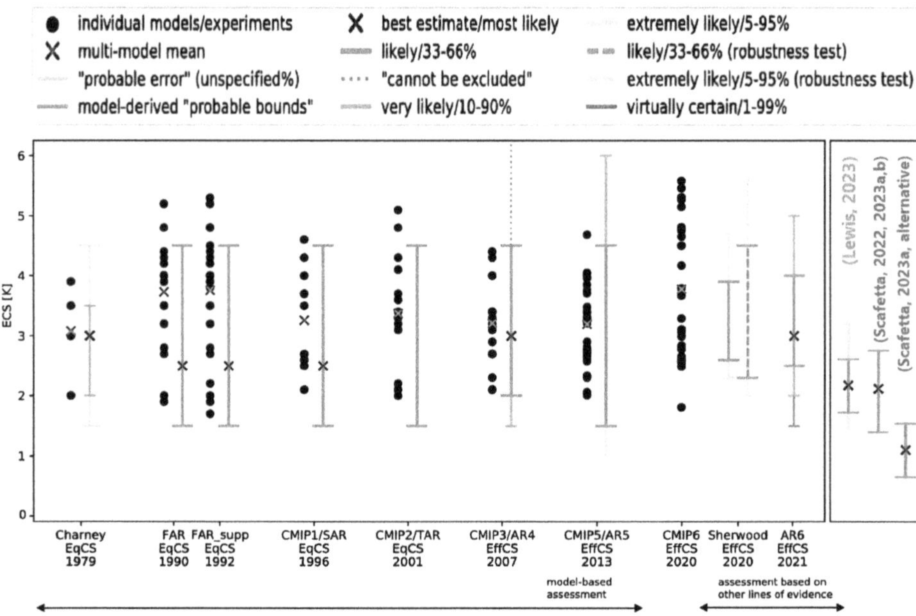

FIGURE 9.2. Historical progression of equilibrium climate sensitivity (ECS) estimates from Charney et al. (1979) to IPCC reports up to AR6 (2021) (adapted from Undorf et al., 2022). On the right are recent estimates by Lewis (2023) and Scafetta (2022, 2023c,b).

counters that lower ECS estimates are more robust when constrained by historical data. Divergent and at times incompatible results are common in the geosciences, where initial conditions, model structures, and data inputs heavily influence the outcomes. Declaring climate science *"settled"* in this context risks oversimplification and intellectual complacency.

***"Consensus"* versus *"empirical verification"*.** Substituting empirical verification with *"consensus"* implies that scientific truth is determined by popularity or authority. Many scientists accept the IPCC summaries without engaging with the underlying uncertainties—a stance that is sociologically understandable but scientifically inadequate. The core issue remains unresolved: the AGWT posits that the warming observed between 1850–1900 and 2011–2020 is entirely attributable to human activity. Yet if, as suggested by Scafetta (2013a, 2024, 2026),[31] the anthropogenic contribution is in fact lower, the CMIP5 and CMIP6 model projections for the 21st century would be overestimated. Such a finding would warrant a recalibration of both risk perception and policy responses.

[31]Scafetta, N., 2013a. Discussion on climate oscillations: CMIP5 general circulation models versus a semi-empirical harmonic model based on astronomical cycles. Earth-Science Reviews, 126, 321–357.
Scafetta, N., 2024. Impacts and risks of "realistic" global warming projections for the 21st century. Geoscience Frontiers, 15(2), 101774.
Scafetta, N., 2025. Detection, attribution, and modeling of climate change: Key open issues. Gondwana Research, DOI: 10.1016/j.gr.2025.05.001.

CHAPTER 10

The impossibility of validating the primary GCM prediction

Scientific progress relies on the formulation of falsifiable hypotheses and on their experimental validation. In climate science, however, this principle confronts the extraordinary complexity and uniqueness of the Earth system, which renders definitive empirical verification of global climate model (GCM) predictions extremely difficult, if not impossible. Yet it is precisely these simulations that underpin the *Anthropogenic Global Warming Theory* (AGWT), and in particular its near-total attribution of the warming observed since 1850–1900 to human activity.

10.1. The scientific method and the non-falsifiability of climate models

As Karl Popper emphasized, the scientific method requires that theories be testable and subject to empirical falsification. If a model fails to reproduce observed data, it must be revised or discarded. Applying this logic to climate science, however, is far from straightforward.

As illustrated in Figure 6.2, the AGWT's central claim—that nearly 100% of the warming between the preindustrial era and the present is anthropogenic—rests on GCM simulations that compare scenarios with and without human influence. These comparisons, however, depend on an unobservable counterfactual: there is no *"twin Earth"* free of CO_2 emissions, deforestation, and industrialization to serve as a control (Figure 10.1).

In essence, GCMs are not subject to the same kind of experimental validation as models in fields such as particle physics or engineering. The interpretation of differing outcomes is derived from model-based assumptions rather than from direct observation. As discussed in Chapter 6, this renders the core prediction of the AGWT non-falsifiable in the traditional scientific sense.

A theory based on models. The AGWT rests on numerical simulations that, however sophisticated, remain simplifications of an extraordinarily complex system. GCMs are built on fundamental fluid dynamics and thermodynamics, but they rely on parameterizations for subsystems such as clouds, vegetation, albedo, and ocean dynamics, all of which introduce uncertainty.

Even paleoclimate reconstructions, frequently cited in support of the AGWT, do not provide definitive validation. The well-known *"hockey stick"* curve by Mann et al. (1999)[1] shows a sharp temperature rise during the 20th century. Yet

[1]Mann, M.E., Bradley, R.S., Hughes, M.K., 1999. Northern hemisphere temperatures during the past millennium: inferences, uncertainties, and limitations. Geophysical Research Letters, 26, 759–762.

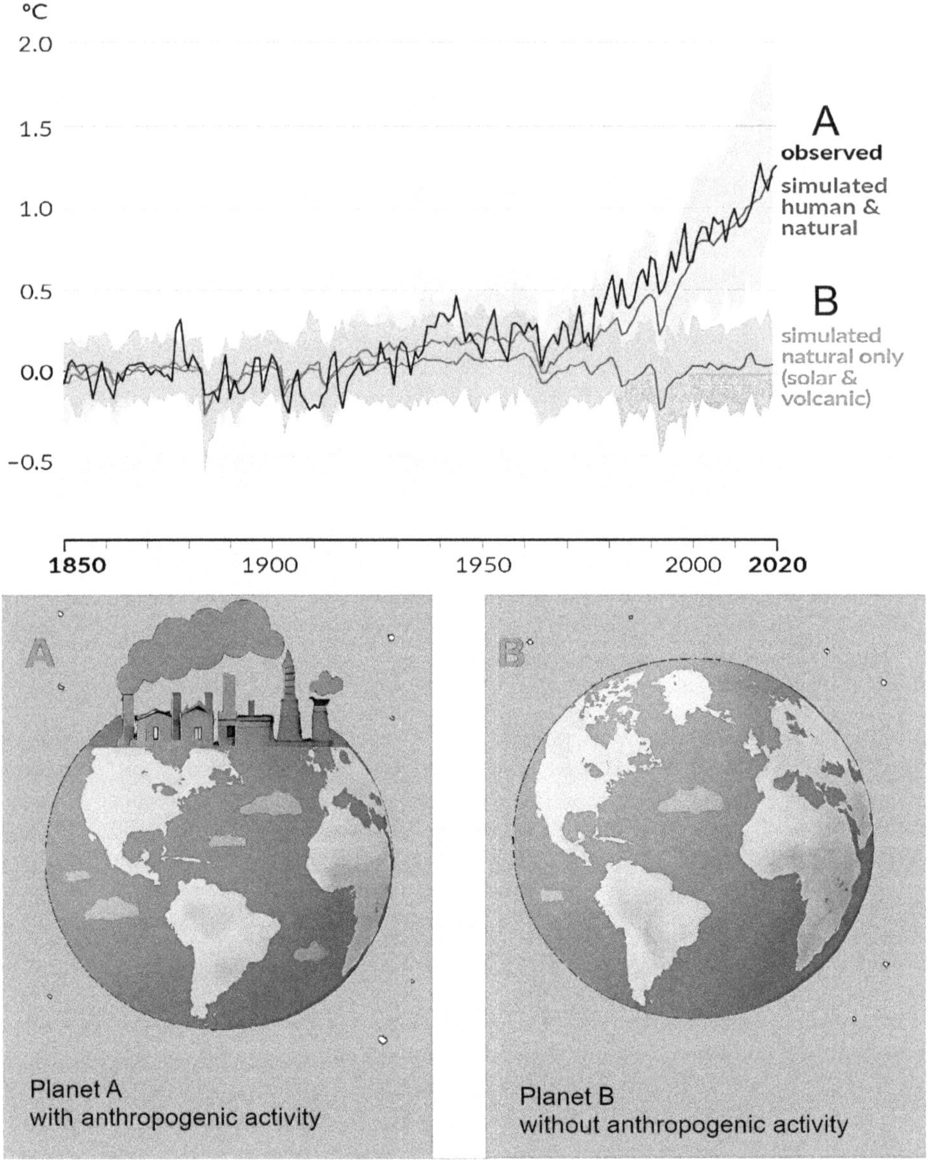

Figure 10.1. Modeling assumption in CMIP6 GCMs. (Adapted from IPCC, 2021). Two hypothetical Earths—one with and one without anthropogenic emissions—are compared. The climate trajectory of the Earth without human activity, shown by the GCM simulations B, cannot be empirically verified. Therefore, the conclusion of the GCM simulations A that observed warming is entirely human-induced cannot be validated through actual observations.

Crowley (2000)[2] estimated that up to 25% of this warming could be attributed to increased solar activity, suggesting that natural factors remain significant.

In short, the exclusive attribution of warming to human activity is not the only interpretation compatible with the existing data. Confidence in the AGWT depends more on the presumed validity of GCMs than on irrefutable empirical evidence.

Structural limitations of GCMs. The design of GCMs entails choices that strongly influence the outcomes. Key limitations include:

- *solar forcings and internal variability*—phenomena such as the Atlantic Multidecadal Oscillation (AMO), the El Niño–Southern Oscillation (ENSO), and chaotic climate variability are often underrepresented;
- *natural aerosols*—volcanic emissions, desert dust, and other natural sources affect the radiative balance, but their cooling effects are difficult to quantify;
- *parameterizations and calibrations*—GCMs contain numerous adjustable parameters, particularly for cloud formation, which are tuned to match historical data; this process more closely resembles engineering optimization than pure physical modeling.

These limitations may lead to an overestimation of future warming and to excessive attribution to anthropogenic influence, particularly if part of the warming observed since 1850–1900 stems from unmodeled natural variability.

Parametric optimization or physical prediction? A central critique of GCMs concerns their calibration. Free parameters are adjusted to optimize agreement with past observations, which can yield simulations that align with historical data but do not guarantee predictive accuracy. McSweeney & Hausfather (2018)[3] acknowledge that such optimization is unavoidable in complex systems. However, subsequent studies (Golaz et al., 2019; Lutsko et al., 2022; Ma et al., 2022; Mauritsen et al., 2019)[4] have shown that small changes in model configuration can yield substantially different results, even when agreement with the historical record is preserved.

This raises a critical point: agreement with past data does not confirm the physical accuracy of a model. The issue is compounded by the lack of convergence on the *Equilibrium Climate Sensitivity* (ECS), which, for the GCMs cited in IPCC AR6 (2021), ranges from 1.8 °C to 5.7 °C—still an excessively wide interval after decades of refinement. If models cannot reliably constrain this key

[2]Crowley, T.J., 2000. Causes of climate change over the past 1000 years. Science, 289, 270–277.

[3]McSweeney, R., Hausfather, Z., 2018. Q&A: How do climate models work? Carbon Brief.

[4]Golaz, J.-C., Caldwell, P.M., Van Roekel, L.P., Petersen, M.R., Tang, Q., et al., 2019. The DOE E3SM coupled model version 1: Overview and evaluation at standard resolution. Journal of Advances in Modeling Earth Systems, 11, 2089–2129.

Lutsko, N.J., Luongo, M.T., Wall, C.J., Myers, T.A., 2022. Correlation between cloud adjustments and cloud feedbacks in CMIP6. Journal of Geophysical Research: Atmospheres, 127, e2022JD037486.

Ma, P.-L., Harrop, B.E., Larson, V.E., Neale, R.B., Gettelman, A., et al., 2022. Better calibration of cloud parameterizations and subgrid effects increases the fidelity of the E3SM Atmosphere Model version 1. Geoscientific Model Development, 15, 2881–2916.

Mauritsen, T., Bader, J., Becker, T., Behrens, J., Bittner, M., et al., 2019. Developments in the MPI-M Earth System Model version 1.2 and its response to increasing CO_2. Journal of Advances in Modeling Earth Systems, 11, 998–1038.

variable, it is reasonable to question the confidence placed in their century-scale projections.

10.2. Parameterizations in climate models

Global Climate Models (GCMs) are essential tools for simulating Earth's climate across past, present, and future timescales. Their construction, however, depends on a critical structural component: parameterizations. These are simplified mathematical representations of physical processes that occur on scales too small or too complex to be directly resolved by the model's grid.

Why parameterizations are necessary. The atmosphere and the oceans are governed by turbulent, nonlinear, and highly localized phenomena (e.g., cloud formation, convection, chemical reactions, and radiative fluxes), which typically operate on scales as small as a few meters. These scales lie far below the spatial resolution of GCMs, which generally ranges from 10 to 100 km. To incorporate these sub-grid processes, models rely on empirical or semi-empirical relationships. Parameterizations thus constitute the invisible yet indispensable backbone of climate modeling.[5]

Main categories of parameterizations. The main categories of parameterizations used in climate models are listed below.

- *Radiative transfer*—Models simulate the absorption, emission, and scattering of solar and infrared radiation. Greenhouse gases (CO_2, CH_4, H_2O) interact with specific wavelengths. These interactions, together with aerosol and cloud effects, are simplified, which introduces margins of uncertainty.
- *Cloud microphysics*—Cloud formation involves condensation nuclei and processes such as evaporation, coalescence, and freezing. GCMs employ empirical schemes to estimate cloud cover and precipitation, which can vary substantially across models.
- *Deep convection*—Vertical air motions, which are critical for cyclones, thunderstorms, and the redistribution of heat, are parameterized. These schemes strongly influence climate sensitivity, particularly in tropical regions.
- *Surface–atmosphere interactions*—Models simulate exchanges of energy and gases between land, vegetation, ice, and the atmosphere. Key processes include evapotranspiration, thermal conductivity, seasonal biospheric dynamics, and albedo variability.
- *Atmospheric photochemistry*—The chemical composition of the atmosphere is modeled through simplified reaction networks. These affect estimates of radiative forcing and atmospheric transparency.
- *Precipitation microphysics*—The formation of rain, snow, and hail is simulated through hydrometeor growth, melting, and crystallization.

[5]For further reading on parameterization in GCMs, useful online resources include:
Ouranos – Understanding Climate Concepts: https://www.ouranos.ca/en/understanding-climate-concepts/physical-parameterizations-climate-model;
Wiki – Parametrization: https://en.wikipedia.org/wiki/Parametrization_(atmospheric_modeling);
Carbon Brief – How Do Climate Models Work? https://www.carbonbrief.org/qa-how-do-climate-models-work/#tune

These parameterizations are based on observational data and vary with local conditions.

- *Cloud fraction and humidity thresholds*—Cloud cover is often triggered when relative humidity exceeds a predefined threshold. This empirical approach underscores the extent to which modeling assumptions shape climate projections.
- *Ocean–atmosphere fluxes*—Exchanges of heat, moisture, and salinity between the ocean and the atmosphere, as well as sea-ice dynamics, are modeled through simplified representations of micro- and mesoscale processes.

Parameterizations represent both a strength and a limitation of climate models. They allow simulations to function within computational constraints but also introduce variability and uncertainty across different models. Improving these formulations remains a major frontier in climate science.

10.3. Climate hypersensitivity to water vapor and cloud formation

Among the unresolved challenges in climate modeling, few are as critical as the representation of water vapor, clouds, and their impact on Earth's radiative balance. Despite decades of scientific and technological progress, cloud parameterization remains the leading source of uncertainty in global climate models (GCMs), significantly affecting estimates of the equilibrium climate sensitivity (ECS)—that is, the projected temperature increase associated with a doubling of atmospheric CO_2.

The radiative power of clouds: far beyond CO_2. Clouds play a dual role in Earth's energy budget. They trap longwave (LW) infrared radiation, thereby contributing to the greenhouse effect, while also reflecting shortwave (SW) solar radiation, thereby producing a cooling effect. The net impact is referred to as the *Cloud Radiative Effect* (CRE) and depends on cloud type, altitude, and optical properties. Low, thick clouds tend to cool, whereas high, thin clouds tend to warm.

Satellite data from the *Geophysical Fluid Dynamics Laboratory* (GFDL) show that:[6]

- global average SW CRE ≈ -50 W/m^2 (cooling through reflection);
- global average LW CRE $\approx +30$ W/m^2 (warming through trapping);
- net CRE ≈ -20 W/m^2 (strong net cooling).

These values are more than ten times greater than the estimated effective radiative forcing from anthropogenic sources between 1750 and 2020 (approximately +2.8 W/m^2, IPCC AR6), underscoring the dominant role of clouds in climate modulation (see Figure 10.2).

The sign of the cloud-related climate feedback—that is, whether clouds amplify or dampen warming—nevertheless depends on how cloud properties respond to changes in temperature, circulation, and atmospheric composition. Estimating this feedback requires a deep, process-level understanding that remains elusive.

[6]Geophysical Fluid Dynamics Laboratory (GFDL), NOAA. "*Clouds and Radiation.*" Available at: https://www.gfdl.noaa.gov/cloud-radiative-effect/

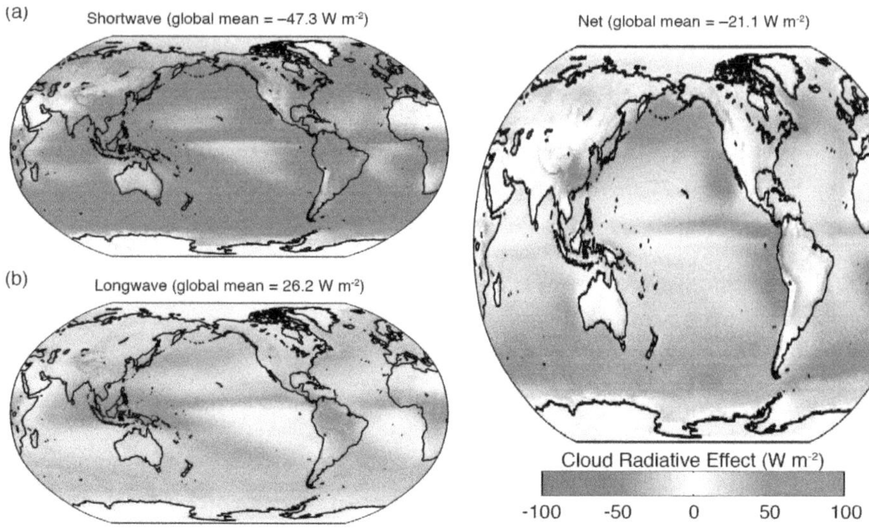

FIGURE 10.2. Distribution of shortwave (a) and longwave (b) cloud radiative effects, and (c) total radiation during the period 2001–2011 from the CERES EBAF dataset. Adapted from the IPCC AR5 report (Chapter 7, Fig. 7.7a–b). Negative values indicate cooling effects from albedo, and positive values indicate warming effects from cloud emissivity.

Unmodeled mechanisms and emerging hypotheses. Some hypotheses suggest that cloud formation may be influenced by mechanisms not currently included in GCMs. For example:

- galactic cosmic rays, modulated by solar activity, may affect cloud nucleation (Svensmark, 2019);[7]
- interplanetary dust cycles, linked to planetary oscillations, may influence atmospheric dynamics (Scafetta et al., 2016, 2020).[8]

If confirmed, these effects would imply that significant natural forcings are currently excluded from the models, which could distort long-term projections.

A scientifically rigorous approach should not dismiss such mechanisms on the grounds of incomplete evidence; rather, it should monitor them and consider their integration into models as empirical support emerges.

Model hypersensitivity to free parameter tuning. The sensitivity of GCMs to cloud formation constitutes a structural limitation. Parametric tuning can dramatically alter both model outcomes and performance.

[7]Svensmark, H., 2019. Force Majeure: The Sun's Role in Climate Change. The Global Warming Policy Foundation. ISBN 978-0993119095.

[8]Scafetta, N., Milani, F., Bianchini, A., Ortolani, S., 2016. On the astronomical origin of the Hallstatt oscillation found in radiocarbon and climate records throughout the Holocene. Earth-Science Reviews, 162, 24–43.

Scafetta, N., Milani, F., Bianchini, A., 2020. A 60-year cycle in the meteorite fall frequency suggests a possible interplanetary dust forcing of the Earth's climate driven by planetary oscillations. Geophysical Research Letters, 47(18), e2020GL089954.

For example, Mauritsen et al. (2019), in developing the ECHAM6.3-HAM2.3 model (used in MPI-ESM1.1 and MPI-ESM1.2), describe how cloud fraction is parameterized through three free parameters (a_1, a_2, a_3) based on relative humidity profiles. The initial values $(a_1 = 0.7, a_2 = 0.9, a_3 = 4.0)$ yielded an ECS of 7.0 °C, far above the IPCC's plausible range. After tuning $(a_1 = 0.75, a_2 = 0.968, a_3 = 1.0)$, the ECS dropped to 2.8 °C, a value considered more acceptable for inclusion in IPCC assessments.

The authors explicitly state that historical warming was also used as a calibration target and that climate sensitivity was deliberately reduced to improve model fidelity. Agreement with observations may therefore reflect successful tuning more than physical realism. Indeed, Mauritsen et al. (2019) also report that their model tends to overestimate warming since the 1990s, which raises further doubts about its predictive reliability.

Conclusion: the structural fragility of GCMs. The sensitivity of GCMs to water vapor and cloud feedback is not a new concern. As early as 1963, Möller (1963)[9] highlighted the uncertainty in the atmospheric water balance and its ambiguous role in climate amplification or damping. His ECS estimates ranged from less than 1 °C to 10 °C, with a central value of approximately 1.5 °C.

Despite advances in resolution and computing power, no model has overcome this epistemological vulnerability. Clouds remain the Achilles' heel of climate simulation. With an estimated influence of 20–30 W/m² on the global radiative balance, their role far exceeds that of anthropogenic forcing; yet it remains the least constrained.

The case examined by Mauritsen et al. (2019) reveals a critical methodological vulnerability: the structural reliance of global climate models (GCMs) on retrospectively calibrated free parameters. Small changes in cloud-related parameters can produce ECS shifts of nearly 4–5 °C. This hypersensitivity points to a deeper methodological issue: models can compensate for missing or poorly understood physics by adjusting other parameters to match historical data. The result is apparent realism—that is, numerical agreement without guaranteed physical accuracy. The fact that an empirical function adjusted through a handful of coefficients can drastically alter the model's estimated ECS raises serious concerns about the robustness of conclusions drawn from such models, particularly when they inform global mitigation policies. As long as the physical behavior of clouds remains poorly constrained, any climate projection derived from GCMs must contend with this epistemological issue, which challenges both the reliability of long-term forecasts and the scientific foundations of climate policy.

10.4. Uncertainty in the absolute temperatures predicted by GCMs

A crucial but often overlooked aspect of evaluating global climate models (GCMs) is their ability to accurately simulate Earth's mean absolute temperature. While most simulations emphasize thermal anomalies—that is, variations relative to a baseline period (e.g., 1850–1900 or 1961–1990)—it remains essential to assess model consistency on the absolute scale as well.

---

[9] Möller, F., 1963. Influence of changes in CO_2 concentration on the radiation balance of Earth's surface and on the climate. Journal of Geophysical Research, 68, 3877–3886.

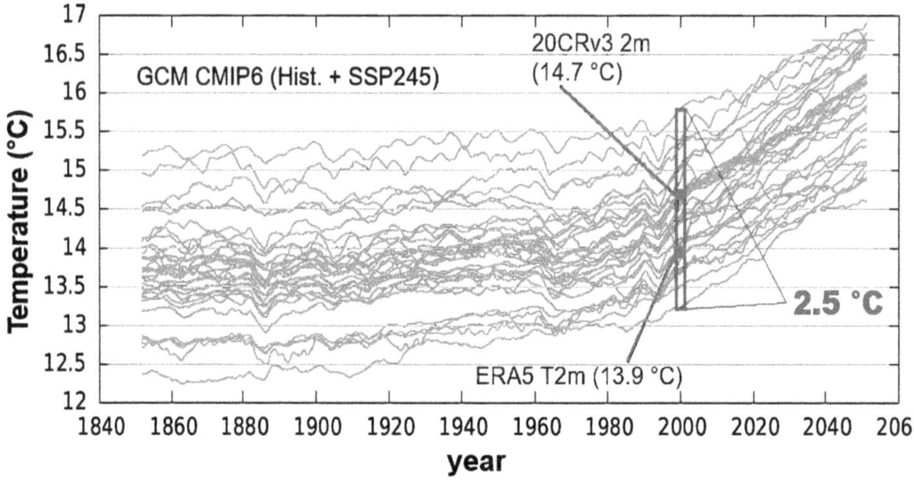

FIGURE 10.3. Comparison of absolute temperatures simulated by 34 CMIP6 CGMs used in the IPCC AR6 (2021). Gray squares show observational estimates for the year 2000 according to NOAA/CIRES/DOE 20th Century Reanalysis (Version 3) and the ERA5 T2m dataset from the Copernicus Climate Change Service. Data available from KNMI Climate Explorer (https://climexp.knmi.nl/).

During the 20th century, the observed global mean temperature ranged between 14 and 15 °C. Figure 10.3 presents absolute temperature outputs from 34 CMIP6 climate models used in IPCC (2021), with the curves representing 48-month moving averages. Gray squares indicate the estimated global mean temperature for the year 2000 according to two datasets: the NOAA/CIRES/DOE 20th Century Reanalysis (Version 3) at 14.7 °C, and ERA5 T2m from the Copernicus Climate Change Service at 13.9 °C.

Although the ensemble range encompasses the observed values, a substantial spread emerges among model outputs. For the year 2000, some GCMs estimate global mean temperatures as low as 13.0–13.5 °C, while others reach 15.5–16.0 °C—a discrepancy spanning approximately 2.5 °C.

This uncertainty is far from trivial. A 2.5 °C error in absolute temperature undermines the precision of climate diagnostics and sensitivity estimates. In climatic terms, such a difference corresponds roughly to a 4–5° shift in latitude, which could distort regional climate projections and feedback assessments. It affects not only the estimation of anomalies relative to a baseline, but also the foundational climatological state from which simulations begin.

Implications and structural causes. Properly addressing the above issue requires:

- improved calibration of models on absolute temperature scales;
- enhanced integration of high-resolution observational datasets;
- critical evaluation of structural inconsistencies in radiative and thermodynamic formulations across GCMs.

This uncertainty is largely attributable to differences in model architecture, including radiative schemes, cloud parameterizations, intrinsic climate sensitivity, and representations of albedo and ocean circulation (Evin et al., 2024).[10] These discrepancies stem from the heterogeneous calibration and tuning strategies adopted by different modeling centers, which lead to divergent climatological baselines.

It is evident that such structural uncertainty can significantly influence the interpretation of climate trends and the estimation of the Earth system's sensitivity to external forcings. Without resolving these inconsistencies, long-term projections risk being built on unstable foundations.

10.5. Planetary albedo and GCMs

Planetary albedo—the fraction of incoming solar radiation reflected back into space—is a fundamental component of Earth's energy balance, directly influencing global warming and cooling (Stephens et al., 2015).[11] Although the average albedo is estimated at approximately 0.30 (PeyrouLauga, 2017),[12] even small variations of just 0.01 can produce radiative effects greater than the current anthropogenic forcing (Datseris & Stevens, 2021).[13]

Climate models diverge significantly from one another both in the absolute value of albedo and in its hemispheric distribution. Satellite data reveal a striking symmetry between the Northern and Southern Hemispheres, despite their geographic differences. This symmetry is attributed to cloud compensation arising from the extratropical storm belts in the Southern Hemisphere (Datseris & Stevens, 2021).

Most CMIP6 models, however, fail to reproduce this balance, exhibiting hemispheric asymmetries that exceed 5 W/m^2—more than twice the radiative forcing attributed to human activity (Rugenstein & Hakuba, 2023).[14] These discrepancies raise important questions about the representation of cloud processes in GCMs and diminish confidence in their ability to project future climate conditions accurately.

10.6. Conclusion: modeling science and observable reality

Global Climate Models (GCMs) are indispensable tools for investigating the complex dynamics of the Earth system. They allow scientists to simulate the interactions among atmospheric, oceanic, and terrestrial processes, offering valuable insights into potential climate trajectories. Their simulated nature, however, demands epistemic caution: GCMs are neither infallible oracles nor substitutes for empirical observation. Their outputs are shaped by assumptions,

---

[10]Evin, G., Ribes, A., Corre, L., 2024. Assessing CMIP6 uncertainties at global warming levels. Climate Dynamics, 62, 8057–8072.

[11]Stephens, G.L., O'Brien, D., Webster, P., Pilewski, P., Kato, S., Li, J.-L., 2015. The albedo of Earth. Reviews of Geophysics, 53(1), 141–163.

[12]Peyrou-Lauga, R., 2017. Using real Earth albedo and Earth IR flux for spacecraft thermal analysis. Proceedings of the 47th International Conference on Environmental Systems (ICES), Charleston, South Carolina, USA.

[13]Datseris, G., Stevens, B., 2021. Earth's albedo and its symmetry. AGU Advances, 2, e2021AV000440.

[14]Rugenstein, M., Hakuba, M., 2023. Connecting hemispheric asymmetries of planetary albedo and surface temperature. Geophysical Research Letters, 50(6), e2022GL101802.

parameterizations, and structural simplifications, all of which influence the narratives they produce.

The most assertive interpretations of the Anthropogenic Global Warming Theory (AGWT) rest on the premise that these models can accurately reconstruct past climate behavior and, by extension, reliably forecast or project future changes under specific conditions. The claim that all observed warming since 1850–1900 is entirely attributable to human activity, however, cannot be experimentally verified. It remains a model-dependent inference, subject to deep epistemological limitations and contingent on the exclusion or underrepresentation of natural forcings.

Scientific rigor demands a critical approach that acknowledges the structural uncertainties inherent in climate modeling and embraces the full spectrum of climatic drivers, both anthropogenic and natural. A mature climate science must move beyond rhetorical consensus and uncritical reliance on simulation outputs; it should be grounded in an open, falsifiable, and iterative process of inquiry, in which competing hypotheses are tested, refined, or discarded on the basis of empirical coherence.

Only by adopting this perspective can climate science evolve into a robust and resilient foundation for informing political decisions and adaptation strategies.

CHAPTER 11

The *"hot model"* problem

In recent years, growing attention has been directed to the limitations of global climate models (GCMs), which must be regarded as indispensable yet imperfect tools for simulating Earth's climate and projecting future scenarios. The structural uncertainties discussed in previous chapters have become more pronounced with the latest generation of CMIP6 models, whose results have sparked debate within both the scientific community and the IPCC itself.

A central concern is the tendency of many models to run *"hot"*—that is, to overestimate the observed global warming, particularly in recent decades. This chapter examines the *"hot model"* issue from several perspectives:

- the wide spread of climate sensitivity values in CMIP6 models;
- the methodological challenges in constraining ECS and TCR;
- the overestimation of tropospheric temperatures in the intertropical region;
- the increasing divergence between model projections and observational data.

11.1. Climate sensitivity and divergences among CMIP6 GCMs

Climate sensitivity—the magnitude of the rise in the global temperature in response to increased greenhouse gases—is a cornerstone of climate modeling. The ECS (*Equilibrium Climate Sensitivity*) refers to the long-term warming following a doubling of CO_2, whereas the TCR (*Transient Climate Response*) captures the short-term response (over approximately 70 years) to a gradual increase in CO_2.

In CMIP6, ECS estimates range from 1.8 °C to 5.7 °C, and TCR from 1.33 °C to 2.79 °C (Hausfather et al., 2022; IPCC, 2021),[1] as shown in Figure 11.1. Although the focus here is on ECS, TCR is also of critical importance, since it is directly linked to mid-century climate projections that inform emissions targets, adaptation strategies, and economic planning. The TCR is roughly correlated with the ECS. In many climate models, the TCR-to-ECS ratio falls between 1/2 and 2/3. For two models with equal ECS, the one with the larger TCR will warm more rapidly in the near term.

These divergences are significant, since they directly affect projections of future climate risks. Figure 6.4 (adapted from Scafetta, 2024),[2] illustrates how the GCM simulations diverge depending on the assumed sensitivity, even under

[1]Hausfather, Z., Marvel, K., Schmidt, G.A., Nielsen-Gammon, J.W., Zelinka, M., 2022. Climate simulations: recognize the 'hot model' problem. Nature, 605(7908), 26–29.

[2]Scafetta, N., 2024. Impacts and risks of "realistic" global warming projections for the 21st century. Geoscience Frontiers, 15(2), 101774.

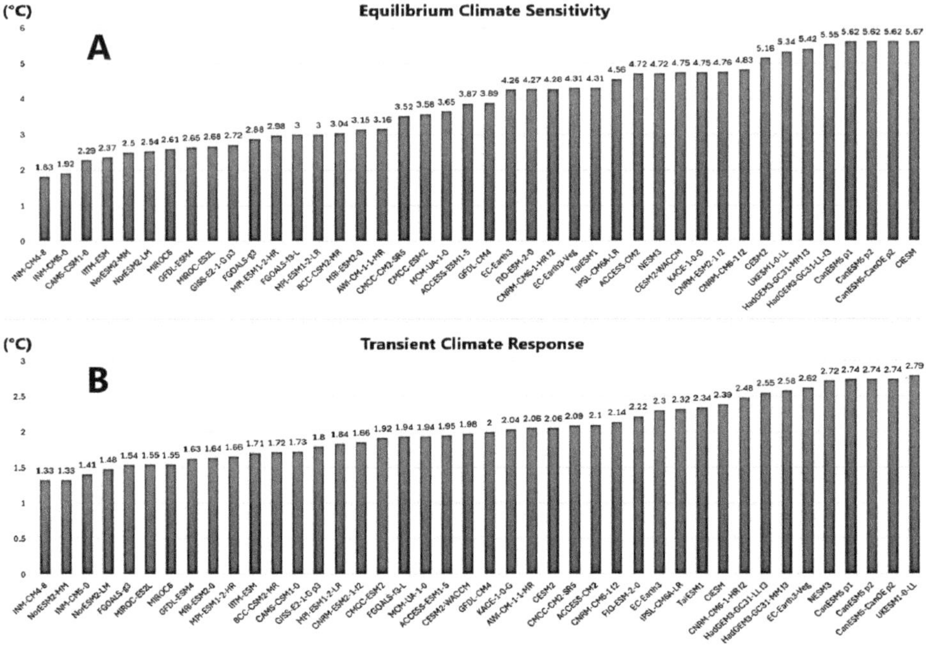

FIGURE 11.1. Values of ECS ("*Equilibrium Climate Sensitivity*") and TCR ("*Transient Climate Response*") predicted by CMIP6 GCMs used by the IPCC. (Adapted from Scafetta, 2026).

identical historical forcing and socioeconomic scenarios (SSPs). This variability reflects an epistemic fragility: small changes in internal parameters can yield radically different outcomes.

The underlying causes include:

- uncertainty in modeling climate feedbacks, especially those associated with clouds, water vapor, and albedo;
- overreliance on free parameters calibrated to historical data, which may prove unreliable for forecasting;
- the lack of a unified methodology for estimating ECS, with approaches ranging from paleoclimate proxies to observational constraints, and often yielding conflicting results.

11.2. Challenges in estimating climate sensitivity to radiative forcing

Quantifying the climate system's response to radiative forcing remains a methodological challenge. Although ECS is conceptually well defined, its practical estimation is fraught with ambiguity. The literature reports values ranging from <1 °C to >10 °C (Knutti et al., 2017),[3] depending on the method employed.

[3] Knutti, R., Rugenstein, M.A., Hegerl, G.C., 2017. Beyond equilibrium climate sensitivity. Nature Geoscience, 10, 727–736.

For IPCC AR6, priority was given to Sherwood et al. (2020),[4] who combined historical and paleoclimate data with process-based modeling to estimate ECS at 3.1 °C, with a likely range of 2.6–3.9 °C. IPCC AR6 proposes:

- a *"likely"* ECS range: 2.5–4.0 °C (17–83% confidence);
- a *"very likely"* ECS range: 2.0–5.0 °C (5–95% confidence).

These estimates, however, are contested.

Lewis (2023)[5] identifies statistical flaws in Sherwood's approach and proposes a median ECS of 2.16 °C, with a range of 1.75–2.7 °C. Other studies suggest even lower values, between 1 and 2 °C (Forster & Gregory, 2006; Lindzen & Choi, 2011; Harde, 2014; Monckton et al., 2015; Scafetta, 2022, 2023c,b,a).

Rugenstein et al. (2023)[6] argue that ECS may not be a fixed value but rather context dependent, varying with timescale, feedback dominance, and initial conditions. The use of a single ECS value may therefore oversimplify a complex, dynamic system.

The *"hot model"* phenomenon. A direct consequence of the high ECS values in CMIP6 models is their tendency to overestimate the warming observed from 1980 to 2022. The *"hot model"* problem is well documented by Hausfather et al. (2022).[7] Scafetta (2022, 2023b)[8] found that models with ECS > 3 °C consistently project temperatures above those recorded in the historical datasets.

Figure 11.2 compares CMIP6 model simulations (grouped by ECS) with observed global temperatures from

- HadCRUT5 (global surface, infilled version) (Morice et al., 2021)[9] and
- the UAH-MSU v6.1 lower troposphere (lt) satellite dataset (Spencer et al., 2017),[10]

both covering the period from 1980 to 2025.

The figure shows that models with ECS > 3 °C significantly overestimate warming. The discrepancy is more pronounced against HadCRUT5 than against UAH, which suggests potential warm biases in the surface climate record.

Figure 11.3 quantifies the warming from 1980–1990 to 2012–2022 (Scafetta, 2022, 2023b):

- HadCRUT5: +0.605 ± 0.02 °C;

[4]Sherwood, S., Webb, M., Annan, J., Armour, K., Forster, P., et al., 2020. An assessment of Earth's climate sensitivity using multiple lines of evidence. Reviews of Geophysics, 58, e2019RG000678.

[5]Lewis, N., 2023. Objectively combining climate sensitivity evidence. Climate Dynamics, 60, 3139–3165.

[6]Rugenstein, M., Zelinka, M.D., Karnauskas, K.B., Ceppi, P., Andrews, T., 2023. Patterns of surface warming matter for climate sensitivity. Eos, 104.

[7]Hausfather, Z., Marvel, K., Schmidt, G.A., Nielsen-Gammon, J.W., Zelinka, M., 2022. Climate simulations: recognize the 'hot model' problem. Nature, 605(7908), 26–29.

[8]Scafetta, N., 2022. Advanced testing of low, medium, and high ECS CMIP6 GCM simulations versus ERA5-T2m. Geophysical Research Letters, 49(6), e2022GL097716.

Scafetta, N., 2023b. CMIP6 GCM ensemble members versus global surface temperatures. Climate Dynamics, 60, 3091–3120.

[9]Morice, C.P., Kennedy, J.J., Rayner, N.A., Winn, J.P., Hogan, E., et al., 2021. An updated assessment of near-surface temperature change from 1850: the HadCRUT5 data set. Journal of Geophysical Research: Atmospheres, 126, e2019JD032361.

[10]Spencer, R.W., Christy, J.R., Braswell, W.D., 2017. UAH version 6 global satellite temperature products: Methodology and results. Asia-Pacific Journal of Atmospheric Sciences, 53(1), 121–130.

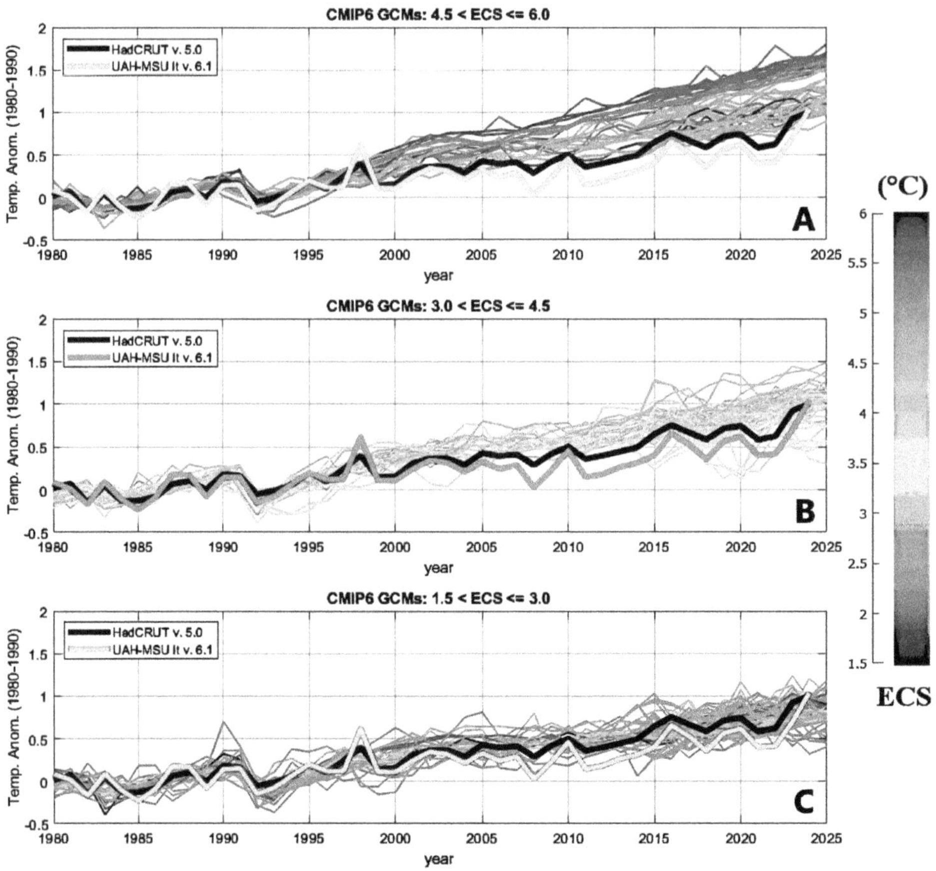

FIGURE 11.2. Comparison between the average simulations of CMIP6 GCMs used by IPCC (2021), divided into three groups by ECS, and global surface temperatures from HadCRUT v5.0 (*"infilled"* version) (Morice et al., 2021), and the satellite temperature dataset UAH-MSU lt v. 6.1 (Spencer et al., 2017). (Adapted from Scafetta, 2026).

- UAH-MSU LT: $+0.427 \pm 0.03$ °C.

If HadCRUT5 is taken as the benchmark, only models with ECS < 3 °C align with the observations. If the UAH-MSU lt record is adopted instead, nearly all CMIP6 models appear *"hot,"* revealing a structural flaw across all GCMs.

The Hunga Tonga–Hunga Ha'apai eruption paradox. Even during the 2014–2024 period, many model simulations remain hotter than the observations, despite the strong global warming peak recorded mainly in 2023–2024. This peak was likely influenced by the Hunga Tonga–Hunga Ha'apai submarine eruption of 11–15 January 2022, together with other natural phenomena such as the 2023–2024 El Niño event. El Niño events enhance deep moist convection in the tropics. Unlike typical volcanic events, which cool the climate for 1–2 years, this eruption injected roughly 10% more water vapor into the stratosphere, thereby

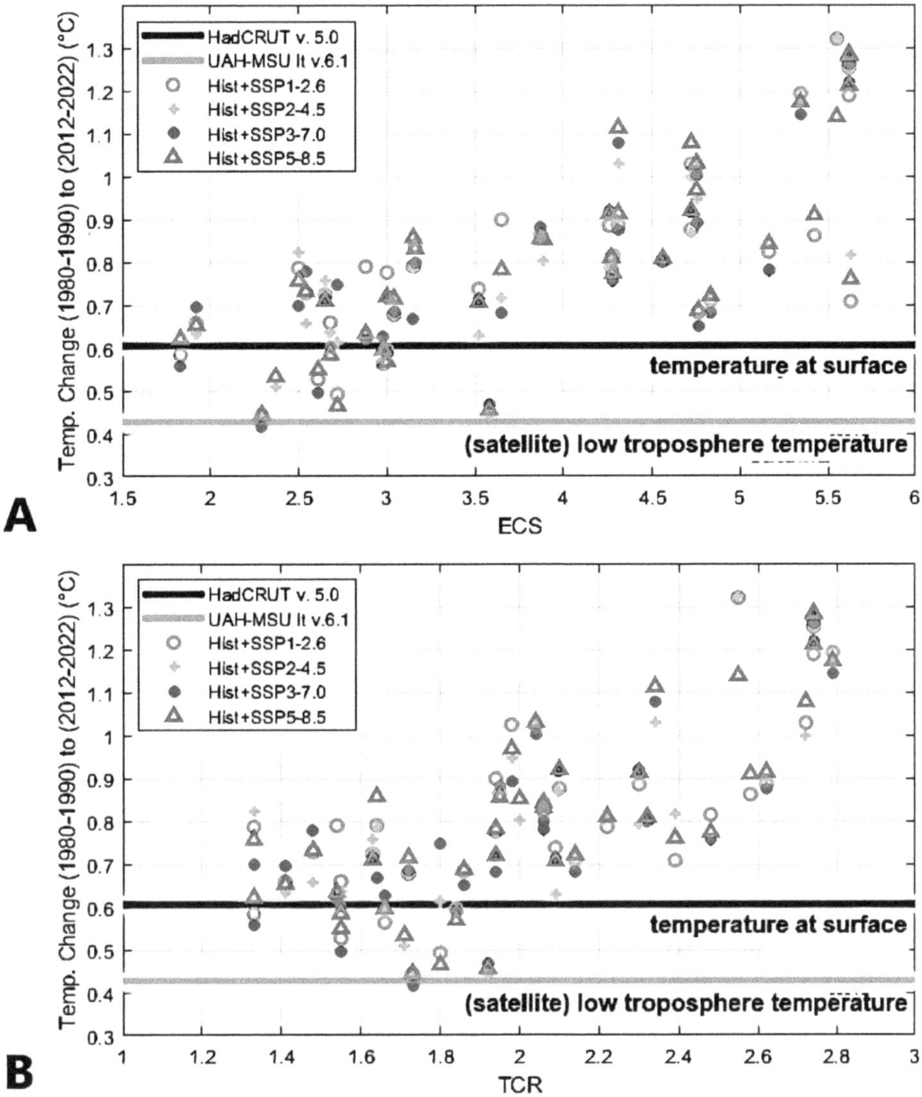

Figure 11.3. Comparison of warming between 1980–1990 and 2012–2022 from average CMIP6 GCM simulations used by IPCC (2021), as a function of their ECS, and global surface temperatures from HadCRUT v5.0 (*"infilled"* version) (Morice et al., 2021), and the satellite temperature dataset UAH-MSU lt v. 6.1 (Spencer et al., 2017). (Adapted from Scafetta, 2026).

temporarily enhancing the radiative forcing (Jenkins et al., 2023; Schoeberl et al., 2024).[11]

[11]Jenkins, S., Smith, C., Allen, M., Grainger, R., 2023. Tonga eruption increases chance of temporary surface temperature anomaly above 1.5 °C. Nature Climate Change, 13(2), 127–129.
Schoeberl, M.R., Wang, Y., Taha, G., Zawada, D.J., Ueyama, R., Dessler, A., 2024. Evolution of the

This anomaly underscores the complexity of climate data and feedbacks, as well as the limitations of current models in capturing rare but high-impact events.

11.3. The GCM tropospheric *"hot spot"* and its embarrassing absence

One of the most persistent predictions of global climate models (GCMs) is the presence of a *"hot spot"* in the upper tropical troposphere—a pronounced warming anomaly between 10 and 12 km of altitude above the equator. This feature, consistently forecast since the earliest CMIP experiments, is prominently visible in the upper-left portion of Figure 11.4 and constitutes a crucial empirical test of the validity of climate simulations.

The theory behind the *"tropical hot spot"*. According to current modeling frameworks, anthropogenic warming should be amplified in the upper tropical troposphere because of:

- *water vapor feedback*: as surface temperatures rise, evaporation increases, raising water vapor concentrations; at higher altitudes, this feedback reduces the efficiency of radiative cooling, thereby intensifying the local warming;
- *tropical convection*: warm, moist air ascends rapidly, condensing and releasing latent heat, which warms the upper troposphere;
- *vertical radiative sensitivity*: the upper troposphere is more responsive to radiative forcing, leading models to predict enhanced warming at these altitudes and latitudes.

On average, GCMs project the warming in this region to be up to three times greater than that at the surface.

The predicted warming is not observed. Despite the consistent expectations of the models, satellite and radiosonde data do not confirm the presence of the tropical *"hot spot"*. This discrepancy was first noted in the *U.S. Climate Change Science Program* report (Karl et al., 2006),[12] and has since been described in the literature as the *"missing hotspot"* (Mitchell et al., 2020; Hudson, 2023; McKitrick & Christy, 2018).[13]

The key findings are as follows:

- satellite datasets (UAH, RSS) and radiosonde measurements show warming below the model predictions, especially over the tropics between 20° S and 20° N;

---

climate forcing during the two years after the Hunga Tonga–Hunga Ha'apai eruption. Journal of Geophysical Research: Atmospheres, 129, e2024JD041296.

[12]Karl, T.R., Hassol, S.J., Miller, C.D., Murray, W.L., 2006. Temperature Trends in the Lower Atmosphere: Steps for Understanding and Reconciling Differences. U.S. Climate Change Science Program and the Subcommittee on Global Change Research. Washington, DC, USA.

[13]Mitchell, D.M., Lo, Y.T.E., Seviour, W.J.M., Haimberger, L., Polvani, L.M., 2020. The vertical profile of recent tropical temperature trends. Environmental Research Letters, 15(10), 104004.
Hudson, R., 2023. The problem of tropospheric temperature trend measurements for global climate models: complex versus direct empiricism. New Technology and the Humanities, 3(2), 117–123.
McKitrick, R., Christy, J., 2018. A test of the tropical 200–300-hPa warming rate in climate models. Earth and Space Science, 5, 529–536.

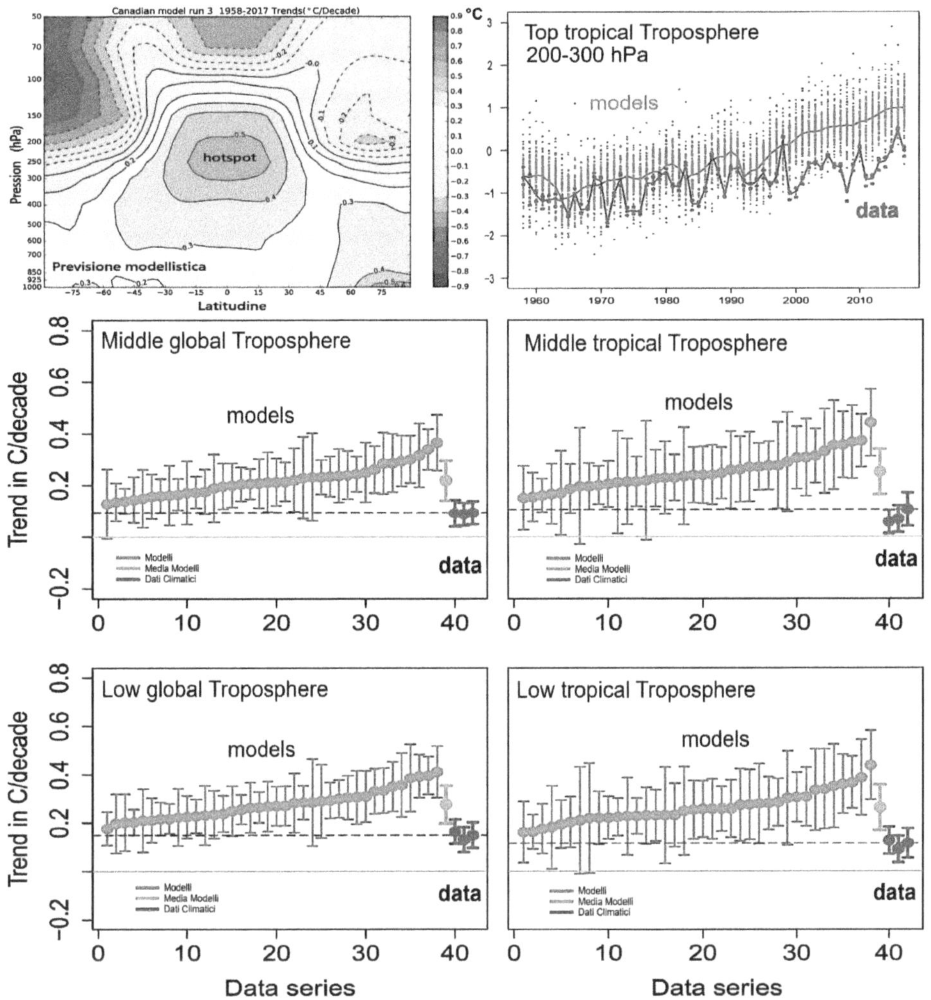

FIGURE 11.4. Top left: model prediction of the *"hot spot"* at 200–300 hPa. Top right: comparison between model outputs and observations. Bottom: warming trends and 95% confidence intervals for individual CMIP6 models (left bars) and their ensemble mean (green bar), and observational datasets (right bars). Dashed line: average trend from satellite and radiosonde data. (Adapted from McKitrick & Christy, 2018, 2020).

- the divergence is systematic, as illustrated in Figures 11.4 and 18.5, and persists across multiple observational platforms;
- IPCC AR6 (its Figure 3.10; cf. Figure 18.5) acknowledges the discrepancy, citing possible causes such as observational error, natural variability, and model limitations.

Figure 11.4 compares warming trends (°C/decade) across 38 CMIP6 models, the ensemble means, and three observational series (McKitrick & Christy, 2020).[14] The dashed line represents the mean observational trend, and clearly shows that most models overestimate tropospheric warming.

A failed test? The absence of the tropical *"hot spot"* is not a minor inconsistency. It calls into question the physical validity of GCMs in simulating vertical atmospheric responses. If models fail to reproduce this dynamic, their reliability in projecting large-scale climate changes must be critically reassessed.

Moreover, this discrepancy casts doubt on the ECS estimates obtained from GCMs. Overstated tropospheric amplification inflates ECS values. McKitrick & Christy (2020) argue that the missing hotspot warrants a downward revision of ECS to approximately 1–2 °C, with significant implications for climate policy and risk assessment.

11.4. Conclusion: toward a critical revision of climate models

The *"hot model"* problem raises foundational questions about the epistemological robustness of these tools as predictive instruments. Divergences in *Equilibrium Climate Sensitivity* (ECS) and *Transient Climate Response* (TCR) estimates, the persistent overestimation of tropospheric temperatures, and the widening discrepancies between simulations and real-world observations all point to the need for a thorough reassessment of climate modeling architecture.

This is not a call to abandon modeling, but to reform it. What is required are models that are:

- transparent in their assumptions, calibration procedures, and parameter choices;
- empirically grounded in high-quality, systematically validated observational datasets;
- structurally aware of their inherent limitations, uncertainties, and simplifications.

Such reform is essential if climate modeling is to fulfill its proper scientific role as a rational and credible framework for long-term decision-making. Models should serve as heuristic instruments—tools for exploring plausible futures, testing hypotheses, and informing policy—rather than as definitive arbiters of truth.

Climate science must embrace methodological pluralism and remain open to alternative modeling approaches, including empirical and semi-empirical frameworks that incorporate natural variability and astronomical forcings. Only through this broader lens can climate science evolve into a discipline that is not only technically sophisticated but also epistemologically sound.

[14] McKitrick, R., Christy, J., 2020. Pervasive warming bias in CMIP6 tropospheric layers. Earth and Space Science, 7, e2020EA001281.

CHAPTER 12

The 50–70-year oscillation and its implications

One of the most underestimated, yet crucial, elements in understanding Earth's climate dynamics is the presence of natural multidecadal oscillations. These long-term fluctuations, often overshadowed by short-term variability and anthropogenic forcings, play a significant role in modulating both regional and global climate patterns. Among them, particular attention is warranted for a cycle of approximately 50–70 years, which is clearly detectable in multiple climate records but poorly reproduced by global climate models (GCMs), including those developed within CMIP3, CMIP5, and CMIP6 (Scafetta, 2012a, 2013a).[1]

As illustrated in Figure 2.2, this oscillation manifests itself as alternating phases of warming and cooling superimposed on a long-term upward trend. The pattern is remarkably consistent across historical temperature reconstructions:

- circa 1850–1880: warming; circa 1880–1910: cooling;
- circa 1910–1940: warming; circa 1940–1970: cooling;
- circa 1970–2000: warming; circa 2000–2030 (?): moderate warming.

This approximately 60-year cycle is corroborated by a wide range of climate indicators (Scafetta, 2014a; Wyatt & Curry, 2014, and authors cited therein).[2] For instance, Lüdecke et al. (2024)[3] identified such a cycle in sunshine hours over Central Europe, suggesting a link between cloud cover variability and surface temperature trends. During the *"Early 20th Century Warming"* (1920–1940), Arctic sea ice anomalies reached down to -1.5 million km^2 in September relative to the climatic average (Semenov et al., 2024).[4] Similarly, Bjørk et al. (2012)[5] documented a rapid retreat of 132 land-terminating glaciers in southeast Greenland during the same period—faster than in the early 2000s—highlighting the influence of air temperature. Figure 26.5 shows an approximately 60-year

---

[1]Scafetta, N., 2012a. Testing an astronomically based decadal-scale empirical harmonic climate model versus the IPCC (2007) general circulation climate models. Journal of Atmospheric and Solar–Terrestrial Physics, 80, 124–137.
Scafetta, N., 2013a. Discussion on climate oscillations: CMIP5 general circulation models versus a semi-empirical harmonic model based on astronomical cycles. Earth-Science Reviews, 126, 321–357.
[2]Scafetta, N., 2014a. Multi-scale dynamical analysis (MSDA) of sea level records versus PDO, AMO, and NAO indexes. Climate Dynamics, 43(1–2), 175–192.
Wyatt, M.G., Curry, J.A., 2014. Role of Eurasian Arctic shelf sea ice in a secularly varying hemispheric climate signal during the 20th century. Climate Dynamics, 42, 2763–2782.
[3]Lüdecke, H.J., Müller-Plath, G., Lüning, S., 2024. Central-European sunshine hours, relationship with the Atlantic Multidecadal Oscillation, and forecast. Scientific Reports, 14, 25152.
[4]Semenov, V. A., Aldonina, T. A., Li, F., Keenlyside, N. S., Wang, L., 2024. Arctic sea ice variations in the first half of the 20th century: A new reconstruction based on hydrometeorological data. Advances in Atmospheric Sciences, 41, 1483–1495.
[5]Bjørk, A.A., Kjær, K.H., Korsgaard, N.J., Khan, S.A., Kjeldsen, K.K., et al., 2012. An aerial view of 80 years of climate-related glacier fluctuations in southeast Greenland. Nature Geoscience, 5, 427–432.

signal in a secular proxy of the Atlantic Multidecadal Oscillation (AMO) (Black et al., 1999)[6] and in historical monsoonal rainfall records from India (Agnihotri & Dutta, 2003).[7] Enzi et al. (2014)[8] identified a 60-year oscillation in a snowfall time-series reconstruction for Italy over the past 300 years, and Ortolani et al. (2019)[9] reported possible 60-year variability in cloud cover at the La Silla Astronomical Observatory in Chile.

Remarkably, volcanic and seismic activity also exhibit a quasi-60-year cycle (Bragato, 2015; Scafetta & Mazzarella, 2015b).[10]

12.1. The Atlantic oscillations: AMO and NAO

A prominent manifestation of the 60-year multidecadal cycle is the Atlantic Multidecadal Oscillation (AMO), which tracks long-term variations in North Atlantic sea surface temperatures. The AMO index is obtained by removing the secular trend from observations in the region [0°N–70°N, 80°W–20°E]. Knudsen et al. (2011)[11] found evidence of similar oscillations extending back at least 8,000 years, underscoring their structural role in the ocean–atmosphere system. This climatic oscillation is also evident in the long-term fluctuations of commercial fish catches over several centuries (Klyashtorin & Lyubushin, 2007).[12]

Millennial climate reconstructions such as that of Mann et al. (1999)[13], together with studies based on marine sediments and historical sources (e.g. Black et al., 1999; Patterson et al., 2004; Mazzarella & Scafetta, 2012)[14], reveal consistent signals of a 50–70-year cycle. Notably, the warming observed between 1910–1940 (~0.4 °C) closely mirrors that of 1970–2000.

Scafetta (2010)[15] demonstrated this similarity by removing the quadratic trend from the HadCRUT3 temperature record and shifting it by 61.5 years.

---

[6]Black, D.E., Peterson, L.C., Overpeck, J.T., Kaplan, A., Evans, M.N., Kashgarian, M., 1999. Eight centuries of North Atlantic Ocean atmosphere variability. Science, 286, 1709–1713.

[7]Agnihotri, R., Dutta, K., 2003. Centennial scale variations in monsoonal rainfall: manifestations of solar variability. Current Science, 85, 459–463.

[8]Enzi, S., Bertolin, C., Diodato, N., 2014. Snowfall time-series reconstruction in Italy over the last 300 years. The Holocene, 24(3), 346–356.

[9]Ortolani, S., Saviane, I., Cavazzani, S. 2019. Cloud coverage and sky brightness at La Silla. *Zenodo,* doi:10.5281/zenodo.3248992

[10]Bragato, P.L., 2015. Italian seismicity and Vesuvius' eruptions synchronize on a quasi 60 year oscillation. Earth and Space Science, 2(10), 453–467.
Scafetta, N., Mazzarella, A. 2015b. Spectral coherence between climate oscillations and the M $\geq$ 7 earthquake historical worldwide record. Natural Hazards, 76, 1807–1829.

[11]Knudsen, M., Seidenkrantz, M.S., Jacobsen, B., Kuijpers, A., 2011. Tracking the Atlantic Multidecadal Oscillation over 8,000 years. Nature Communications, 2, 178.

[12]Klyashtorin, L.B., Lyubushin, A.A., 2007. Cyclic Climate Changes and Fish Productivity. Moscow, VNIRO Publishing.

[13]Mann, M.E., Bradley, R.S., Hughes, M.K., 1999. Northern hemisphere temperatures during the past millennium: inferences, uncertainties, and limitations. Geophysical Research Letters, 26, 759–762.

[14]Black, D.E., Peterson, L.C., Overpeck, J.T., Kaplan, A., Evans, M.N., Kashgarian, M., 1999. Eight centuries of North Atlantic Ocean atmosphere variability. Science, 286, 1709–1713.
Patterson, R.T., Prokoph, A., Chang, A., 2004. Late Holocene sedimentary response to solar and cosmic ray activity influenced climate variability in the NE Pacific. Sedimentary Geology, 172, 67–84.
Mazzarella, A., Scafetta, N., 2012. Evidences for a quasi 60-year North Atlantic Oscillation since 1700 and its meaning for global climate change. Theoretical and Applied Climatology, 107, 599–609.

[15]Scafetta, N., 2010. Empirical evidence for a celestial origin of the climate oscillations and its implications. Journal of Atmospheric and Solar–Terrestrial Physics, 72, 951–970.

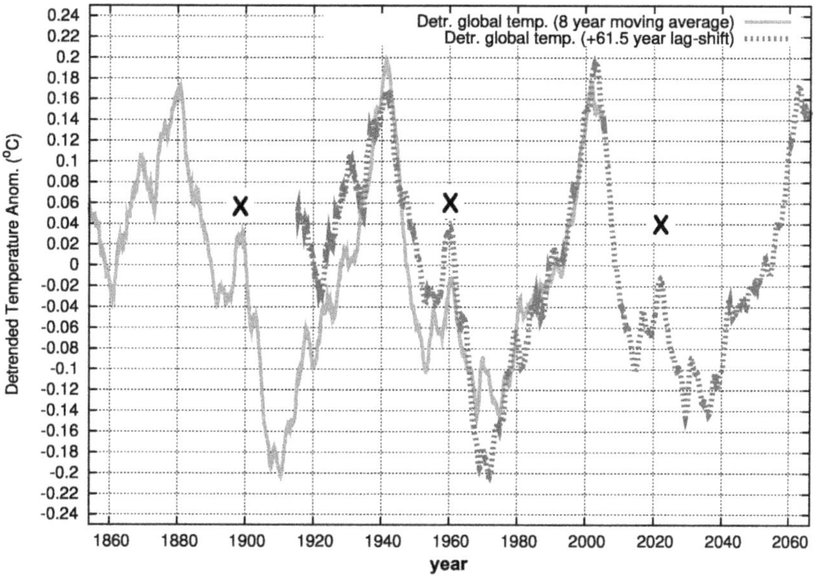

FIGURE 12.1. Comparison of the HadCRUT3 temperature record with its trend removed and the same record shifted by 61.5 years. (Adapted from Scafetta, 2010).

Figure 12.1 shows that the 1880–1940 cycle is nearly replicated in the 1940–2000 period. Secondary temperature peaks around 1900, 1961, and a predicted peak between 2020 and 2025 are marked with "X". This temperature peak is confirmed by recent warming trends (cf. Figure 2.2).

Similar periodicities appear in the North Atlantic Oscillation (NAO) index, indicating spectral coherence across independent climate variables. Scafetta (2014a)[16] applied a Multi-Scale Acceleration Analysis (MSAA) to long-term records such as global sea level (Jevrejeva et al., 2008)[17] and the NAO index (Luterbacher et al., 2002),[18] revealing a persistent 55–70-year cycle that has been active since 1700 (Figure 12.2). Chambers et al. (2012)[19] showed that a quasi-60-year cycle is present in several sea-level records.

This cyclic behavior enables a remarkably accurate reconstruction of climate variability since 1850 by means of a simple harmonic model superimposed on a secular trend (Scafetta, 2010; Loehle & Scafetta, 2011).[20]

---

[16]Scafetta, N., 2014a. Multi-scale dynamical analysis (MSDA) of sea level records versus PDO, AMO, and NAO indexes. Climate Dynamics, 43(1–2), 175–192.

[17]Jevrejeva, S., Moore, J.C., Grinsted, A., Woodworth, P.L., 2008. Recent global sea level acceleration started over 200 years ago? Geophysical Research Letters, 35, L08715.

[18]Luterbacher, J., Xoplaki, E., Dietrich, D., Jones, P.D., Davies, T.D., et al., 2002. Extending North Atlantic Oscillation reconstructions back to 1500. Atmospheric Science Letters, 2(1–4), 114–124.

[19]Chambers, D.P., Merrifield, M.A., Nerem, R.S., 2012. Is there a 60-year oscillation in global mean sea level? Geophysical Research Letters, 39, L18607.

[20]Loehle, C., Scafetta, N., 2011. Climate change attribution using empirical decomposition of climatic data. The Open Atmospheric Science Journal, 5, 74–86.

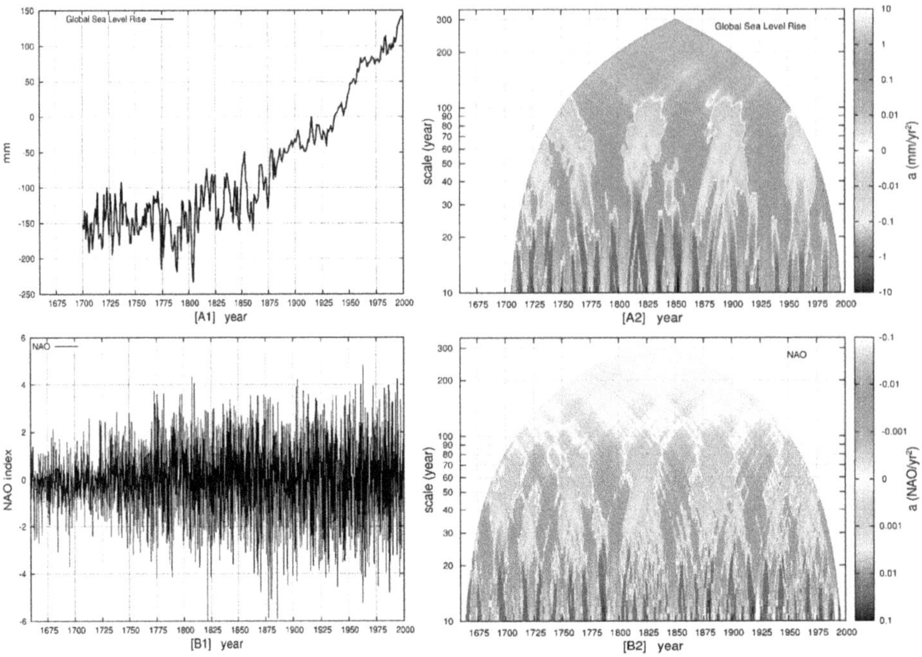

FIGURE 12.2. [A] (left) Historical mean sea level reconstruction (left) and MSA from Jevrejeva et al. (2008); (right) MSAA diagram alternating colored bands indicate a persistent 55–70 year oscillation within the 30–110 year scale.
[B] NAO index series (left) and corresponding MSAA diagram (right) from Luterbacher et al. (2002), showing cyclicity consistent with sea level trends.
(Adapted from Scafetta, 2014a).

12.2. A cycle not modeled by GCMs

Global climate models (GCMs) must attribute the warming episode of 1910 to the 1940s to natural variability, yet they fail to reproduce it. The warming from the 1970s to the 2000s is instead explained in anthropogenic terms. This interpretive asymmetry points to a possible structural inconsistency. Indeed, Figure 12.3 compares CMIP6 GCM simulations with observed sea surface temperature trends (HadSST4, NOAA Reanalysis) in the AMO region. While the models simulate a monotonic warming from 1880 to 1960, the observational data reveal an oscillatory pattern with amplitudes of up to approximately 0.4 °C.

This divergence suggests that, while GCMs tend to mimic the tendencies of known external forcings (see Figure 6.2), real-world observations reflect a more complex internal variability. The inability of GCMs to reproduce the observed multidecadal oscillation indicates that they fail to simulate the natural mechanisms responsible for large, empirically evident climate oscillations.

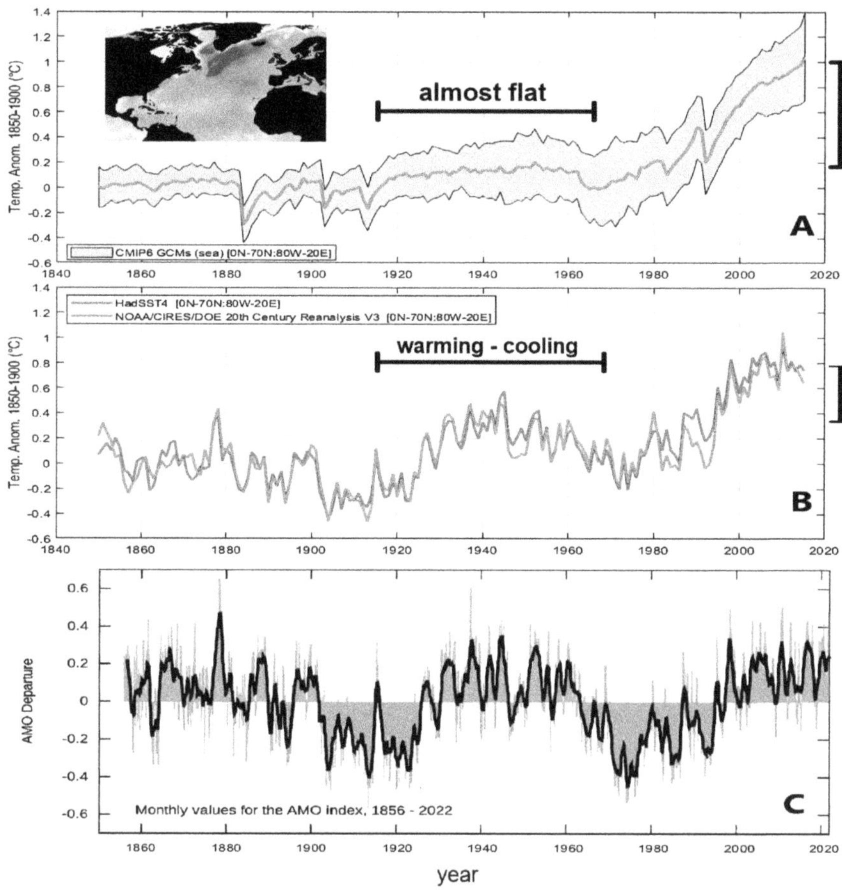

FIGURE 12.3. [A] CMIP6 model simulations of North Atlantic surface temperature ([0°N–70°N, 80°W–20°E]).
[B] Observational series from HadSST4 (Kennedy et al., 2019) and NOAA/CIRES/DOE Reanalysis V3.
[C] AMO index derived from detrended Atlantic SST anomalies (1856–2022).
(Adapted from Scafetta, 2026).

12.3. Further evidence: North Atlantic hurricanes

In the early 2000s, the apparent increase in the frequency and intensity of hurricanes in the North Atlantic was widely interpreted as a signal of anthropogenic climate change. A landmark study by Webster et al. (2005)[21], co-authored by Judith Curry, documented an increase in category 4 and 5 tropical cyclones between 1970 and 2004, suggesting a link to ocean surface warming driven by greenhouse gas emissions.

[21] Webster, P.J., Holland, G.J., Curry, J.A., Chang, H.-R., 2005. Changes in tropical cyclone number, duration, and intensity in a warming environment. Science, 309(5742), 1844–1846.

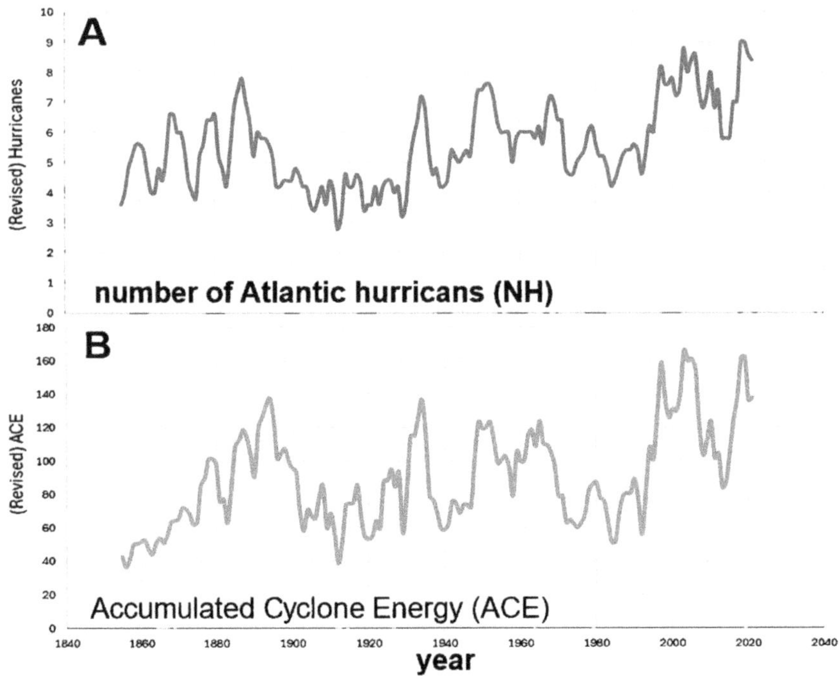

FIGURE 12.4. [A] Number of hurricanes in the North Atlantic. [B] Accumulated Cyclone Energy (ACE) index, reflecting strength, duration, and frequency (5-year moving averages; NOAA, 2023).

This interpretation, however, was subsequently reassessed. Vecchi & Knutson (2011)[22] pointed out that the analyzed period coincided with a rising phase of the AMO, which follows a natural ~60-year cycle. Figure 12.4 shows that both hurricane counts and the Accumulated Cyclone Energy (ACE) index from 1853 to 2023 exhibit a clear multidecadal oscillation, in phase with the AMO patterns (Figure 12.3).

Vecchi & Knutson (2011) also noted that limitations in pre-satellite data (especially over 1878–1965) likely led to the undercounting of hurricanes in the historical records. They concluded that natural variability and incomplete data could account for much of the observed long-term increase, thereby challenging the notion that it is primarily anthropogenic.

Indeed, cyclonic activity in the North Atlantic appears to follow a cyclical pattern correlated with ocean surface temperature changes that are consistent with the AMO oscillations (Figure 12.3), rather than behaving as a direct response to anthropogenic warming, as GCMs often suggest.

Judith Curry herself gradually revised her position in light of the emerging evidence, distancing herself from earlier conclusions on the anthropogenic attribution of intense hurricanes. This shift did not represent a rejection of

[22] Vecchi, G., Knutson, T., 2011. Estimating annual numbers of Atlantic hurricanes missing from the HURDAT database (1878–1965) using ship track density. Journal of Climate, 24(6), 1736–1746.

climate science, but rather a critical reassessment of the methodologies used to interpret data and construct projections.

Curry emphasized that the complexity of the climate system, together with its intrinsic long-term natural variability—particularly that linked to the AMO—makes it difficult to isolate the anthropogenic signal with certainty. She also highlighted the presence of a 60-year cycle observable across multiple climate subsystems, which she and Wyatt (2014)[23] described as generating a *"stadium wave"* within the climate system (Wyatt & Curry, 2014).[24]

In public testimony before the U.S. Congress, Curry emphasized the difficulties of attributing hurricane trends to global warming, citing model uncertainties and gaps in the historical data. She subsequently acknowledged that the conclusions of Webster et al. (2005) had been shaped by a simplified framework that overlooked multidecadal natural climate oscillations.

On the policy front, Curry has cautioned against adopting radical climate measures based on highly uncertain projections. In her book *Climate Uncertainty and Risk* (2023),[25] she advocates a paradigm grounded in systematic uncertainty assessment and robust scientific evidence.

Her intellectual evolution has sparked debate and contributed to the polarization of discussions on hurricane attribution. Today, Judith Curry stands as a prominent example of a scientist who, while initially aligned with the IPCC consensus, has developed a critical and independent perspective driven by a deepening awareness of climate complexity and of the epistemological limits of current modeling approaches.

12.4. Implications for the Anthropogenic Global Warming Theory

Knight et al. (2009)[26] argue that climate models should not be expected to simulate temperature fluctuations in fine detail over short timescales. They emphasize, however, that GCMs must be able to reproduce climate trends over periods longer than 10–15 years. The persistent inability to simulate the 50–70-year climate oscillation thus reveals a structural flaw that cannot be ignored.

This limitation has direct implications that call the AGWT into question. If the post-1970s warming is partially attributable to the positive phase of a natural multidecadal oscillation, then GCMs necessarily overestimate the anthropogenic contribution to climate change by ascribing all of that warming to human activity.

Several authors have shown that the failure of CMIP models to reproduce such oscillations compromises the ability to correctly separate the natural and anthropogenic components of the observed warming (e.g.: Scafetta, 2010; Gervais, 2016; Scafetta, 2013a)[27]. This misattribution affects not only temperature trends but also the broader interpretation of climate dynamics.

[23] Wyatt, M.G., Curry, J.A., 2014. Role of Eurasian Arctic shelf sea ice in a secularly varying hemispheric climate signal during the 20th century. Climate Dynamics, 42, 2763–2782.

[24] Wyatt, M.G., Curry, J.A., 2014. Role of Eurasian Arctic shelf sea ice in a secularly varying hemispheric climate signal during the 20th century. Climate Dynamics, 42, 2763–2782.

[25] Curry, J., 2023. Climate Uncertainty and Risk: Rethinking Our Response. Anthem Press.

[26] Knight, J., Kennedy, J.J., Folland, C., Harris, G., Jones, G.S., et al., 2009. Do global temperature trends over the last decade falsify climate predictions? Bulletin of the American Meteorological Society, 90(8), S22–S23.

[27] Scafetta, N., 2010. Empirical evidence for a celestial origin of the climate oscillations and its

Moreover, GCMs struggle to accurately simulate other key climate parameters, including precipitation variability, extreme rainfall events, and hydrological cycles (Li et al., 2022; Zhang et al., 2025a; Hellmuth et al., 2025). Connolly et al. (2019),[28] for example, compared the observed snow-cover trends in the Northern Hemisphere (1967–2018) with CMIP5 simulations, revealing significant seasonal discrepancies and a poor match between models and observations.

Figure 12.3 illustrates a notable discrepancy between GCM projections and empirical data for the global warming between 1940 and 2020:

- GCMs estimate a temperature increase of approximately +0.8 °C;
- observational data indicate a warming closer to +0.4 °C.

This substantial overestimation cannot be explained by internal variability or statistical uncertainty alone. It points to an inflated climate sensitivity and to a lack of realism in the modeling of natural cyclic climate components.

A key reason for this modeling failure also lies in the IPCC's claim that oceanic and atmospheric oscillations—such as the AMO and PDO—are manifestations of random internal variability. The IPCC AR6 report (2021, p. 197) states that the AMO and the PDO are unpredictable beyond a few years, simply because current GCMs cannot simulate them. The IPCC thus resolves the dilemma by rhetorically renaming the "*Atlantic Multidecadal Oscillation*" (AMO) as the "*Atlantic Multidecadal Variability*" (AMV), and the "*Pacific Decadal Oscillation*" (PDO) as the "*Pacific Decadal Variability*" (PDV), among others.

Although the IPCC (AR6 2021, p. 114) claims that the observed variance in Pacific and Atlantic climatic records does not show significant changes, numerous peer-reviewed studies—e.g., Ghil et al. (2002),[29] Scafetta (2010), Mantua et al. (1997)[30] and Gray et al. (2004)[31]—have demonstrated that the oscillations apparent in the AMO, PDO, ENSO, and global mean surface temperature (GMST) records are statistically significant at the 95% level. These findings challenge the assumption that such variability is merely stochastic noise and suggest, on the contrary, that it plays a meaningful role in shaping long-term climate trends.

12.5. The decline in tropical cyclone frequency

Regarding the secular trend uncertainty in cyclonic frequency and intensity, one of the most compelling hypotheses linking global warming to changes

implications. Journal of Atmospheric and Solar–Terrestrial Physics, 72, 951–970.
Scafetta, N., 2013a. Discussion on climate oscillations: CMIP5 general circulation models versus a semi-empirical harmonic model based on astronomical cycles. Earth-Science Reviews, 126, 321–357.
Gervais, F., 2016. Anthropogenic CO_2 warming challenged by 60-year cycle. Earth-Science Reviews, 155, 129–135.

[28]Connolly, R., Connolly, M., Soon, W., Legates, D.R., Cionco, R.G., Velasco Herrera, V.M., 2019. Northern hemisphere snow-cover trends (1967–2018): A comparison between climate models and observations. Geosciences, 9(3), 135.

[29]Ghil, M., Allen, M.R., Dettinger, M.D., Ide, K., Kondrashov, et al., 2002. Advanced spectral methods for climatic time series. Reviews of Geophysics, 40(1), 3.1–3.41.

[30]Mantua, N.J., Hare, S.R., Zhang, Y., Wallace, J.M., Francis, R.C., 1997. A Pacific interdecadal climate oscillation with impacts on salmon production. Bulletin of the American Meteorological Society, 78, 1069–1080.

[31]Gray, S.T., Graumlich, L.J., Betancourt, J.L., Pederson, G.T., 2004. A tree-ring based reconstruction of the Atlantic Multidecadal Oscillation since 1567 A.D. Geophysical Research Letters, 31, L12205.

in tropical cyclone (TC) activity concerns the modification of large-scale atmospheric pressure gradients. As anthropogenic warming progresses, polar regions—particularly the Arctic—are expected to experience accelerated temperature increases relative to the tropics, a phenomenon known as polar amplification. This differential warming reduces the meridional (equator-to-pole) temperature and pressure gradients, which are fundamental drivers of Earth's general circulation.

A weakened gradient implies a reduction in the strength of the Hadley and Walker circulations, both of which play critical roles in the genesis and maintenance of tropical cyclones. In particular, the Hadley circulation governs vertical motion and convective activity in the tropics. Its attenuation produces a less favorable thermodynamic environment for TC formation, characterized by diminished low-level convergence, reduced mid-tropospheric humidity, and altered vertical wind-shear profiles.

This theoretical framework is consistent with the findings of Chand et al. (2022),[32] who reported a statistically significant decline in global TC frequency over the twentieth century (Figure 12.5). Using a reconstruction based on the Twentieth Century Reanalysis dataset, the authors attributed the observed trend to less favorable tropospheric dynamics arising from weakened large-scale circulations. Their detection algorithm, which requires deep tropospheric curvature, sustained moisture, and persistence over 48 hours, revealed a robust downward trend in TC occurrence.

Emanuel (2023)[33], however, challenged the robustness of this result, arguing that the decline may be an artifact of the increasing density of assimilated pressure observations in the reanalysis data. He proposed an alternative proxy based on the Okubo–Weiss–Zeta parameter, which did not exhibit a comparable trend. In response, Chand et al. (2023)[34] defended their methodology, demonstrating that the decline persists even in reanalysis products that do not assimilate sea-level pressure.

Complementing these observational and methodological debates, Knutson et al. (2020)[35] provided a synthesis of climate model projections under anthropogenic forcing. Their assessment of CMIP5 and CMIP6 simulations indicates that, while the intensity of tropical cyclones is expected to increase, their frequency may remain stable or decline modestly. This model-based evidence reinforces the notion that a warming climate may suppress cyclone genesis through dynamical constraints, even as the thermodynamic potential for intensification rises.

Taken together, these findings suggest that the weakening of the equator-to-pole pressure gradient under global warming may contribute to a decline in

[32]Chand, S.S., Tory, K.J., Walsh, K.J.E., Camargo, S.J., Kuleshov, Y., Ye, H., 2022. Declining tropical cyclone frequency under global warming. Nature Climate Change, 12, 655–660.

[33]Emanuel, K., 2023. Limitations of reanalyses for detecting tropical cyclone trends. Nature Climate Change, 13, 379–380.

[34]Chand, S.S., Tory, K.J., Walsh, K.J.E., Camargo, S.J., Kuleshov, Y., Ye, H., 2023. Reply to: Limitations of reanalyses for detecting tropical cyclone trends. Nature Climate Change, 13, 381–383.

[35]Knutson, T.R., Camargo, S.J., Chan, J.C.L., Emanuel, K., Ho, C.H., et al., 2020. Tropical cyclones and climate change assessment: Part II. Projected response to anthropogenic warming. Bulletin of the American Meteorological Society, 101(3), E303–E322.

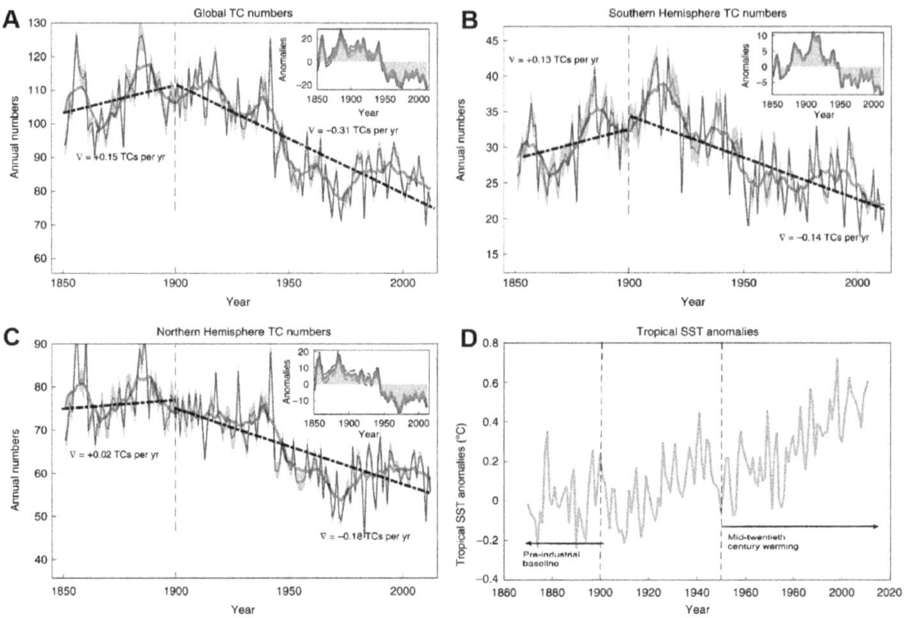

FIGURE 12.5. [A–C] Time series of ensemble-mean tropical cyclone (TC) counts for the global domain (a), Southern Hemisphere (b), and Northern Hemisphere (c), shown as annual values (blue solid line) with 95% confidence intervals (blue shading). Long-term linear trends for the preindustrial and twentieth-century periods are indicated by black dashed lines, with associated gradients (∇), and the five-year running mean is plotted in red. Top-right inserts display TC count anomalies relative to the full-period mean.
[D] Time series of tropical summertime sea surface temperature (SST) anomalies relative to the preindustrial baseline, highlighting the onset and progression of warming since the mid-twentieth century. Vertical dashed lines in panels a–c mark the end of the preindustrial reference period (circa 1900). (Adapted from Chand et al., 2022).

TC frequency, even as the thermodynamic potential for cyclone intensification increases. This nuanced dynamic underscores the importance of distinguishing between cyclone genesis and intensification mechanisms, and highlights the need for physically grounded metrics in climate model evaluation.

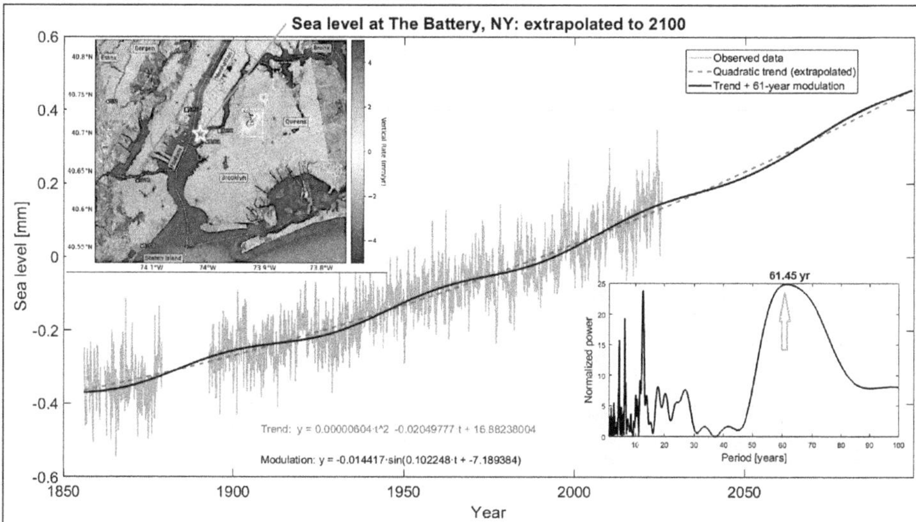

FIGURE 12.6. **Relative Sea Level at the Battery (New York City).** Local relative sea level (RSL) trends are measured by tide gauges, which record sea level with respect to a fixed benchmark on land. Consequently, RSL reflects a combination of actual sea level change and local vertical land motion. RSL data for numerous locations are available from NOAA (https://tidesandcurrents.noaa.gov/), PSMLS (https://psmsl.org/) and other sources.

At the Battery in New York City, records extend back to 1856. Here, RSL has risen almost linearly at a rate of 2.95 mm/year, with only a slight acceleration. The periodogram of the data also reveals a ∼60-year oscillation correlated with the Atlantic Multidecadal Oscillation (AMO) (Scafetta, 2013b, 2014a). Assuming that the quadratic trend persists throughout the 21st century, RSL is projected to rise by approximately 40 cm between 2000 and 2100. New York City, however, is undergoing land subsidence of approximately 1.5–2.0 mm/year, likely due to the city's structural load (Boretti, 2025; Buzzanga et al., 2023, https://www.jpl.nasa.gov/images/pia25527-map-of-new-york-city-subsidence-and-uplift/). The actual geocentric sea-level rise has therefore been closer to 1.0–1.5 mm/year, implying a projected geocentric sea-level rise of 20–25 cm from 2000 to 2100.

The presence of a ∼60-year oscillation in relative sea level (RSL) complicates the reliable detection of acceleration in records shorter than one full cycle. Scafetta (2014a) demonstrated that datasets spanning approximately 110 years or more are required to effectively disentangle a long-term accelerating trend from the influence of this multidecadal oscillation.

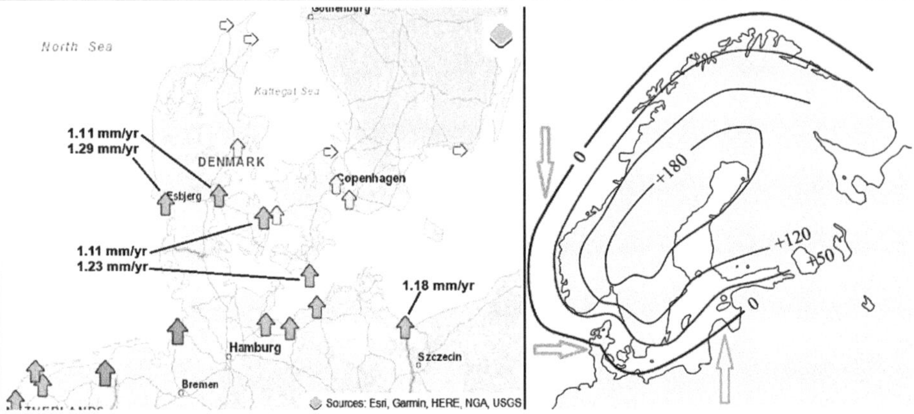

Figure 12.7. **Relative Sea Level in Central Denmark.** Rates of relative sea-level rise of approximately 1.0–1.5 mm/year are observed in several locations.
Notably, the central region of Denmark—in particular, the area surrounding the island of Funen (Fyn)—marks the approximate hinge line where vertical land motion is near zero. This geodynamically stable zone lies between the uplifting terrains of northern Denmark—where post-glacial rebound, driven by the melting of the Fennoscandian Ice Sheet over Norway and Sweden, causes the land to rise—and the subsiding regions of the south, which are gradually sinking owing to sediment compaction and residual isostatic adjustment. Because vertical land motion is minimal in this central corridor, it provides a robust reference frame for isolating the geocentric component of sea-level rise. Tide-gauge records from this region are particularly valuable for validating satellite altimetry and for benchmarking relative sea-level trends across Northern Europe. Data from PSMLS (https://psmsl.org/).

Other coastal locations where the vertical motion of the land is negligible include:
Honolulu, Hawaii, RSL = 1.54 mm/yr;
Auckland, New Zealand, RSL = 1.29 mm/yr;
Fremantle, Australia, RSL = 1.84 mm/yr;
Sydney, Australia, RSL = 1.15 mm/yr.

In general, during Phanerozoic icehouse periods, when large continental ice sheets were present, orbital-scale glacio-eustatic fluctuations could exceed 100 meters, comparable to the Quaternary glacial cycles. During greenhouse periods, with little or no permanent ice, sea-level changes were much smaller—typically 20–40 meters or less—and in some intervals were nearly absent (Van der Meer et al., 2025).

CHAPTER 13

The *"Roman Optimum"* and the *"Medieval Warm Period"*

Over the past millennium, Earth's climate has exhibited significant variability, marked by two anomalous periods in terms of intensity and duration: the Medieval Warm Period (MWP) and the Little Ice Age (LIA). The MWP, dated approximately between 900 and 1300 AD, was characterized by relatively high temperatures across many regions of the globe. The LIA, spanning roughly from 1400 to 1850, was a period of widespread cooling. This thermal contrast has been documented since the 1960s (Lamb, 1965)[1] and confirmed by a wide array of paleoclimatic and historical sources.

The climatic significance of the MWP has been the subject of intense debate, particularly following the publication of the *"hockey stick"* graph by Mann et al. (1999),[2] which depicted a relatively stable climate during the centuries preceding modern warming, the latter being described as *"unprecedented."* Numerous empirical studies, however, have challenged this narrative by showing that the MWP was not only widespread but, in several regions, likely featured temperatures comparable to, or even exceeding, those of today (Soon & Baliunas, 2003; Soon et al., 2003a; von Storch et al., 2004; Lasher & Axford, 2019; Ge et al., 2017, and many others).[3]

13.1. Geographic distribution of the MWP

A central question in the debate is whether the MWP was a global or a regional phenomenon. While some researchers, such as Mann et al. (2009),[4]

[1]Lamb, H.H., 1965. The early medieval warm epoch and its sequel. Palaeogeography, Palaeoclimatology, Palaeoecology, 1, 13–37.
[2]Mann, M.E., Bradley, R.S., Hughes, M.K., 1999. Northern hemisphere temperatures during the past millennium: inferences, uncertainties, and limitations. Geophysical Research Letters, 26, 759–762.
[3]Soon, W., Baliunas, S., 2003. Proxy climatic and environmental changes of the past 1000 years. Climate Research, 23(2), 89–110.
Soon, W., Baliunas, S., Idso, C., Idso, S., Legates, D.R., 2003a. Reconstructing climatic and environmental changes of the past 1000 years: A reappraisal. Energy and Environment, 14(2–3), 233–296.
von Storch, H., Zorita, E., Jones, J.M., Dimitriev, Y., González-Rouco, F., Tett, S.F.B., 2004. Reconstructing past climate from noisy data. Science, 306(5696), 679–682.
Lasher, G.E., Axford, Y., 2019. Medieval warmth confirmed at the Norse Eastern Settlement in Greenland. Geology, 47(3), 267–270.
Ge, Q.S., Liu, H.L., Ma, X., Zheng, J.Y., Hao, Z.X., 2017. Characteristics of temperature change in China over the last 2000 years. Advances in Atmospheric Sciences, 34, 941–951.
[4]Mann, M.E., Zhang, Z., Rutherford, S., Bradley, R.S., Malcolm, K., et al., 2009. Global signatures and dynamical origins of the Little Ice Age and Medieval Climate Anomaly. Science, 326(5957), 1256–1260.

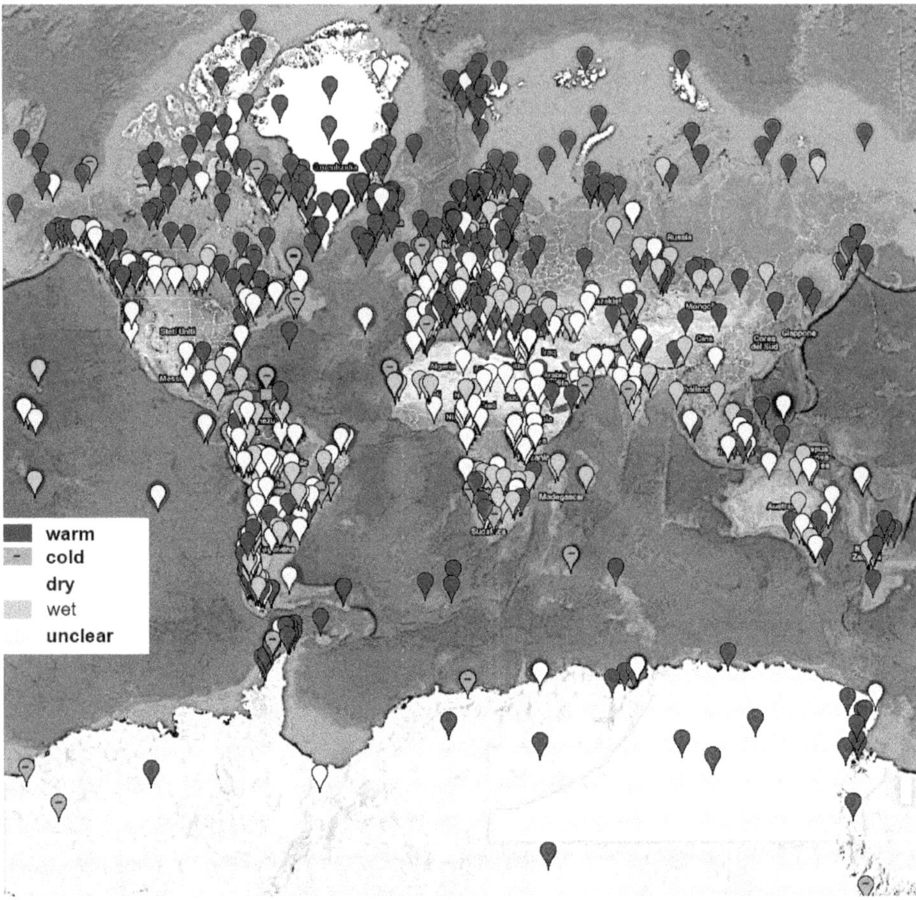

FIGURE 13.1. Climate reconstructions from 1,274 studies on the MWP (1000–1200 AD), with each point representing a study location. Colors indicate climate conditions: warm, cold, dry, wet, and unclear. In the grayscale map, cold regions are distinguished from wet ones by the use of the "-" symbol. Based on Lüning's "Climate reconstructions of the 'Medieval Warm Period' 1000–1200 AD" map (Lüning, 2022, http://t1p.de/mwp). (Adapted from Scafetta, 2026).

argue that it was confined to the North Atlantic, the *"Medieval Warm Period (1000–1200 AD)"* project by Lüning (2022)[5] analyzed nearly 1,300 scientific publications and found evidence of warming in most regions of the planet. Figure 13.1 shows a predominance of positive thermal anomalies, indicating that the MWP was of near-global extent, including Antarctica (Forte et al., 2025).[6]

[5]Lüning, S., 2022. Climate reconstructions of the 'Medieval Warm Period' 1000–1200 AD. Available at: http://t1p.de/mwp

[6]Forte, E., Azzaro, M., Cannone, N., Longhi, A., Santin, I, Guglielmin, M., 2025. A warming pulse in the Antarctic continent changed the landscape during the Middle Ages. Commun Earth Environ 6, 281 (2025).

13.2. The MWP versus global climate models

Multiproxy paleoclimate reconstructions based on tree rings, ice cores, lake sediments, and corals carry inherent uncertainties (IPCC, 2013, see their Figure 5.7). Yet a systematic discrepancy emerges between these reconstructions and GCM simulations, particularly in the representation of past warm periods such as the MWP.

Figures 13.2A and 13.2B—taken from the IPCC AR5 (2013)[7] and the IPCC AR6 (2021)[8], respectively—highlight the gap between proxy data and CMIP5/CMIP6 model averages. The pairs of arrows on the left side of the graphs indicate that the CMIP6 GCMs do not reproduce the MWP and flatten the preindustrial climate variability, simulating a nearly stationary climate from 850 to 1900 AD apart from brief volcanic cooling episodes.

This flattening is further illustrated in Figure 13.2C, where the semi-empirical reconstruction of Scafetta (2013a, 2021b)[9] (based on Moberg et al. (2005)[10] and calibrated against HadCRUT post-1850 data) diverges substantially from the multimodel GCM mean. The failure of CMIP5 and CMIP6 models to reproduce medieval warmth is attributed to their limited treatment of solar variability. These models often assume a near-constant solar forcing, which restricts their ability to simulate past warm episodes such as the MWP.

Even IPCC (AR6, 2021, p. 433) acknowledges *"greater disagreement between models and temperature reconstructions"* prior to 1300 AD, thereby implicitly conceding the limitations of current simulations. Notably, AR6 emphasizes the PAGES-2K Consortium (2019)[11] reconstruction, which substantially dampens the MWP signal in a manner similar to the original *"hockey stick"* graph by Mann et al. (1999). This choice by the IPCC has, however, been criticized by Esper et al. (2024a),[12] who argue that AR6 has oversimplified Holocene climate variability, neglecting the diversity of available evidence.

13.3. The millennial climate cycle

The *Medieval Warm Period* (MWP) is but one manifestation of a broader pattern of cyclical natural climate variability. Over the past 9,000 years of the Holocene, a quasi-millennial climate cycle has been consistently observed. Ice core data from Greenland's GISP2 project (Alley, 2000),[13] together with numerous paleoclimate reconstructions (Bond et al., 2001; Kerr, 2001; Neff et al., 2001;

---

[7] IPCC, 2013. Climate Change 2013: The Physical Science Basis. Cambridge University Press.

[8] IPCC, 2021. Climate Change 2021: The Physical Science Basis. Cambridge University Press.

[9] Scafetta, N., 2013a. Discussion on climate oscillations: CMIP5 general circulation models versus a semi-empirical harmonic model based on astronomical cycles. Earth-Science Reviews, 126, 321–357. Scafetta, N., 2021b. Reconstruction of the interannual to millennial scale patterns of the global surface temperature. Atmosphere, 12, 147.

[10] Moberg, A., Sonechkin, D.M., Holmgren, K., Datsenko, N.M., Karlén, W., 2005. Highly variable Northern Hemisphere temperatures reconstructed from proxy data. Nature, 433, 613–617.

[11] PAGES-2K Consortium, 2019. Consistent multi-decadal variability in global temperature reconstructions and simulations over the Common Era. Nature Geoscience, 12(8), 643–649.

[12] Esper, J., Smerdon, J.E., Anchukaitis, K.J., Cook, E.R., D'Arrigo, R., et al., 2024. The IPCC's reductive Common Era temperature history. Communications Earth & Environment, 5, 222.

[13] Alley, R.B., 2000. The Younger Dryas cold interval as viewed from central Greenland. Quaternary Science Reviews, 19, 213–226.

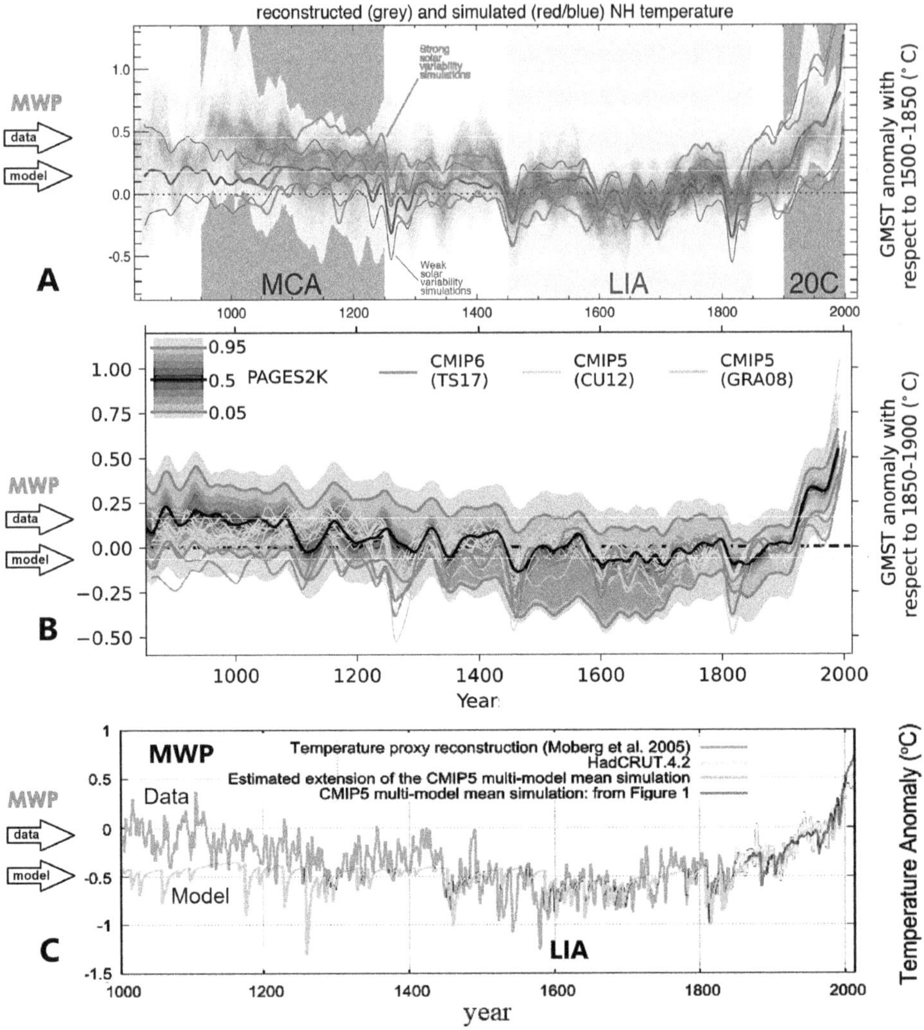

FIGURE 13.2. [A] Simulated vs. reconstructed temperature anomalies in the Northern Hemisphere over the last millennium, including the MWP, LIA, and 20th-century warming (adapted from IPCC, 2013, Fig. 5.8).
[B] Global temperature anomalies from the PAGES2K multiproxy model (baseline 1850–1900), alongside CMIP5 and CMIP6 simulations (adapted from IPCC, 2021, Fig. 3.2).
[C] Comparison between CMIP model mean, Moberg et al. (2005) proxy reconstruction, and HadCRUT data post-1850. Arrows indicate MWP temperature estimates from proxy data vs. model simulations.
(Adapted from Scafetta, 2013a; 2021b).

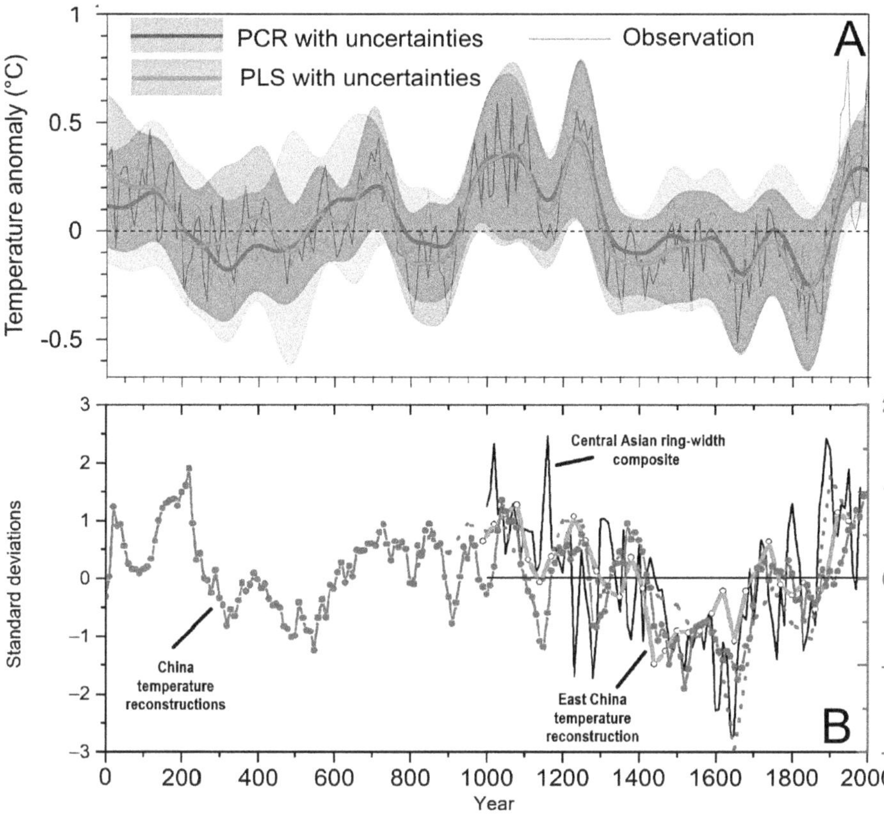

Figure 13.3. Temperature reconstruction [A] in China and [B] climate proxy trends in arid regions of Central Asia. (Adapted from Ge et al., 2017; Yang et al., 2009).

Kutschera et al., 2020),[14], reveal a regular alternation of warm and cold phases, each lasting approximately 900–1,000 years.

For instance, the MWP and the Roman Warm Period (RWP) are separated by approximately 900–1,000 years. Both are clearly evident in temperature reconstructions from China (Ge et al., 2017)[15] and from arid regions of Central

[14]Bond, G., Kromer, B., Beer, J., Muscheler, R., Evans, M.N., et al., 2001. Persistent solar influence on North Atlantic climate during the Holocene. Science, 294(5549), 2130–2136.
Kerr, R.A., 2001. A variable sun paces millennial climate. Science, 294(5546), 1431–1433.
Neff, U., Burns, S.J., Mangini, A., Mudelsee, M., Fleitmann, D., Matter, A., 2001. Strong coherence between solar variability and the monsoon in Oman between 9 and 6 kyr ago. Nature, 411, 290–293.
Kutschera, W., Patzelt, G., Schaefer, J.M., Schlüchter, C., Steier, P., Wild, E.M., 2020. The movements of Alpine glaciers throughout the last 10,000 years as sensitive proxies of temperature and climate changes. EPJ Web of Conferences, 232, 02002.
[15]Ge, Q.S., Liu, H.L., Ma, X., Zheng, J.Y., Hao, Z.X., 2017. Characteristics of temperature change in China over the last 2000 years. Advances in Atmospheric Sciences, 34, 941–951.

Asia (Yang et al., 2009),[16] which also indicate that both periods may have been climatically comparable to the present warm phase. See Figure 13.3.

Further evidence comes from Mediterranean sea surface temperature (SST) records. Figure 13.4 presents high-resolution SST reconstructions from the Sicily Channel (Margaritelli et al., 2020),[17] the Alboran Sea (Rodrigo-Gámiz et al., 2014),[18] and the Minorca Basin (Cisneros et al., 2016),[19] compared with Northern Hemisphere temperature reconstructions (Ljungqvist, 2010)[20] and the NAO index (Faust et al., 2016).[21] These studies show that Mediterranean SSTs during the Roman period (1–500 AD) were up to 2 °C higher than the average of the subsequent centuries, including the 20th century, with a further warm phase during the MWP (1000–1300 AD). Such conditions may have contributed to the stability and expansion of the Roman Empire. The Mediterranean subsequently experienced gradual cooling and increased climatic instability (cf. Camuffo, 2010).[22]

Figure 13.5 compares Northern Hemisphere temperature reconstructions (Ljungqvist, 2010; Luterbacher, 2016; Matskovsky & Helama, 2014)[23] with total solar irradiance (TSI) derived from radiocarbon ^{14}C data (Steinhilber et al., 2012).[24] The Roman, Medieval, and Modern warm periods exhibit comparable amplitudes and durations, whereas the intervening cold phases differ in their intensity.

Figure 15.2 provides further support for this millennial climatic oscillation, showing Holocene climate reconstructions from Greenland ice cores (Alley,

---

[16]Yang, B., Wang, J., Bräuning, A., Dong, Z., Esper, J., 2009. Late Holocene climatic and environmental changes in arid central Asia. Quaternary International, 194(1–2), 68–78.

[17]Margaritelli, G., Cacho, I., Català, A., Barra, M., Bellucci, L. G., Lubritto, C., Rettori, R., Lirer, F., 2020. Persistent warm Mediterranean surface waters during the Roman period. Scientific Reports, 10, 10430.

[18]Rodrigo-Gámiz, M., Martínez-Ruiz, F., Rampen, S.W., Schouten, S., Sinninghe Damsté, J.S., 2014. Sea surface temperature variations in the western Mediterranean Sea over the last 20 kyr: A dual-organic proxy (UK′37 and LDI) approach. Paleoceanography, 29, 87–98.

[19]Cisneros, M., Cacho, I., Frigola, J., Canals, M., Masqué, P., et al., 2016. Sea surface temperature variability in the central-western Mediterranean Sea during the last 2700 years: A multi-proxy and multi-record approach. Climate of the Past, 12(4), 849–869.

[20]Ljungqvist, F.C., 2010. A new reconstruction of temperature variability in the extra-tropical northern hemisphere during the last two millennia. Geografiska Annaler: Series A, Physical Geography, 92, 339–351.

[21]Faust, J.C., Fabian, K., Milzerc, G., Giraudeau, J. Kniesa, J., 2016. Norwegian fjord sediments reveal NAO related winter temperature and precipitation changes of the past 2800 years. Earth and Planetary Science Letters, 435, 84–93.

[22]Camuffo, D., Bertolin, C., Barriendos, M., Dominguez-Castro, F., Cocheo, C., et al., 2010. 500-year temperature reconstruction in the Mediterranean Basin using documentary data and instrumental observations. Climatic Change, 101, 169–199.

[23]Luterbacher, J., Werner, J.P., Smerdon, J.E., Fernández-Donado, L., González-Rouco, F.J., et al., 2016. European summer temperatures since Roman times. Environmental Research Letters, 11, 024001. Matskovsky, V.V., Helama, S., 2014. Testing long-term summer temperature reconstruction based on maximum density chronologies obtained by reanalysis of tree-ring data sets from northernmost Sweden and Finland. Climate of the Past, 10(4), 1473–1487.

[24]Steinhilber, F., Abreu, J.A., Beer, J., Brunner, I., Christl, M., et al., 2012. 9,400 years of cosmic radiation and solar activity from ice cores and tree rings. Proceedings of the National Academy of Sciences of the United States of America, 109(16), 5967–5971.

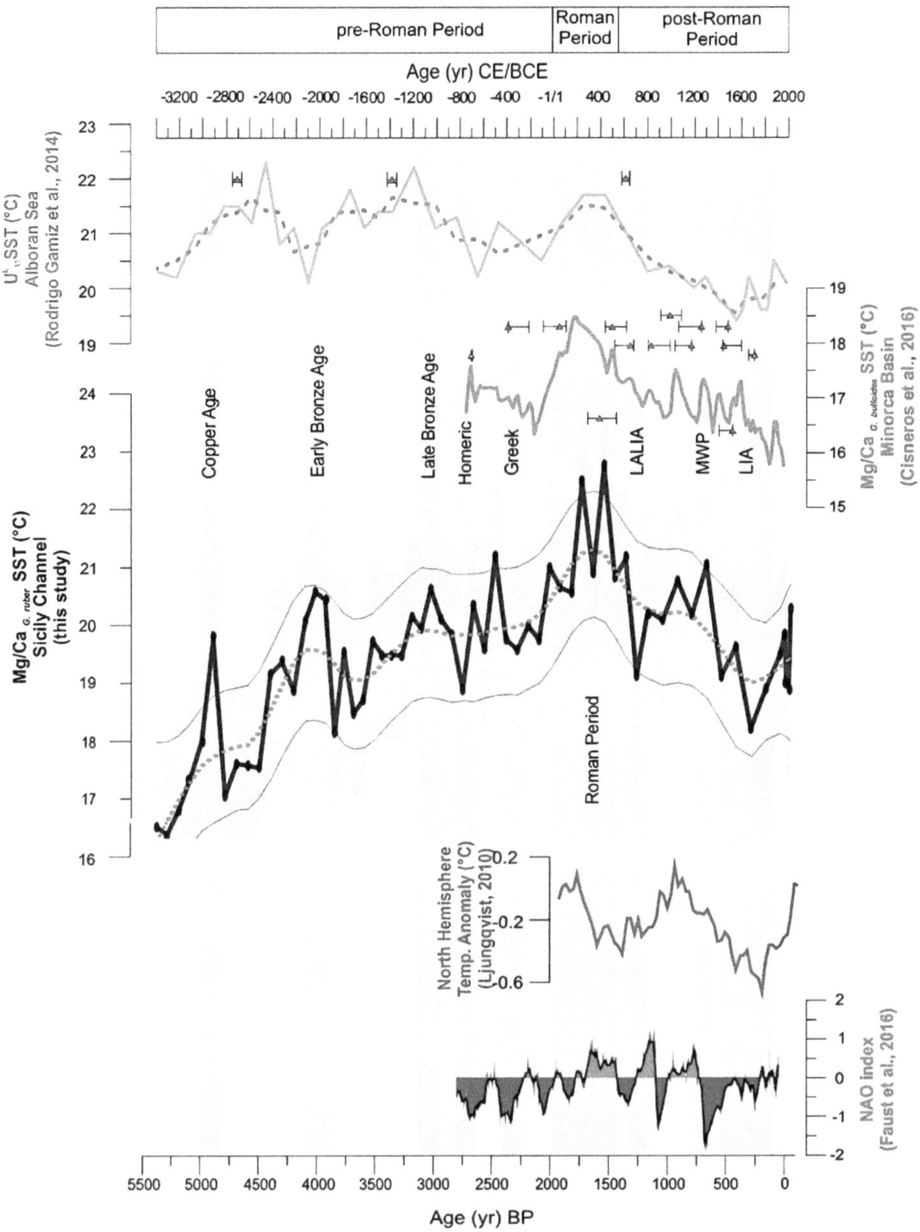

FIGURE 13.4. Reconstruction of Mediterranean Sea surface temperatures. Comparison of SST records from the Sicily Channel (Mg/Ca) (Margaritelli et al., 2020), Alboran Sea (Rodrigo-Gámiz et al., 2014), Minorca Basin (Cisneros et al., 2016) with Northern Hemisphere temperature reconstruction (Ljungqvist, 2010) and NAO index (Faust et al., 2016). (Adapted from Margaritelli et al., 2020).

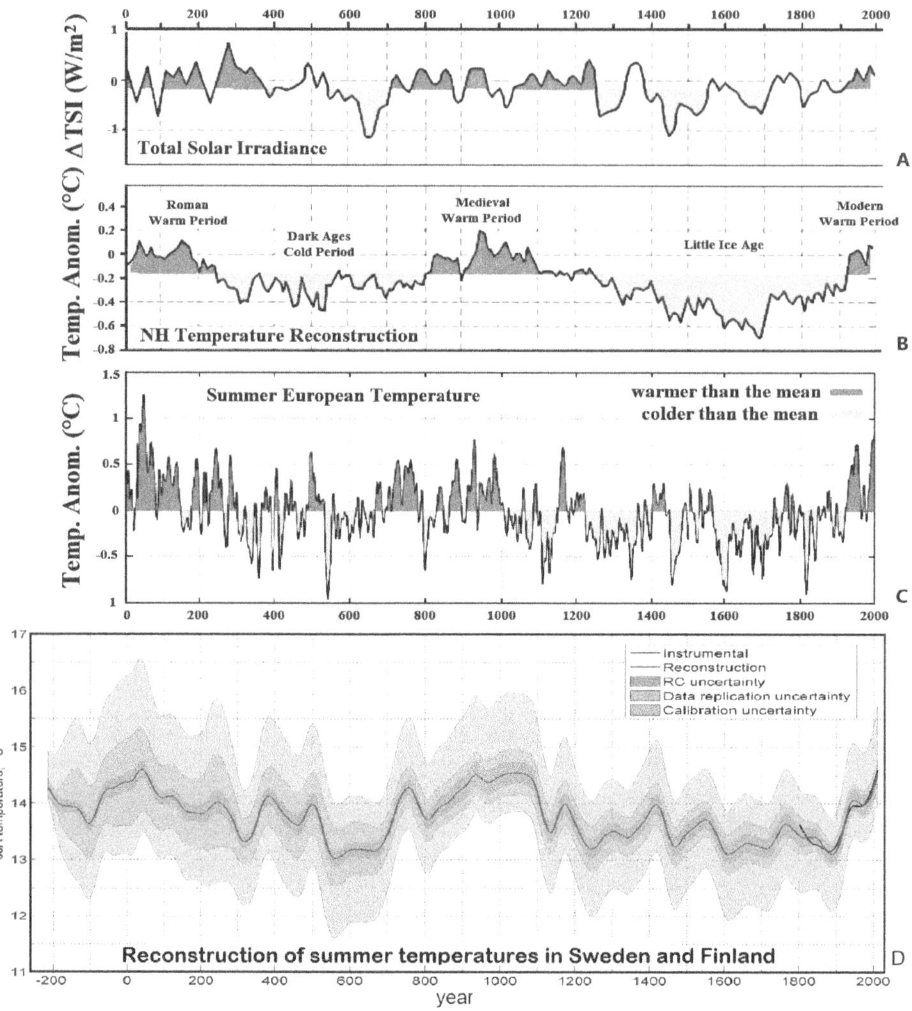

FIGURE 13.5. Quasi-millennial synchronous cycles observed in solar activity and climate proxies: [A] a reconstruction of total solar irradiance (Steinhilber et al., 2012); [B] a reconstruction of Northern Hemisphere surface temperatures (Ljungqvist, 2010); [C] a reconstruction of European summer temperatures (Luterbacher, 2016); [D] summer (JJA) temperature reconstruction in Sweden and Finland (Adapted from Matskovsky & Helama, 2014).

2000)[25] and proxies from Slovakia (Melo et al., 2022),[26] in which both the RWP and MWP are again clearly evident.

[25]Alley, R.B., 2000. The Younger Dryas cold interval as viewed from central Greenland. Quaternary Science Reviews, 19, 213–226.

[26]Melo, M., Lapin, M., Pecho, J., 2022. Climate in the Past and Present in the Slovak Landscapes—The Central European Context. In: Lehotský, M., Boltižiar, M. (eds.) Landscapes and Landforms of

This recurring pattern cannot be explained solely by changes in atmospheric greenhouse gas concentrations. Some studies link this quasi-millennial cycle to modulations of solar activity and to astronomical cycles (e.g.: Scafetta, 2012b; Scafetta & Bianchini, 2023).[27] These mechanisms are explored further in Figures 23.1, 23.2, 36.2, and 36.3 in Part 6 of this book. The next peak of this millennial cycle is expected in the second half of the 21st century, suggesting that the warming observed since the 17th century may be at least partially solar-driven.

The existence of a millennial cycle driven by solar activity challenges a core claim of the AGWT: namely, that the warming from 1850–1900 to 2011–2020 is entirely (~100%) attributable to human activity. The historical recurrence of natural thermal phases on secular and millennial timescales instead calls for a revision of the GCMs.

13.4. Model–data mismatch in past climate reconstructions

The CMIP6 global climate models (GCMs), when forced solely by natural influences such as solar and volcanic activity, simulate near-total climate stability between 800 AD and the modern era (Figure 6.2). This result stands in stark contrast to paleoclimatic and historical evidence. All major multiproxy reconstructions (e.g.: IPCC, 2013; PAGES-2K Consortium, 2019; Esper et al., 2024a) confirm a cooling trend between the Medieval Warm Period (MWP) and the Little Ice Age (LIA), although the magnitude of this cooling varies across datasets.

The inability of current GCMs to reproduce past natural warm periods, such as the Roman and Medieval epochs, raises serious concerns about their reliability in diagnosing and attributing the causes of modern warming. If the models cannot accurately reconstruct preindustrial climate variability, it is scientifically reasonable to hypothesize that part of the post-1850 warming may be attributable to natural mechanisms, including solar variability, oceanic oscillations, or internal climate feedbacks not adequately captured by current models.

A robust attribution of recent global warming must therefore be grounded in a comprehensive understanding of the natural cycles that shaped Earth's climate prior to the industrial era. The systematic underestimation of natural variability—particularly in the case of the RWP and the MWP—is not merely a technical shortcoming, but a deeper epistemological vulnerability within the theoretical framework that underpins contemporary climate attribution.

Slovakia, Springer.

[27] Scafetta, N., Bianchini, A., 2023. Overview of the spectral coherence between planetary resonances and solar and climate oscillations. Climate, 11(4), 77.

Pass	Modern Region / Border	Roman-Era Use	Current Physical Condition	Archaeological / Historical Evidence
Schnidejoch Pass	Bernese Oberland (Switzerland)	Used in Roman and prehistoric times as a high-altitude crossing between the Bernese Oberland and the Rhône Valley	Seasonal glacier, rapidly retreating; pass typically snowbound most of the year	Numerous finds from Neolithic, Bronze Age, and Roman periods: arrows, leather, textiles, containers
Theodul Pass	Aosta Valley (Italy) – Zermatt (Switzerland)	Probable Roman transalpine route connecting Gaul and Raetia; minor compared to Great St Bernard	Partially covered by Theodul Glacier; only seasonally accessible	Roman coins, leather shoes, animal remains, and artifacts exposed by glacial melt
Monte Moro Pass	Piedmont (Italy) – Valais (Switzerland)	Probable Roman-era route linking the Anzasca Valley to the Rhône Valley; not a major consular road	Glacier and permanent snowfields; difficult and dangerous transit	Limited but plausible finds of metal and organic material; historical mention of early crossings
Ambin Pass	Susa Valley (Italy) – Maurienne (France)	Possible secondary Roman route toward Gaul, complementing the Montgenèvre route	Perennial snow and seasonal ice; remote terrain	Sparse archaeological traces; some historical toponyms suggesting Roman activity
Theodulejoch (col adjacent to Theodul)	Near Theodul Pass, Zermatt sector	Possibly part of the same ancient transalpine network	Covered by glacier year-round	Occasional finds of human remains and artifacts of Roman or post-Roman date

FIGURE 13.6. **Roman Alpine transit routes vs modern glacial barriers.** In 218 BC, Hannibal crossed the Alps with his army and war elephants, likely via the *Col du Clapier* (2,500 m) or the *Col de la Traversette* (2,950 m) (Mahaney et al., 2017); today those high-altitude routes are partially obstructed by glaciers. Several high-altitude transalpine routes, once viable for Roman-era mobility, have become increasingly difficult to traverse owing to modern glacial expansion or residual ice cover. Archaeological and historical evidence confirms substantial human activity across these passes, particularly at sites such as Schnidejoch and Theodul Pass, where recent glacial retreat has revealed exceptionally well-preserved organic and metallic artifacts, including textiles, leather goods, and tools (Hafner & Schwörer, 2018; Harriss, 1971). These discoveries not only confirm the historical plausibility of Roman alpine crossings but also provide tangible data for reconstructing paleoclimatic conditions. The alignment between archaeological finds and historical accounts of Roman transalpine logistics (Lamb, 1995; Migliario, 2022) highlights the critical role of interdisciplinary approaches that combine glaciology, archaeology, and climatology in elucidating the dynamic interplay between climate variability and ancient mobility networks.

CHAPTER 14

The temperature–CO_2 conundrums of the Holocene and Phanerozoic

The Holocene, spanning the last 11,700 years since the end of the last glaciation, is a pivotal epoch for understanding natural climate evolution and its relationship with the rise of human civilization; see Figure 15.2. The analysis of this period reveals one of the most persistent inconsistencies between global climate model (GCM) simulations and paleoclimatic evidence: the so-called *"Holocene Temperature Conundrum,"* as defined by Liu et al. (2014).[1] This paradox remains unresolved and poses a significant challenge to the reliability of climate models in reconstructing the past and projecting the future.

The transition to the Holocene was preceded by a sequence of abrupt climatic events. As Earth emerged from the Last Glacial Maximum, Greenland ice cores record a strong warming during the Bølling interstadial (approximately 14,700 years ago), followed by a brief cooling known as the Older Dryas (~14,000 years ago). A subsequent warm phase, the Allerød interstadial (~13,900–12,900 years ago), was then abruptly interrupted by the intense cooling of the Younger Dryas (~12,900–11,700 years ago). Finally, around 11,700 years ago, a rapid and dramatic warming marked the onset of the Holocene interglacial, when regional temperatures in Greenland rose by several degrees within only a few decades. Twenty-five similar Dansgaard–Oeschger (D-O) events occurred during the preceding glacial period (Capron et al., 2021).[2] These events reveal repeated patterns of sudden warming followed by gradual cooling, underscoring the inherent instability of the climate system. Compared with the modern rate of ~1 °C over the past century, these past events highlight that natural variability has at times produced changes of greater speed and magnitude, reminding us that today's warming must be understood within the broader context of Earth's dynamic climate history.

14.1. Model limitations and the strength of data

Throughout the Holocene, natural greenhouse gas concentrations (in particular CO_2 and CH_4) have gradually increased (cf. Monnin et al., 2004; Bereiter

[1] Liu, Z., Zhu, J., Rosenthal, Y., Timm, O.E., 2014. The Holocene temperature conundrum. Proceedings of the National Academy of Sciences, 111(34), E3501–E3505.

[2] Capron, E., Rasmussen, S.O., Popp, T.J., Erhardt, T., Fischer, H., et al., 2021. The anatomy of past abrupt warmings recorded in Greenland ice. Nature Communications, 12, 2106.

et al., 2015).[3] According to GCMs, this should have produced a steady warming trend across the epoch. Indeed, model simulations reflect this expectation, as shown by the rising curve in Figure 14.1.

Empirical data from paleoclimatic archives derived from ice cores, lake sediments, fossil pollen, and isotopic markers, however, tell a different story. They reveal a pronounced warm phase during the early and middle Holocene, known as the Holocene Thermal Maximum (HTM) or Holocene Climatic Optimum (HCO), followed by a gradual cooling that culminated in the Little Ice Age (LIA). This observational framework is captured in the global temperature reconstruction by Kaufman et al. (2020),[4] shown in Figure 14.1.

The Holocene is traditionally subdivided into five climatic epochs:

- *Preboreal*: rapid deglaciation and warming;
- *Boreal*: temperate climate with seasonal extremes;
- *Atlantic*: warm and humid, with maximum forest expansion;
- *Subboreal*: more variable, with a cooling trend;
- *Subatlantic*: cooler and unstable, encompassing the LIA and the present period.

These phases align with ~2300-year Bray–Hallstatt cycles (see Figure 40.2), suggesting a structured rhythm in Holocene climate evolution.

14.2. Regional evidence supporting the HTM

The HTM is not a localized anomaly, since it is supported by a wide range of regional studies:

- *Eastern Greenland*: summer temperatures during the early Holocene were at least 3 °C higher than today (Westhoff et al., 2022).[5]
- *Central Europe*: warm anomalies confirmed by Zander et al. (2024)[6] and Melo et al. (2022);[7]
- *Eastern Mediterranean*: pollen-based reconstructions show sustained warmth (Cruz-Silva et al., 2023);[8]

[3]Monnin, E., Steig, E.J., Siegenthaler, U., Kawamura, K., Schwander, J., et al., 2004. Evidence for substantial accumulation rate variability in Antarctica during the Holocene, through synchronization of CO_2 in the Taylor Dome, Dome C and DML ice cores. Earth and Planetary Science Letters, 224, 45–54.

Bereiter, B., Eggleston, S., Schmitt, J., Nehrbass-Ahles, C., Stocker, T.F., et al., 2015. Revision of the EPICA Dome-C CO_2 record from 800 to 600 kyr before present. Geophysical Research Letters, 42, 542–549.

[4]Kaufman, D., McKay, N., Routson, C., Erb, M., Dätwyler, C., et al., 2020. Holocene global mean surface temperature: a multi-method reconstruction. Scientific Data, 7, 201.

[5]Westhoff, J., Sinnl, G., Svensson, A., Freitag, J., Kjær, H.A., et al., 2022. Melt in the Greenland EastGRIP ice core reveals Holocene warm events. Climate of the Past, 18, 1011–1034.

[6]Zander, P.D., Böhl, D., Sirocko, F., Auderset, A., Haug, G.H., Martínez-García, A., 2024. Reconstruction of warm-season temperatures in central Europe during the past 60,000 years from lacustrine branched glycerol dialkyl glycerol tetraethers (brGDGTs). Climate of the Past, 20, 841–864.

[7]Melo, M., Lapin, M., Pecho, J., 2022. Climate in the Past and Present in the Slovak Landscapes—The Central European Context. In: Lehotský, M., Boltižiar, M. (eds.) Landscapes and Landforms of Slovakia, Springer.

[8]Cruz-Silva, E., Harrison, S.P., Prentice, I.C., Marinova, E., Bartlein, P.J., et al., 2023. Pollen-based reconstructions of Holocene climate trends in the eastern Mediterranean region. Climate of the Past, 19, 2093–2108.

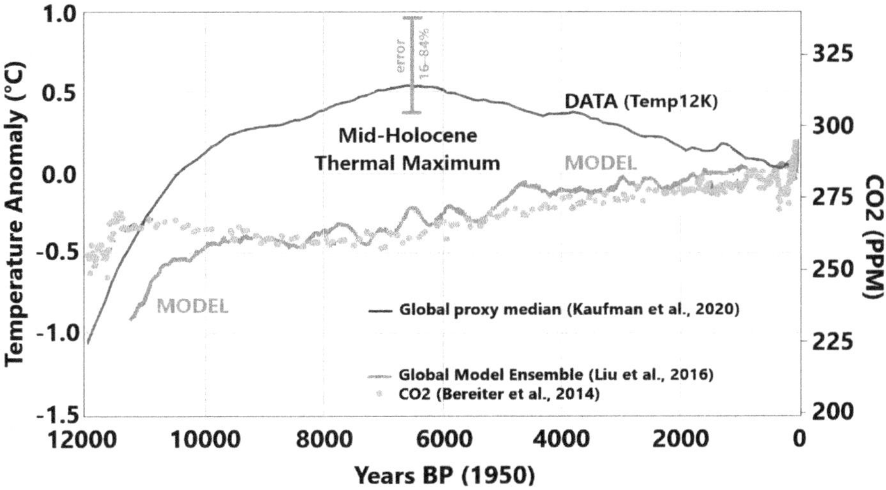

Figure 14.1. The Holocene Temperature Conundrum: (black line) global median temperature proxy from Kaufman et al. (2020); (warming line) ensemble of global climate models from Liu et al. (2014); (dots) EPICA Dome-C CO_2 record from Bereiter et al. (2015).

- *Japan*: isotope profiles reveal elevated Holocene temperatures (Murata et al., 2025);[9]
- *Antarctica*: recent reinterpretations suggest the presence of the HTM (Jones et al., 2023).[10]

Indirect indicators further reinforce the HTM narrative:

- *Mid-Holocene sea levels*: global mean sea level was likely higher than today by a few meters (Creel et al., 2024; Zhang et al., 2025b).[11] Following the Last Glacial Maximum, sea levels rose rapidly, at rates of up to 10 mm/year, until ∼6000–7000 years ago, after which they stabilized with minor fluctuations linked to Holocene climate oscillations. The behavior of sea level is shown in Figure 14.2, which is based on the reconstruction by Fairbridge (1961).[12] These changes, driven by eustatic, isostatic, and tectonic processes (cf. Murray-Wallace &

[9]Murata, A., Mori, T., Kato, H., Hu, H.-M., Shen, C.-C., et al., 2025. Holocene Temperature Trend Inferred From Oxygen and Carbonate Clumped Isotope Profiles. Island Arc, 34, e70002.

[10]Jones, T.R., Cuffey, K.M., Roberts, W.H.G., Markle, B.R., Steig, E.J., et al., 2023. Seasonal temperatures in West Antarctica during the Holocene. Nature, 613, 292–297.

[11]Creel, R.C., Austermann, J., Kopp, R.E., Khan, N.S., Albrecht, T., Kingslake, J., 2024. Global mean sea level likely higher than present during the Holocene. Nature Communications, 15, 10731.
Zhang, L., Lü, T., Xue, L., Mo, W., Wang, C., Zhao, X., Hu, D., 2025. An overview of the Holocene high sea level around the South China sea: age, height, and mechanisms. Atmosphere, 16(8), 993.

[12]Fairbridge, R.W., 1961. Eustatic changes in sea level. Physics and Chemistry of the Earth, 4, 99–185.

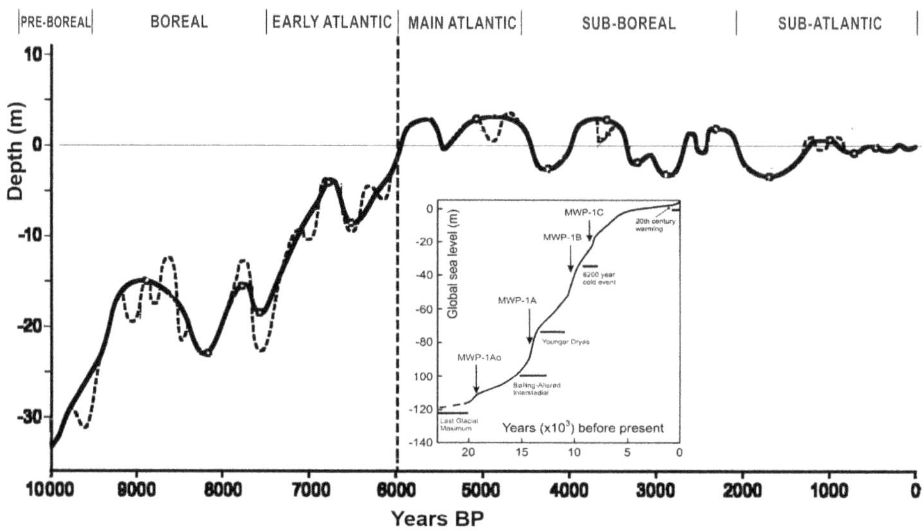

FIGURE 14.2. Sea level variations from the Last Glacial Maximum (about 20,000 years ago) to the present (adapted from Fairbridge, 1961). The insert shows the sea level rise during the last deglaciation with its Meltwater Pulses.[14]

Woodroffe, 2014),[13] reshaped coastlines and influenced human migration and biodiversity. In particular, Mörner et al. (2004)[14] show that the Maldives present morphological evidence of sea level fluctuations of ±1 m over the past 5000 years and emphasize the resilience of the islands during Late Quaternary time. Before the Holocene, sea level rose intermittently and usually very rapidly (insert of Figure 14.2).[15]

- *African Humid Period* (14,800–5500 years ago): the Sahara was transformed into a green landscape rich in lakes, driven by strong tropical

[13]Murray-Wallace, C.V., Woodroffe, C.D., 2014. Quaternary Sea-Level Changes: A Global Perspective. Cambridge University Press.

[14]Mörner, N.-A., Tooley, M., Possnert, G., 2004. New perspectives for the future of the Maldives. Global and Planetary Change, 40(1–2), 177–182.

[15]During the last deglaciation before the Holocene, sea level rose by about 120–130 meters. However, the Earth experienced several abrupt episodes of sea level rise known as Meltwater Pulses (MWPs), which reflect intense phases of ice sheet collapse and freshwater input into the oceans. As shown in the insert of Figure 14.2, the most prominent, MWP-1A, occurred around 14,600 years ago, raising global sea level by approximately 12–22 meters over a few centuries, with peak rates exceeding 40 mm/year. This event is attributed to rapid melting of the Laurentide, Fennoscandian, and possibly West Antarctic ice sheets. MWP-1B, dated to around 11,300 years ago, was smaller and more debated, with estimates ranging from 4 to 7 meters of sea level rise, possibly linked to final collapses of Northern Hemisphere ice masses. MWP-1C, occurring near 8,200 years ago, coincides with the so-called 8.2 ka event, which was a cooling episode triggered by freshwater discharge from proglacial lakes in North America. Though its sea level contribution was modest (likely 1–2 meters), it had significant climatic repercussions. Together, these pulses illustrate the non-linear nature of postglacial sea level rise and provide paleoclimatic benchmarks for evaluating modern trends.

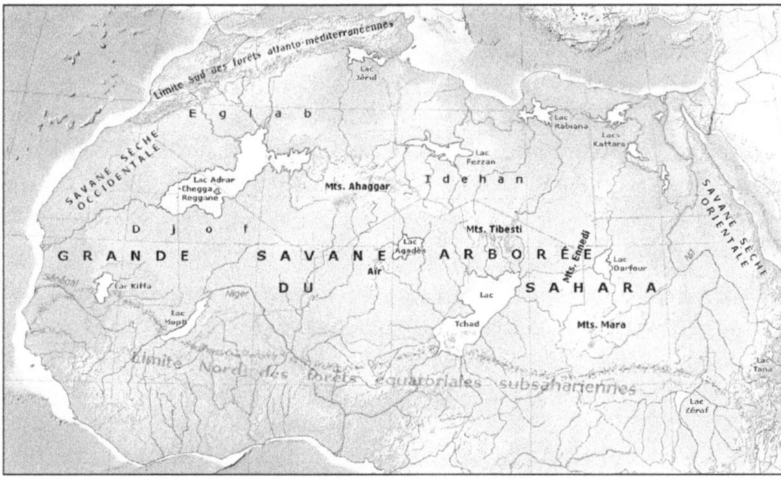

FIGURE 14.3. Map of North Africa about 8000 years ago, at the peak of the African Humid Period during the Holocene Optimum.

monsoons and elevated global temperatures (de Menocal & Tierney, 2012; Kaufman et al., 2020).[16] Figure 14.3 illustrates this dramatic shift.
- *Tree trunks under glaciers*: numerous discoveries of subfossil tree trunks in proximity to retreating glaciers (Figure 40.3) provide compelling evidence of episodic warmth and ecological shifts throughout the Holocene. In particular, Hormes et al. (2001)[17] documented eight distinct Holocene phases of reduced glacier extent in the Central Swiss Alps.

These data are difficult to reconcile with current GCM simulations, which predict climatic stasis and gradual warming throughout the Holocene. The mismatch between model outputs and paleoclimate data underscores a fundamental limitation in our ability to simulate long-term natural variability.

14.3. Orbits, seasons, and proxies: interpretive keys

A compelling interpretive framework for the Holocene thermal profile lies in the evolution of mean daily solar insolation at high latitudes (e.g., 65°N), governed by Earth's orbital cycles as described by Laskar et al. (2011).[18] These variations, rooted in Milankovitch theory, drive the alternation between glacial and interglacial periods and likely influenced Holocene climate dynamics as well.

---

[16] de Menocal, P.B., Tierney, J.E., 2012. Green Sahara: African humid periods paced by Earth's orbital changes. Nature Education Knowledge, 3(10), 12.
Kaufman, D., McKay, N., Routson, C., Erb, M., Dätwyler, C., et al., 2020. Holocene global mean surface temperature: a multi-method reconstruction. Scientific Data, 7, 201.

[17] Hormes, A., Müller, B. U., Schlüchter, C., 2001. The Alps with little ice: Evidence for eight Holocene phases of reduced glacier extent in the Central Swiss Alps. The Holocene, 11(3), 255–265.

[18] Laskar, J., Fienga, A., Gastineau, M., Manche, H., 2011. La2010: a new orbital solution for the long-term motion of the Earth. Astronomy and Astrophysics, 532, A89.

Some studies suggest that paleoclimate proxies predominantly reflect summer temperatures rather than annual means. Bova et al. (2021)[19] hypothesized that this seasonal bias may explain part of the discrepancy between proxy-based reconstructions and model simulations. This interpretation, however, has been challenged (Cartapanis et al., 2022; Dong et al., 2022; Kaufman & Broadman, 2023):[20] the presence of the Holocene Thermal Maximum (HTM) is also confirmed in annual temperature reconstructions, which once again confirms the divergence between observations and model predictions.

Implications for climate modeling. The persistent mismatch between GCM simulations and paleoclimate reconstructions throughout the Holocene suggests that models tend to overestimate the influence of greenhouse gases while underrepresenting other natural drivers, including solar variability and ocean–atmosphere oscillations. This limitation is particularly significant given that these models underpin modern climate attribution and future projections.

The Holocene conundrum fits within the broader context discussed in Chapter 13, where CMIP5 and CMIP6 models fail to reproduce the quasi-millennial oscillations documented by Alley (2000),[21] Bond et al. (2001),[22] and Kutschera et al. (2020).[23] Nor do they accurately simulate the transition from the Medieval Warm Period to the Little Ice Age, despite robust evidence from paleoclimate archives and historical records.

14.4. The interglacial period paradox over the last 500,000 years

A long-term comparison of global temperature and atmospheric CO_2 over the past 500,000 years (Figure 14.4) reveals a consistent pattern: during the interglacial periods, temperatures rise first, followed by a delayed increase in CO_2 concentrations, typically lagging by several centuries (Koutsoyiannis, 2024).[24] This behavior is consistent with Henry's law, which states that warmer oceans release more CO_2 into the atmosphere. These large climate oscillations are correlated with changes in Earth's orbital eccentricity (Figure 32.3).

Notably, all past interglacials recorded temperatures significantly higher than today's, despite CO_2 levels remaining below 300 ppm—much less than

[19]Bova, S., Rosenthal, Y., Liu, Z., Yan, M., 2021. Seasonal origin of the thermal maxima at the Holocene and the last interglacial. Nature, 589, 548–553.

[20]Cartapanis, O., Jonkers, L., Moffa-Sanchez, P., Jaccard, S.L., de Vernal, A., 2022. Complex spatio-temporal structure of the Holocene Thermal Maximum. Nature Communications, 13, 5662.
Dong, Y., Wu, N., Li, F., Zhang, D., Zhang, Y., et al., 2022. The Holocene temperature conundrum answered by mollusk records from East Asia. Nature Communications, 13, 5153.
Kaufman, D.S., Broadman, E., 2023. Revisiting the Holocene global temperature conundrum. Nature, 614, 425–435.

[21]Alley, R.B., 2000. The Younger Dryas cold interval as viewed from central Greenland. Quaternary Science Reviews, 19, 213–226.

[22]Bond, G., Kromer, B., Beer, J., Muscheler, R., Evans, M.N., et al., 2001. Persistent solar influence on North Atlantic climate during the Holocene. Science, 294(5549), 2130–2136.

[23]Kutschera, W., Patzelt, G., Schaefer, J.M., Schlüchter, C., Steier, P., Wild, E.M., 2020. The movements of Alpine glaciers throughout the last 10,000 years as sensitive proxies of temperature and climate changes. EPJ Web of Conferences, 232, 02002.

[24]Koutsoyiannis, D., 2024. Stochastic assessment of temperature–CO_2 causal relationship in climate from the Phanerozoic through modern times. Mathematical Biosciences and Engineering, 21(7), 6560–6602.

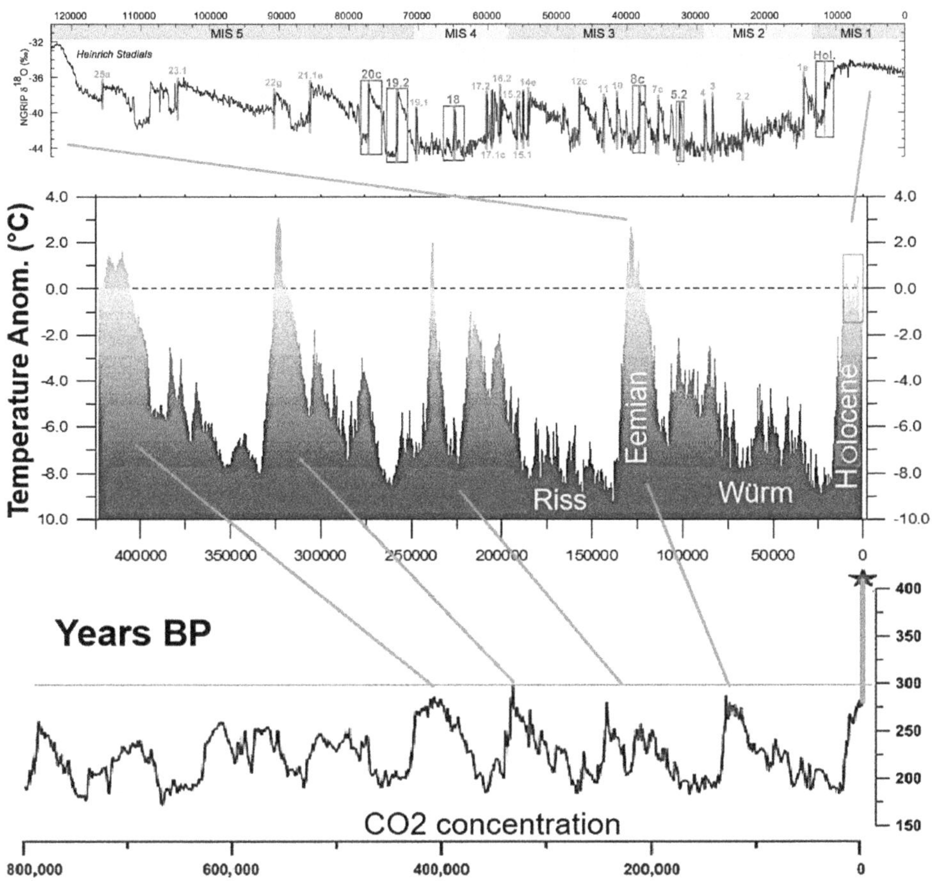

FIGURE 14.4. Top: Reconstruction of global temperature over the last 420,000 years, based on ice core data from Vostok, Antarctica with the 25 Dansgaard–Oeschger (D-O) events occurred during the last glacial period (Petit, 2001; Capron et al., 2021). The graph covers four glaciations and five interglacial phases, including the current one. The horizontal line represents modern temperature.
Bottom: Atmospheric CO_2 concentration; the thick vertical line on the right highlights the CO_2 increase observed since 1900.

current levels (∼425 ppm). This discrepancy suggests that other natural factors, such as orbital variations, solar activity, and ocean circulation, played a dominant role in past climate warming periods.

The Eemian period (Marine Isotope Stage 5e), occurring ∼130,000 to 115,000 years ago, exemplifies this paradox. Global temperatures during the Eemian were likely ∼3 °C higher than today, especially at high latitudes. Ice sheet retreat in Greenland and West Antarctica caused sea levels to rise by 6 to 9, or even 11, meters above present levels (Antonioli et al., 2018).[25] See Figure 14.5.

[25]Antonioli, F., Ferranti, L., Stocchi, P., Deiana, G., Lo Presti, V., et al., 2018. Morphometry and

Evidence from ice cores, marine sediments, and fossil pollen reveals dramatic ecological shifts:

- forests extended into the Arctic, with oaks and hazelnuts in regions now classified as tundra;
- European fauna included elephants, hippos, and lions even at high latitudes, as in areas such as London;
- the climate was more humid and stable, although regionally variable.

Despite these conditions, atmospheric CO_2 remained at around 280 ppm, similar to preindustrial levels (Petit, 2001).[26] The high temperatures of the Eemian thus point to other natural forcings as primary drivers of interglacial warmth. Yet climate models struggle to reproduce these warm interglacial periods. Capron et al. (2017)[27] concludes that GCMs fail to simulate the early evolution of the Eemian climate when forced solely with orbital and greenhouse gas parameters, which suggests the existence of additional processes that are still poorly understood.

14.5. Temperature leads CO_2 throughout the Phanerozoic

The climatic evolution of the Phanerozoic Eon reveals a system governed by multi-scale interactions among tectonic, astronomical, and biogeochemical processes. Understanding the causal relationship between global temperature and atmospheric CO_2 across these timescales remains a central challenge in paleoclimatology. This appendix develops a monographic synthesis that integrates causality analyses, long-term temperature reconstructions, and modern instrumental evidence from CO_2–temperature comparisons (Mazzarella et al., 2013; Humlum et al., 2013; Davis, 2017; Koutsoyiannis, 2024).[28] By examining evidence from the full Phanerozoic, the Cenozoic, the Quaternary, and the instrumental era, it argues for a scale-dependent causal structure in which temperature variations typically precede CO_2 changes on geological and millennial timescales. The modern anthropogenic era represents a more complex situation, because CO_2 changes produced by natural feedbacks are superimposed on those arising from anthropogenic emissions.

---

elevation of the last interglacial tidal notches in tectonically stable coasts of the Mediterranean Sea. Earth-Science Reviews, 185, 600–623.

[26]Petit, J.R., et al., 2001. Vostok Ice Core Data for 420,000 Years. IGBP PAGES/World Data Center for Paleoclimatology Data Contribution Series #2001-076. NOAA/NGDC Paleoclimatology Program, Boulder CO, USA.

[27]Capron, E., Govin, A., Stone, E.J., Masson-Delmotte, V., Mulitza, S., Otto-Bliesner, B., Rasmussen, T.L., 2017. Temporal and spatial structure of multi-millennial temperature changes at high latitudes during the Last Interglacial. Quaternary Science Reviews, 103, 116–133.

[28]Mazzarella, A., Giuliacci, A., Scafetta, N., 2013. Quantifying the Multivariate ENSO Index (MEI) coupling to CO_2 concentration and to the length of day variations. Theor Appl Climatol 111, 601–607.
Humlum, O., Stordahl, K., Solheim, J.-E., 2013. The phase relation between atmospheric carbon dioxide and global temperature. Global and Planetary Change, 100, 51–69.
Davis, W.J., 2017. The Relationship between Atmospheric Carbon Dioxide Concentration and Global Temperature for the Last 425 Million Years. Climate, 5, 76.
Koutsoyiannis, D., 2024. Stochastic assessment of temperature–CO_2 causal relationship in climate from the Phanerozoic through modern times. Mathematical Biosciences and Engineering, 21, 6560–6602.

FIGURE 14.5. Wave-cut notch about 7–8 meters above present sea level, corresponding to sea level during the Eemian interglacial (about 125,000 years ago). The photo shows a section of the Monte Orlando cliff visible from Serapo beach in Gaeta, Italy. (See also other locations shown in Antonioli et al., 2018).

The following sections examine Phanerozoic climate variability and the associated causality problem across temporal scales ranging from hundreds of millions of years to intra-annual fluctuations.

14.6. Phanerozoic climate variability and the causality problem

The climatic history of the Phanerozoic Eon, spanning approximately the last 540 million years as shown in Figures 14.6 and 14.7, is characterized by alternating greenhouse and icehouse states, punctuated by abrupt transitions and long-term secular trends. The figures synthesize multiple paleotemperature proxies, including oxygen-isotope measurements from marine carbonates, sedimentological indicators of glaciation, and paleogeographic reconstructions. Despite the uncertainties inherent in deep-time proxies, the figure robustly captures first-order climatic transitions such as the warm early Paleozoic, the Late Paleozoic Ice Age, the Mesozoic greenhouse, and the progressive cooling of the Cenozoic culminating in the Quaternary glacial cycles (Veizer et al., 2000; Royer et al., 2004; Scotese et al., 2021).[29]

[29] Veizer, J., Godderis, Y., Francois, L.M., 2000. Evidence for decoupling of atmospheric CO_2 and global climate during the Phanerozoic eon. Nature, 408, 698–701.
Royer, D.L., Berner, R.A., Montañez, I.P., Tabor, N.J., Beerling, D.J., 2004. CO_2 as a primary driver of Phanerozoic climate. Nature, 446, 530–532.
Scotese, C.R., Song, H., Mills, B.J.W., van der Meer, D.G., 2021. Phanerozoic paleotemperatures: The earth's changing climate. Earth-Science Reviews, 215, 103503.

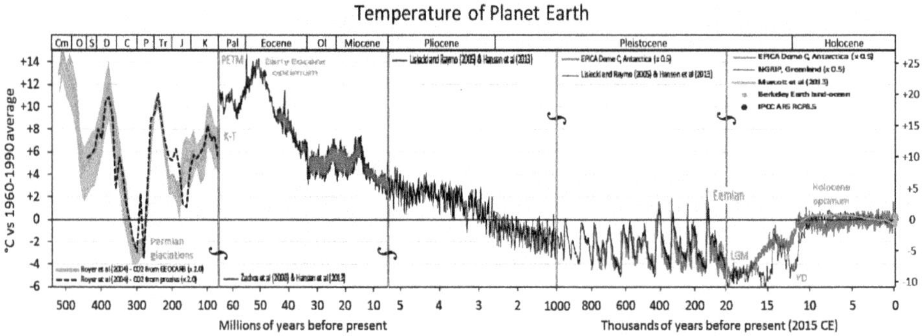

FIGURE 14.6. Composite of estimates of Earth's global average surface air temperature over the ~540 My of the Phanerozoic Eon, since the first major proliferation of complex life forms on our planet. Adapted from https://commons.wikimedia.org/wiki/File:All_palaeotemps.png.

The observed large-scale climatic variations raise a fundamental question: what governs climate change and, in particular, the direction of causality between temperature and atmospheric CO_2 on multimillion-year timescales? Traditional interpretations often assume that CO_2 acts as the primary driver of long-term climate change. The geological record, however, frequently reveals temporal offsets between temperature and CO_2, suggesting a more complex causal structure.

This issue can be reframed by analyzing the causality framework applied to multiple paleoclimate datasets. Rather than presupposing a dominant forcing, the analysis evaluates whether changes in temperature or in CO_2 systematically precede the other. This approach is particularly suited to the Phanerozoic, where tectonic, orbital, and astronomical forcings interact with the carbon cycle in nontrivial ways.

14.7. Long-period climate cycles and external forcing

A prominent feature of the palaeotemperature synthesis is the presence of quasi-periodic climate oscillations on timescales of approximately 150 million years, as evidenced in Figure 14.7. These oscillations have been interpreted within two principal frameworks.

The first relates them to the Wilson cycle of supercontinent assembly and breakup, which modulates global climate through changes in volcanic outgassing, continental configuration, ocean circulation, and silicate weathering rates (Berner, 1994; Berner & Kothavala, 2001; Nance et al., 2014).[30] These tectonic processes can initiate major climatic transitions independently of atmospheric CO_2, or in advance of the CO_2 response.

[30]Berner, R.A., 1994. GEOCARB II: A revised model of atmospheric CO_2 over Phanerozoic time. American Journal of Science, 294, 56–91.
Berner, R.A., Kothavala, Z., 2001. GEOCARB III: A revised model of atmospheric CO_2 over Phanerozoic time. American Journal of Science, 301, 182–204.
Nance, R.D., Murphy, J.B., Santosh, M., Pisarevsky, S.A., Stern, R.J., 2014. The supercontinent cycle. Geoscience Frontiers, 5, 395–411.

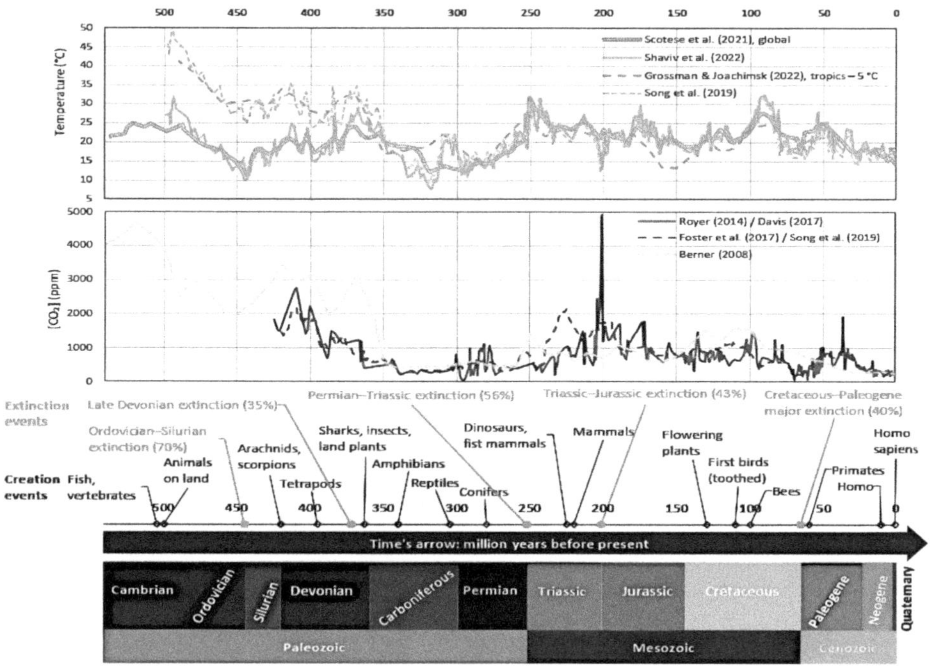

FIGURE 14.7. Recent reconstructions of atmospheric temperature and CO_2 concentrations over the 541-million-year Phanerozoic Eon. Temperature curves are derived from datasets by Scotese et al. (2021), Shaviv & Veizer (2003), Grossman & Joachimski (2022) (adjusted by –5 °C for consistency), and Song et al. (2019). CO_2 reconstructions come from Royer et al. (2004), Foster et al. (2012), Song et al. (2019), and Berner (1994). The chart also includes the standard subdivision of the Phanerozoic into three eras and twelve periods, together with major biological origination and extinction events, including genus-level extinction percentages. Adapted from Koutsoyiannis (2024).

A second interpretive framework explores the possible role of astronomical forcing on galactic scales, as illustrated in Figure 26.3. Shaviv (2003)[31] and Shaviv & Veizer (2003)[32] proposed that the passage of the Solar System through the Milky Way's spiral arms could modulate the flux of galactic cosmic rays (GCRs), with potential indirect effects on Earth's climate through changes in cloud microphysics and planetary albedo.

Some correlations between inferred spiral-arm crossings and broad Phanerozoic climate states have been reported, suggesting a possible coincidence between enhanced GCR exposure and cooler climatic intervals. The physical mechanisms linking cosmic rays to cloud formation and climate, however, remain uncertain, and the magnitude of any resulting climatic effect is still debated

[31] Shaviv, N.J., 2003. The spiral structure of the Milky Way, cosmic rays, and ice age epochs on Earth. New Astronomy, 8, 39–77.
[32] Shaviv, N. J., Veizer, J., 2003. Celestial driver of Phanerozoic climate? GSA Today, 13, 4–10.

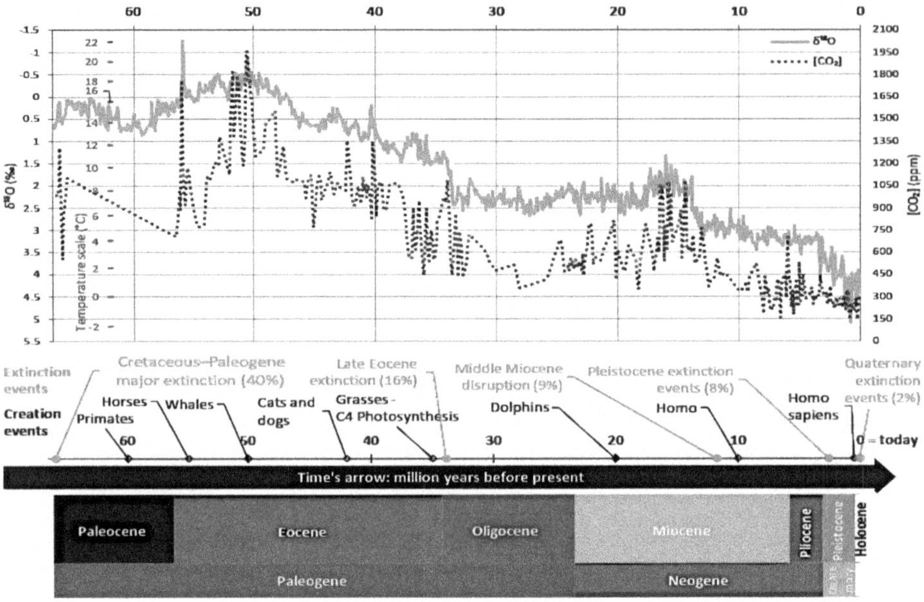

FIGURE 14.8. Recent reconstructions of the $\delta^{18}O$ temperature proxy and atmospheric CO_2 concentrations over the 66-million-year Cenozoic. Temperature data come from Westerhold et al. (2020), while CO_2 estimates combine alkenone- and boron-isotope–based records from Rae et al. (2021). The chart also includes the standard subdivision of the Cenozoic into its periods and epochs, together with major biological origination and extinction events, including genus-level extinction percentages. Adapted from Koutsoyiannis (2024).

within the scientific community. Nevertheless, astronomical forcing provides a plausible explanation for temperature variations that precede CO_2 changes (Svensmark, 2007).[33]

14.8. Phanerozoic evidence

Figure 14.7 juxtaposes multiple independent reconstructions of temperature and CO_2 across the Phanerozoic. Temperature proxies include isotopic and sedimentological datasets (Grossman & Joachimski, 2022; Song et al., 2019; Scotese et al., 2021).[34] CO_2 reconstructions are derived from GEOCARB-type models and proxy-based estimates (Berner, 1994; Berner & Kothavala, 2001; Royer et al., 2004).

---

[33] Svensmark, H., 2007. Cosmoclimatology: a new theory emerges. Astronomy & Geophysics, 48(1), 1.18–1.24.

[34] Grossman, E.L., Joachimski, M.M., 2022. Oxygen isotope stratigraphy of the Phanerozoic. Annual Review of Earth and Planetary Sciences, 50, 1–35.

Song, H., Scotese, C.R., Mills, B.J.W., van der Meer, D.G., 2019. Phanerozoic climate change reflected in global marine oxygen-isotope data. Gondwana Research, 67, 172–187.

Scotese, C.R., Song, H., Mills, B.J.W., van der Meer, D.G., 2021. Phanerozoic paleotemperatures: The earth's changing climate. Earth-Science Reviews, 215, 103503.

The comparison reveals broad coevolution but also notable discrepancies in timing and amplitude, including weak or negative correlations, in line with Davis (2017), whose findings that CO_2 changes did not drive ancient temperature variations are consistent with the stochastic causality analysis of Koutsoyiannis (2024), which identifies temperature as generally leading CO_2 on geological timescales. For example, the Late Ordovician glaciation around 445 Ma is clearly expressed in temperature proxies, yet CO_2 reconstructions often display delayed or muted responses. Similarly, during the Late Paleozoic Ice Age, the temperature decline precedes the major CO_2 drawdown. These patterns suggest that CO_2 was not the dominant driver of climate over most of the Phanerozoic, but acted primarily as a feedback to temperature changes initiated by tectonic or astronomical forcings.

14.9. Cenozoic refinement

The Cenozoic Era provides a more detailed and better constrained record for evaluating the coupling between temperature and atmospheric CO_2. High-resolution benthic $\delta^{18}O$ records document a long-term cooling trend from the early Paleogene greenhouse to the Neogene icehouse (Zachos et al., 2001, 2008),[35] while proxy-based CO_2 reconstructions indicate a broadly declining trend with substantial uncertainty and temporal variability (Foster et al., 2012).[36]

Figure 14.8 presents recent reconstructions of benthic $\delta^{18}O$ and atmospheric CO_2 over the past 66 million years (Westerhold et al., 2020; Rae et al., 2021).[37] The comparison highlights intervals in which temperature and CO_2 exhibit broadly synchronous long-term trends, as well as phases in which changes in one variable appear to precede or lag the other within the limits of proxy resolution. Detailed comparison, however, also reveals several transitions in which temperature changes appear to precede CO_2 variations.

The Eocene–Oligocene transition ($\sim$34 Ma) represents a key example of this complex coupling. Tectonically driven changes in ocean gateways and the isolation of Antarctica are widely regarded as essential preconditions for Antarctic glaciation, while declining atmospheric CO_2 likely acted as a reinforcing feedback that stabilized the emerging icehouse state (Cramer et al., 2009).[38] Given proxy uncertainties, the current evidence does not uniquely constrain the precise temporal ordering of temperature and CO_2 changes across this transition, but supports a view in which multiple forcings and feedbacks acted in concert.

[35]Zachos, J., Pagani, M., Sloan, L., Thomas, E., Billups, K., 2001. Trends, rhythms, and aberrations in global climate 65 Ma to present. Science, 292, 686–693.
Zachos, J.C., Dickens, G.R., Zeebe, R.E., 2008. An early Cenozoic perspective on greenhouse warming and carbon-cycle dynamics. Nature, 451, 279–283.

[36]Foster, G.L., Lear, C.H., Rae, J.W.B., 2012. The evolution of pCO_2 over the past 65 Myr. Earth and Planetary Science Letters, 341–344, 243–254.

[37]Westerhold, T., Marwan, N., Drury, A. J., Liebrand, D., Agnini, C., et al., 2020. An astronomically dated record of Earth's climate and its predictability over the last 66 million years. Science, 369(6509), 1383–1387.
Rae, J.W.B., Burke, A., Robinson, L.F., Adkins, J.F., Chen, T., et al., 2021. CO_2 over the past 66 million years from marine archives. Annual Review of Earth and Planetary Sciences, 49, 599–631.

[38]Cramer, B.S., Miller, K.G., Barrett, P.J., Wright, J.D., 2009. Late Cretaceous–Neogene trends in deep ocean temperature and continental ice volume. Science, 327, 1187–1191.

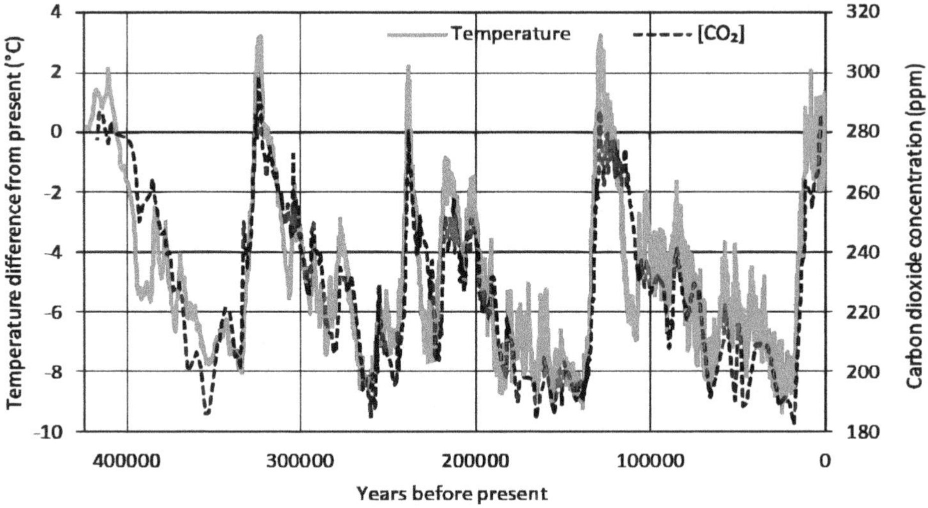

FIGURE 14.9. Time series of temperature and CO_2 concentration from the Vostok ice core. Temperature changes precede CO_2 changes. Adapted from Koutsoyiannis (2024).

14.10. Quaternary constraints

The Vostok ice-core record provides direct measurements of atmospheric CO_2 together with high-resolution temperature proxies across several glacial–interglacial cycles (Petit et al., 1999; Jouzel et al., 2007).[39] Figure 14.9 shows that temperature increases precede the rises in CO_2 by several centuries during glacial terminations. This behavior is consistent with orbital forcing theory, in which initial warming driven by insolation changes and ice-albedo feedbacks triggers the release of CO_2 from the oceans (Imbrie et al., 1992).[40] The stochastic causality analysis formalizes this interpretation, identifying a dominant *orbital forcing → temperature changes → CO_2 changes* causal direction.

14.11. Millennial variability

Koutsoyiannis (2024) extends the analysis to the past two millennia, comparing several temperature reconstructions with CO_2 data from ice cores and stomatal proxies. In the preindustrial period, temperature variability often precedes CO_2 changes, reflecting ocean–atmosphere carbon exchange responding to climate fluctuations. Only in the industrial era does CO_2 become a primary forcing as a result of fossil fuel combustion, marking a reversal of the long-standing geological pattern.

[39]Petit, J.R., Jouzel, J., Raynaud, D., Barkov, N.I., Barnola, J.-M., et al., 1999. Climate and atmospheric history of the past 420,000 years from the Vostok ice core. Nature, 399, 429–436.
Jouzel, J., Masson-Delmotte, V., Cattani, O., Dreyfus, G., Falourd, S., et al., 2007. Orbital and millennial Antarctic climate variability over the past 800,000 years. Science, 317, 793–796.
[40]Imbrie, J., Boyle, E.A., Clemens, S.C., Duffy, A.C., Howard, W.R., et al., 1992. On the structure and origin of major glaciation cycles. Paleoceanography, 7, 701–738.

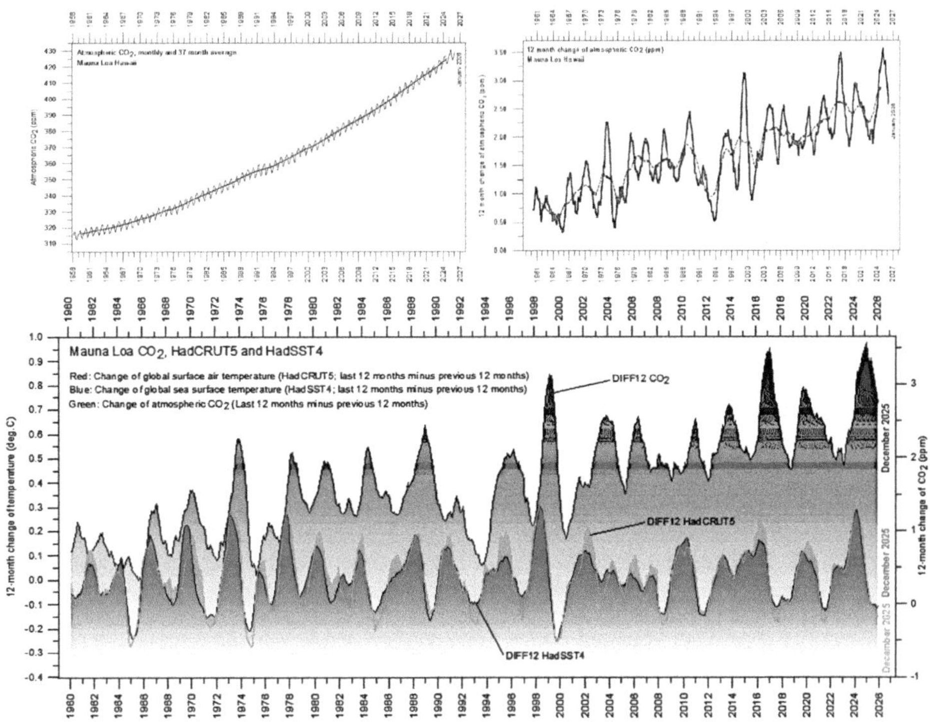

FIGURE 14.10. [A] Monthly amount of atmospheric CO_2 since March 1958, measured at the Mauna Loa Observatory, Hawaii. [B] Annual (12 month) growth rate (ppm) of atmospheric CO_2 since 1959, calculated as the average amount of atmospheric CO_2 during the last 12 months, minus the average for the preceding 12 months. [C] 12-month change of global atmospheric CO_2 concentration (Mauna Loa), global sea surface temperature (HadSST4) and global surface air temperature (HadCRUT5). Temperature changes precede CO_2 changes by a few months. (Humlum et al., 2013, https://www.climate4you.com/).

Beck (2022)[41] proposes an alternative reconstruction of atmospheric CO_2 concentrations from 1826 to 1960 based on approximately 100,000 historical chemical measurements. His record is controversial and is not accepted in mainstream climate science, since it contradicts the ice-core CO_2 record, which is very smooth and shows CO_2 levels below 310 ppm. In any case, Beck's record exhibits a close correlation with global sea surface temperature with a one-year lag, including a pronounced CO_2 peak of approximately 380 ppm from 1940 to 1945, interpreted as a response to the temperature maximum during the same period (Figure 12.3).

[41]Beck, E.-G., 2022. Reconstruction of atmospheric CO_2 background levels since 1826 from direct measurements near ground. Science of Climate Change, 2(2), 148–211.

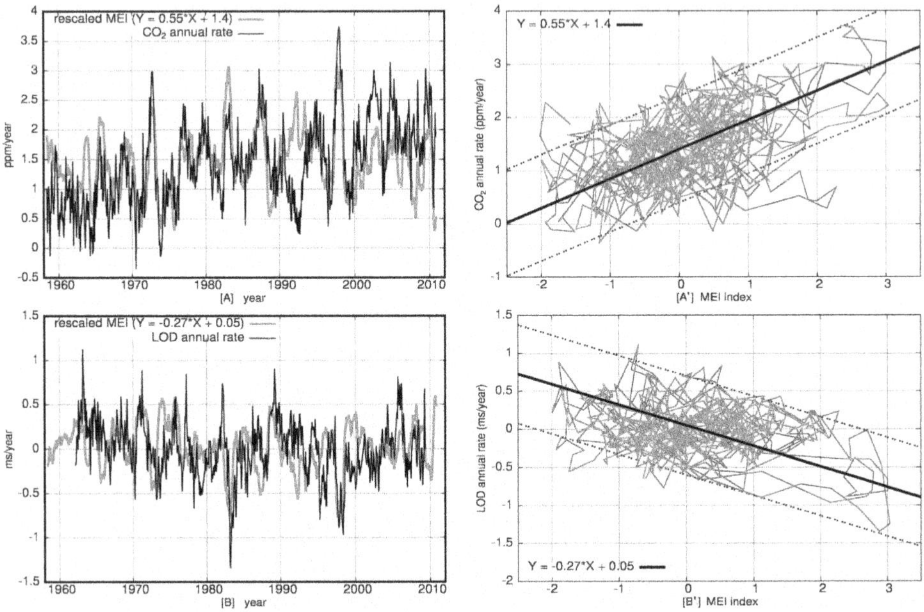

FIGURE 14.11. [A] Time plot of rescaled MEI index (red line) and of CO_2 annual rate variation. The CO_2 curve is shifted 3 months back for best correlation. [B] Time plot of MEI (red line) and of LOD annual rate, which is shifted 3 months back for best correlation. The right panel show the relative scatter-plots. Temperature changes precede CO_2 changes by a few months. Adapted from Mazzarella et al. (2013).

14.12. Modern observational constraints from CO_2–temperature data

Modern instrumental observations provide important constraints on the short-term interactions between atmospheric CO_2 and global temperature. Several studies have examined phase relationships and correlations on interannual to decadal timescales, highlighting the strong influence of internal climate variability—and in particular of the El Niño–Southern Oscillation (ENSO)—on both temperature and the short-term growth rate of atmospheric CO_2 (Mazzarella et al., 2013; Humlum et al., 2013; Koutsoyiannis, 2024).

On annual to decadal timescales, the correlation between atmospheric CO_2 concentration and global mean surface temperature is relatively weak. Since the onset of continuous CO_2 measurements in 1958, atmospheric CO_2 has increased almost monotonically, whereas global temperature has exhibited pronounced interannual and multidecadal variability. Periods such as the mid-20th-century cooling (approximately 1940–1975) and the early-21st-century slowdown in surface warming illustrate that short-term temperature variability can be dominated by internal climate dynamics, ocean–atmosphere interactions, and other forcings, rather than by the direct radiative effect of CO_2 alone.

The seasonal cycle of atmospheric CO_2 further illustrates this decoupling on short timescales. Its dominant amplitude reflects biospheric uptake and release in the Northern Hemisphere and is largely independent of long-term global temperature trends. This behavior underscores the sensitivity of natural carbon sinks and sources to biological and climatic processes operating on seasonal to interannual timescales.

Further insight into short-term climate–carbon coupling is provided by the analysis of the annual CO_2 growth rate (ΔCO_2), commonly defined as the difference between the mean atmospheric CO_2 concentration over consecutive 12-month periods. Figures 14.10 and 14.11 show that variations in ΔCO_2 exhibit coherent oscillations with global and sea-surface temperature, ENSO indices such as the Multivariate ENSO Index (MEI), and related geophysical indicators including the Length of Day (LOD), often with temperature leading CO_2 by a few months. This lagged relationship is consistent with the response of terrestrial ecosystems and the ocean carbon cycle to climate variability, particularly during ENSO events.

These short-term lead–lag relationships, however, do not directly constrain the role of atmospheric CO_2 as a radiative forcing on multidecadal to centennial timescales. Since the mid-20th century, the sustained increase in atmospheric CO_2 driven primarily by fossil-fuel combustion and land-use change has produced an independent positive radiative forcing. While internal variability and other forcings modulate the year-to-decade evolution of global temperature, they do not negate the long-term warming influence associated with anthropogenic CO_2 emissions. At the same time, the presence of clearly identifiable climate–carbon feedback mechanisms, operating in response to both natural and anthropogenic forcings, calls into question the attribution of the entire observed CO_2 increase solely to direct human emissions, as done by the IPCC.

Accordingly, the modern instrumental record is best interpreted as reflecting a superposition of climate–carbon feedbacks operating across a range of timescales—from interannual to multidecadal and longer—through which temperature variability can influence the rate of atmospheric CO_2 change, together with a dominant anthropogenic CO_2 forcing that contributes to long-term warming. This perspective reconciles the observed temperature–CO_2 phase relationships on short timescales with the established role of CO_2 as a key driver of long-term climate change, while acknowledging the bidirectional coupling between climate and the carbon cycle.

14.13. Conclusion

The Holocene temperature conundrum, together with analogous discrepancies recurring throughout the past 540 million years of Earth's history, underscores persistent structural limitations in mainstream climate reconstructions and modeling frameworks. These long-term mismatches between proxy evidence and model expectations suggest that key climatic drivers, feedbacks, and boundary conditions remain only partially understood or incompletely represented. Consequently, climate policies grounded primarily on GCM outputs, under the assumption that climate change strongly depends on CO_2 variations, warrant renewed and rigorous scrutiny to ensure that their assumptions, uncertainties, and predictive constraints are fully acknowledged and resolved before being translated into societal or regulatory action.

The interpretation of deep-time climate proxies entails substantial uncertainties. Oxygen-isotope records are influenced by diagenesis, seawater chemistry, and paleolatitude. Sedimentological indicators provide qualitative constraints but lack temporal precision. CO_2 proxies—including paleosols, stomatal indices, and boron isotopes—each carry methodological limitations and calibration uncertainties. GEOCARB-type models depend on assumptions regarding volcanic outgassing, weathering rates, and biological evolution.

Despite these uncertainties, the integration of paleoclimatic and historical records, together with causality analyses (e.g. Mazzarella et al., 2013; Humlum et al., 2013; Davis, 2017; Koutsoyiannis, 2024), reveals a recurrent pattern: across much of geological time, temperature changes tend to precede variations in atmospheric CO_2. This behavior is consistent with a climate system in which CO_2 often acts as a feedback to climate changes initiated by other forcings, rather than as the sole or primary driver.

The available evidence supports a hierarchical climate system in which different processes dominate at different timescales. On multimillion-year scales, tectonic and, potentially, astronomical factors modulate Earth's energy balance and initiate major climatic transitions. On million-year scales, CO_2 variations commonly act as an amplifying feedback, reinforcing temperature changes. On millennial scales, orbital forcing and internal climate variability play a dominant role, with carbon-cycle responses closely coupled to temperature evolution.

In the modern era, anthropogenic CO_2 emissions represent a perturbation of the climate–carbon system that is without precedent in the geological record in terms of rate and magnitude. Nevertheless, it remains essential to distinguish anthropogenic contributions from climate-driven variability in atmospheric CO_2. The observed time lags between temperature and CO_2 changes—documented in paleoclimate archives and detectable during the instrumental period—demonstrate that climate–carbon feedbacks operate alongside direct human emissions. This distinction matters for the interpretation of climate sensitivity estimates, particularly those derived from concentration-driven climate model experiments in which the radiative forcing from post-1850 CO_2 changes is classified as solely anthropogenic. For example, the IPCC (AR6, 2021, chapter 5) states that "*it is unequivocal that the increases in atmospheric carbon dioxide* (CO_2), *methane* (CH_4) *and nitrous oxide* (N_2O) *since the preindustrial period are caused by human activities,*"[42] which is likely erroneous. Indeed, Figures 14.6 and 14.9 suggest that the ~7 °C cooling from the peak of the Eemian to the Last Glacial Minimum corresponded to a CO_2 decrease of about 120 ppm. This implies that, on average, a 1 °C change in global temperature might correspond to a CO_2 change of approximately 15–20 ppm. If a portion of the observed CO_2 evolution reflects feedback responses to temperature across multiple timescales, then sensitivity estimates that neglect this bidirectional coupling may be warm-biased and warrant careful reassessment.

This perspective situates the greenhouse effect within a multi-scale, dynamically coupled climate system. The modern period departs from the long-term natural patterns, since a substantial share of CO_2 now acts as direct radiative forcing. Quantifying the various interactions and feedbacks is essential for refining estimates of climate sensitivity across timescales.

[42]Available at: https://www.ipcc.ch/report/ar6/wg1/chapter/chapter-5/

CHAPTER 15

Assessment of GCMs and the social cost of carbon

In light of the natural climate variability observed across multiple timescales, the simulations produced by global climate models (GCMs) offer a valuable lens through which to critically assess the limitations of the *Anthropogenic Global Warming Theory* (AGWT). As discussed in previous chapters, GCMs exhibit structural weaknesses in simulating key natural dynamics, such as the 50–70-year climate cycle, and consistently fail to reproduce the secular warm periods documented throughout history. Despite these limitations, GCMs remain central to attribution studies and to the projection of future climate scenarios, even when they are used to calibrate simpler and more tractable models—e.g., the FaIR (*Finite Amplitude Impulse Response*) model—adopted for assessing the social cost of carbon (cf. EPA, 2023).[1]

According to the AGWT, as formulated in IPCC (2021),[2] Earth's climate would have remained substantially stable between the preindustrial period (1850–1900) and the present (2011–2020) in the absence of anthropogenic forcings. This hypothesis, however, is not empirically verifiable, since no direct observational counterfactual exists. In this context, GCMs play a pivotal role in the attribution of climate change. This very reliance, however, raises a fundamental question: if the models cannot accurately reconstruct past natural warm periods, how reliable are they in diagnosing the causes of recent warming?

15.1. GCMs' failure to reproduce historical climate variations

Climate models from CMIP3, CMIP5, and CMIP6 have consistently failed to simulate several well-documented climate episodes, including:

- the Holocene Thermal Maximum (HTM), about 9000–5000 years ago;
- the Roman Warm Period (RWP), about 0–300 AD;
- the Medieval Warm Period (MWP), about 800–1300 AD;
- millennial-scale oscillations linking warm phases to cold intervals, such as the Dark Ages Cold Period (DACP) and the Little Ice Age (LIA).

These events, extensively documented in the scientific literature (Ljungqvist, 2010; Christiansen & Ljungqvist, 2012; Klimenko et al., 2014; Kutschera et al., 2020; Li et al., 2023),[3] reveal a natural climate variability far more pronounced than that simulated by the available GCMs.

[1]EPA, 2023. Technical Support Document: Social Cost of Greenhouse Gases. U.S. Environmental Protection Agency.

[2]IPCC, 2021. Climate Change 2021: The Physical Science Basis. Cambridge University Press.

[3]Ljungqvist, F.C., 2010. A new reconstruction of temperature variability in the extra-tropical northern hemisphere during the last two millennia. Geografiska Annaler: Series A, Physical Geography, 92,

This failure raises two critical questions:

(1) do current models incorporate appropriate natural radiative forcings?
(2) do they adequately represent the solar, oceanic, and astronomical mechanisms that drive long-term climatic oscillations?

The available evidence suggests that they do not. GCMs fail to capture the millennial-scale variability of the Holocene, severely limiting their ability to distinguish between natural and anthropogenic contributions to the warming observed from 1850 to 2025, which also coincides with a millennial warm phase.

Even at regional scales, model performance is often poor. For example, in the Corn Belt states of the United States, GCMs have substantially overestimated warming trends (Spencer, 2023).[4] Palmer & Stevens (2019)[5] conclude that, for many local and global applications, the current generation of GCMs is simply not fit for purpose.

Implications and conclusion. These considerations do not negate the role of anthropogenic activity in contemporary climate change. Rather, they call for a more cautious and methodologically rigorous approach. In particular, the exclusive attribution of modern warming to human emissions appears scientifically premature, at least until the following conditions are met:

- reliable reproduction of past warm and cold climate periods that occurred without significant anthropogenic influence;
- integration of cyclic natural forcings, especially solar and astronomical drivers, into climate models, where they are likely underrepresented;
- empirical validation of the models' ability to distinguish between anthropogenic and natural components, based not only on simulations but on systematic comparisons with observational data.

In the absence of a comprehensive representation of natural processes and long-term variability, GCMs appear to overestimate the anthropogenic contribution to post-19th-century warming. This bias arises from structural limitations within the models themselves, which often omit or oversimplify critical climate dynamics. As a result, theories and policies derived from these simulations rest on scientifically fragile foundations, calling into question the robustness of the *Anthropogenic Global Warming Theory* (AGWT).

339–351.

Christiansen, B., Ljungqvist, F.C., 2012. The extra-tropical Northern Hemisphere temperature in the last two millennia: Reconstructions of low-frequency variability. Climate of the Past, 8, 765–786.

Klimenko, V.V., Matskovsky, V.V., Dahlmann, D., 2014. Multi-archive temperature reconstruction of the Russian Arctic for the past two millennia. Geography, Environment, Sustainability, 4(12), 84–95.

Kutschera, W., Patzelt, G., Schaefer, J.M., Schlüchter, C., Steier, P., Wild, E.M., 2020. The movements of Alpine glaciers throughout the last 10,000 years as sensitive proxies of temperature and climate changes. EPJ Web of Conferences, 232, 02002.

Li, Y., Wu, D., Wang, T., Chen, L., Zhang, C., Guo, S., 2023. Late Holocene temperature and precipitation variations in an alpine region of the northeastern Tibetan Plateau. Palaeogeography, Palaeoclimatology, Palaeoecology, 615, 111442.

[4]Spencer, R.W., 2023. Epic fail in America's Heartland: climate models greatly overestimate Corn Belt warming. Available at: https://www.drroyspencer.com

[5]Palmer, T., Stevens, B., 2019. The scientific challenge of understanding and estimating climate change. Proceedings of the National Academy of Sciences, 116(49), 24390–24395.

15.2. Attribution of the *"climate impact drivers"*

The concept of *Climate Impact Drivers* (CIDs), introduced in Chapter 12 of IPCC AR6 (2021), refers to physical conditions of the climate system—such as long-term averages, discrete events, and extremes—that directly affect ecosystems and human societies. These drivers may exert negative, neutral, or even beneficial effects depending on the regional context and the sensitivity of the system. For instance, the *"Hot and Cold"* CID category includes variables such as mean air temperature, heatwaves, cold spells, and frost events.

Despite growing concern over anthropogenic climate influence, the emergence of clear human-induced signals in CID trends remains limited. According to Table 12.12 of IPCC AR6, only 5 out of 33 meteorological impact categories exhibit an anthropogenic signal with high confidence, and another 4 with medium confidence. Most of these are related to mean temperature or ocean chemistry, rather than to extreme weather events. Notably, AR6 does not attribute significant human influence to phenomena such as mean wind, violent storms, tropical cyclones, dust storms, drought conditions, or fire-prone environments.

Even under the high-emission RCP8.5/SSP5-8.5 scenario—now widely regarded as implausible (Hausfather & Peters, 2020; Meinshausen et al., 2020)[6]—CMIP6 global climate models do not project the emergence of anthropogenic signals for most of these phenomena either by 2050 or even by 2100 (cf. Alimonti et al., 2022).[7] Several researchers have criticized the continued use of RCP8.5/SSP5-8.5, noting that it no longer reflects current or projected emission trajectories. Others (cf. Palmer & Stevens, 2019; Nissan et al., 2019)[8] have highlighted the limitations of current GCMs in capturing regional dynamics, particularly in contexts relevant to international development and disaster preparedness.

Natural variability and the limits of attribution. A key challenge in attribution lies in the dominance of natural variability in the modeling of extreme events. Simplistic causal claims are often undermined by regional heterogeneity, trend reversals, and non-climatic confounders, including land-use changes and evolving reporting practices. For example, Alimonti & Mariani (2023)[9] analyzed EM-DAT disaster data and identified three distinct phases in global natural disaster records: linear growth until the mid-20th century, a rapid increase until around 2000, and a marked decline from 2002 to 2022. Their findings suggest

[6]Hausfather, Z., Peters, G.P., 2020. Emissions—the 'business as usual' story is misleading. Nature, 577(7792), 618–620.
Meinshausen, M., Nicholls, Z.R.J., Lewis, J., Gidden, M.J., Vogel, E., et al., 2020. The shared socio-economic pathway (SSP) greenhouse gas concentrations and their extensions to 2500. Geoscientific Model Development, 13(8), 3571–3605.

[7]Alimonti, G., Mariani, L., Prodi, F., Ricci, R.A., 2022. A critical assessment of extreme events trends in times of global warming. European Physical Journal Plus, 137, Article 112.

[8]Palmer, T., Stevens, B., 2019. The scientific challenge of understanding and estimating climate change. Proceedings of the National Academy of Sciences, 116(49), 24390–24395.
Nissan, H., Goddard, L., Coughlan de Perez, E., Furlow, J., Baethgen, W., Thomson, M.C., Mason, S.J., 2019. On the use and misuse of climate change projections in international development. WIREs Climate Change, 10(6), e579.

[9]Alimonti, G., Mariani, L., 2023. Is the number of global natural disasters increasing? Environmental Hazards, 22(4), 345–368.

Climatic Impact-driver Type	Climatic Impact-driver Category	Already Emerged in Historical Period	Emerging by 2050 at Least for RCP8.5/SSP5-8.5	Emerging Between 2050 and 2100 for at Least RC8.5/SSP5-8.5
Heat and Cold	Mean air temperature	++	++	++
	Extreme heat	++	++	++
	Cold spell	-	-	-
	Frost			
Wet and Dry	Mean precipitation		++ -	++ -
	River flood			
	Heavy precipitation and pluvial flood			+
	Landslide			
	Aridity			
	Hydrological drought			
	Agricultural and ecological drought			
	Fire weather			
Wind	Mean wind speed			
	Severe wind storm			
	Tropical cyclone			
	Sand and dust storm			
Snow and Ice	Snow, glacier and ice sheet		-	-
	Permafrost	-	-	-
	Lake, river and sea ice	--	--	--
	Heavy snowfall and ice storm			
	Hail			
	Snow avalanche			
Coastal	Relative sea level		+	+
	Coastal flood			
	Coastal erosion			
Open Ocean	Mean ocean temperature	++	++	++
	Marine heatwave			
	Ocean acidity		++	++
	Ocean salinity	+	+	+
	Dissolved oxygen	-	-	-
Other	Air pollution weather			
	Atmospheric CO_2 at surface	++	++	++
	Radiation at surface			

--	-		+	++
High confidence of decrease	Medium confidence of decrease	Low confidence in direction of change	Medium confidence of increase	High confidence of increase

FIGURE 15.1. Emergence of anthropogenic signals in historical data, and projections for 2050 and 2100 under the extreme RCP8.5/SSP5-8.5 scenario. White: low confidence in direction of change. "+" (medium) & "++" (high) "confidence of increase", and "−" (medium) & "− −" (high) "confidence of decrease". Adapted from IPCC (2021, Chapter 12, Table 12.12).

that improved event reporting, rather than a real increase in disaster frequency, accounts for much of the observed trend, thereby challenging assertions that link global warming directly to rising disaster rates.

Building on this empirical foundation, Alimonti & Mariani (2025)[10] propose a quantitative framework for defining the "climate crisis" through Response

[10] Alimonti, G., & Mariani, L., 2025. Quantifying the climate crisis: A data-driven framework using

Indicators (RINDs) derived from CIDs. Their analysis of time series data using Mann–Kendall trend tests reveals significant increases in global surface temperature (1951–2024), atmospheric CO_2 levels (1959–2024), and precipitation anomalies (1951–2018), alongside significant decreases in Arctic ice volume (1979–2024), flooded area (2002–2018), and wildfire extent (2002–2023). By contrast, indicators such as tropical cyclone frequency and energy (1980–2024), flood-related deaths (2000–2024), and deadly flood events in Italy (1951–2024) display no statistically significant trends. Socio-economic metrics, including global GDP (1960–2023), total crop harvest (1961–2022), and crop yields (1961–2023), exhibit strong positive trajectories, suggesting increased resilience and adaptive capacity.

This body of evidence supports the authors' call for a dynamic "crisis index" that replaces rhetorical urgency with localized, data-driven adaptation strategies. By anchoring climate governance in statistically validated indicators, Alimonti & Mariani (2025) offer a compelling alternative to alarmist framings, one that emphasizes methodological transparency, regional specificity, and innovation as a pathway to long-term sustainability.

Summary and implication. The attribution of climate impacts to human activity is likely overestimated, while the role of natural variability remains underappreciated. Even the IPCC acknowledges that most extreme weather phenomena do not currently exhibit clear anthropogenic signals, and that such signals are not expected to emerge within the 21st century, even under scenarios now considered unrealistic. This has direct implications for economic modeling, particularly for estimates such as the social cost of carbon, which depend on accurate attribution of damages. If climate models cannot reliably distinguish between natural and anthropogenic drivers, then economic projections based on their simulations risk being inflated, misdirected, or both.

15.3. Climate change, economics, and the social cost of carbon

The climate change debate today is often dominated by alarmist narratives and model-driven projections which, while scientifically valuable, risk overshadowing a more empirical and rational assessment of the actual impacts. In particular, the intersection of climate and economics has been extensively analyzed, but it has also been found to be subject to simplifications, distortions, and methodological fragility (e.g. IEA, 2024, 2025).[11] This section critically examines the concept of the *Social Cost of Carbon* (SCC), its modeling foundations, and its political implications, in light of historical evidence and persistent uncertainties.

Climate as a secondary economic factor. Historically, climate has been viewed as a secondary driver of economic growth. Thomas Schelling (1992)[12] argued that the effects of climate change on the U.S. economy would be *"small compared to many other changes,"* noting that sectors such as finance, education, and healthcare are *"little affected by climate."* This perspective was echoed in

---

response indicators for evidence-based adaptation policies. Environmental Hazards, Advance online publication. https://doi.org/10.1080/17477891.2025.2571708

[11]International Energy Agency, 2024. World Energy Outlook 2024.
International Energy Agency, 2025. CO_2 Emissions: Global Energy Review 2025.

[12]Schelling, T.C., 1992. Some economics of global warming. American Economic Review, 82(1), 1–14.

IPCC AR5 (Arent et al., 2014),[13] which concluded that *"for most economic sectors, the impact of climate change will be small compared to other factors."*

More recent studies, such as Mohaddes et al. (2023),[14] show that heat-related shocks have minimal economic effects, whereas cold-related shocks are more significant. This suggests that a warmer climate may even yield net benefits—particularly in temperate regions such as Europe and the United States—challenging the assumption that every degree of warming is inherently harmful.

Global growth and historical resilience. Despite a global temperature increase of approximately 1.1 °C from 1850–1900 to 2011–2020, humanity has flourished:

- life expectancy rose from 32 to 72 years;
- per-capita GDP increased seven-fold;
- mortality from extreme climate events dropped fiftyfold (O'Neill, 2023).[15]

Economic losses from climate-related disasters, as a share of global GDP, have declined (Pielke, 2018, 2020),[16] and insurance payouts have increased only in proportion to the growth of asset values (Hu & McKitrick, 2015).[17] These trends suggest that the adaptive capacity of modern societies has thus far exceeded the negative impacts of climate change.

IAM models and the mitigation dilemma. The SCC is calculated using *Integrated Assessment Models* (IAMs), such as the DICE model (Nordhaus, 1993, 2018).[18] These models simulate future scenarios and estimate the economic damages associated with CO_2 emissions. Simulations show, however, that aggressive mitigation policies often entail costs that exceed their projected benefits. For example, in the *"Optimal"* DICE scenario, warming reaches +3.5 °C with lower total costs than in a scenario targeting +2.5 °C. Nordhaus concludes that *"most CO_2 emissions are less harmful than the policies required to reduce them."*

Empirical uncertainty and methodological controversies. The literature on the economic impact of global warming is fragmented and often contradictory:

- Dell et al. (2012)[19] found significant damage in poor countries;

[13]Arent, D.J., Tol, R., Faust, E., Hella, J., Kumar, S., et al., 2014. Key economic sectors and services. In: Climate Change 2014: Impacts, Adaptation, and Vulnerability. IPCC WGII AR5.

[14]Mohaddes, K., Ng, S., Raissi, M., 2023. Climate Change and Macroeconomic Forecasting. IMF Working Paper No. WP/23/45. International Monetary Fund.

[15]O'Neill, B.C., 2023. Global development trends and climate resilience. Journal of Climate Policy, 23(2), 145–162.

[16]Pielke Jr., R., 2018. Normalized disaster losses in the United States: 1900–2017. Environmental Hazards, 17(2), 123–136.

Pielke Jr., R., 2020. Tracking progress on disaster risk reduction. Nature Sustainability, 3(9), 706–709.

[17]Hu, Y., McKitrick, R., 2015. Insurance payouts and asset exposure in climate-sensitive regions. Climate Economics Review, 9(1), 33–49.

[18]Nordhaus, W.D., 1993. Rolling the 'DICE': An optimal transition path for controlling greenhouse gases. Resource and Energy Economics, 15(1), 27–50.

Nordhaus, W.D., 2018. Projections and uncertainties about climate change in an era of minimal climate policies. American Economic Journal: Economic Policy, 10(3), 333–360.

[19]Dell, M., Jones, B.F., Olken, B.A., 2012. Temperature shocks and economic growth: Evidence from the last half century. American Economic Journal: Macroeconomics, 4(3), 66–95.

- Burke et al. (2015)[20] identified negative effects above 13 °C;
- Zhao et al. (2018)[21] and Berg et al. (2023)[22] reported economic benefits below 16 °C;
- Newell et al. (2021)[23] concluded that, given model uncertainty, the effect of warming on global growth is *"too uncertain to be distinguished from zero."*

The social cost of carbon: a conditional tool. The SCC represents the estimated monetary damage caused by emitting one ton of CO_2. The SCC, however, is not a direct empirical measure but a model-dependent construct based on conditional assumptions. Key variables include:

- *discount rate*: higher rates lower the present value of future damages;
- *climate sensitivity* (ECS): a lower ECS reduces the SCC (Dayaratna et al., 2017, 2023);[24]
- *damage function*: FUND includes CO_2 fertilization benefits; DICE excludes them (McKitrick, 2025);[25]
- *emission scenarios*: higher projected emissions increase the SCC.

The EPA (2023) quintupled its SCC estimate compared with that of a decade earlier—not because of new data, but because of revised assumptions, some of which have been contested (Moore et al., 2017; McKitrick, 2025).[26]

Critical points and climate bifurcations. A debated issue is the role of *"tipping points,"* which fall into two categories:

- *forced discontinuities*: abrupt changes triggered by threshold crossings (e.g., ice sheet collapse);
- *bifurcations*: transitions between stable climate states, even in the absence of external forcing (Crawford, 1991; Shen et al., 2023).[27]

---

[20]Burke, M., Hsiang, S.M., Miguel, E., 2015. Global non-linear effect of temperature on economic production. Nature, 527(7577), 235–239.

[21]Zhao, Z., Deng, Y., Zhang, J., 2018. Temperature and economic performance in developing countries. World Development, 105, 314–326.

[22]Berg, M., Kuralbayeva, K., Venables, A.J., 2023. Climate change and economic development: Revisiting the evidence. Oxford Economic Papers, 75(1), 1–24.

[23]Newell, R.G., Prest, B.C., Sexton, S., 2021. The social cost of carbon: Advances in long-term probabilistic modeling. Brookings Papers on Economic Activity, spring 2021, 1–65.

[24]Dayaratna, K., McKitrick, R., Kreutzer, D., 2017. Empirical climate sensitivity and the social cost of carbon. Climate Change Economics, 8(2), 1750006.
Dayaratna, K., McKitrick, R., Kreutzer, D., 2023. Updated estimates of climate sensitivity and implications for policy. Heritage Foundation Technical Report.

[25]McKitrick, R., 2025. Extended crop yield meta-analysis data do not support upward SCC revision. Scientific Reports, 15, 5575.

[26]Moore, S., Smith, J., Taylor, L., 2017. Reassessing the EPA's social cost of carbon. Regulation, 40(3), 14–20.

[27]Crawford, J., 1991. Climate bifurcations and nonlinear transitions. Climatic Dynamics, 6(3), 145–158.
Shen, X., Liu, Y., Wang, H., 2023. Stable climate states and bifurcation thresholds. Earth System Dynamics, 14(1), 89–104.

According to the IPCC (2021), most tipping points are assigned a low probability. Even when included in SCC models, their impact on cost estimates is modest (Dietz, 2021).[28]

Alternatives to SCC and policy implications. Some economists advocate abandoning the *Social Cost of Carbon* (SCC) metric on account of its variability and sensitivity to assumptions (Thoung, 2017).[29] Alternative policies (e.g., those mandated for electric vehicles or energy efficiency) often entail substantially higher costs per ton of CO_2 avoided, and fail basic cost-benefit tests. Moreover, a significant share of "green" technologies exhibits faster-than-expected degradation (e.g.: Mishnaevsky, 2022; Tang et al., 2026).[30] In this context, the SCC risks becoming a rhetorical device rather than a reliable operational tool, employed to justify policies that may be disproportionate or ineffective.

Estimated U.S. GDP loss due to climate change. The joint report of the Council of Economic Advisers and the Office of Management and Budget presents projected U.S. GDP losses under various warming scenarios (CEA-OMB, 2023, Figure 1).[31] Even with several degrees Celsius of warming above preindustrial levels, the estimated GDP loss is expected to remain modest, on the order of a few percentage points.

Since the U.S. economy grows at $\sim$1–2% annually, the projected impact of climate change on GDP appears relatively negligible. This reinforces the view that, at least for advanced economies such as the United States and Europe, the economic effects of global warming may be manageable without drastic interventions.

Conclusion. Climate change is a complex and multifaceted reality, but its translation into economic terms demands prudence, rigor, and transparency. Although the Social Cost of Carbon (SCC) can be a useful tool for policy analysis, it remains heavily conditioned by modeling assumptions and empirical uncertainties. Historical evidence demonstrates the remarkable adaptive capacity of modern societies, suggesting that climate policies must weigh not only future risks but also present costs and viable alternatives.

Global spending on climate-related initiatives from 2011 to 2018 exceeded $3.6 trillion, with more than half directed toward wind and solar energy (ÓhAiseadha et al., 2020).[32] Despite this massive investment, these technologies have contributed only modestly to global energy supply, rising from 0.5% to 3%.

---

[28]Dietz, S., Rising, J., Stoerk, T., Wagner, G., 2021. Economic impacts of tipping points in the climate system. Proceedings of the National Academy of Sciences, 118(34), e2103081118.

[29]Thoung, T., 2017. Why we should abandon the social cost of carbon. Energy Policy, 105, 512–520.

[30]Mishnaevsky, L. Jr., 2022. Root causes and mechanisms of failure of wind turbine blades: overview. Materials, 15, 2959.

Tang, Y., Poddar, S., Kay, M., Rougieux, F.E., 2026. Understanding and reducing the risk of extreme photovoltaic degradation. IEEE Journal of Photovoltaics, 16(1), 150–159.

[31]CEA-OMB, Council of Economic Advisers & Office of Management and Budget. Methodologies and Considerations for Integrating the Physical and Transition Risks of Climate Change into Macroeconomic Forecasting for the President's Budget. White Paper, March 13, 2023.

[32]ÓhAiseadha, C., Quinn, G., Connolly, R., Connolly, M., Soon, W., 2020. Energy and climate policy—an evaluation of global climate change expenditure 2011–2018. Energies, 13(18), 4839.

The authors argue that such a disproportion calls into question the strategic efficiency of current climate policies, and urges a more balanced and results-driven approach to the energy transition.

A balanced assessment must be grounded in observable data and open to methodological scrutiny. This approach is essential to avoid ideological drift and to promote a sustainable, proportionate, and realistic transition.

15.4. Editorial controversies and methodological limitations

Economic analyses of climate change are particularly delicate, often entangled with political interests and subject to intense editorial scrutiny. The scientific community continues to question how certain studies gain acceptance in high-impact journals, especially when they appear to bypass established methodological standards.

A prominent case is the publication of *The Economic Commitment of Climate Change* by Kotz, Levermann, and Wenz (2024)[33] in *Nature*, which projected a 19% reduction in global income by 2050 due to already "committed" climate impacts. The study drew criticism for the insufficient treatment of spatial correlations between regions—a methodological limitation that undermines the statistical robustness of its findings. In a formal reply, Christof Schötz (2025)[34] argued that the analysis *"can not provide a sufficiently solid empirical basis to inform climate policy"* unless spatial dependencies are properly addressed, showing that some results deviated by up to a factor of three. The controversy intensified when *KlimaNachrichten*[35] reported that peer reviewers had already flagged these issues during the review process, yet the article was published nonetheless, raising concerns about editorial judgment and the tension between narrative impact and scientific integrity.

Despite its methodological limitations, the findings of Kotz et al. (2024) have influenced financial debates. The *Network for Greening the Financial System* (NGFS), which brings together central banks such as the Federal Reserve and the European Central Bank, discussed the study in the context of revising damage functions for climate stress test scenarios. While not adopted as a standalone model, its estimates contributed to shaping assessments of income risk, portfolio exposure, and macroeconomic stability. This chain of influence—rooted in a contested analytical foundation—highlights the urgent need to ensure that flawed models do not become institutionalized in financial policymaking.

---

[33] Kotz, M., Levermann, A., Wenz, L., 2024. The economic commitment of climate change. Nature, 628, 551–557.

[34] Schötz, C., 2025. Spatial correlation in economic analysis of climate change. Nature, 644, E27–E30.

[35] KlimaNachrichten Redaktion. (2025, August 12). Wenn Studien falsch sind. KlimaNachrichten. Available at: https://klimanachrichten.de/2025/08/12/wenn-studien-falsch-sind/

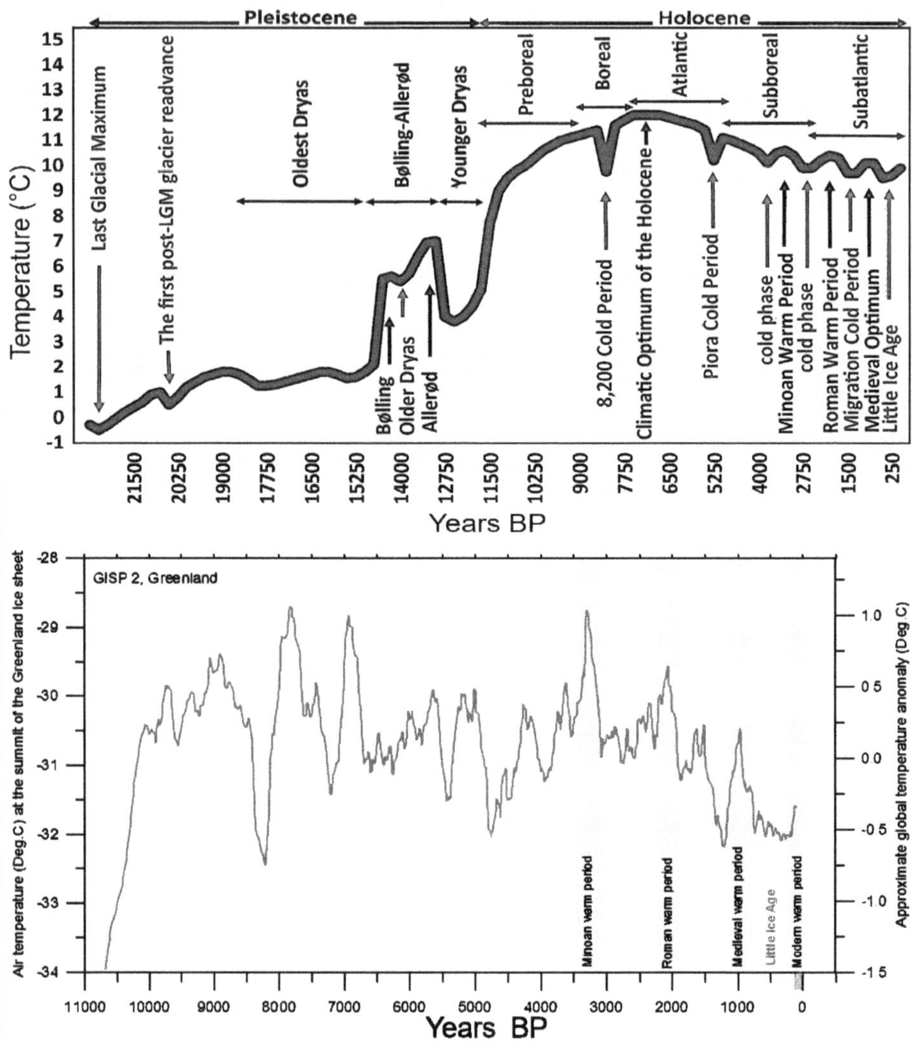

Figure 15.2. **Holocene temperature reconstructions.** (Top) Reconstructed mean annual air temperature at Hurbanovo, Slovakia, spanning the last 22,500 years, adapted from. (Recreated from Melo et al., 2022). (Bottom) Comparative record of temperature and atmospheric CO_2 concentrations derived from GISP2 ice cores in Greenland (Alley, 2000). The data highlight the Holocene Thermal Maximum and reveal several millennial-scale climatic oscillations. Notably, the thermal peaks of the Holocene Optimum, and of these cycles, were likely warmer than present-day conditions, despite lower atmospheric CO_2 levels.

Part 3

New Perspectives for Quantifying Climate Change

CHAPTER 16

Uncertainty in global surface temperature records

In the Sixth Assessment Report, the *Intergovernmental Panel on Climate Change* (IPCC, 2023)[1] states with high confidence that global surface temperature increased by 1.09 °C (range: 0.95–1.20 °C) during the period 2011–2020 relative to the 1850–1900 baseline. The warming was more pronounced over land (+1.59 °C) than over the oceans (+0.88 °C). However, a critical analysis of the reliability of the underlying temperature records raises important methodological questions. The precise magnitude of global warming is far from straightforward, as it depends heavily on technical and statistical choices made in constructing the datasets used to study climate change.

16.1. The foundations of historical temperature series

Global surface temperature anomalies are not derived from direct, continuous measurements but rather from a complex statistical process. Raw temperature data originate from thousands of land-based meteorological stations and marine platforms (buoys, ships, and fixed installations), collected with diverse instruments, techniques, and sampling frequencies that have evolved over time.

Daily maximum and minimum temperatures are averaged monthly and then annually. These averages are normalized against a reference period (e.g., 1961–1990 or 1850–1900) to produce *"anomalies,"* that is, deviations from the local climatic norm. The anomalies are then spatially aggregated to estimate global climate variation. It is therefore essential to recognize that the "global surface temperature" is not a direct empirical measurement, but rather a modeled and interpolated construct subject to multiple layers of uncertainty.

16.2. Spatial and temporal coverage issues

One of the most significant limitations is the uneven spatial distribution of observations. Historically, meteorological networks have been concentrated in densely populated regions of the Northern Hemisphere—North America, Europe, and East Asia—while vast areas such as Antarctica, the Arctic, the Saharan and sub-Saharan regions of Africa, the Southern Ocean, rainforests, and deserts remain sparsely monitored, especially prior to 1950 (Lawrimore, 2011; Menne et al., 2018).[2]

---

[1]IPCC, 2023. Climate Change 2023: Synthesis Report. IPCC, Geneva, Switzerland.

[2]Lawrimore, J.H., Menne, M.J., Gleason, B.E., Williams, C.N., Wuertz, D.B., et al., 2011. An overview of the Global Historical Climatology Network monthly mean temperature data set, version 3. Journal of Geophysical Research: Atmospheres, 116, D19121.
Menne, M.J., Williams, C.N., Gleason, B.E., Rennie, J.J., Lawrimore, J.H., 2018. The Global Historical Climatology Network monthly temperature dataset, version 4. Journal of Climate, 31(24), 9835–9854.

To address these gaps, gridded averaging techniques are employed, dividing Earth's surface into regular cells (e.g., $0.25° \times 0.25°$) and interpolating the available data. However, when entire regions lack observations, interpolation can introduce systematic errors. Temporal coverage is also inconsistent: the number of active stations has fluctuated over time, with closures and relocations compromising data continuity, particularly in developing countries.

The principal sources of uncertainty include:

- instrumental changes: upgrades in thermometers, reading systems, or automation;
- station relocations: even small shifts can alter the local environmental conditions;
- data gaps: interruptions, transcription errors, or accidental losses.

These factors complicate the reconstruction of long-term climate trends.

16.3. Local effects and systematic distortions

Meteorological stations are designed primarily for short-term weather monitoring, not for long-term climate analysis. Over time, the surrounding environment may change dramatically: urbanization, deforestation, vegetation growth, or land-use shifts can alter the local microclimate, introducing variations unrelated to global climate change.

To ensure reliability, stations must meet strict technical and environmental criteria established by organizations such as the World Meteorological Organization (WMO). These criteria include proper sensor placement (e.g., thermo-hygrometers positioned 1.5–2 m above grassy ground, anemometers at 10 m in open terrain), siting in rural or lightly urbanized areas, avoidance of artificial surfaces, the use of certified instrumentation, and continuous data collection with quality control and cross-validation.

In practice, these ideal conditions are rarely met. Urbanization introduces the well-documented *Urban Heat Island* (UHI) effect: artificial surfaces and anthropogenic heat sources raise temperatures in urban areas relative to rural ones. If not properly corrected, such local phenomena can distort global temperature averages and exaggerate warming trends.

As noted by D'Aleo (2016)[3] and in the report *Corrupted Climate Stations* by Anthony Watts (2022),[4] approximately 96% of U.S. stations used for climate monitoring fail to meet NOAA's siting standards, and are often located near heat sources such as air-conditioning units. Additional issues include physical degradation: older station enclosures lose their reflective coating and darken over time, altering thermal readings. According to Watts, these combined factors introduce a systemic thermal bias that may overstate warming trends in the United States and potentially worldwide.

A recent study by Spencer, Christy, and Braswell (2025)[5] quantified the UHI effect in U.S. summer temperature data from 1895 to 2023. By analyzing millions

[3]D'Aleo, J.S., 2016. A critical look at surface temperature records. In: Easterbrook, D.J. (Ed.) Evidence-Based Climate Science, 2nd Edition. Elsevier, pp. 11–48.

[4]Watts, A., 2022. Corrupted climate stations: The official U.S. temperature record remains fatally flawed. The Heartland Institute.

[5]Spencer, R.W., Christy, J.R., Braswell, W.D., 2025. Urban heat island effects in U.S. summer surface temperature data, 1895–2023. Journal of Applied Meteorology and Climatology, 64(7), 717–728.

FIGURE 16.1. Illustration: warming induced by the *Urban Heat Island* effect. A rural station (on the right) records lower temperatures than an urban station (on the left). The gap widens as urbanization intensifies.

of raw observations and correlating them with population density, the authors found that UHI contributed approximately:

- 65% of the observed warming in urban and suburban areas;
- 8% in rural zones;
- 22% as the national average impact on summer warming trends.

The fastest warming occurred in low-density areas undergoing urbanization, while in already densely populated cities the effect tended to plateau.

Implications for dataset integrity. Many stations in the United States, and globally, exhibit systematic anomalies due to siting and environmental context, raising legitimate concerns about the quality of long-term temperature records.

Although agencies such as NOAA, NASA, and HadCRUT apply corrections to account for non-climatic distortions, the transparency and effectiveness of these adjustments remain debated. Updated versions of datasets often show more pronounced warming trends than earlier iterations, fueling skepticism about the objectivity of the correction methodologies.

16.4. Conclusion: a necessary caution

The assertion that global temperature has risen by approximately 1.09 °C over the past 150 years rests on complex statistical reconstructions derived from fragmented and heterogeneous datasets, each subject to significant uncertainty.

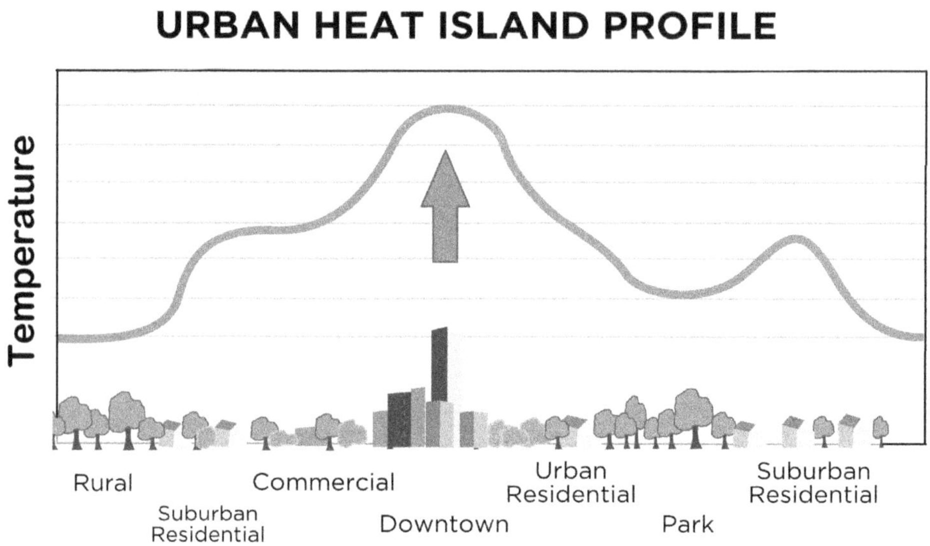

FIGURE 16.2. Illustration: temperature profile around an urban heat island: a city surrounded by suburban and rural areas.

While multiple lines of evidence—such as glacier retreat and sea level rise—clearly support the occurrence of global warming since 1850–1900, the precision with which this warming is quantified, and the implications drawn from it, demand careful scrutiny.

In a context where methodological choices can yield divergent estimates, and where environmental changes and instrumental anomalies may distort climate signals, maintaining a critical and transparent approach is essential. The reliability of global mean temperatures as indicators of long-term climate trends should not be assumed but continuously evaluated through methodological review and interdisciplinary comparison.

Historical temperature series remain a valuable yet inherently imperfect tool. Their interpretation—particularly when used to inform global policy decisions—requires a clear understanding of the technical and structural limitations that accompany them. Only through such caution can climate science maintain both its credibility and its utility in guiding responsible action.

CHAPTER 17

The *"Urban Heat Island"* (UHI) effect

17.1. Global population growth and urban expansion

Cities are not only hubs of economic, cultural, and political activity—they are also sources of artificial heat. The phenomenon known as the *Urban Heat Island* (UHI) describes the systematic tendency of urbanized areas to record higher temperatures than the surrounding rural zones. This effect, documented and studied for decades (McKitrick & Michaels, 2007; Mohajerani, 2017; Stewart & Oke, 2012),[1] results from a combination of factors linked to the anthropogenic transformation of the environment.

Urban construction materials—such as asphalt, concrete, brick, and glass—have a high thermal capacity and low reflectivity, absorbing and slowly releasing heat. The vertical structure of cities, with buildings that obstruct air circulation, gives rise to thermal canyons. Soil sealing reduces evaporative cooling, while sparse vegetation limits evapotranspiration. To this is added the continuous input of thermal energy from vehicles, infrastructure, and human activity. Together, these elements generate persistent localized warming, making the UHI a structural feature of the contemporary urban climate.

Figures 17.1A and 17.1B illustrate the uneven global distribution of population and the rapid growth of major urban centers over time. In Asia, the Americas, and Africa, urban expansion has been especially pronounced—for example, Beijing's population grew from approximately 1.5 million in 1950 to about 20 million in 2020—compared to Europe, where cities are older and have expanded more gradually.

17.2. The influence of urbanization on climate data

The UHI effect is not merely a local meteorological phenomenon—it directly affects the reliability of global climate datasets. Meteorological stations, which form the basis of historical temperature series, are often located in urban or peri-urban environments. Although urbanized areas cover less than 4% of Earth's

[1]McKitrick, R.R., Michaels, P.J., 2007. Quantifying the influence of anthropogenic surface processes and inhomogeneities on gridded global climate data. Journal of Geophysical Research, 112, D24S09.
Mohajerani, A., Bakaric, J., Jeffrey-Bailey, T., 2017. The urban heat island effect, its causes, and mitigation, with reference to the thermal properties of asphalt concrete. Journal of Environmental Management, 197, 522–538.
Stewart, I.D., Oke, T.R., 2012. Local climate zones for urban temperature studies. Bulletin of the American Meteorological Society, 93(12), 1879–1900.

A

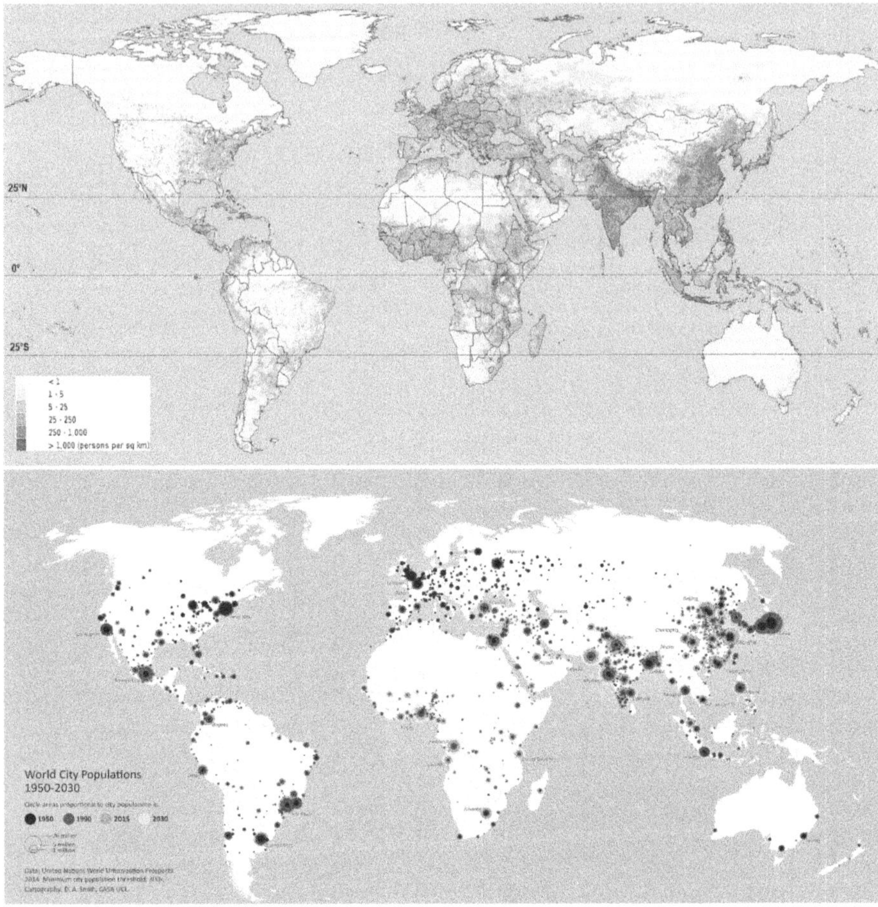

B

FIGURE 17.1. (A) Global population density in 2015. Available at: https://sedac.ciesin.columbia.edu/data/collection/gpw-v4).
(B) Growth of the global urban population in 1950, 1990, 2015, and projected for 2030.
Available at: https://population.un.org/wup/.

surface, a disproportionate number of stations in the Global Historical Climatology Network (GHCN) are situated in areas affected by UHI (Lawrimore, 2011).[2] In many cases, originally rural stations have been absorbed by expanding urban zones (Scafetta, 2021a).[3]

This systemic pattern introduces a potential structural distortion in historical climate records. Given the rapid global urbanization since 1950, it is

[2]Lawrimore, J.H., Menne, M.J., Gleason, B.E., Williams, C.N., Wuertz, D.B., et al., 2011. An overview of the Global Historical Climatology Network monthly mean temperature data set, version 3. Journal of Geophysical Research: Atmospheres, 116, D19121.

[3]Scafetta, N., 2021a. Detection of non-climatic biases in land surface temperature records by comparing climatic data and their model simulations. Climate Dynamics, 56(9–10), 2959–2982.

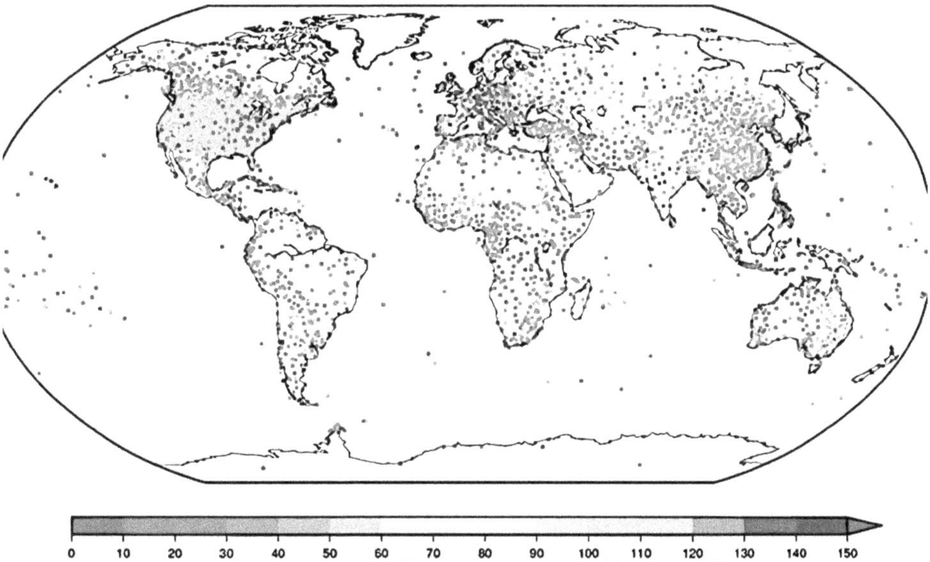

FIGURE 17.2. Global Historical Climatology Network – Monthly (GHCN-M), Version 3. Color indicates the length in years of available data (Lawrimore, 2011). https://www.ncei.noaa.gov/metadata/geoportal/rest/metadata/item/gov.noaa.ncdc:C00839/html

plausible that part of the warming observed in global climate datasets reflects the intensification of UHI rather than purely global climate change.

17.3. The IPCC's controversial interpretations

Several studies (Jones et al., 1990; Peterson et al., 1999; Parker, 2006; Wickham et al., 2013)[4] have attempted to assess UHI contamination by comparing warming trends between urban and rural stations. However, it remains uncertain whether these methods are capable of detecting the full UHI bias when present (cf. McKitrick, 2013).[5]

Before the Fourth Assessment Report (AR4, IPCC, 2007)[6], several studies argued that UHI contributed significantly (30–50%) to the observed warming and was not adequately accounted for in climate models. Their conclusions

[4]Jones, P.D., Groisman, P.Ya., Coughlan, M., Plummer, N., Wang, W.-C., Karl, T.R., 1990. Assessment of urbanization effects in time series of surface air temperature over land. Nature, 347(6289), 169–172.
Peterson, T.C., Gallo, K.P., Lawrimore, J., Owen, T.W., Huang, A., McKittrick, D., 1999. Global rural temperature trends. Geophysical Research Letters, 26(3), 329–332.
Parker, D.E., 2006. A Demonstration That Large-Scale Warming Is Not Urban. Journal of Climate, 19(12), 2882–2895.
Wickham, C., Rohde, R., Muller, R.A., Wurtele, J., Curry, J., et al., 2013. Influence of Urban Heating on the Global Temperature Land Average using Rural Sites Identified from MODIS Classifications. Geoinformatics & Geostatistics: An Overview, 1(2), doi:10.4172/2327-4581.1000104.

[5]McKitrick, R., 2013. Encompassing Tests of Socioeconomic Signals in Surface Climate Data. Climatic Change, 120(1–2), 95–107.

[6]IPCC, 2007. Climate Change 2007: The Physical Science Basis. Cambridge University Press.

were based on correlations between warming trends and areas of socioeconomic development (de Laat & Maurellis, 2006; McKitrick & Michaels, 2007).[7]

AR4 dismissed these findings, attributing the correlations to natural atmospheric circulation patterns without providing explicit evidence. This approach was criticized for its lack of justification. In the Fifth Assessment Report (IPCC, 2013),[8] the panel acknowledged that AR4 had failed to substantiate its dismissal and conceded that land temperature data were likely contaminated by UHI. Nevertheless, it maintained that the effect accounted for less than 10% of the observed warming and issued no warnings regarding the use of land-based datasets.

In Chapter 2 of the Sixth Assessment Report (IPCC, 2021),[9] the IPCC again recognized contamination from UHI and other land-transformation processes but downplayed their impact. It concluded—with *"very high confidence"*—that urbanization has a negligible effect on annual global warming.

This conclusion has been challenged in the recent literature. Helbling & Meierrieks (2023),[10] analyzing data from 118 countries between 1960 and 2016, found a strong correlation between temperature increases and urbanization rates, suggesting that the UHI influence may be underestimated. The temporal overlap between postwar urban growth and the observed warming trends raises questions that cannot easily be dismissed.

Spencer, Christy, and Braswell (2025)[11] examined UHI effects in U.S. summer temperature data from 1895 to 2023, using millions of raw observations. By segmenting the data by population density and time interval, they showed that the fastest warming occurs in low-density areas undergoing urbanization, whereas warming stabilizes in already urbanized zones. Their study estimates that the UHI accounts for:

- approximately 8% of the warming in rural areas;
- up to approximately 65% in urban and suburban zones;
- approximately 22% as the national average impact on summer warming trends.

These findings challenge the accuracy of the official temperature trends, which rely on homogenized data that may not effectively remove urbanization effects. Katata et al. (2023)[12] further demonstrated evidence of urban blending in homogenized temperature records in Japan and the United States.

[7]de Laat, A.T.J., Maurellis, A.N., 2006. Evidence for influence of anthropogenic surface processes on lower tropospheric and surface temperature trends. International Journal of Climatology, 26(7), 897–913.

[8]IPCC, 2013. Climate Change 2013: The Physical Science Basis. Cambridge University Press.

[9]IPCC, 2021. Climate Change 2021: The Physical Science Basis. Cambridge University Press.

[10]Helbling, M., Meierrieks, D., 2023. Global warming and urbanization. Journal of Population Economics, 36, 1187–1223.

[11]Spencer, R.W., Christy, J.R., Braswell, W.D., 2025. Urban heat island effects in U.S. summer surface temperature data, 1895–2023. Journal of Applied Meteorology and Climatology, 64(7), 717–728.

[12]Katata, G., Connolly, R., O'Neill, P., 2023. Evidence of urban blending in homogenized temperature records in Japan and the United States. Journal of Applied Meteorology and Climatology, 62(8), 1095–1114.

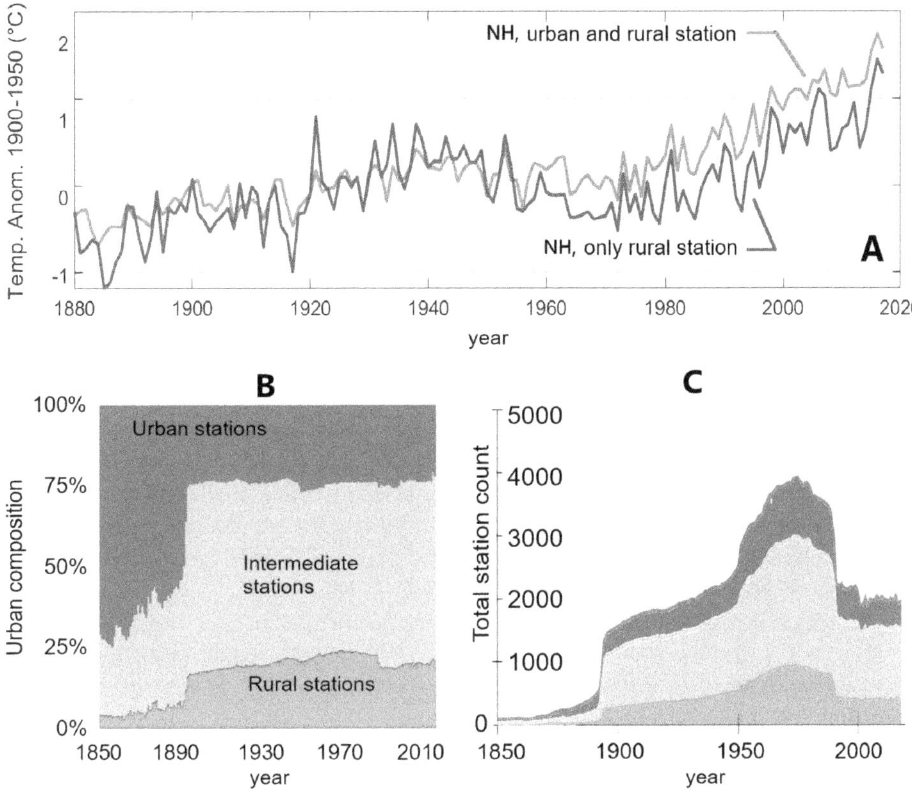

FIGURE 17.3. Observed divergence in Northern Hemisphere surface air temperature estimates:
[A] Comparison between reconstructions using both urban and rural stations vs. rural-only stations.
[B, C] Percentage and total count of stations classified as urban, intermediate, or rural. (Adapted from Soon et al., 2023).

17.4. Comparison between urban and rural stations

In support of this interpretation, Soon et al. (2023)[13] compared two reconstructions of land surface temperatures in the Northern Hemisphere: one incorporating both urban and rural stations, and another based exclusively on verified rural stations. The results were striking: the mixed dataset yielded a warming rate of 0.89 °C per century, compared with 0.55 °C per century in the rural-only dataset. This difference of nearly 40%, illustrated in Figure 17.3A,

[13]Soon, W., Connolly, R., Connolly, M., Akasofu, S.-I., Baliunas, S., Berglund, J., Bianchini, A., Briggs, W.M., Butler, C.J., Cionco, R.G., Crok, M., Elias, A. G., Fedorov, V.M., Gervais, F., Harde, H., Henry, G.W., Hoyt, D.V., Humlum, O., Legates, D.R., Lupo, A.R., Maruyama, S., Moore, P., Ogurtsov, M., ÓhAiseadha, C., Oliveira, M.J., Park, S.-S., Qiu, S., Quinn, G., Scafetta, N., Solheim, J.-E., Steele, J., Szarka, L., Tanaka, H.L., Taylor, M.K., Vahrenholt, F., Velasco Herrera, V.M., Zhang, W., 2023. The detection and attribution of Northern Hemisphere land surface warming (1850–2018) in terms of human and natural factors: Challenges of inadequate data. Climate, 11(9), 179.

underscores the critical importance of sample selection in constructing historical climate series.

Figures 17.3B and 17.3C provide further insight into the composition and distribution of the stations analyzed, demonstrating how local environmental conditions can significantly influence observed trends and affect the integrity of global temperature estimates.

17.5. Conclusion

The *Urban Heat Island* (UHI) effect exemplifies how local environmental factors can distort the measurement of global climate phenomena. Underestimating these influences—particularly in official climate narratives—risks undermining the quality of scientific discourse and shaping public policy on uncertain foundations.

If urbanization and other surface-transformation processes have significantly distorted the warming trends reported in official climate records, it becomes imperative to reassess not only the data but also the mitigation strategies derived from models calibrated on them. In this context, climate science must confront the complexity of its observational tools and methodologies, rather than obscure it behind overly simplified or reassuring estimates.

If not properly corrected, the UHI effect risks introducing systematic distortions into the datasets used for global climate assessments. Homogenization techniques, intended to mitigate such distortions, are not always effective and may, in some cases, introduce new biases.

In the light of this evidence, several researchers have called for a critical and systematic review of global temperature datasets and of the methodologies used to adjust them. Greater emphasis should be placed on analyzing long-term stable rural stations and on expanding observational networks in less urbanized regions, in order to produce a more representative and unbiased picture of global warming.

CHAPTER 18

Dendrochronological, satellite, and instrumental data

One of the most delicate and controversial issues in modern climate science concerns the consistency among the various metrics used to estimate surface temperature variations on Earth. The following sections examine key discrepancies between alternative climatic records.

18.1. Dendrochronological proxies versus instrumental temperatures

A representative example is shown in Figure 18.1, which compares instrumental summer temperatures (June–August) between 30°N and 70°N with temperatures reconstructed from tree rings (dendrochronological proxies) in the same latitudinal band (Esper et al., 2018).[1] The figure shows that, beginning around 1980, a marked divergence emerges: the instrumental data display accelerated warming, whereas the proxy data indicate much weaker warming, especially from 1970 onward. This phenomenon is known as the *"divergence problem"* in dendroclimatology.

The gap raises questions about the reliability of either modern instrumental data or paleoclimatic reconstructions, or both. As noted by Büntgen et al. (2021)[2] and Cai & Lu (2023),[3] it remains unclear whether instrumental datasets overestimate the recent warming or whether tree-ring proxies underestimate it owing to biological limitations. Moberg et al. (2005)[4] had already observed that tree rings are more sensitive to short-term fluctuations but tend to dampen long-term trends, since physiological adaptations may obscure decadal- or centennial-scale changes.

Notably, this divergence parallels the discrepancy observed between urban and rural station data (Figure 17.3), reinforcing the hypothesis that instrumental datasets may be influenced by non-climatic factors such as the intensification of urban heat island effects, land-use changes, and interpolation techniques.

18.2. Climate reconstructions in small isolated northern settlements

Figure 18.2 presents temperature trends from several Arctic locations—presumably small settlements where urbanization has had minimal impact.

---

[1]Esper, J., St. George, S., Anchukaitis, K., D'Arrigo, R., Ljungqvist, F.C., et al., 2018. Large-scale, millennial-length temperature reconstructions from tree rings. Dendrochronologia, 50, 81–90.

[2]Büntgen, U., Kirdyanov, A.V., Krusic, P.J., Shishov, V.V., Esper, J., 2021. Arctic aerosols and the 'divergence problem' in dendroclimatology. Dendrochronologia, 67, 125837.

[3]Cai, M., Lu, J., 2023. Research progress on the divergence problem of tree ring growth-climate relationship. Journal of Sichuan Forestry Science and Technology, 44(6), 7–13.

[4]Moberg, A., Sonechkin, D.M., Holmgren, K., Datsenko, N.M., Karlén, W., 2005. Highly variable Northern Hemisphere temperatures reconstructed from proxy data. Nature, 433, 613–617.

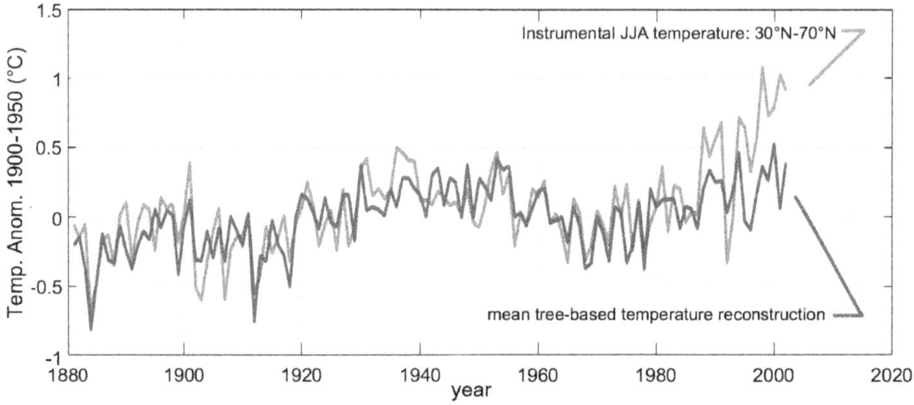

FIGURE 18.1. Divergence between instrumental summer temperature averages (JJA) and tree-ring-based reconstructions for land areas between 30°N and 70°N, relative to the 1930–1960 reference period. A notable divergence appears from 1970 onward. (Adapted from Esper et al., 2018).

On average, the temperatures recorded around the 1940s are comparable to those observed after 2000. These two warm phases, together with cold intervals in the early 20th century and in the 1970s, suggest the presence of a multi-decadal climatic oscillation of approximately 60 years.

This oscillatory pattern is consistent with the Atlantic Multidecadal Oscillation (AMO) (Figure 12.3) and with dendrochronological reconstructions (Figure 17.3) from forests free of anthropogenic influence.

Figure 18.3 compares temperatures in Iceland and Greenland—sparsely populated regions—using CRU-TS 4 data (Harris et al., 2020)[5] and CMIP6 climate simulations (144 simulations).[6] The models fail to reproduce the cooling of approximately 1 °C observed between the 1940s and the 1970s, and they overestimate the warming between the 1940s and 2000–2020 (1 °C vs. 0.5 °C). In addition, the spread in the simulated mean absolute temperatures exceeds 6–7 °C, indicating substantial model uncertainty.

18.3. Comparison between surface temperature and satellite data

Significant discrepancies also emerge when surface temperature records are compared with satellite-based measurements of the lower troposphere. Figure 18.4 compares the HadCRUT5, CRUTEM5, and HadSST datasets (Morice et al., 2021; Osborn et al., 2021; Kennedy et al., 2019)[7] with satellite data from UAH

[5]Harris, I., Osborn, T.J., Jones, P., Lister, D., 2020. Version 4 of the CRU TS monthly high-resolution gridded multivariate climate dataset. Scientific Data, 7, 109.

[6]Data available from the "*KNMI Climate Change Atlas*" at https://climexp.knmi.nl/plot_atlas_form.py

[7]Morice, C.P., Kennedy, J.J., Rayner, N.A., Winn, J.P., Hogan, E., et al., 2021. An updated assessment of near-surface temperature change from 1850: the HadCRUT5 data set. Journal of Geophysical Research: Atmospheres, 126, e2019JD032361.

Osborn, T.J., Jones, P.D., Lister, D.H., Morice, C.P., Simpson, I.R., et al., 2021. Land surface air temperature variations across the globe updated to 2019: the CRUTEM5 dataset. Journal of Geophysical

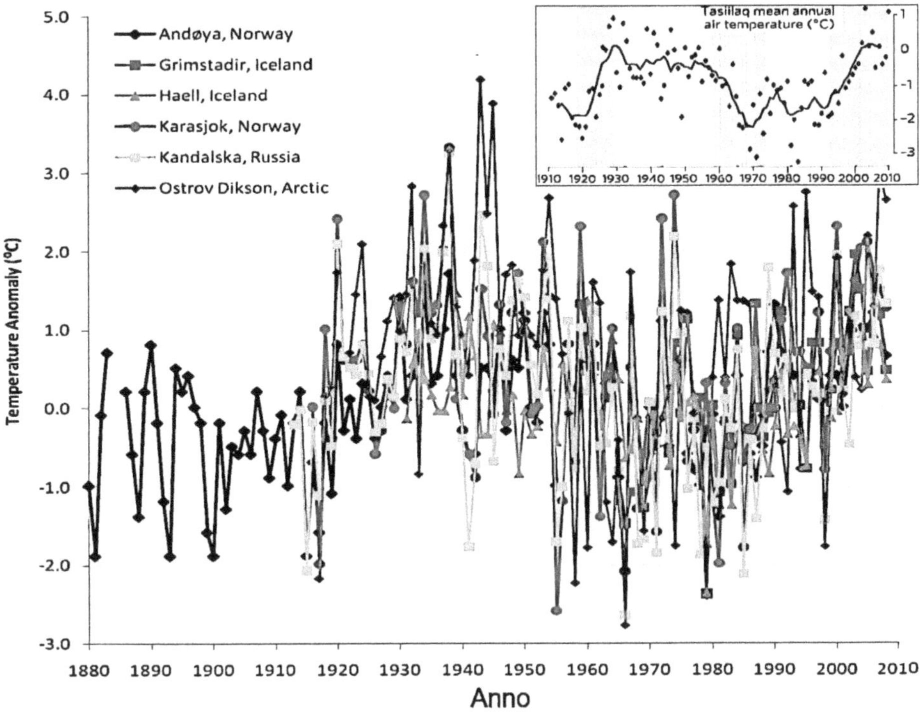

FIGURE 18.2. Temperature trends in small Arctic settlements. Warm periods around the 1940s and post-2000 suggest multi-decadal climate oscillations. (cf. Bjørk et al., 2012).

MSU v6.1 (Spencer et al., 2017),[8] available since 1978, and NOAA-STAR v5.0 (Zou et al., 2023).[9]

Contrary to the climate model simulations—which predict greater warming in the troposphere, especially in the upper tropical band (see Figure 11.4 and IPCC AR6 Figure 3.10[10])—the satellite data show a lower warming trend than the surface records. Between the decades 1981–1990 and 2014–2024, the surface datasets are approximately +0.15 °C warmer, that is, 21% higher than the satellite estimates.

An anomaly between hemispheres: north vs. south. Even more pronounced is the hemispheric discrepancy: in the Northern Hemisphere, the surface data show excess warming relative to the satellite data—33% over land

Research: Atmospheres, 126, e2019JD032352.

Kennedy, J.J., Rayner, N.A., Atkinson, C.P., Killick, R.E., 2019. An ensemble dataset of sea-surface temperature change from 1850. Journal of Geophysical Research: Atmospheres, 124, 7719–7763.

[8]Spencer, R.W., Christy, J.R., Braswell, W.D., 2017. UAH version 6 global satellite temperature products: Methodology and results. Asia-Pacific Journal of Atmospheric Sciences, 53(1), 121–130.

[9]Zou, C.-Z., Xu, H., Hao, X., Liu, Q., 2023. Mid-tropospheric layer temperature record derived from satellite microwave sounder observations with backward merging approach. Journal of Geophysical Research: Atmospheres, 128, e2022JD037472.

[10]Available at: https://www.ipcc.ch/report/ar6/wg1/figures/chapter-3/figure-3-10/

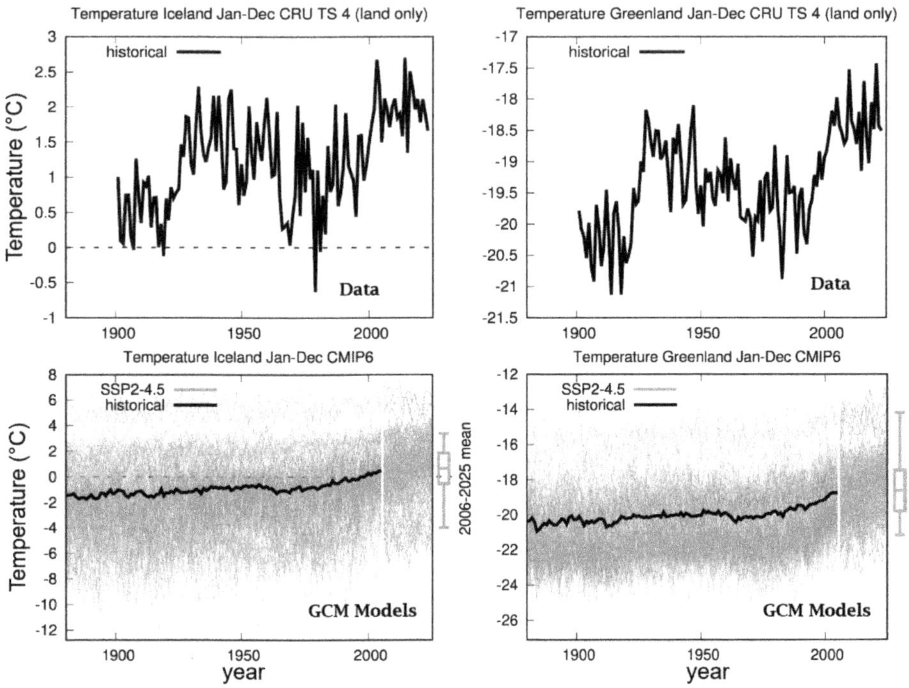

FIGURE 18.3. Comparison of temperatures in Iceland and Greenland from CRU-TS 4 and CMIP6 simulations. Note the models' inability to reproduce the mid-century cooling from the 1940s to 1970s and their overestimation of recent warming.

and 23% over ocean—whereas in the Southern Hemisphere the two datasets agree within their uncertainty margins (Scafetta, 2026).[11]

This suggests that the Northern Hemisphere, being more urbanized and densely populated, is more susceptible to local interferences that distort the surface measurements. By contrast, the Southern Hemisphere—with its larger oceanic extent and lower urban density—more accurately reflects the atmospheric dynamics.

The issue of the "*tropical hot spot*"—a region in the upper tropical troposphere predicted by the models to exhibit intensified warming—adds further complexity. Satellite observations do not confirm the existence of this hot spot (McKitrick & Christy, 2018; Mitchell et al., 2020),[12] highlighting yet another divergence between model projections and empirical data. This discrepancy is evident in Figures 11.4 and 18.5 between 100 and 500 hPa, which compare the observed and simulated mean tropical temperature trends using both CMIP5 and CMIP6 GCMs.

[11]Scafetta, N., 2025. Detection, attribution, and modeling of climate change: Key open issues. Gondwana Research, DOI: 10.1016/j.gr.2025.05.001.

[12]McKitrick, R., Christy, J., 2018. A test of the tropical 200–300-hPa warming rate in climate models. Earth and Space Science, 5, 529–536.
Mitchell, D.M., Lo, Y.T.E., Seviour, W.J.M., Haimberger, L., Polvani, L.M., 2020. The vertical profile of recent tropical temperature trends. Environmental Research Letters, 15(10), 104004.

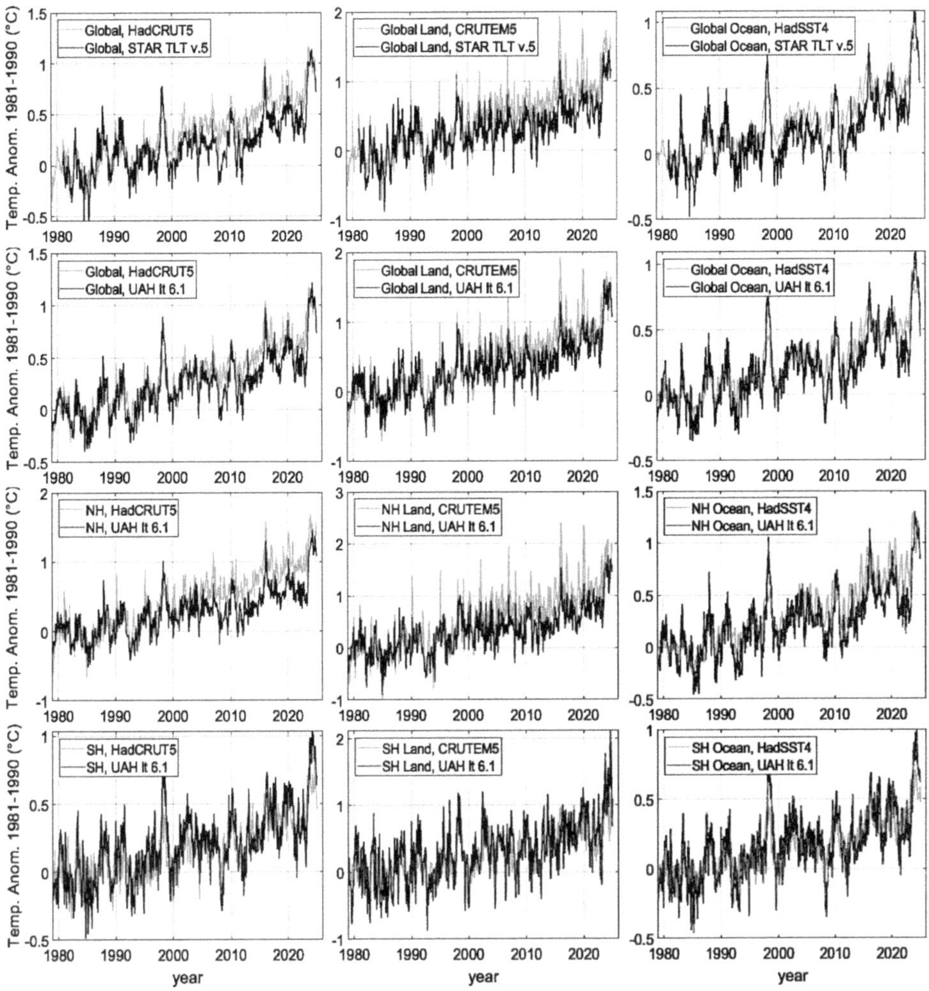

FIGURE 18.4. Comparison between surface temperature records—HadCRUT5 (global), CRUTEM5 (land), and HadSST (ocean) (upper curves) (Morice et al., 2021; Osborn et al., 2021; Kennedy et al., 2019)—and lower troposphere temperature records based on satellite NOAA-STAR v5.0 (Zou et al., 2023) and UAH MSU v. 6.1 (lower curves) (Spencer et al., 2017). The surface temperature records (lighter curves) consistently warm faster than the satellite lower troposphere temperature records (darker curves), particularly for the Northern Hemisphere land. (Adapted from Scafetta, 2026).

Revisions and controversies in satellite data. The debate over satellite-derived temperature records intensified in 2014, when Remote Sensing Systems (RSS) updated its dataset to better align with surface-based estimates (Mears and Wentz, 2016),[13] resulting in a version that exhibited more pronounced

[13]Mears, C.A., Wentz, F.J., 2016. Sensitivity of satellite-derived tropospheric temperature trends to

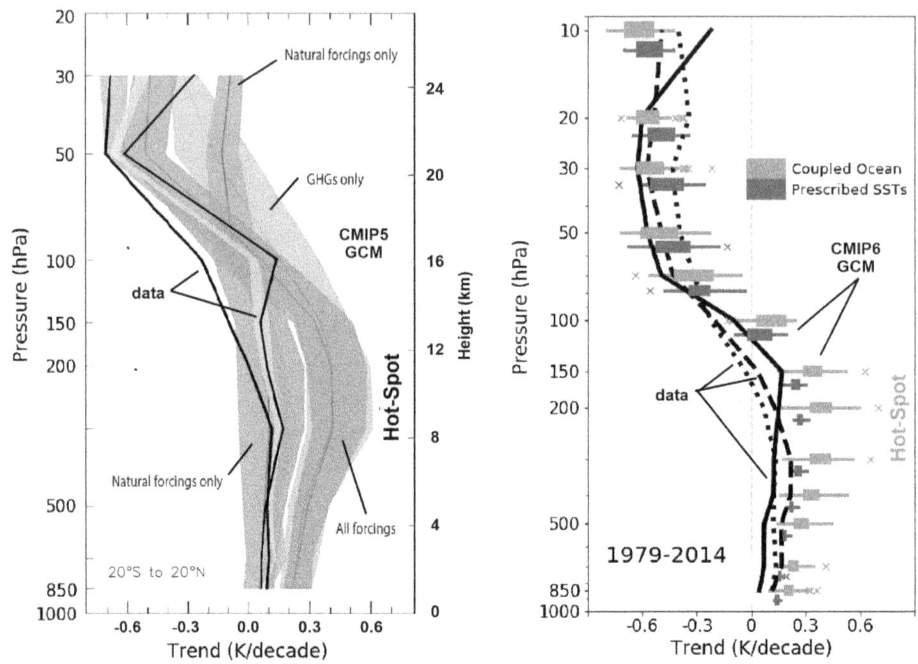

FIGURE 18.5. (Left) Observed (black curves) and simulated mean tropical temperature trends from 1961 to 2010 for CMIP5 simulations under various forcing assumptions; radiosonde observations from HadAT2, RAOBCORE, and RICH (adapted from IPCC, 2013, Figure 10.SM.1).
(Right) A similar result using the CMIP6 GCMs (adapted from Mitchell et al., 2020; IPCC, 2021, its figure 3.10).

warming. By contrast, the UAH MSU v6.1 dataset showed greater consistency with IGRA radiosonde data and with several independent reanalyses (Christy et al., 2018).[14]

Although NOAA-STAR v4.0 initially indicated warming trends similar to those of RSS, subsequent revisions—culminating in version 5.0 (Zou et al., 2023)[15]—corrected earlier errors and produced global warming trends slightly lower than those reported by UAH MSU. This evolution strengthens the hypothesis that the UAH MSU satellite data may be among the most physically consistent, while surface-based climate records—often derived from averaging and homogenizing meteorological station data—may be affected by local contamination and statistical artifacts that exaggerate the warming.

the diurnal cycle adjustment. Journal of Climate, 29(10), 3629–3646.

[14]Christy, J.R., Spencer, R.W., Braswell, W.D., Junod, R., 2018. Examination of space-based bulk atmospheric temperatures used in climate research. International Journal of Remote Sensing, 39, 3580–3607.

[15]Zou, C.-Z., Xu, H., Hao, X., Liu, Q., 2023. Mid-tropospheric layer temperature record derived from satellite microwave sounder observations with backward merging approach. Journal of Geophysical Research: Atmospheres, 128, e2022JD037472.

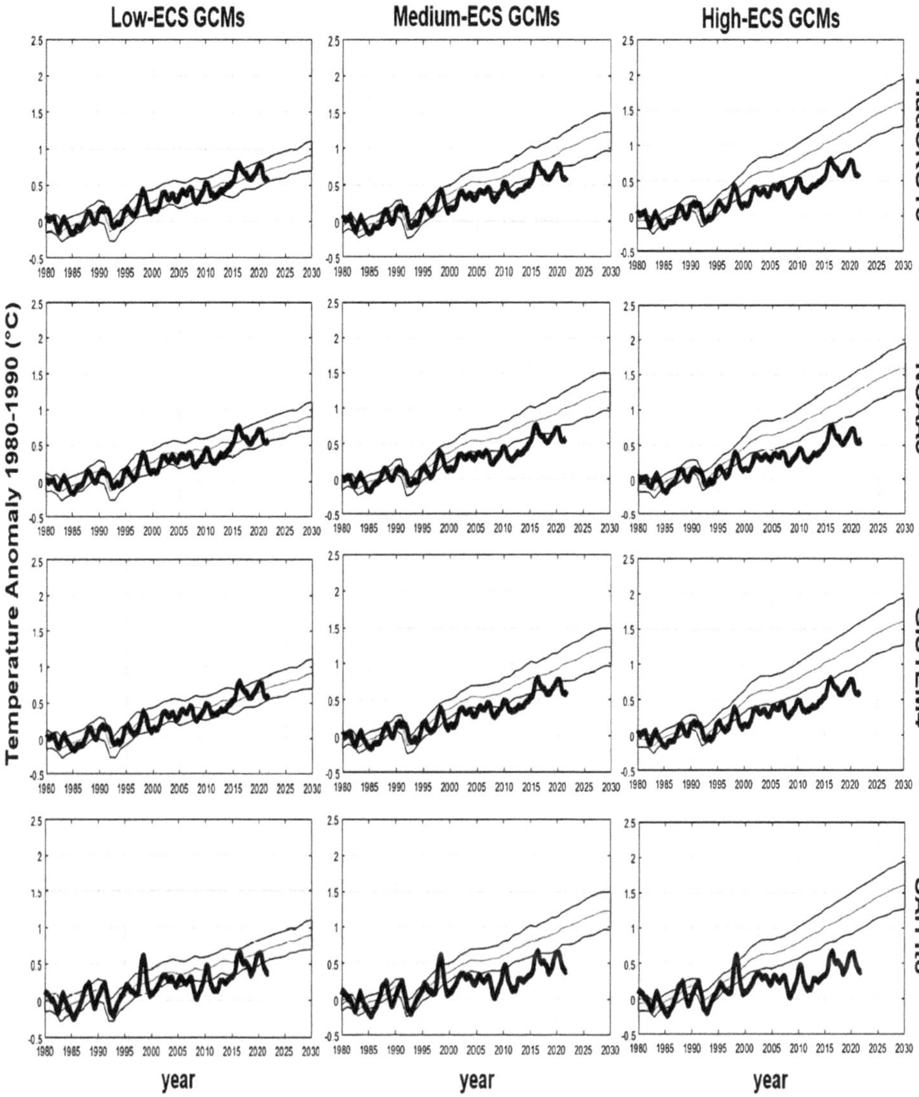

FIGURE 18.6. GCM global surface temperature ensembles (yellow area) versus HadCRUT5 (infilled data), GISTEMP v4, NOAAGlobTemp v5, and UAH-MSU-lt v6 temperature records (black, 12-month moving average). The climate models are grouped into low, medium, and high ECS. (Adapted from Scafetta, 2023b).

Implications for climate modeling. The divergence between satellite and surface temperature datasets not only casts doubt on the reliability of modern instrumental records but also raises concerns about the robustness of the predictive models that inform global climate policy. If satellite reconstructions more accurately represent the physical state of the troposphere, many current projections may be overly pessimistic, potentially misguiding mitigation strategies and economic planning. In fact, the global climate models (GCMs) developed under the CMIP6 project incorporate numerous free parameters, many of which are tuned to replicate observed warming trends. As discussed in Chapter 10 and noted by Mauritsen et al. (2019),[16] this tuning process can lead to an overestimation of climate sensitivity and to exaggerated projections of future warming if the climatic data are warm-biased.

The divergence between satellite and surface temperature datasets, on the one hand, and the CMIP6 GCM predictions, on the other, is illustrated in Figure 18.6, which compares global surface temperature ensembles from GCMs with the following observational datasets: HadCRUT5 (infilled), GISTEMP v4, NOAAGlobTemp v5, and UAH-MSU-lt v6 (shown in black as a 12-month moving average). The models are grouped by equilibrium climate sensitivity (ECS) into low, medium, and high categories. The figure suggests that only the low-ECS models (ECS $< 3°C$) are consistent with the observed surface temperature records. However, even this agreement breaks down when satellite-based lower-troposphere datasets are considered, since they show significantly less warming than the surface-based records (Scafetta, 2023b).[17]

18.4. Conclusion: data, models, and climate truth

In summary, the discrepancies among dendrochronological proxies, satellite observations, and instrumental surface records underscore the persistent uncertainties in observational climatology. Datasets considered "*canonical*" by institutions such as the IPCC—e.g., HadCRUT5 (Met Office & CRU), GISTEMP v4 (NASA GISS), NOAAGlobalTemp v6.0 (NOAA), and Berkeley Earth Temperature—appear vulnerable to systematic biases. Meanwhile, satellite and radiosonde measurements—though not without limitations—may offer greater physical consistency.

In this context, a critical reassessment of the data sources used in climate evaluations is essential. Only through such scrutiny can climate science maintain its credibility and effectively guide policy in a complex and evolving world.

---

[16]Mauritsen, T., Bader, J., Becker, T., Behrens, J., Bittner, M., et al., 2019. Developments in the MPI-M Earth System Model version 1.2 and its response to increasing CO_2. Journal of Advances in Modeling Earth Systems, 11, 998–1038.

[17]Scafetta, N., 2023b. CMIP6 GCM ensemble members versus global surface temperatures. Climate Dynamics, 60, 3091–3120.

CHAPTER 19

Contamination in the homogenization of meteorological data

In observational climatology, data homogenization is a necessary practice for correcting non-climatic discontinuities in meteorological records. These discontinuities may arise from instrumental changes, station relocations, shifts in observation protocols, or environmental transformations such as urbanization and land-use change. Without appropriate corrections, historical temperature series risk reflecting artificial trends that misrepresent the actual climate variability.

Datasets such as the Global Historical Climatology Network (GHCN) provide both raw and homogenized versions of temperature data. The homogenized datasets—designated as *"corrected"*—are used by the major climate agencies (NASA, NOAA, Hadley Centre) to produce official global temperature records. The algorithms (Menne et al., 2018)[1] aim to identify and adjust for *"break points"* and other statistical discontinuities attributed to non-climatic causes in the raw time series.

Although the theoretical framework is sound, the practical application of these algorithms raises several concerns.

19.1. *"Break points"* and algorithmic instability

One major challenge lies in the automatic detection of break points. Statistical algorithms often rely on generic assumptions and may overlook the environmental context of individual stations. This can result in misclassifications—whereby genuine climate variations are "corrected" as spurious, or actual discontinuities go undetected.

Squintu et al. (2020)[2] highlighted inconsistencies in break-point detection between geographically proximate stations under similar conditions, suggesting that algorithmic outcomes can vary significantly with the data configuration.

Even more problematic is the dynamic instability of the homogenized datasets. As noted by O'Neill et al. (2022),[3] each update—whether arising from new data or from minor corrections—can retroactively alter the entire

[1] Menne, M.J., Williams, C.N., Gleason, B.E., Rennie, J.J., Lawrimore, J.H., 2018. The Global Historical Climatology Network monthly temperature dataset, version 4. Journal of Climate, 31(24), 9835–9854.

[2] Squintu, A.A., van der Schrier, G., Štěpánek, P., Zahradníček, P., Tank, A.K., 2020. Comparison of homogenization methods for daily temperature series against an observation-based benchmark dataset. Theoretical and Applied Climatology, 140, 285–301.

[3] O'Neill, P., Connolly, R., Connolly, M., Soon, W., Chimani, B., et al., 2022. Evaluation of the homogenization adjustments applied to European temperature records in the Global Historical Climatology Network dataset. Atmosphere, 13(2), 285.

homogenized sequence, even changing historical trends for the same station. This undermines scientific reproducibility, a cornerstone of empirical research.

19.2. *"Urban blending"*: thermal contamination between stations

A significant side effect of homogenization is *"urban blending,"* which denotes the unintended contamination of rural data by urban warming signals. In theory, urban stations should be corrected by reference to nearby rural stations. In practice, however, the algorithms employed may reverse this logic, especially when uncontaminated rural stations are statistically interpreted as outliers relative to nearby, more urbanized stations.

Katata et al. (2023)[4] documented how spurious warming trends from urban heat islands (UHI) are sometimes redistributed to surrounding rural stations during homogenization, effectively corrupting them. The algorithms tend to apply symmetric adjustments between nearby stations, even when the underlying causes are asymmetric and localized, such as urban development. This leads to a dilution of the UHI signal across the network, rather than to its removal.

Given that most surface stations are located in urban or peri-urban areas, an accurate correction of UHI effects is difficult to achieve without targeted comparisons between carefully selected rural stations.

19.3. A systemic distortion: artificial convergence of stations

Urban blending becomes particularly insidious in the presence of accelerating urbanization. As more stations are enveloped by anthropogenic environments, the risk grows that the entire meteorological network becomes contaminated by local non-climatic effects, which are then normalized at the global level.

This may explain several discrepancies observed in the literature:

- nighttime minimum (T_{min}) temperatures warming much faster than daytime maximum (T_{max}) temperatures;
- surface temperature data warming faster than satellite and radiosonde lower-troposphere measurements;
- similar conflicts with independent climate proxies (e.g., tree rings, ice cores) and stable rural stations.

Figures 17.3, 18.1, and 18.4 illustrate these divergences, suggesting that the global climate average may reflect a statistical artifact rather than a faithful representation of climate evolution.

Case study: U.S. Historical Climatology Network (USHCN). Katata et al. (2023) highlight a possible homogenization bias by comparing climate reconstructions for the United States based on USHCN stations (excluding Alaska and the non-continental territories), analyzing three versions of the data:

- raw meteorological data;
- TOB-adjusted data (corrected for observation-time bias);
- fully homogenized data (PHA + TOB-adjusted).

[4]Katata, G., Connolly, R., O'Neill, P., 2023. Evidence of urban blending in homogenized temperature records in Japan and the United States. Journal of Applied Meteorology and Climatology, 62(8), 1095–1114.

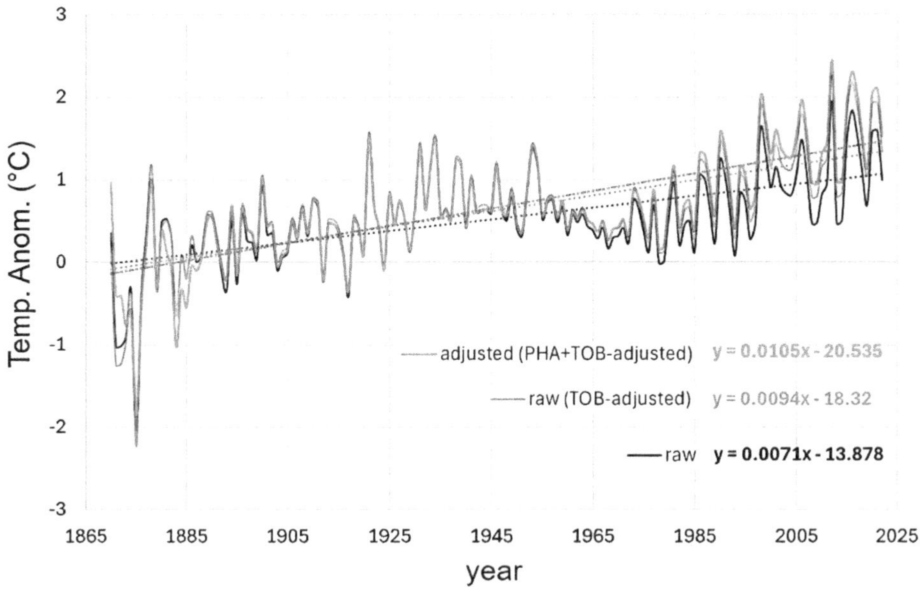

FIGURE 19.1. Comparison of climate reconstructions for the United States using stations from the U.S. Historical Climatology Network (USHCN) from 1870 to 2022 in three versions: (1) raw meteorological data show a warming of 1.08 °C; (2) data adjusted for observation time error (TOB-adjusted) show a warming of 1.43 °C; (3) fully homogenized climate record (PHA+TOB-adjusted) shows a warming of about 1.60 °C. Thermal anomalies relative to 1870–1900. (Adapted from O'Neill et al., 2022).

Figure 19.1 shows that:

- the raw data suggest a warming of about 1.08 °C from 1870 to 2022;
- the TOB-adjusted data show a warming of about 1.43 °C;
- the fully homogenized data indicate a warming of about 1.60 °C.

This result appears paradoxical: the strongest warming trend emerges from the most *"corrected"* dataset, despite the expectation that urbanization should have contributed to additional warming. Instead, the homogenized data imply that cities may have cooled the climate record—an implausible outcome.

The complexity of station distribution and temporal coverage may not fully account for this anomaly. It is reasonable to hypothesize that the homogenization algorithm may have corrupted, rather than corrected, the original series, especially the rural ones, by transferring to them urban warming through statistical blending.

Alternative dataset comparison. Roy Spencer (2012)[5] conducted a critical analysis of USHCN data from 1973 to 2012, comparing:

- HadCRUT3;

[5]Spencer, R.W., 2012. USHCN Surface Temperatures, 1973–2012: Dramatic Warming Adjustments,

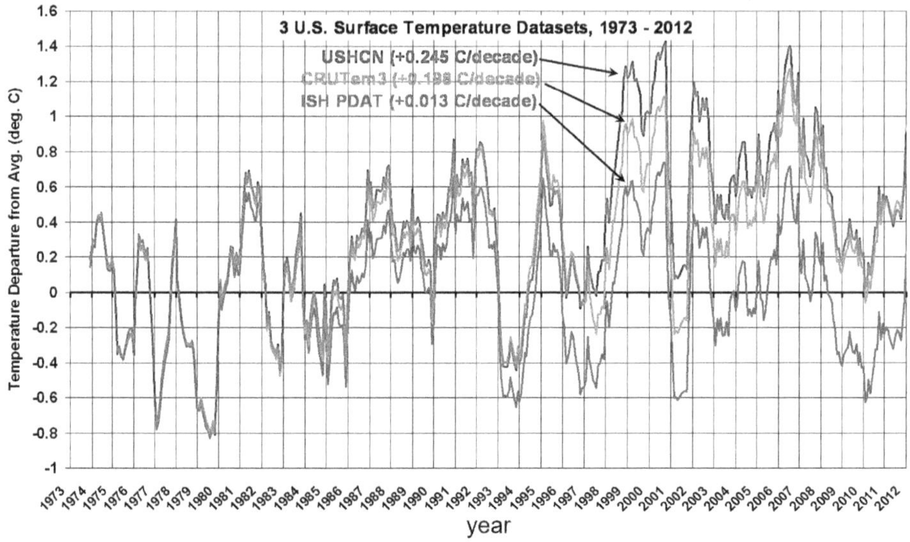

FIGURE 19.2. Comparison between the corrected (homogenized) version of USHCN published by NOAA/NCDC, HadCRUT3, and the alternative dataset "*International Surface Hourly*" (ISH), which represents raw data corrected based on population density. (Adapted from Spencer, 2012).

- NOAA/NCDC homogenized USHCN data;
- the ISH (International Surface Hourly) dataset.

The ISH dataset, derived from hourly observations at airport and meteorological stations, was adjusted using population density as a proxy for the Urban Heat Island (UHI) effect (see Figure 19.2). Spencer identified a modest warming trend of +0.013 °C per decade, in stark contrast to the +0.245 °C per decade trend found in the homogenized USHCN data. This discrepancy suggests that much of the warming in the corrected dataset may have been introduced by statistical algorithms rather than directly observed.

Spencer (2012) also noted that HadCRUT3 exhibited a more moderate global warming trend than later datasets such as HadCRUT4 and the updated NOAA records. The divergence between HadCRUT3 and its successors, together with the inconsistency between the ISH and USHCN data, raises concerns that homogenization procedures may have artificially inflated warming trends. Although these algorithms are designed to correct for discontinuities caused by changes in instrumentation or by station relocations, they may have overcompensated, spreading localized warming anomalies across multiple stations. Critically, the most substantial adjustments occurred between 1995 and 1997, coinciding with a sudden rise in the warming trend of the corrected data—an increase that is not present in either the raw or the hourly datasets. Taken together, the comparison among HadCRUT3, homogenized USHCN data, and raw ISH data supports the

Noisy Trends. Available at: https://www.drroyspencer.com/2012/04/ushcn-surface-temperatures-1973-2012-dramatic-warming-adjustments-noisy-trends/

hypothesis that part of the warming reflected in the official climate series may result from methodological artifacts rather than from actual climatic changes.

19.4. Conclusion: a methodological paradox

The stated goal of homogenization is to filter out non-climatic noise and reveal the authentic signal of climate change. In principle, this process aims to correct for discontinuities caused by changes in instrumentation, station relocations, or observational practices. However, the algorithmic nature of homogenization can introduce new layers of uncertainty—subtle and often invisible to non-specialists, yet critical when reconstructing long-term temperature trends.

The issues raised in this chapter highlight the urgent need for a more robust and transparent approach to data homogenization, including:

- greater methodological transparency, with full disclosure of the adjustment algorithms, calibration criteria, and decision thresholds;
- independent verification, also using proxies (e.g., tree rings, ice cores, historical records) and alternative datasets not subject to the same correction procedures;
- rigorous screening of meteorological stations, taking into account geographic settings, urbanization effects, and local anthropogenic influences;
- a reassessment of the scientific weight assigned to potentially contaminated or heavily adjusted series, especially when they dominate regional or global averages.

If the observational datasets that underpin climate projections are subject to unstable, opaque, or poorly validated modifications, then the conclusions derived from them—however widely accepted—must be critically re-evaluated. This is particularly important when such conclusions serve as the foundation for global policy decisions, economic planning, and societal transformation.

Climate science must commit to methodological integrity to ensure that its empirical foundations are as solid and transparent as its theoretical frameworks. Only then can it retain public trust and fulfill its role as a guide for responsible, evidence-based action.

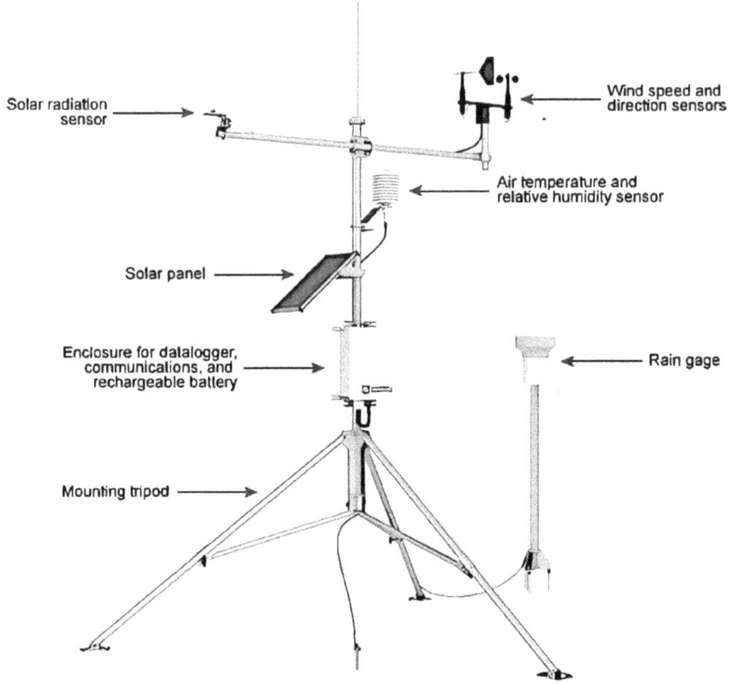

FIGURE 19.3. **Digital automated weather station.** Illustration of a complete digital weather station designed for continuous atmospheric monitoring. The components are mounted on a stable tripod and arranged so as to minimize interference between sensors. At the top, the solar radiation sensor measures the incoming solar energy, essential for climate analysis and energy-balance studies. Nearby, the wind speed and direction sensors (anemometer and vane) capture air movement, which is crucial for meteorological modeling and forecasting. At an intermediate level, the combined air temperature and relative humidity sensor provides data for calculating the heat index and for assessing atmospheric stability. The station is powered by a solar panel that charges an internal battery housed within a protective enclosure, which also contains the data logger and the communication modules. This system enables real-time data recording and remote transmission. At the base, the rain gauge accurately measures precipitation, supporting hydrological assessments and water-resource management. The entire setup is engineered to withstand harsh weather conditions and to ensure reliable measurements in remote or exposed environments. This configuration represents a standard for automated meteorological data collection, widely used in climatology, precision agriculture, and environmental research.

CHAPTER 20

Nighttime temperatures rising faster than daytime

A promising method for detecting non-climatic distortions in surface temperature records was first proposed by Scafetta & Ouyang (2019),[1] and later refined by Scafetta (2021a).[2] The initial study focused on identifying Urban Heat Island (UHI) biases in the Chinese climate network by analyzing the divergence between nighttime minimum temperatures (T_{min}) and daytime maximum temperatures (T_{max}). The subsequent work extended the methodology to global datasets, again comparing observed temperature trends with climate model simulations in order to detect non-climatic biases in land surface records.

This approach relies on the differential behavior of T_{min} (typically reached around sunrise) and T_{max} (usually recorded one to two hours after solar noon). By examining their respective warming trends in homogenized global land datasets such as CRU-TS4 (Harris et al., 2020),[3] the method helps to isolate artificial influences from genuine climatic signals.

Global Climate Models (GCMs) predict that, in a warming world, T_{min} should increase slightly more than T_{max} owing to nighttime heat retention near the surface. This means that the *diurnal temperature range* (DTR)—the difference between the daily maximum temperature (T_{max}) and the daily minimum temperature (T_{min})—diminishes, while the *"negative" diurnal temperature range,*

$$\mathrm{NDTR}(t) = -\mathrm{DRT}(t) = T_{min}(t) - T_{max}(t),$$

increases. A positive NDTR(t) trend indicates that T_{min} warms faster than T_{max}.

Observations show, however, that in many regions—especially across the Northern Hemisphere and near clusters of urban centers—nighttime temperatures are not merely warming moderately more than daytime temperatures, but have been warming significantly more than predicted by the GCMs. This discrepancy suggests that non-climatic surface factors, such as urbanization and land-use changes, may have disproportionately affected T_{min} measurements, thereby distorting the climate signal derived from the daily averages of T_{min} and T_{max}.

---

[1]Scafetta, N., Ouyang, S., 2019. Detection of UHI bias in China climate network using Tmin and Tmax surface temperature divergence. Global and Planetary Change, 181, 102989.

[2]Scafetta, N., 2021a. Detection of non-climatic biases in land surface temperature records by comparing climatic data and their model simulations. Climate Dynamics, 56(9–10), 2959–2982.

[3]Harris, I., Osborn, T.J., Jones, P., Lister, D., 2020. Version 4 of the CRU TS monthly high-resolution gridded multivariate climate dataset. Scientific Data, 7, 109.
(CRU-TS4 = Climatic Research Unit Time Series version 4; it covers only land temperature).

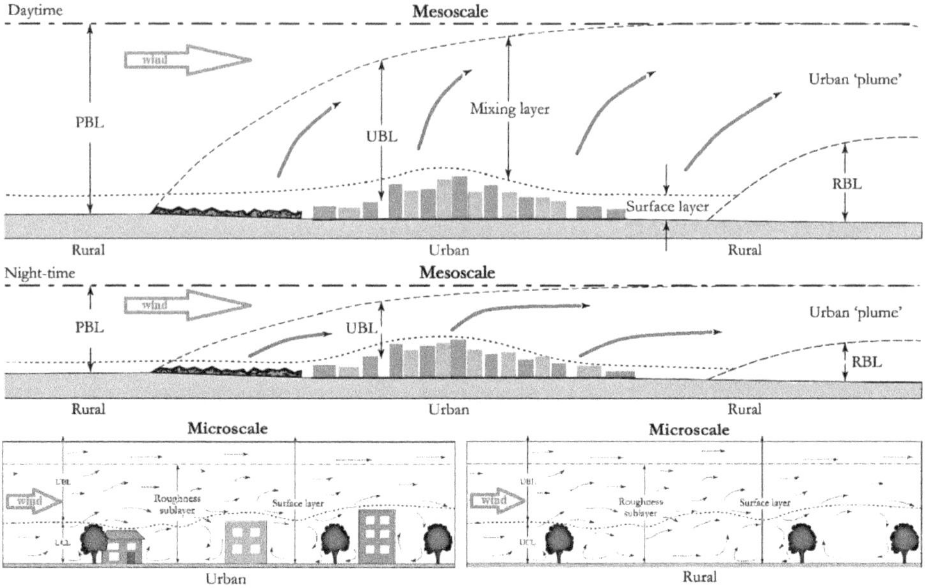

FIGURE 20.1. Schematic structure of the atmospheric boundary layer over urban and rural areas. During the daytime, the boundary layers are higher owing to solar heating, enhanced turbulence, and urban heat dissipation; at night, they become lower and more stratified. Horizontal wind flow transports heat from urban centers to the surrounding rural zones, contributing to localized warming effects by enlarging the UHI. The key layers include: the Planetary Boundary Layer (PBL), which encompasses the full turbulent zone influenced by surface processes; the Urban Boundary Layer (UBL), which reflects urban-specific thermal and dynamical structures; and the Urban Canopy Layer (UCL), confined within the built environment and directly shaped by buildings, streets, and vegetation. (Adapted from Scafetta, 2021a).

20.1. Atmospheric mechanisms behind the phenomenon

To understand the excess nighttime warming, it is essential to examine the dynamics of the near-surface atmosphere. During the day, solar radiation induces vertical convection, dispersing heat into higher atmospheric layers and reducing the influence of overheated surfaces.

At night, the ground cools rapidly through radiative loss, and the thermal-inversion boundary layer descends. Heat becomes trapped within the first $\sim$100 m of the atmosphere—precisely where urban surfaces exert the strongest influence. Materials such as asphalt and concrete release stored heat slowly, raising local nighttime temperatures. This heat then disperses horizontally, expanding the urban heat island (UHI) footprint owing to the "cap" effect created by the lowered inversion layer.

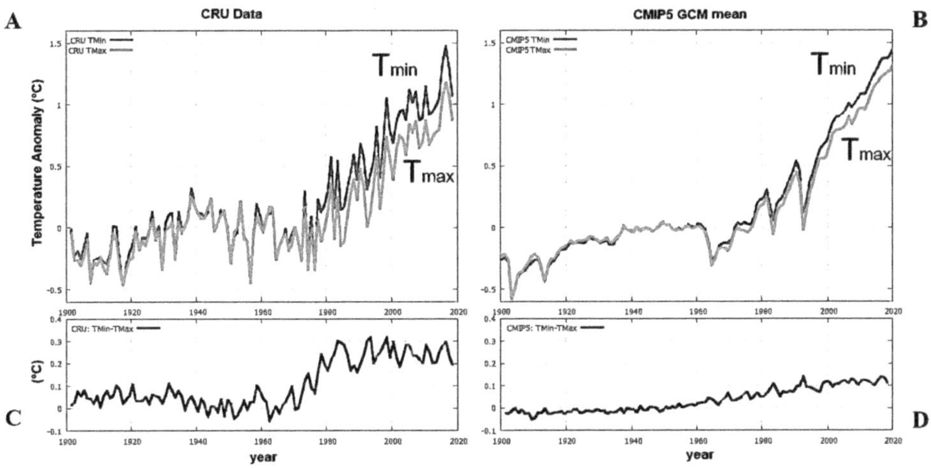

FIGURE 20.2. Comparison of CRU-TS4 T_{min} and T_{max} trends with CMIP5 multi-model averages. The observed increase in NDTR is significantly larger than that simulated by the models. (Adapted from Scafetta, 2021a).

Figure 20.1 schematically shows the difference in altitude of the atmospheric boundary layer during the day (typically higher) and at night (typically lower). Urban expansion disproportionately affects T_{min}, while T_{max} remains less impacted, thereby amplifying the UHI effect at night. Kim & Christy (2022)[4] observed the same phenomenon in the San Joaquin Valley of California.

20.2. Spatial divergence between T_{min} and T_{max} values

Figure 20.2 compares global land-surface T_{min} and T_{max} trends from CRU-TS4 with CMIP5 model projections. The lower panels show:

- (a) CMIP5 ensemble simulations for T_{min};
- (b) CMIP5 simulations for T_{max};
- (c–d) the differences between observed and simulated trends.

Both models and observations show T_{min} warming faster than T_{max}. However, the CRU-TS4 data reveal that, from 1900 to 2020, the observed divergence—$\text{NDTR} = -\text{DTR} = T_{min} - T_{max}$—is nearly twice as large as predicted by the CMIP5 GCMs, as illustrated in the lower panels of Figure 20.2. The most pronounced divergence occurs between 1960 and 1990, coinciding with rapid global urbanization.

Figure 20.3 maps the variation in $\text{NDTR} = T_{min} - T_{max}$ across global regions between the decade 1945–1955 and the decade 2013–2023. The anomaly shown between the two decades is

$$\Theta_{1950-2017} = \text{NDTR}_{2013-2023} - \text{NDTR}_{1945-1955}.$$

Regions exhibiting values of $\Theta_{1950-2017}$ significantly higher than those theoretically predicted by the GCMs indicate a surplus in T_{min} warming that is not

[4]Kim, D., Christy, J.R., 2022. Detecting impacts of surface development near weather stations since 1895 in the San Joaquin Valley of California. Theoretical and Applied Climatology, 149, 1223–1238.

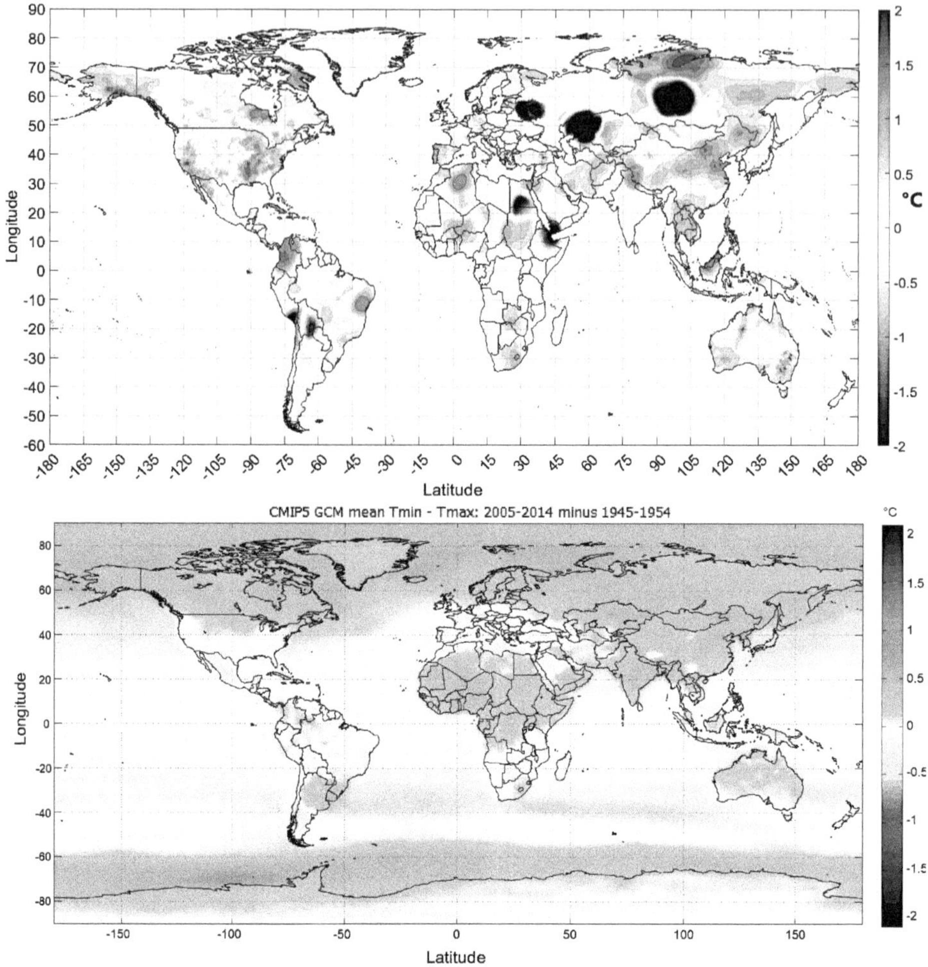

FIGURE 20.3. Top: Global distribution of $\Theta_{1950-2017}$ calculated as the difference between the 1945–1954 and 2013–2023 periods, based on the CRU-TS4 dataset (adapted from Scafetta, 2026). Bottom: Global distribution of the same function between 1945–1954 and 2005–2014, derived from CMIP5 GCM projections (adapted from Scafetta, 2021a).

captured by the models. This discrepancy may be attributed to residual UHI effects and other surface thermal anomalies that are insufficiently corrected by the homogenization algorithms.

The upper-panel map of Figure 20.3 is computed using the CRU TS monthly data (Harris et al., 2020).[5] It displays several regional strong positive values of Θ, mostly across the Northern Hemisphere, with zones indicating NDTR increases ranging from +0.5 °C to +2 °C, often clustered around urban centers.

[5]Harris, I., Osborn, T.J., Jones, P., Lister, D., 2020. Version 4 of the CRU TS monthly high-resolution gridded multivariate climate dataset. Scientific Data, 7, 109.

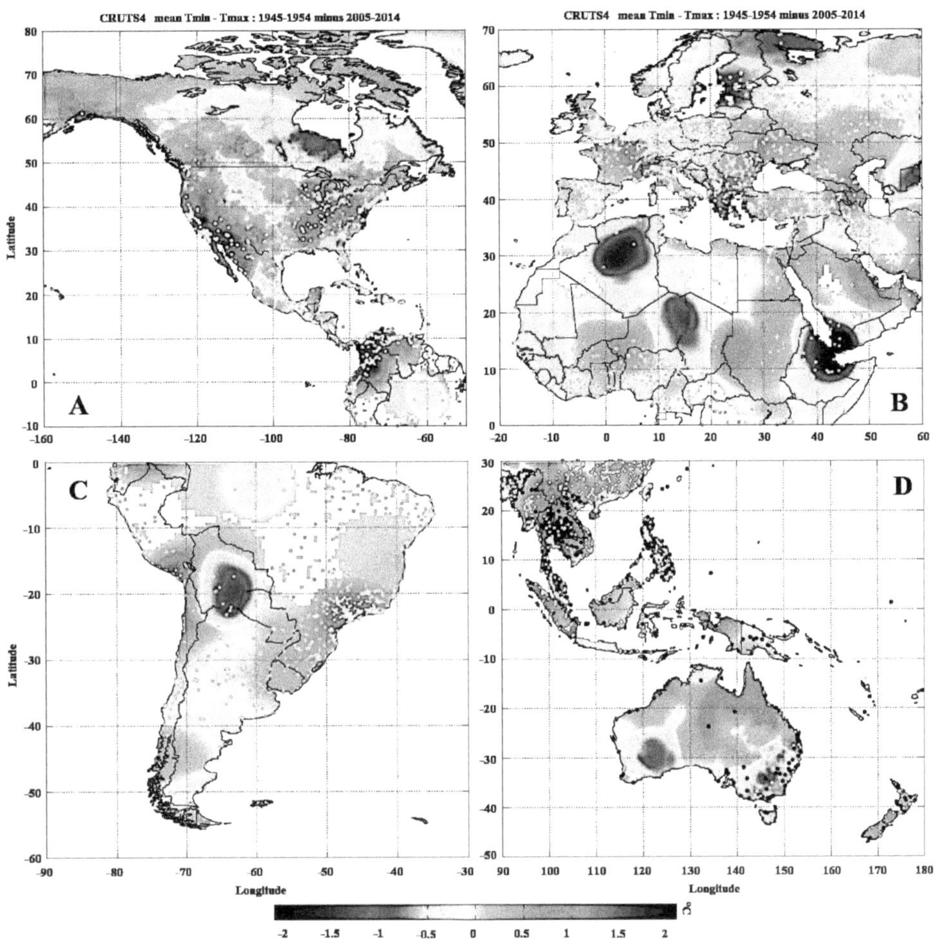

FIGURE 20.4. Global distribution of Θ between 1945–1954 and 2005–2014 based on the CRU-TS4 dataset. Colored dots indicate cities with populations exceeding 50,000; black dots in the lower right panel represent cities with more than 10,000 inhabitants. (Adapted from Scafetta, 2021a).

By contrast, the lower-panel map shows the CMIP5 GCM-based theoretical prediction, characterized by more homogeneous values of Θ, typically ranging between –0.3 °C and +0.5 °C.

Even small urban agglomerations can produce measurable UHI effects, as demonstrated by Cardoso et al. (2017)[6] and Pinho & Orgaz (2000),[7] indicating that the distortion is not limited to large metropolitan areas. By contrast, the Southern Hemisphere shows a much smaller $\Theta_{1950-2017}$ anomaly, consistent

[6]Cardoso, R.D.S., Dorigon, L.P., Teixeira, D.C.F., Amorim, M.C.d.C.T., 2017. Assessment of urban heat islands in small- and mid-sized cities in Brazil. Climate, 5(1), 14.

[7]Pinho, O., Orgaz, M., 2000. The urban heat island in a small city in coastal Portugal. International Journal of Biometeorology, 44, 198–203.

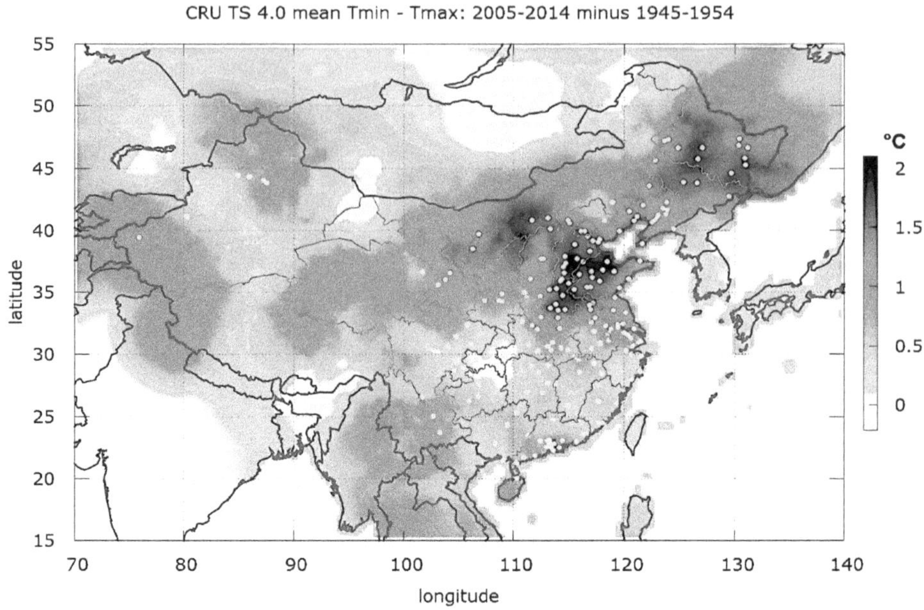

FIGURE 20.5. Global distribution of Θ between 1945–1954 and 2005–2014, based on the CRU-TS4 dataset. Dots indicate Chinese cities with populations exceeding 50,000. (Adapted from Scafetta & Ouyang, 2019).

with its lower urbanization and greater oceanic coverage Scafetta (2026).[8] Some dark areas show negative Θ values, possibly linked to *"Urban Cool Island"* (UCI) effects, especially in arid zones where cities are built near oases or water bodies. Other contributing factors may include deforestation, irrigation, permafrost changes, or differences in soil emissivity (Scafetta & Ouyang, 2019; Scafetta, 2021a).

Figure 20.4 magnifies the same type of map shown in Figure 20.3, comparing it with the distribution of cities with more than 50,000 inhabitants and, in some cases, with urban centers of more than 10,000 inhabitants. Figure 20.5 does the same for China. The darker areas are often characterized by clusters of numerous cities, as in China, the United States, Colombia, the region surrounding São Paulo in Brazil, and others.

Figure 20.6 compares summer and winter T_{max} and T_{min} temperatures from the CRU-TS4 dataset in China with their corresponding CMIP5 GCM simulations (Scafetta & Ouyang, 2019). A key feature is the magnitude of warming observed between the periods 1940–1950 and 2000–2010, which encompass two distinct warming peaks. The empirical data consistently show that the T_{min} records exhibit greater warming than the T_{max} records, and that both winter T_{min} and winter T_{max} trends are stronger than their summer counterparts.

The observed differences among these warming trends are, however, significantly larger than those simulated by the CMIP5 models. Notably, the summer

[8]Scafetta, N., 2025. Detection, attribution, and modeling of climate change: Key open issues. Gondwana Research, DOI: 10.1016/j.gr.2025.05.001.

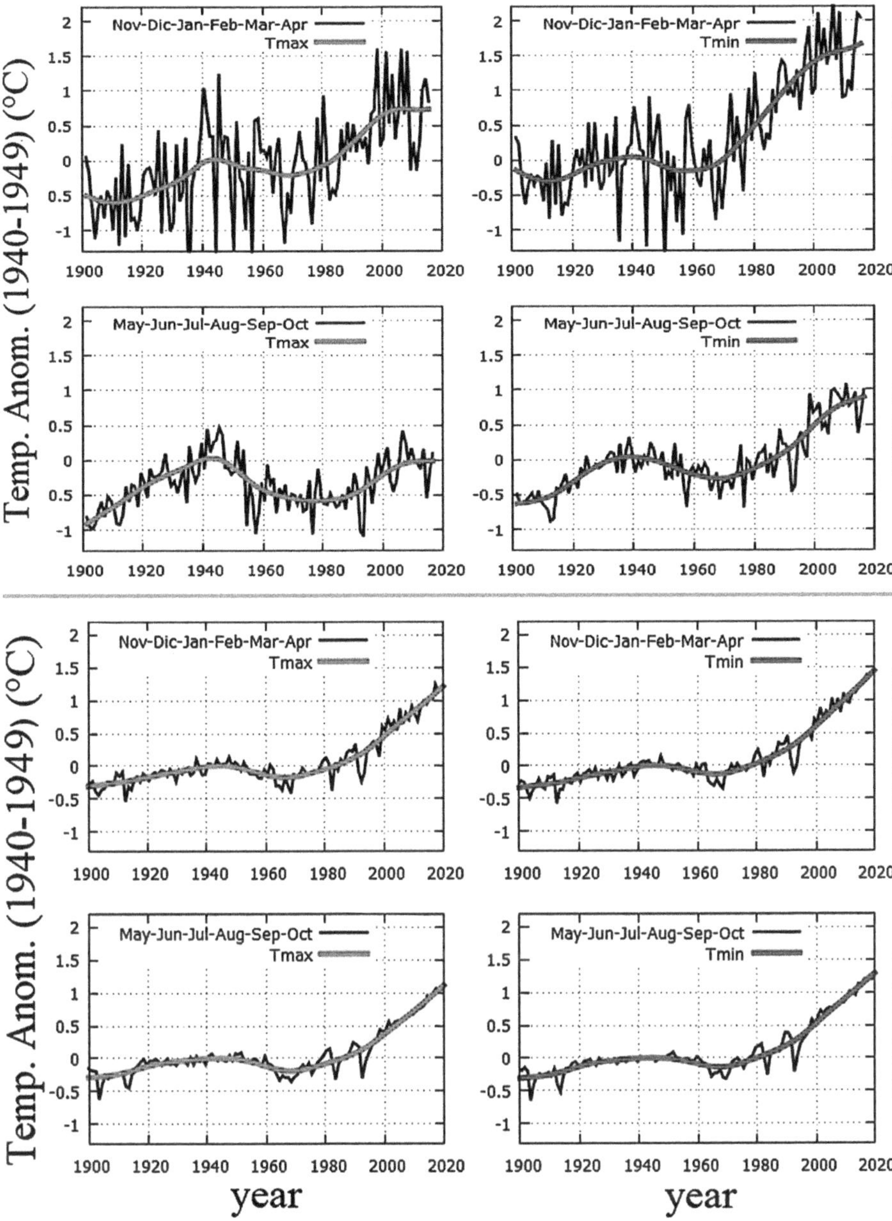

FIGURE 20.6. Observed summer and winter T_{max} and T_{min} temperatures in China (top), compared with the corresponding simulations from CMIP5 Global Climate Models (bottom). (Adapted from Scafetta & Ouyang, 2019).

T_{max} record shows no appreciable warming between 1940–1950 and 2000–2010, a result that is consistent with the Northern Hemisphere tree-ring-based temperature reconstruction shown in Figure 18.1. This empirical finding stands in stark contrast to the CMIP5 GCM simulations, which indicate pronounced warming in the summer T_{max} record. Such a data–model discrepancy strongly suggests substantial contamination from urban heat island (UHI) effects in the observational data.

20.3. Implications for data quality and the perception of warming

These findings suggest that local anthropogenic changes have selectively influenced T_{min}, and that these anomalies have not been adequately corrected by the homogenization algorithms. Since global temperature averages are computed from the daily means of T_{min} and T_{max}, an excessive distortion of T_{min} can artificially inflate global warming estimates.

The observed divergence between nighttime minimum temperatures (T_{min}) and daytime maximum temperatures (T_{max}) thus serves as a powerful diagnostic tool for evaluating the integrity of surface climate records. This asymmetry not only reveals the influence of local anthropogenic factors—such as urbanization and land-use change—but also helps to explain the persistent gap between surface temperature trends and those observed in the lower troposphere, as discussed in connection with Figure 18.4 and throughout the preceding chapters.

20.4. Conclusion: the thermometer measures only the local microclimate

The question inevitably resurfaces:

> *Do surface thermometers truly measure global climate, or are they capturing a signal increasingly shaped by local thermal alterations—independent of the radiative forcings modeled in GCMs?*

In this light, the thermometer becomes less a window onto planetary climate dynamics and more a lens onto microclimatic evolution, shaped by human infrastructure, environmental transformation, and urban sprawl. Recognizing this distinction is essential to ensuring that climate assessments remain grounded in physical reality rather than in statistical artifacts.

CHAPTER 21

The 2000–2014 temperature *"pause"* controversy

Between approximately 2000 and 2014, a notable slowdown in the rate of global warming was observed. This phenomenon has been widely referred to in the literature as the *hiatus*, the *thermal plateau*, or the *pause*. During this period, the rise in surface temperatures was significantly lower than projected by the global climate models (GCMs), raising serious questions about the physical accuracy of the models.

In its Fifth Assessment Report, the IPCC (2013)[1] explicitly acknowledged this pause, describing it as a significant discrepancy between observations and model outputs for the period 1998–2012. Numerous independent studies (cf. Scafetta, 2012a; Fyfe et al., 2013)[2] confirmed the flattening of global mean temperatures during this fifteen-year interval.

21.1. The HadCRUT records: upgrades or revisionism?

One of the principal datasets used to assess global temperature trends is HadCRUT, produced by the Met Office Hadley Centre (Had) in collaboration with the Climatic Research Unit (CRU) at the University of East Anglia. In the acronym, "T" stands for *"temperature."* HadCRUT records combine both land and ocean data. They have undergone multiple revisions:

- HadCRUT3 (version 3, discontinued in 2014; Brohan et al., 2006[3]);
- HadCRUT4 (version 4, discontinued in 2021; Morice et al., 2012[4]);
- HadCRUT5 (version 5, the current version; Morice et al., 2021[5]).

Each update has significantly altered the reported values, especially in the post-2000 period. Figure 21.1 compares HadCRUT3, HadCRUT4, and HadCRUT5,

---

[1]IPCC, 2013. Climate Change 2013: The Physical Science Basis. Cambridge University Press.

[2]Scafetta, N., 2012a. Testing an astronomically based decadal-scale empirical harmonic climate model versus the IPCC (2007) general circulation climate models. Journal of Atmospheric and Solar–Terrestrial Physics, 80, 124–137.

Fyfe, J., Gillett, N., Zwiers, F., 2013. Overestimated global warming over the past 20 years. Nature Climate Change, 3, 767–769.

[3]Brohan, P., Kennedy, J.J., Harris, I., Tett, S.F.B., Jones, P.D., 2006. Uncertainty estimates in regional and global observed temperature changes: A new dataset from 1850. Journal of Geophysical Research: Atmospheres, 111, D12106.

[4]Morice, C.P., Kennedy, J.J., Rayner, N.A., Jones, P.D., 2012. Quantifying uncertainties in global and regional temperature change using an ensemble of observational estimates: the HadCRUT4 dataset. Journal of Geophysical Research: Atmospheres, 117, D08101.

[5]Morice, C.P., Kennedy, J.J., Rayner, N.A., Winn, J.P., Hogan, E., et al., 2021. An updated assessment of near-surface temperature change from 1850: the HadCRUT5 data set. Journal of Geophysical Research: Atmospheres, 126, e2019JD032361.

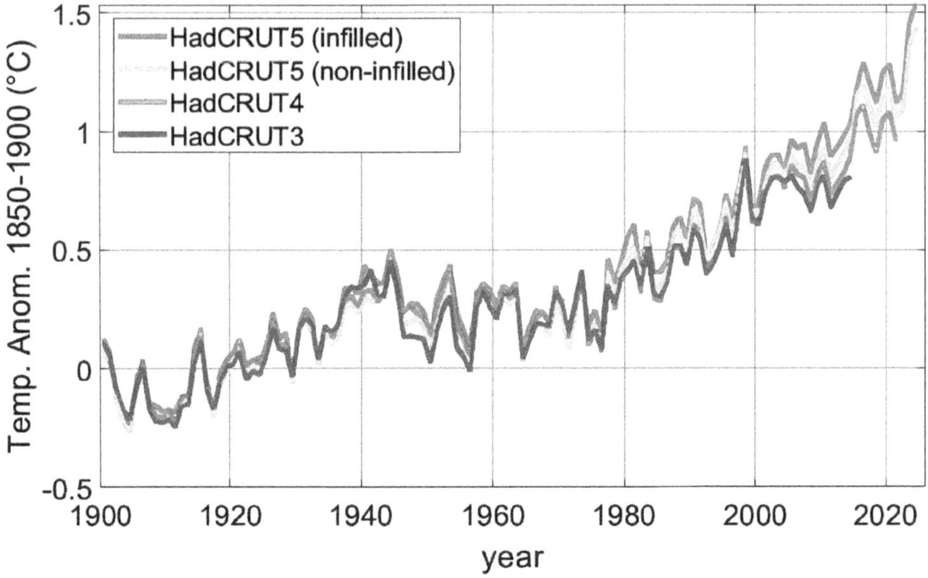

FIGURE 21.1. Comparison of global temperature records: HadCRUT3 (discontinued in 2014; Brohan et al., 2006), HadCRUT4 (retired in 2021; Morice et al., 2012), and HadCRUT5, including the *filled* and *non-filled* versions (Morice et al., 2021).

showing that the warming trend from 2000 to 2014 increased by up to +20% across versions, effectively transforming a *pause* into a *warming trend.*

It is also relevant to note that HadCRUT5 is published in two formats:

- *non-infilled data*: based solely on homogenized direct observations; covers approximately 80% of Earth's surface;
- *infilled data*: incorporates statistical modeling to estimate anomalies in data-sparse regions; covers 100% of Earth's surface.

The infilled version, which includes synthetic data, exhibits the most pronounced warming, nearly matching the trend observed from 1970 to 2000. These changes are typically attributed to:

- the inclusion of new observational data from under-sampled regions;
- updates to the homogenization algorithms;
- model-based interpolations, especially in the Arctic.

Although these adjustments are presented as methodologically justified, they have retroactively reshaped the climate curve, effectively erasing an inconvenient anomaly relative to the model projections.

Figure 21.2 shows the surface coverage of the HadCRUT3 (Brohan et al., 2006) and HadCRUT4 (Morice et al., 2012) datasets for two selected months. White areas indicate regions without data. The table shown in Figure 21.3 reports the percentage of surface coverage in the HadCRUT datasets from version 1 to version 5. It shows that only 80% of Earth's surface is covered by direct measurements. HadCRUT v.5 (*infilled*) covers 100% of the surface, but does so by filling the gaps with synthetic data.

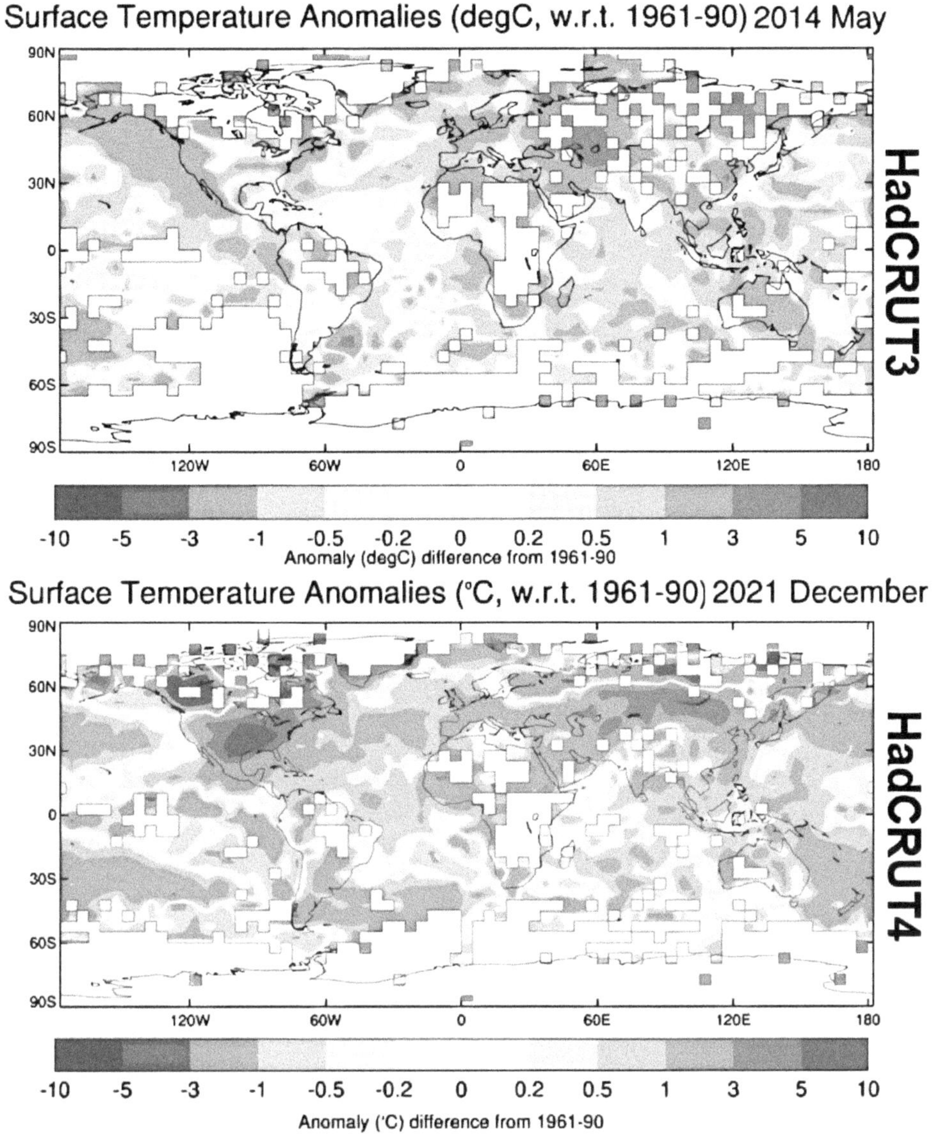

FIGURE 21.2. Surface coverage in the HadCRUT3 dataset (Brohan et al., 2006) and HadCRUT4 (Morice et al., 2012) for the indicated months and years. White areas are not covered.

21.2. The persistence of the pause in alternative climate data

Despite the apparent disappearance of the *hiatus* in the updated surface datasets, this phase of slowed warming remains evident in the satellite records and, to some extent, in the oceanic data. Sea surface temperatures (SSTs) and satellite measurements of the lower troposphere (e.g., UAH, NOAA-STAR)

HadCRUT Version	Reference Period	Estimated Land-Surface Area Coverage	Main Notes
HadCRUT1	1881–1993 (esteso a 1856–2002)	~65%	First combined dataset; 5×5° grid, no interpolation; limited coverage.
HadCRUT2	1856–2005	~70%	Improvements in bias and uncertainty; still no interpolation.
HadCRUT3	1850–2012	~75%	Introduction of a more complete uncertainty model; improved coverage but not global.
HadCRUT4	1850–2020	~80%	Addition of new data and corrections in marine biases; still no interpolation.
HadCRUT5 (non infilled)	1850–oggi	~80%	Traditional version; includes only cells with direct observations, similar to HadCRUT4.
HadCRUT5 (infilled)	1850–oggi	~100%	Interpolated version ("analyzed"); uses Gaussian Process Regression to estimate temperatures even in areas without data.

FIGURE 21.3. Percentage of surface coverage in HadCRUT datasets from version 1 to version 5.

continue to display a significant attenuation of warming from 2000 to 2014 (Spencer et al., 2017; Zou et al., 2023),[6] as shown in Figure 21.4.

[6]Spencer, R.W., Christy, J.R., Braswell, W.D., 2017. UAH version 6 global satellite temperature products: Methodology and results. Asia-Pacific Journal of Atmospheric Sciences, 53(1), 121–130. Zou, C.-Z., Xu, H., Hao, X., Liu, Q., 2023. Mid-tropospheric layer temperature record derived from satellite microwave sounder observations with backward merging approach. Journal of Geophysical Research: Atmospheres, 128, e2022JD037472.

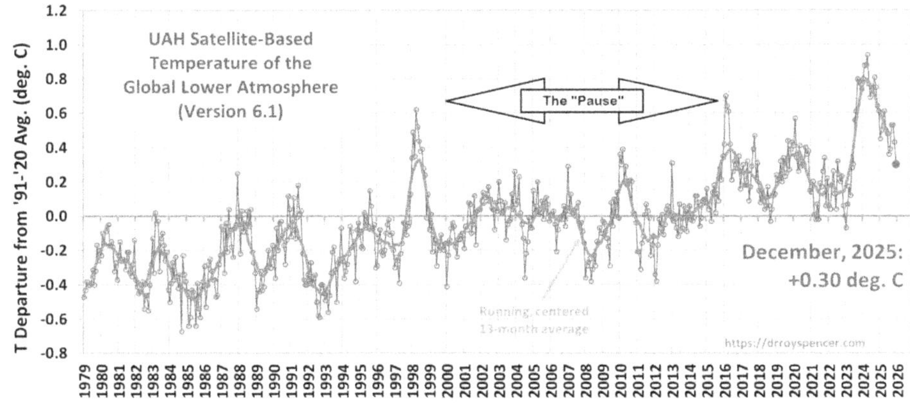

FIGURE 21.4. Global satellite temperature estimates of the lower troposphere, UAH MSU v6.0 lower troposphere (lt) (Spencer et al., 2017), continue to show the *pause* from 2000 to 2014. Note also the strong global warm peak recorded mostly in 2023–2024, likely influenced by the Hunga Tonga–Hunga Ha'apai submarine eruption of 11–15 January 2022 and the 2023–2024 El Niño event. From April 2024 to December 2025 the temperature anomaly (relative to the 1991–2020 average) decreased from 0.94 °C to 0.30 °C.

This behavior is consistent with earlier datasets such as HadCRUT3 and reinforces the suspicion that the removal of the pause—particularly in the HadCRUT5 infilled data—may be due in part to homogenization procedures designed to restore consistency with the GCM projections.

21.3. The role of the Arctic and under-sampled regions

As noted by Scafetta (2023a),[7] the inclusion of synthetic data in regions lacking direct observations—especially the Arctic—may have allowed the introduction of a warm anomaly that has skewed the global trends during the *pause* period. Many temperature values in these remote zones are derived from model-based interpolation rather than from direct measurement. This practice, common in climate reanalysis, becomes especially problematic when applied to climatically extreme regions such as the Arctic.

During the *pause*, CMIP6 GCMs project intense Arctic warming that far exceeds the observed data, suggesting a systematic overestimation (Scafetta, 2021c, 2022).[8] The use of synthetic estimates in official reconstructions has likely exaggerated the apparent warming, generating trends not reflected in the empirical observations.

[7]Scafetta, N., 2023a. CMIP6 GCM validation based on ECS and TCR ranking for 21st-century temperature projections and risk assessment. Atmosphere, 14, 345.

[8]Scafetta, N., 2021c. Testing the CMIP6 GCM simulations versus surface temperature records from 1980–1990 to 2011–2021: High ECS is not supported. Climate, 9, 161.

Scafetta, N., 2022. Advanced testing of low, medium, and high ECS CMIP6 GCM simulations versus ERA5-T2m. Geophysical Research Letters, 49(6), e2022GL097716.

21.4. A necessary revision of IPCC estimates?

According to Scafetta (2021a),[9] the actual global warming—once corrected for urban heat effects, forced interpolations, and algorithmic biases—could be 15–25% lower than the estimates reported in the IPCC's Assessment Reports.

This discrepancy is critical, since future climate projections and policy decisions assume that the observational data are accurate and independent of the models. However, if the datasets have been retroactively modified to align with the GCM outputs, an epistemological feedback loop emerges: models are validated by data that, in part, derive from similar models or share their theoretical assumptions.

In essence, there is a risk of conflating theoretical output with observational reality, undermining the scientific integrity of climate assessments and the effectiveness of the policies built upon them.

21.5. Overview of the climate reanalyses

Climate reanalyses have become a cornerstone of modern meteorology, offering the promise of reconstructing Earth's atmospheric history with high consistency and spatio-temporal detail. These datasets combine real-world observations—from weather stations, satellites, buoys, and radiosondes—with numerical atmospheric models that assimilate the data using advanced techniques.

This process of retrospective simulation generates a continuous, three-dimensional time series of key atmospheric variables (temperature, pressure, wind, humidity), uniformly distributed across global or regional grids. It functions as a kind of scientific time machine: for each day in the past, the meteorological conditions of any location are hindcast using all the available data.

Methodologically, reanalyses rely on:

- data assimilation: the integration of observations to guide and correct model outputs;
- physical and dynamical consistency: ensuring that the relationships between variables obey atmospheric laws;
- temporal homogeneity: producing uniform data even as observational coverage evolves over time.

Reanalyses are widely used in climatology, model validation, climate change research, agriculture, hydrology, and urban planning. Prominent datasets include:

- ERA5 (ECMWF ReAnalysis version 5) (Copernicus Climate Change Service: Hersbach et al., 2020);[10]
- MERRA-2 (NASA Global Modeling and Assimilation Office: Gelaro et al., 2017);[11]

---

[9]Scafetta, N., 2021a. Detection of non-climatic biases in land surface temperature records by comparing climatic data and their model simulations. Climate Dynamics, 56(9–10), 2959–2982.

[10]Hersbach, H., Bell, B., Berrisford, P., et al., 2020. The ERA5 global reanalysis. Quarterly Journal of the Royal Meteorological Society, 146, 1999–2049.
(ECMWF = European Centre for Medium-Range Weather Forecasts).

[11]Gelaro, R., McCarty, W., Suárez, M. J., et al., 2017. The Modern-Era Retrospective Analysis for Research and Applications, Version 2 (MERRA-2). Journal of Climate, 30, 5419–5454.

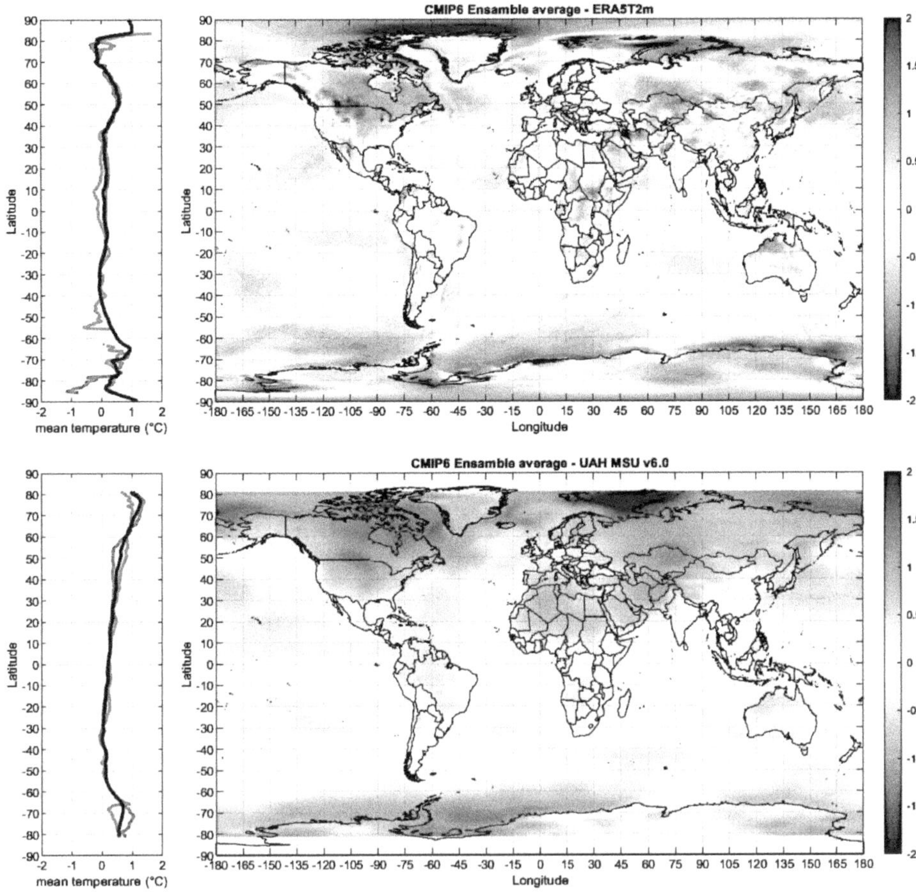

FIGURE 21.5. Difference in warming from 1980–1990 to 2011–2021 predicted by the ensemble mean of CMIP6 GCM simulations and recorded by ERA5-T2m and UAH MSU v6.0 lower troposphere (lt) temperature datasets. (Adapted from Scafetta, 2021c). Refer to the original paper for the color versions of the figures.

- JRA-55 (Japan Meteorological Agency: Kobayashi et al., 2015).[12]

These datasets span multiple decades with hourly or daily resolution, transforming sparse and imperfect observations into a coherent and scientifically robust climate narrative.

The reliability of reanalyses for detailed climate analysis must, however, be critically assessed. Despite their strengths, they are built on numerical models that can be affected by:

- systematic model biases;

(MERRA-2 = Modern-Era Retrospective analysis for Research and Applications version 2).

[12]Kobayashi, S., Ota, Y., Harada, Y., et al., 2015. The JRA-55 Reanalysis: General Specifications and Basic Characteristics. Journal of the Meteorological Society of Japan, 93(1), 5–48.
(JRA-55 = Japanese ReAnalysis – 55-year version).

- limited spatial resolution;
- variable data quality, especially in poorly monitored regions (e.g., polar zones, remote oceans);
- artificial discontinuities introduced by evolving instrumentation.

These limitations are particularly relevant for periods preceding the satellite era. A critical reading of reanalyses is therefore essential, always taking into account their uncertainty and the specific scientific application.

Importantly, reanalyses also enable direct comparison with GCM outputs and can reveal model deficiencies. For instance, Figure 21.5 shows the difference in warming from 1980–1990 to 2011–2021 as predicted by the CMIP6 ensemble simulations versus that recorded by ERA5-T2m and UAH MSU v6.0. Large positive areas—especially over land in the Northern Hemisphere and around Antarctica—highlight where, and to what extent, GCM predictions diverge from the observed data by predicting too much warming.

In particular, Scafetta & Mazzarella (2015a)[13] showed that the Antarctic sea-ice extent increased from 1980 to 2015, in stark contrast to the projections of the general circulation models (GCMs), which predicted pronounced regional warming and, consequently, a reduction in sea-ice coverage analogous to the observed decline in the Arctic.

21.6. Conclusions: observational science under pressure

The disappearance of the 2000–2014 climate *pause* in the updated datasets raises a fundamental question: how stable and independent are historical temperature series? If a stationary phase can be removed through methodological revisions, the boundary between observation and construction becomes blurred.

Without denying the reality of long-term warming, these concerns call for:

- rigorous verification of observational climate sequences;
- transparent comparison between historical and updated versions;
- cross-validation with independent sources (e.g., satellites, climate proxies);
- caution in the political use of climate data potentially altered by synthetic reconstructions.

The progress of climate science depends on reconstructing the past through the integrity of observations, not through alignment with forecasts. Its epistemological credibility rests on a clear separation between empirical data and interpretive intent, ensuring that what is observed remains independent of what is simulated.

[13] Scafetta, N., Mazzarella, A., 2015a. The Arctic and Antarctic sea-ice area index records versus measured and modeled temperature data. Advances in Meteorology, 481834, 1–8.

CHAPTER 22

Assessing non-climatic systematic biases in global warming

The arguments developed in the preceding chapters suggest that modern terrestrial climate archives may contain significant systematic distortions attributable to non-climatic factors. This chapter presents a methodology for estimating the artificial component of global warming—the portion of the reported trend that is not driven by actual climate forcings, but rather by instrumental biases, particularly in land temperature records. This methodology was proposed by Scafetta (2021a).[1]

22.1. Comparison between climate change over land and ocean

The method begins with a direct comparison between the observed surface temperature trends over land and ocean (SST) and the corresponding simulations from CMIP5 climate models.

The climatic data are drawn from HadCRUT4.6 (Morice et al., 2012).[2] A comparison is proposed between the temperature records over the continents and those over the ocean (sea surface temperature, SST), relative to the 1940–1960 baseline. Figure 22.1 shows these comparisons.

The rationale is that SSTs, measured in remote and largely non-urbanized environments, are less affected by distortions such as:

- urban heat island effects;
- land-use changes;
- station relocations or instrumental drift.

Kennedy et al. (2011),[3] Kennedy (2014)[4] and D'Aleo (2016)[5] have discussed several sources of uncertainty related to marine sampling methods, especially

[1]Scafetta, N., 2021a. Detection of non-climatic biases in land surface temperature records by comparing climatic data and their model simulations. Climate Dynamics, 56(9–10), 2959–2982.

[2]Morice, C.P., Kennedy, J.J., Rayner, N.A., Jones, P.D., 2012. Quantifying uncertainties in global and regional temperature change using an ensemble of observational estimates: the HadCRUT4 dataset. Journal of Geophysical Research: Atmospheres, 117, D08101.

[3]Kennedy, J.J., Rayner, N.A., Smith, R.O., Parker, D.E., Saunby, M., 2011. Reassessing biases and other uncertainties in sea surface temperature observations measured in situ since 1850: 2. Biases and homogenization. Journal of Geophysical Research: Atmospheres, 116, D14104.

[4]Kennedy, J.J., 2014. A review of uncertainty in situ measurements and data sets of sea surface temperature. Reviews of Geophysics, 52, 1–32.

[5]D'Aleo, J.S., 2016. A critical look at surface temperature records. In: Easterbrook, D.J. (Ed.) Evidence-Based Climate Science, 2nd Edition. Elsevier, pp. 11–48.

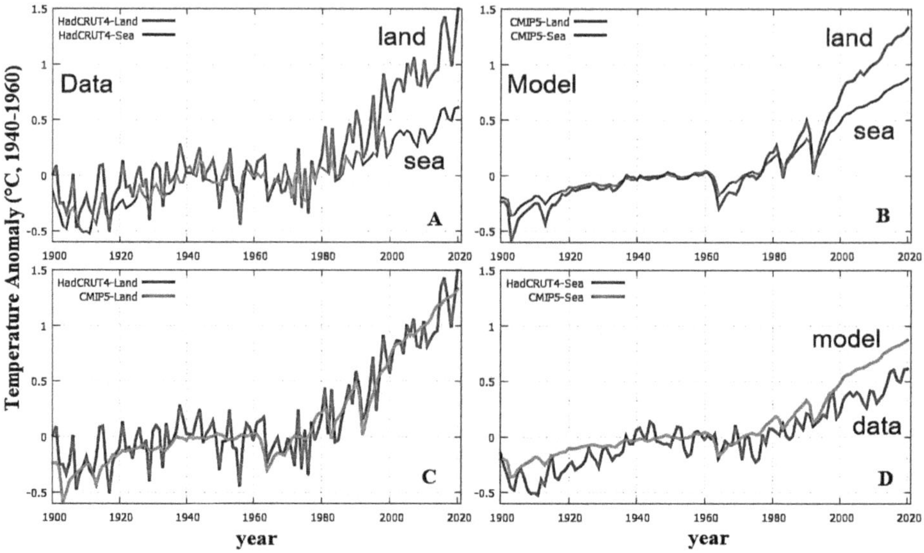

FIGURE 22.1. Comparison between HadCRUT4.6 records of land surface temperature and oceanic SST, and the respective ensemble mean simulations generated by the GCMs of CMIP5. Anomalies are relative to the 1940–1960 average. (Adapted from Scafetta, 2021a).

during the early twentieth century. Moreover, Davis (2019)[6] highlighted that the SST trend during the 1950–1975 period depends strongly on the type of dataset used. However, although SSTs are not immune to error, they are considered sufficiently reliable from 1940 onward for the present analysis.

22.2. Divergence between observations and models

Global climate models (GCMs) predict that land surfaces warm more rapidly than the oceans, owing to the lower heat capacity of soil relative to water. This behavior is indeed observed in both the data and the simulations, as shown in Figures 22.1A and 22.1B. However, a significant discrepancy emerges: the CMIP5 GCM simulations accurately reproduce the trajectory of land temperature (Figure 22.1C), but significantly overestimate ocean warming (Figure 22.1D).

- Land warming (2000–2020):
  - Observed: 0.97 ± 0.05 °C
  - CMIP5: 1.03 ± 0.04 °C
- Ocean warming (SST):
  - Observed: 0.41 ± 0.03 °C
  - CMIP5: 0.69 ± 0.03 °C

[6]Davis, L.L.B., Thompson, D.W.J., Kennedy, J.J., Kent, E.C., 2019. The importance of unresolved biases in twentieth-century sea surface temperature observations. Bulletin of the American Meteorological Society, 100, 621–629.

The GCMs' overestimation of ocean warming—which accounts for about 70% of Earth's surface—suggests that the models are overestimating surface warming, and that the apparent agreement between model simulations and land data is inflated by non-climatic factors warm-biasing the land climatic record.

22.3. Empirical calibration of CMIP5 models

This result can be used to estimate the warming bias in the observational record as follows. Let us assume that the SST record provides a reliable estimate of the actual climate warming. It can then be used to retroactively calibrate the CMIP5 GCM ensemble simulation. Specifically, the ensemble model output is rescaled to match the observed ocean warming.

The procedure assumes that:

(1) the CMIP5 GCMs correctly simulate the land/ocean warming ratio;
(2) the climate sensitivity (ECS) varies within a plausible range (1–6 °C), with CMIP5 averaging $\sim$3 °C (Knutti et al., 2017; IPCC, 2013).[7] Rescaling the CMIP5 outputs by up to one third is therefore theoretically defensible.

22.4. Estimating the non-climatic systematic deviation

The step-by-step calculations are as follows:

(1) to reproduce the SST warming, the climate models must be rescaled by 0.41 / 0.69 $\approx$ 0.59;
(2) the same scaling factor is then applied to the GCM land warming: 1.03 $\times$ 0.59 $\approx$ 0.61 °C;
(3) the comparison between the modeled and the reported land warming yields 0.61 / 0.97 $\approx$ 0.63;
(4) thus, the observed land record needs to be rescaled to 0.97 $\times$ 0.63 $\approx$ 0.61 $\pm$ 0.07 °C.

The difference of 0.36 $\pm$ 0.04 °C suggests that 25–45% of the land warming may be due to non-climatic biases occurring at the surface.

22.5. A new estimate of the corrected global temperature

Combining the corrected land temperatures (30% of the surface) with the unmodified SSTs (70%) yields:

$$T_{corr} = 0.3 \cdot T_{land,corr} + 0.7 \cdot T_{SST}.$$

The results of the rescaling for the period from 1940–1960 to 2000–2020 are:

- original HadCRUT4.6 record: 0.59 $\pm$ 0.03 °C;
- UAH MSU v6.0 lower troposphere temperature record: 0.44 $\pm$ 0.03 °C;
- corrected temperature record: 0.48 $\pm$ 0.03 °C;
- CMIP5 ensemble mean: 0.78 $\pm$ 0.03 °C.

The corrected global surface temperature record is thus found to align closely with the satellite estimates from UAH MSU v6.0.

[7]Knutti, R., Rugenstein, M.A., Hegerl, G.C., 2017. Beyond equilibrium climate sensitivity. Nature Geoscience, 10, 727–736.
IPCC, 2013. Climate Change 2013: The Physical Science Basis. Cambridge University Press.

22.6. Conclusion

The corrected global temperature record is presented in Figure 22.2, and shows statistical consistency with three independent lines of evidence:

- satellite measurements from the UAH MSU v6.0 dataset (Spencer et al., 2017),[8] which provide a direct and homogenized view of lower-tropospheric temperature trends since 1979;
- rural station datasets, as analyzed by Soon et al. (2023),[9] which minimize urban heat island effects and other anthropogenic biases by focusing on minimally disturbed locations;
- the methodological corrections proposed by (Scafetta, 2021a), which address structural inconsistencies and potential overestimations in mainstream surface temperature reconstructions.

Together, these approaches converge in suggesting that the official global surface temperature records—such as HadCRUT5 (infilled)—may contain spurious warming signals over land, accounting for approximately 40% of the observed warming, as illustrated in Figure 17.3.

This discrepancy is far from trivial: it implies that a significant portion of the reported warming may stem from non-climatic influences, including urbanization, station siting issues, and homogenization artifacts.

Moreover, the CMIP5 and CMIP6 GCM ensembles tend to systematically overestimate land warming, pointing to deeper issues in model calibration and tuning. These biases may arise from the use of surface datasets that embed non-climatic distortions, leading to inflated sensitivity estimates and exaggerated projections.

The analysis indicates that non-climatic surface thermal distortions—including measurement biases, land-use changes, and data-processing artifacts—may account for 15–25% of the total global warming (land + sea) reported between 1940–1960 and 2000–2020. This is a substantial fraction, with important implications for both scientific interpretation and policy formulation.

If surface temperature datasets are used to calibrate climate models or to justify far-reaching mitigation strategies, it is essential that their integrity be rigorously assessed. A more cautious and transparent approach is warranted, one that distinguishes genuine climatic signals from methodological noise and ensures that policy decisions are grounded in robust, empirically validated evidence.

---

[8]Spencer, R.W., Christy, J.R., Braswell, W.D., 2017. UAH version 6 global satellite temperature products: Methodology and results. Asia-Pacific Journal of Atmospheric Sciences, 53(1), 121–130.

[9]Soon, W., Connolly, R., Connolly, M., Akasofu, S.-I., Baliunas, S., Berglund, J., Bianchini, A., Briggs, W.M., Butler, C.J., Cionco, R.G., Crok, M., Elias, A. G., Fedorov, V.M., Gervais, F., Harde, H., Henry, G.W., Hoyt, D.V., Humlum, O., Legates, D.R., Lupo, A.R., Maruyama, S., Moore, P., Ogurtsov, M., ÓhAiseadha, C., Oliveira, M.J., Park, S.-S., Qiu, S., Quinn, G., Scafetta, N., Solheim, J.-E., Steele, J., Szarka, L., Tanaka, H.L., Taylor, M.K., Vahrenholt, F., Velasco Herrera, V.M., Zhang, W., 2023. The detection and attribution of Northern Hemisphere land surface warming (1850–2018) in terms of human and natural factors: Challenges of inadequate data. Climate, 11(9), 179.

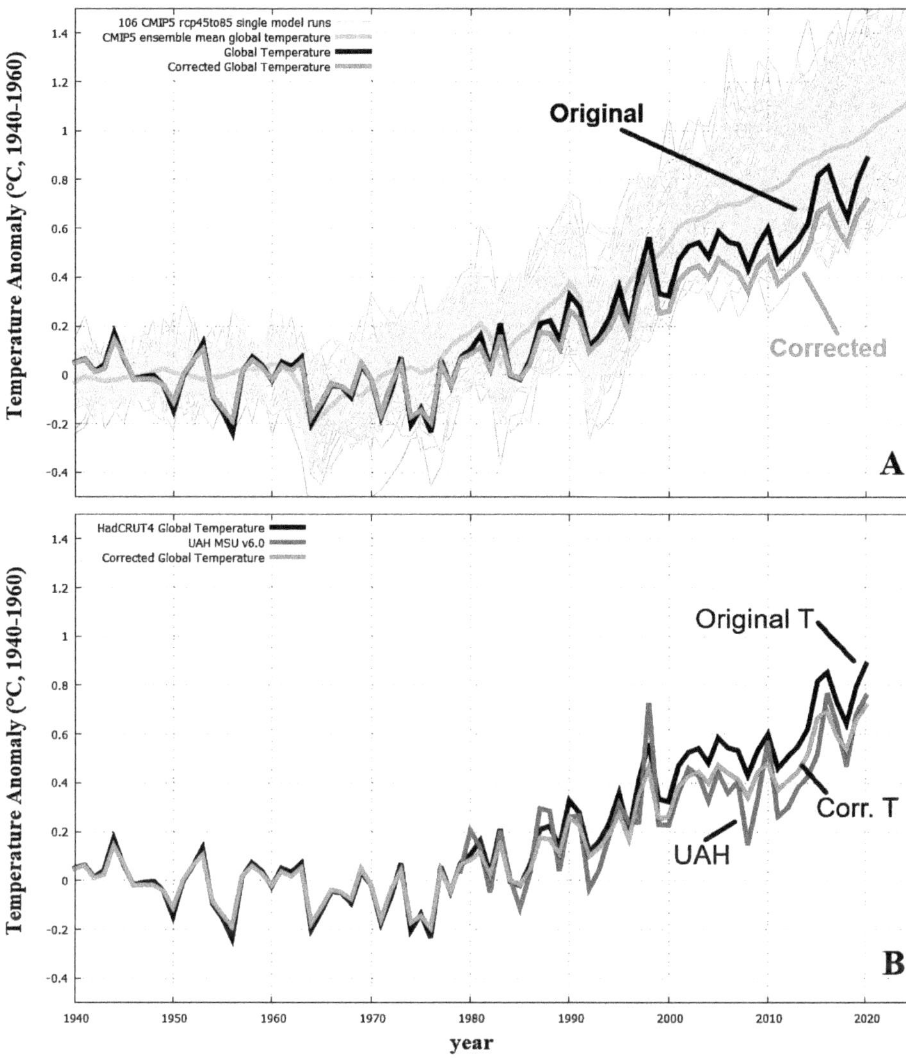

FIGURE 22.2. A) Comparison between the original HadCRUT4.6 record, the proposed corrected record, the CMIP5 model ensemble mean, and the 106 individual simulations.
B) Comparison of the corrected record with the UAH MSU v6.0 satellite dataset for the lower troposphere, using the same 1979–1985 baseline.
All data are anomalies relative to the 1940–1960 period. (Adapted from Scafetta, 2021a).

FIGURE 22.3. **Frozen Venice, Italy: the Epiphany of 1709.** On the nights of January 5 and 6, 1709, Venice experienced one of the coldest episodes in its recorded history. Amid the broader European deep freeze known as the *Great Winter of 1709*—linked to the solar Maunder Minimum and the Little Ice Age—the Venetian Lagoon froze solid. This rare phenomenon allowed citizens to walk, skate, and even set up temporary stalls on the ice, transforming the aquatic city into a surreal winter landscape. Temperatures plunged to –17.5 °C, driven by fierce bora winds that paralyzed navigation and commerce. The event caused widespread damage to agriculture: olive trees, vineyards, and fruit orchards perished across the Veneto region and beyond. Gabriele Bella's painting captures this extraordinary moment, depicting figures crossing the frozen lagoon, bundled in heavy garments, surrounded by a city suspended in silence. The image stands as both artistic testimony and climatic evidence of solar-induced cooling during the Maunder Minimum. It reminds us that Venice, a city defined by water, once became a city of ice.
Across Europe, the same cold wave devastated crops, froze rivers such as the Seine and the Thames, and led to famine and social unrest. In France, thousands died from exposure and hunger, while in Germany and Central Europe entire forests cracked under the frost. The winter of 1709 remains one of the most dramatic examples of natural climate variability in historical memory.

Part 4

Rethinking Climate Attribution: The Solar Perspective

CHAPTER 23

The changing Sun and Earth's climate

23.1. Correlation between solar and climate records

Since the earliest investigations into climate, the Sun has been recognized as the primary regulator of Earth's energy balance. The hypothesis that variations in solar activity influence climate, either directly or indirectly, has deep roots in paleoclimatology and has gained increasing empirical support since the mid-twentieth century.

Pioneering studies by Bray (1968)[1] and Eddy (1976, 1977)[2] identified long-term solar cycles using cosmogenic isotopes such as carbon-14 (^{14}C) and beryllium-10 (^{10}Be), which serve as proxies for solar activity. These records exhibit strong correlations with Holocene climate fluctuations.

Figures 23.1 A–C (adapted from Eddy, 1977) illustrate how solar activity inferred from ^{14}C records aligns with periods of advance and retreat of Alpine (Le Roy Ladurie, 1967)[3] and global glaciers (Denton & Karlén, 1973),[4] regional temperature shifts, and documented episodes of severe winters in Western Europe (Lamb, 1972).[5]

In his seminal article *The Maunder Minimum* (1976), Eddy characterized the period from 1645 to 1715 as one of exceptionally low solar activity, marked by a near-absence of sunspots. This interval coincided with the coldest phase of the Little Ice Age (approximately 1300–1850), which featured harsh winters and extreme climatic conditions across Europe, North America, and China. By contrast, the Medieval Warm Period (approximately 950–1300) was associated with heightened solar activity, as evidenced by ^{14}C records and historical naked-eye sunspot observations, as shown in the lower panel of Figure 23.1.

Eddy's integration of historical data, astronomical records, and isotope analysis laid the foundation for a renewed debate on the Sun's role in long-term climate variability. Subsequent studies have reinforced the hypothesis that multi-decadal, centennial, and millennial climate fluctuations may be largely modulated by solar variability.

---

[1]Bray, J.R., 1968. Glaciation and solar activity since the fifth century BC and the solar cycle. Nature, 220, 672–674.

[2]Eddy, J.A., 1976. The Maunder Minimum. Science, 192, 1189–1202.

Eddy, J.A., 1977. Climate and the changing Sun. Climatic Change, 1, 173–190.

[3]Le Roy Ladurie, E., 1967. Histoire du Climat depuis l'an mil. Flammarion, Paris (trans. B. Bray, Doubleday & Co., 1971).

[4]Denton, G.H., Karlén, W., 1973. Holocene climatic variations—their pattern and possible cause. Quaternary Research, 3, 155–205.

[5]Lamb, H.H., 1972. Climate: Present, Past and Future. Vol. 1. Methuen.

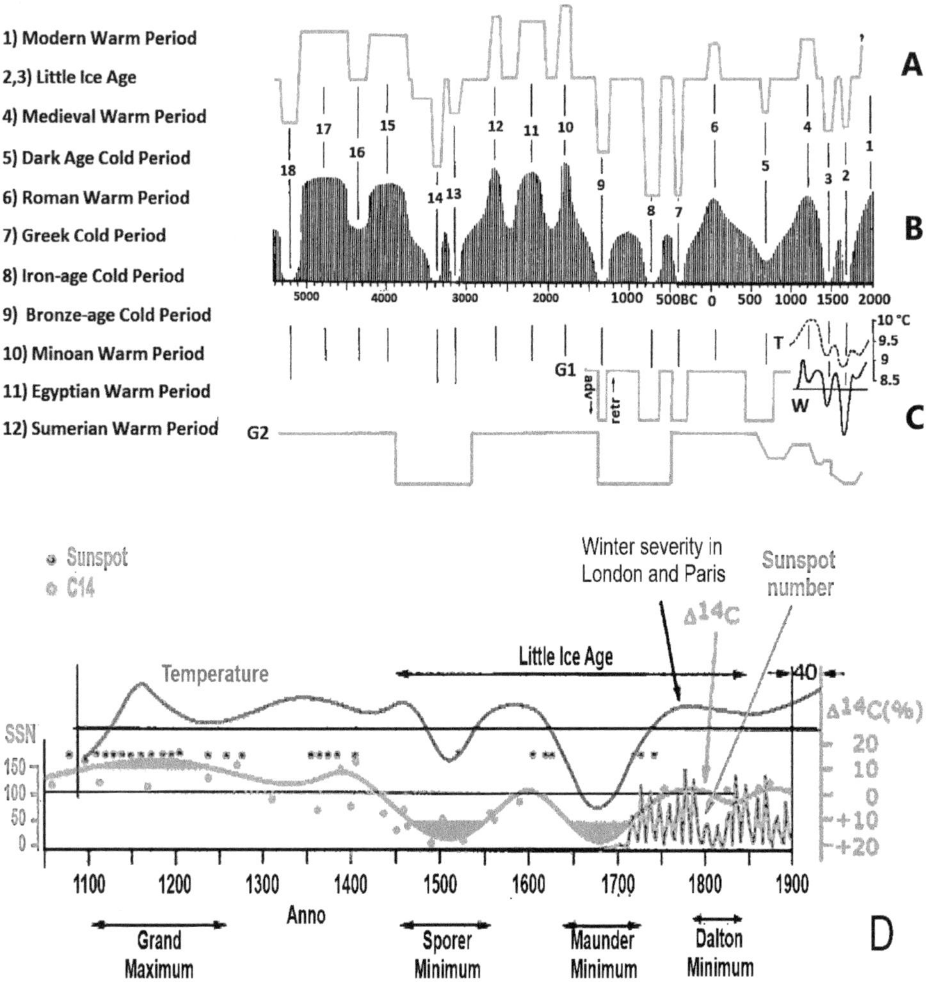

FIGURE 23.1. Correlation between solar and climate data.
[A] Schematic variation of the solar proxy based on ^{14}C (adapted from Lin et al., 1975).
[B] Interpretation of curve [A] as representing the long-term envelope of solar activity.
[C] Four estimates of climate change during the Holocene: (G1) periods of advance and retreat of Alpine glaciers (adapted from Le Roy Ladurie, 1967); (G2) global glacier fluctuations (adapted from Denton & Karlén, 1973); (T) estimates of mean annual temperature in central England (adapted from Lamb, 1965); (W) winter severity index for the Paris–London region (adapted from Lamb, 1972). The upper left side lists documented warm and cold periods in human history.
[D] Comparison between the solar proxy based on ^{14}C, sunspot records, and winter climate severity in London and Paris. (Adapted from Eddy, 1977).

23.2. Climate and civilization: a link with solar cycles?

One of the most striking insights to emerge from this line of research is the apparent synchrony between major Holocene climate oscillations —including the Roman Warm Period, the Medieval Warm Period (MWP), and the Little Ice Age (LIA)—and long-term solar activity cycles (see Figure 13.5). This correlation suggests that variations in solar irradiance may have played a more substantial role in shaping historical climate regimes than is currently acknowledged in mainstream climate models.

Historical and paleoclimatic studies (Le Roy Ladurie, 1967; Lamb, 1972; Denton & Karlén, 1973; Fagan, 2008)[6] point to a compelling connection between climate variability and the rise and fall of civilizations. Periods of prosperity, agricultural expansion, and demographic growth often coincide with solar maxima, whereas episodes of crisis—marked by famines, migrations, and sociopolitical instability—tend to align with grand solar minima.

A notable example is the Viking settlement in Greenland, which flourished during the MWP, a period characterized by elevated solar activity and relatively mild northern temperatures. These communities were eventually abandoned during the onset of the LIA, an interval marked by declining solar irradiance and harsher climatic conditions (Eddy, 1977; Lasher & Axford, 2019).[7] Similar patterns are observable in the timing of historical disruptions across Europe and Asia, many of which coincide with multi-decadal solar minima such as the Maunder Minimum (1650–1715) and the Dalton Minimum (1790–1830).

These correlations do not imply a simplistic causality, but they underscore the need to reevaluate the role of solar variability as a potential driver of long-term climate dynamics. Disregarding these patterns in favor of exclusively anthropogenic explanations risks overlooking a fundamental component of Earth's climatic history—one that has shaped ecosystems, societies, and civilizations for millennia.

23.3. New confirmations from proxies and modern climate series

In recent decades, a growing body of research has strengthened the statistical and dynamical link between solar variability and climate change across secular and multi-decadal timescales. Studies by Hoyt & Schatten (1997),[8] Baliunas & Soon (1991),[9] Bond et al. (2001),[10] Neff et al. (2001),[11] Kirkby (2007),[12] Tung

[6]Fagan, B., 2008. The Great Warming: Climate Change and the Rise and Fall of Civilizations. Bloomsbury Press.

[7]Lasher, G.E., Axford, Y., 2019. Medieval warmth confirmed at the Norse Eastern Settlement in Greenland. Geology, 47(3), 267–270.

[8]Hoyt, D.V., Schatten, K.H., 1997. The Role of the Sun in Climate Change. Oxford University Press.

[9]Baliunas, S.L., Soon, W.H., 1991. Are variations in the length of the activity cycle related to changes in brightness in solar-type stars? The Astrophysical Journal, 374, 463–470.

[10]Bond, G., Kromer, B., Beer, J., Muscheler, R., Evans, M.N., et al., 2001. Persistent solar influence on North Atlantic climate during the Holocene. Science, 294(5549), 2130–2136.

[11]Neff, U., Burns, S.J., Mangini, A., Mudelsee, M., Fleitmann, D., Matter, A., 2001. Strong coherence between solar variability and the monsoon in Oman between 9 and 6 kyr ago. Nature, 411, 290–293.

[12]Kirkby, J., 2007. Cosmic rays and climate. Surveys in Geophysics, 28(5–6), 333–375.

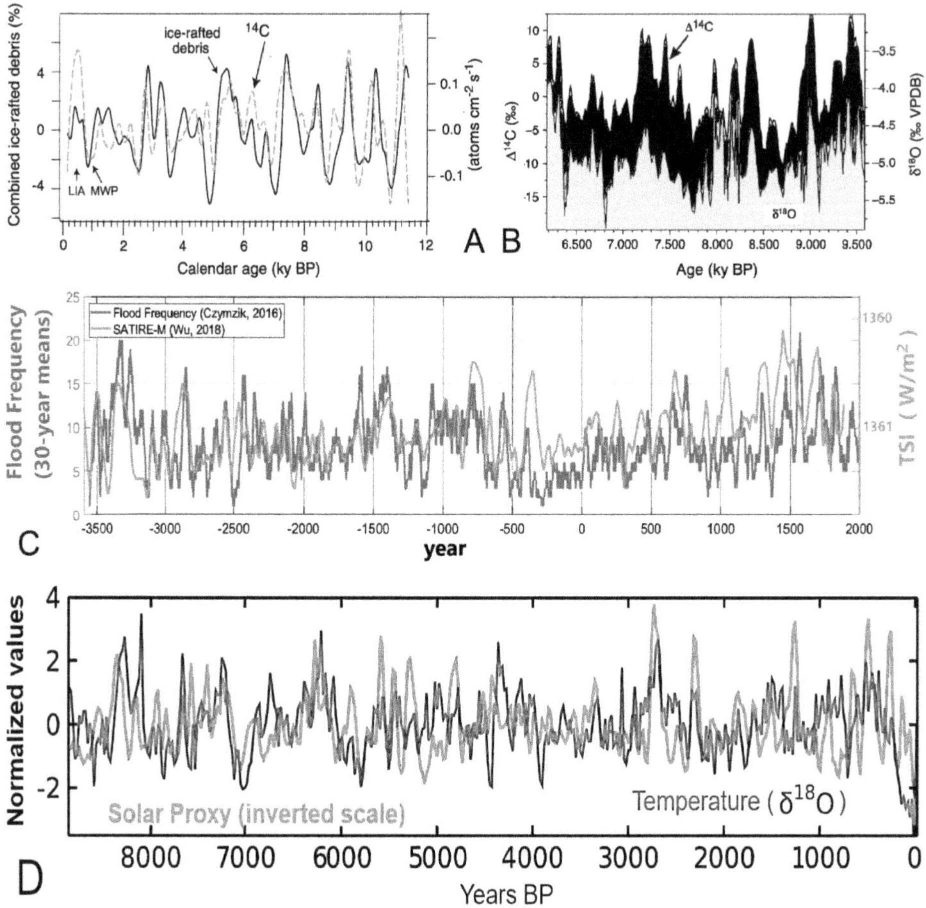

FIGURE 23.2. [A, B] Comparison between the solar proxy based on ^{14}C and two climate records (adapted from Bond et al., 2001; Neff et al., 2001; Kirkby, 2007). [C] Correlation between reconstructed total solar irradiance (recreated from Wu et al., 2018) and flood frequency in Central Europe (30-year moving window) (Czymzik et al., 2016: $r = -0.4$, $p < 0.0001$). [D] Comparison between the solar proxy based on ^{14}C and a climate series from Dongge Cave, China, representing Asian climate and likely monsoon variability (adapted from Steinhilber et al., 2012).

& Camp (2008),[13] Scafetta (2009),[14] Steinhilber et al. (2012),[15] Czymzik et al.

[13] Tung, K.K., Camp, C.D., 2008. Solar cycle warming at the Earth's surface in NCEP and ERA-40 data: A linear discriminant analysis. Journal of Geophysical Research: Atmospheres, 113, D05114.
[14] Scafetta, N., 2009. Empirical analysis of the solar contribution to global mean air surface temperature change. Journal of Atmospheric and Solar–Terrestrial Physics, 71, 17–18.
[15] Steinhilber, F., Abreu, J.A., Beer, J., Brunner, I., Christl, M., et al., 2012. 9,400 years of cosmic radiation and solar activity from ice cores and tree rings. Proceedings of the National Academy of Sciences of the United States of America, 109(16), 5967–5971.

(2016),[16] and Scafetta (2023c),[17] have proposed robust interpretive frameworks and presented compelling empirical evidence in support of this connection. In recent years, significant correlations have been identified between solar activity, climate change, length of day (LOD), and large earthquakes worldwide (Scafetta & Mazzarella, 2015b; Marchitelli et al., 2020).[18]

Figures 23.2 and 23.3 highlight strong temporal correlations between solar cycles and observed climate oscillations on millennial, centennial, and decadal timescales. These findings reinforce the view that solar activity is not merely a background variable but a dynamic driver of terrestrial climate variability.[19]

Recent climate data confirm that solar variability can significantly influence key atmospheric parameters, including surface temperature, precipitation patterns, and broader meteorological dynamics. This view is further supported by recent studies (e.g., Connolly et al., 2023, 2024; Scafetta & Bianchini, 2023; Xiao et al., 2024),[20] which examine the challenges of attributing observed climate trends solely to anthropogenic factors.

For example, Scafetta (2023c) proposes three balanced multi-proxy models of total solar activity (TSA), integrating diverse solar proxies to assess their influence on global and sea surface temperatures. His methodology, based on differential multi-linear regression and energy balance modeling, suggests that the Sun's impact may be substantially greater than is currently recognized in CMIP6 GCMs.

---

[16]Czymzik, M., Muscheler, R., Brauer, A., 2016. Solar modulation of flood frequency in central Europe during spring and summer on interannual to multi-centennial timescales. Climate of the Past, 12, 799–805.

[17]Scafetta, N., 2023c. Empirical assessment of the role of the Sun in climate change using balanced multi-proxy solar records. Geoscience Frontiers, 14(6), 101650.

[18]Scafetta, N., Mazzarella, A. 2015b. Spectral coherence between climate oscillations and the M ≥ 7 earthquake historical worldwide record. Natural Hazards, 76, 1807–1829.

Marchitelli, V., Harabaglia, P., Troise, C., De Natale, G., 2020. On the correlation between solar activity and large earthquakes worldwide. Scientific Reports, 10, 11495.

[19]Baliunas, S.L., Soon, W.H., 1991. Are variations in the length of the activity cycle related to changes in brightness in solar-type stars? The Astrophysical Journal, 374, 463–470.

Hoyt, D.V., Schatten, K.H., 1993. A discussion of plausible solar irradiance variations, 1700–1992. Journal of Geophysical Research: Space Physics, 98(A11), 18895–18906.

Eichler, A., Olivier, S., Henderson, K., Laube, A., Beer, J., Papina, T., Gäggeler, H. W., Schwikowski, M., 2009. Temperature response in the Altai region lags solar forcing. Geophysical Research Letters, 36, L01808.

Svensmark, H., Friis-Christensen, E., 2007. Reply to Lockwood and Fröhlich—The persistent role of the Sun in climate forcing. Danish National Space Center Scientific Report 3/2007.

Shaviv, N.J., 2008. Using the oceans as a calorimeter to quantify the solar radiative forcing. Journal of Geophysical Research: Space Physics, 113, A11101.

[20]Connolly, R., Soon, W., Connolly, M., Baliunas, S., Berglund, J., et al., 2023. Challenges in the detection and attribution of Northern Hemisphere surface temperature trends since 1850. Research in Astronomy and Astrophysics, 23, 105015.

Connolly, R., Soon, W., Connolly, M., Cionco, R.G., Elias, A.G., et al., 2024. Multiple new or updated satellite total solar irradiance (TSI) composites (1978–2023). The Astrophysical Journal, 975, 102.

Scafetta, N., Bianchini, A., 2023. Overview of the spectral coherence between planetary resonances and solar and climate oscillations. Climate, 11(4), 77.

Xiao, Z., Zhao, L., Zhou, L., Huo, W., Mironova, I.A., Miyahara, H., 2024. Editorial: Impact of solar activities on weather and climate. Frontiers in Earth Science, 11, 1338416.

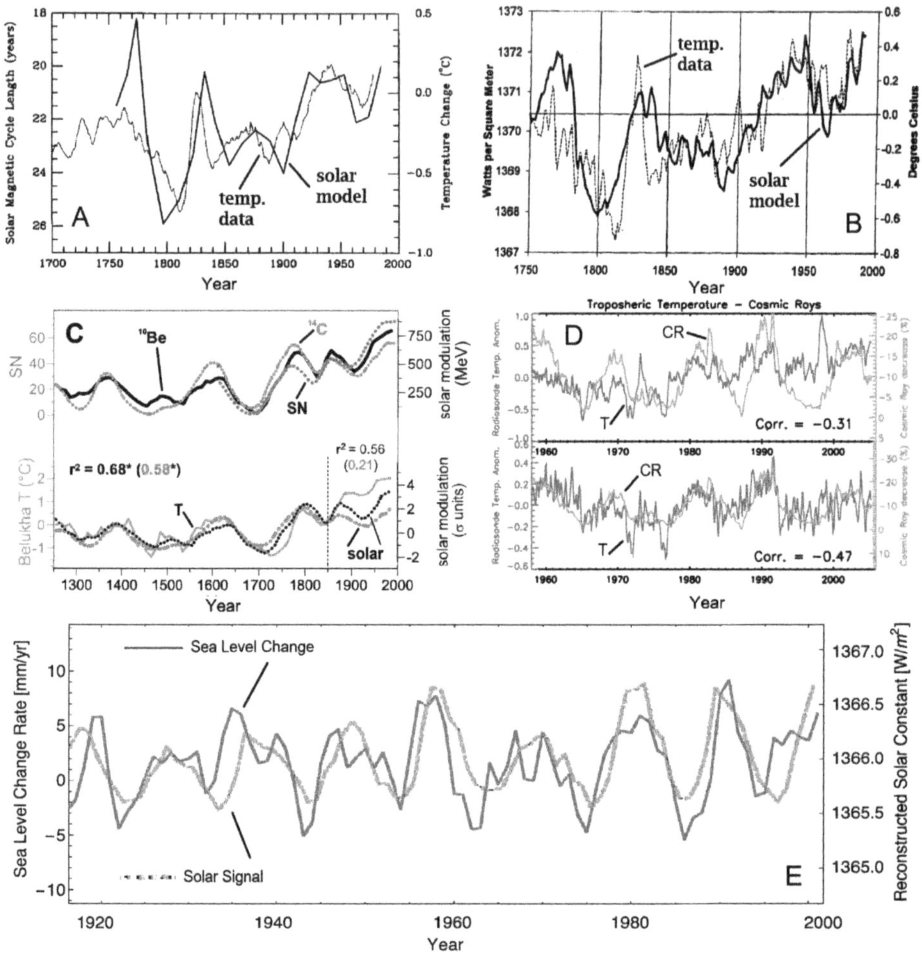

FIGURE 23.3. Examples of correlations between solar and climate sequences: [A] Comparison between the length of the Hale solar cycle (two consecutive Schwabe cycles) and a reconstruction of Northern Hemisphere temperatures (adapted from Baliunas & Soon, 1991). [B] Comparison between a reconstruction of Northern Hemisphere temperatures and the TSI proxy model by Hoyt & Schatten (1993, 1997). [C] Comparison between the Belukha temperature reconstruction and solar activity proxies from ^{14}C and ^{10}Be (adapted from Eichler et al., 2009). [D] The top panel shows temperature observations and cosmic rays. The bottom panel shows the correlation after removing the effects of El Niño, the North Atlantic Oscillation, volcanic aerosols, and a linear trend (0.14 ± 0.4 °C/decade) (adapted from Svensmark & Friis-Christensen, 2007). [E] Comparison between solar oscillation (dashed line) and the rate of sea-level rise (adapted from Shaviv, 2008).

23.4. Missing forcings and mechanisms in CMIP6 climate models

Despite a vast body of paleoclimatic and archaeoclimatic literature emphasizing the Sun as the primary long-term climate driver, the IPCC Sixth Assessment Report IPCC (AR6, 2021)[21] —in alignment with CMIP6 climate model simulations— asserts that the solar contribution to global warming from 1850–1900 to 2010–2020 is negligible (see Figures 6.2 and 6.3). Nearly all modern warming is attributed to anthropogenic greenhouse gas emissions.

This position raises a fundamental question:

> *How can the marginalization of solar influence in CMIP GCMs be reconciled with paleoclimatic records that consistently identify the Sun as a decisive force in shaping Earth's climate over millennia?*

If solar variability drove significant climate oscillations in the past, why has its impact become "irrelevant" in the modern era?

A possible explanation lies in the epistemological limitations of current climate models, particularly in their treatment of solar forcing. Critical studies (e.g., Scafetta et al., 2019a; Scafetta, 2023c; Connolly et al., 2023, 2024),[22] have shown that CMIP6 GCMs adopt the solar forcing derived from the total solar irradiance (TSI) model proposed by Matthes et al. (2017),[23] which exhibits extremely low secular variability.

This choice effectively minimizes the Sun's climatic role within the models. Yet many researchers contend that this reconstruction fails to capture the full range of solar variability documented in paleoclimate proxies, such as cosmogenic isotopes and geological records, leading to a systematic underestimation of the solar influence by at least an order of magnitude.

Moreover, CMIP6 GCMs focus almost exclusively on the radiative component of solar forcing (TSI and SSI), while neglecting other potentially significant mechanisms, including:

- solar–cosmic ray interactions, which may influence cloud nucleation and albedo (Kirkby, 2007; Svensmark et al., 2016; Svensmark, 2019, 2022);[24]

[21]IPCC, 2021. Climate Change 2021: The Physical Science Basis. Cambridge University Press.
[22]Scafetta, N., Willson, R.C., Lee, J.N., Wu, D.L., 2019. Modeling quiet solar luminosity variability from TSI satellite measurements and proxy models during 1980–2018. Remote Sensing, 11(21), 2569.
Scafetta, N., 2023c. Empirical assessment of the role of the Sun in climate change using balanced multi-proxy solar records. Geoscience Frontiers, 14(6), 101650.
Connolly, R., Soon, W., Connolly, M., Baliunas, S., Berglund, J., et al., 2023. Challenges in the detection and attribution of Northern Hemisphere surface temperature trends since 1850. Research in Astronomy and Astrophysics, 23, 105015.
Connolly, R., Soon, W., Connolly, M., Cionco, R.G., Elias, A.G., et al., 2024. Multiple new or updated satellite total solar irradiance (TSI) composites (1978–2023). The Astrophysical Journal, 975, 102.
[23]Matthes, K., Funke, B., Andersson, M.E., Barnard, L., Beer, J., et al., 2017. Solar forcing for CMIP6 (v3.2). Geoscientific Model Development, 10, 2247–2302.
[24]Svensmark, J., Enghoff, M.B., Shaviv, N.J., Svensmark, H., 2016. The response of clouds and aerosols to cosmic ray decreases. Journal of Geophysical Research: Space Physics, 121(9), 8152–8181.
Svensmark, H., 2019. Force Majeure: The Sun's Role in Climate Change. The Global Warming Policy Foundation. ISBN 978-0993119095.
Svensmark, H., 2022. Supernova rates and burial of organic matter. Geophysical Research Letters, 49(1), e2021GL096376.

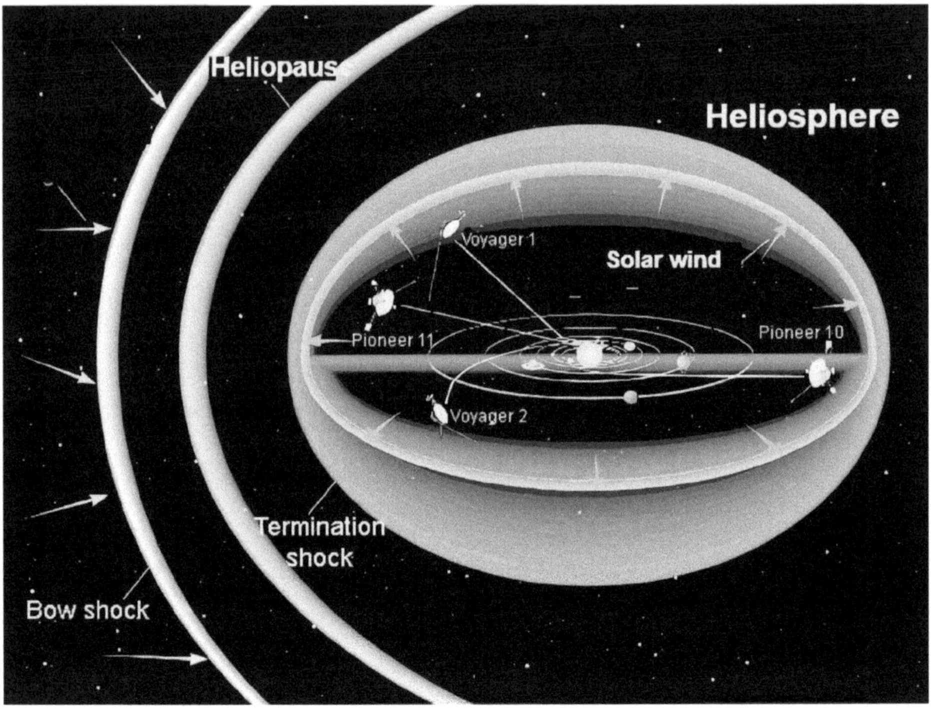

FIGURE 23.4. The Solar System and the heliosphere generated by the Sun.

- solar magnetic cycles, which affect atmospheric circulation and particle fluxes;
- spectral variations, particularly in the UV band, which modulate stratospheric dynamics;
- nonlinear ocean–atmosphere coupling, potentially synchronized with solar cycles.

The omission or simplification of these processes may help explain persistent discrepancies between model simulations and observed climate data, particularly on multi-decadal and longer timescales.

23.5. Beyond the anthropocentric paradigm

The *"neutralization"* of the Sun in official climate narratives, often achieved through selective reconstructions of solar activity, reflects a deeply anthropocentric paradigm. Natural drivers of climate variability are systematically subordinated to human-induced forcings, reinforcing a model-centric view in which anthropogenic emissions dominate the explanatory landscape.

While this approach may serve policy objectives by offering a clear and actionable narrative, it risks oversimplifying the inherent complexity of Earth's climate system. Climate is not a linear response to a single variable, but a dynamic interplay of multiple forcings—solar, volcanic, oceanic, atmospheric, and anthropogenic—operating across a wide range of spatial and temporal scales. Marginalizing solar variability, especially when paleoclimatic and proxy data consistently indicate its historical significance, undermines the integrity of climate attribution and narrows the scope of scientific inquiry.

To move beyond this partial and potentially reductive view, a more balanced and inclusive scientific approach is needed—one that embraces:

- a critical reassessment of solar activity reconstructions used in climate models, including transparent evaluation of methodologies, assumptions, and uncertainties;
- the integration of emerging physical mechanisms—including solar–cloud interactions, cosmic-ray modulation, and magnetospheric coupling—even when these processes are not yet fully understood or readily parameterizable;
- the reopening of the scientific debate on the Sun's role in recent climate change, informed by historical records, natural proxies (e.g., ^{10}Be and ^{14}C isotopes), and modern satellite observations.

Such an approach does not reject anthropogenic influence, but rather seeks to contextualize it within a broader framework of natural variability. Only by restoring the Sun to its rightful place in the climate narrative can we hope to develop models and policies that reflect the full complexity of the Earth system and remain open to revision as new evidence emerges.

23.6. Conclusion: an open mystery at the heart of the climate system

The uncertainty surrounding solar activity remains one of the most profound and unresolved questions in contemporary climate science. Despite its central role in modulating Earth's energy balance, solar variability is often marginalized in mainstream climate models, which tend to prioritize anthropogenic forcings. Yet a vast body of paleoclimatic evidence, historical documentation, and natural proxies—from ice cores and tree rings to sediment layers and solar cycle reconstructions—continues to affirm that solar activity has long acted, and likely still acts, as a major driver of Earth's climate dynamics. For example, during the Middle Holocene, oscillations in the Indian–Omani monsoon exhibit a strong correlation with solar activity: periods of reduced solar output correspond to weaker monsoon precipitation, whereas intervals of enhanced solar activity coincide with stronger monsoon rainfall (Fleitmann & Matter, 2009).[25]

The disconnect between empirical evidence and model representation raises important epistemological concerns. Neglecting solar influences in favor of computational simulations, however sophisticated, risks producing distorted projections and informing environmental policies built on fragile or incomplete assumptions. As Eddy (1976) presciently warned, ignoring solar variability is tantamount to excluding one of the central protagonists in the history of planetary climate evolution.

The principal challenge is conceptual: to reintegrate solar physics into climate science in a manner that respects both empirical complexity and theoretical rigor. Doing so requires a reassessment of attribution frameworks, a more nuanced treatment of natural forcings, and an openness to interdisciplinary approaches that bridge climate science, astrophysics, and geophysics.

[25] Fleitmann, D., Matter, A., 2009. The speleothem record of climate variability in Southern Arabia. Comptes Rendus Geoscience, 341 (8–9), 633–642.

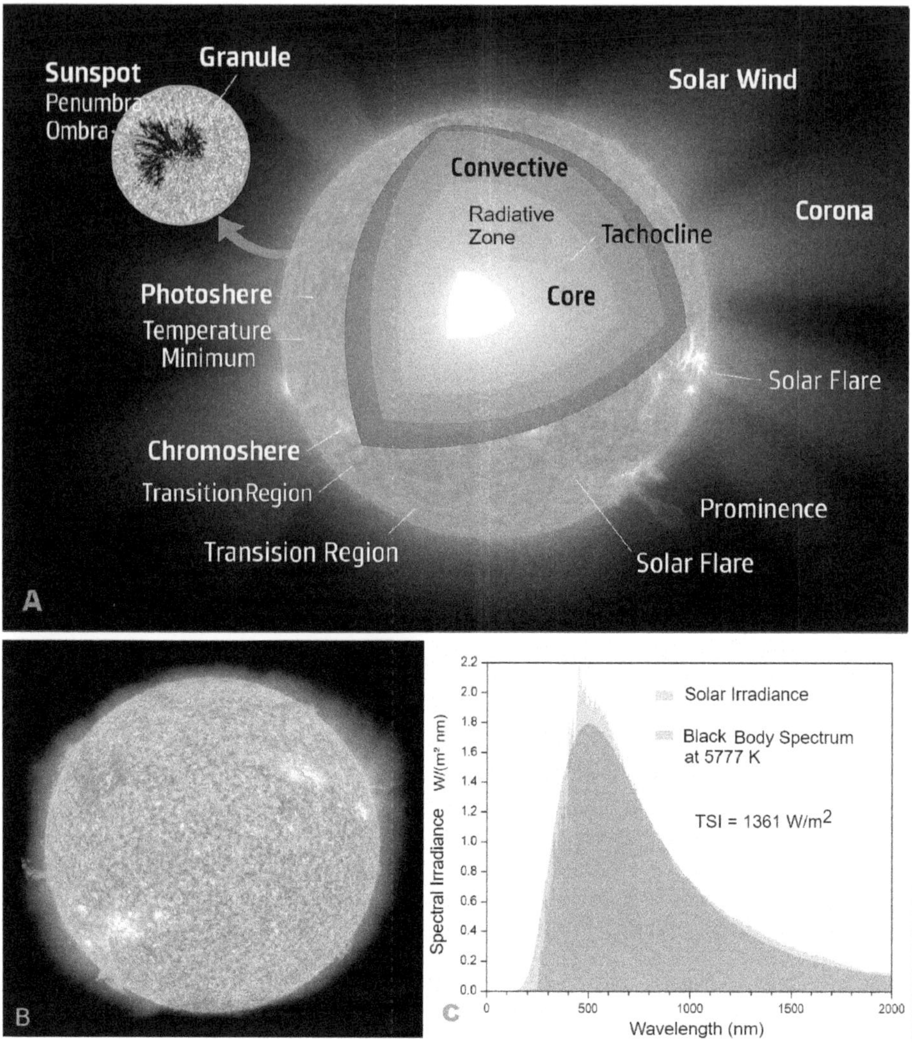

Figure 23.5. **Structure of the Sun.** [A–B] The Sun has a layered structure composed of both internal and external zones. Internally, it consists of the core, where nuclear fusion generates energy; the radiative zone, where energy moves outward via photon diffusion; and the convective zone, where hot plasma circulates in turbulent motions. Externally, the Sun comprises the photosphere (visible surface), the chromosphere, and the corona, its outer atmosphere, which extends into space as the solar wind. Each layer plays a crucial role in solar dynamics, from energy production to electromagnetic emissions. [C] The Sun's emission spectrum (yellow area) compared with that of a black body of the same size emitting the same amount of radiant energy (gray area).

CHAPTER 24

TSI satellite composites: a controversy spanning four decades

Total solar irradiance (TSI) represents the total amount of solar energy received by Earth per unit time and per unit surface area. It is measured outside Earth's atmosphere at the average Earth–Sun distance (1 AU $\approx$ 149.6 million km). Expressed in watts per square meter (W/m^2), TSI is the fundamental input to Earth's climate energy balance and is commonly known as the "*solar constant*," although it varies with solar luminosity.

Although short-term TSI variations, such as those associated with the 11-year solar cycle, are relatively small (about 0.1%, or slightly more than 1 W/m^2 since 1980), their climatic impact may be amplified through direct, indirect, and nonlinear feedbacks involving the atmosphere and oceans. Moreover, secular and multi-decadal solar variability may have contributed significantly to climate fluctuations in both the preindustrial and modern eras (Scafetta, 2023c; Soon et al., 2023).[1]

24.1. Discrepancies in TSI measurements

Direct TSI measurements began with the satellite era. In November 1978, the "*Earth Radiation Budget*" (ERB) radiometer aboard Nimbus-7 inaugurated space-based TSI monitoring, followed in 1980 by ACRIM ("*Active Cavity Radiometer Irradiance Monitor*") I aboard the Solar Maximum Mission (SMM).

Subsequent missions include:

- ACRIM-II (aboard UARS);
- ACRIM-III (on ACRIMSAT);
- ERBE/ERBS (Earth Radiation Budget Experiment);
- VIRGO (on the SoHO satellite);
- TIM/SORCE (Solar Radiation and Climate Experiment);
- TSIS-1 (Total and Spectral Solar Irradiance Sensor-1, operating on the International Space Station since 2017).

SORCE/TIM and TSIS-1 (Woods & Leibacher, 2023)[2] also employ an alternative cross-calibration methodology.

---

[1]Scafetta, N., 2023c. Empirical assessment of the role of the Sun in climate change using balanced multi-proxy solar records. Geoscience Frontiers, 14(6), 101650.
Soon, W., Connolly, R., Connolly, M., Akasofu, S.-I., Baliunas, S., et al., 2023. The detection and attribution of Northern Hemisphere land surface warming (1850–2018) in terms of human and natural factors: Challenges of inadequate data. Climate, 11, 179.
[2]Woods, T.N., Leibacher, J.W., 2023. The Solar Radiation and Climate Experiment (SORCE) mission: Final calibrations and data products. Solar Physics, 298, 25.

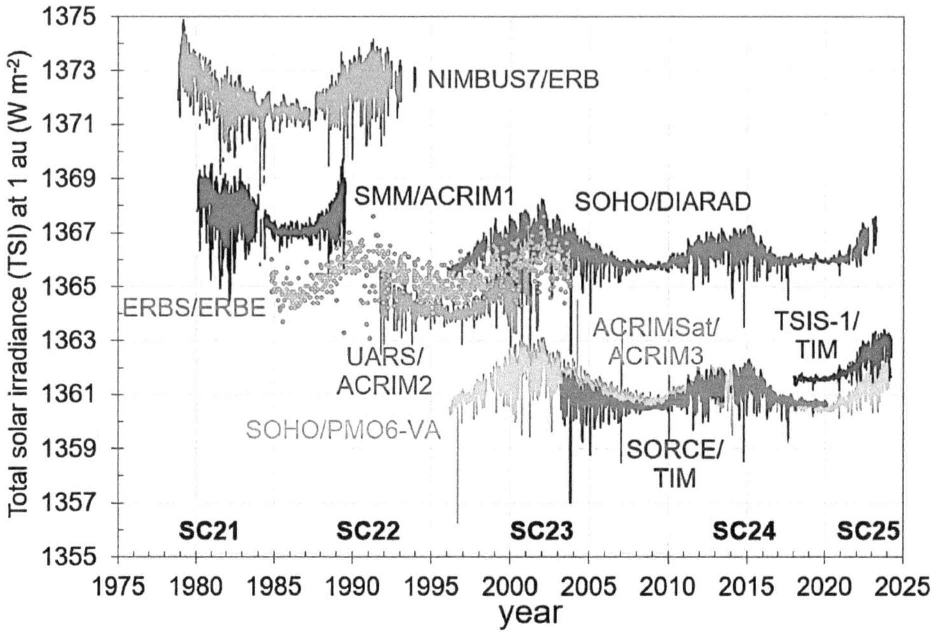

FIGURE 24.1. Original daily satellite TSI data provided by mission teams used to generate TSI composites. (Adapted from Connolly et al., 2024).

The launch of SORCE/TIM in 2003 and TSIS-1 in 2017 marked a leap in precision, with TIM achieving an uncertainty below ±0.03% (Kopp & Lean, 2011),[3] leading to a revised TSI baseline of approximately 1361 W/m². Earlier estimates had placed the value near 1367 W/m² (see Figure 24.1).

Beyond TSI, TSIS-1 introduced high-resolution solar spectral irradiance (SSI) measurements, enabling wavelength-specific analysis of solar effects on atmospheric chemistry, particularly stratospheric ozone.

However, no single mission has provided continuous, uninterrupted TSI coverage from 1978 to the present. Figure 24.1 clearly displays the 11-year cycle, but discrepancies among instruments—arising from calibration and sensitivity differences—have led to inconsistencies.

To bridge the observational gaps, several research groups have developed TSI composites by empirically cross-calibrating the records from different instruments:

- ACRIM Composite (Willson & Mordvinov, 2003)[4]: based on ACRIM I, II, III, and ERB data; uses originally published records without modification;

[3]Kopp, G., Lean, J.L., 2011. A new, lower value of total solar irradiance: Evidence and climate significance. Geophysical Research Letters, 38, L01706.
[4]Willson, R.C., Mordvinov, A.V., 2003. Secular total solar irradiance trend during solar cycles 21–23. Geophysical Research Letters, 30(5), 1199.

- PMOD Composite (Fröhlich & Lean, 1998; Fröhlich, 2012)[5]: developed by PMOD in Davos; uses modified versions of the original satellite data based on various models and assumptions;
- RMIB Composite (Chapman et al., 2013; Dewitte & Clerbaux, 2017)[6]: developed by the Royal Meteorological Institute of Belgium; uses an alternative cross-calibration methodology.

Figure 24.2 compares the original satellite-based TSI records (NIMBUS7/ERB, ACRIM1, and ERBS/ERBE) with the modified versions produced by Fröhlich and infilled by RMIB. The adjustments introduced by PMOD to the NIMBUS7/ERB and ERBS/ERBE datasets are explicitly shown in Panels "c," "f," and "i," and are substantial, effectively lowering the TSI minimum observed in 1996 to match the level recorded in 1985.

Figure 24.3 compares the TSI satellite composites proposed by ACRIM and PMOD. According to ACRIM, the TSI level in 1996 was approximately 0.4 W/m^2 higher than in 1985, whereas the PMOD composite indicates that the two TSI minima were nearly equal.

The divergence between the ACRIM and PMOD composites during this gap has profound implications for solar variability reconstructions and for the estimated solar contribution to global warming.

24.2. The critical issue: The ACRIM-gap and PMOD's revision

The ACRIM and PMOD TSI composites differ in how they treat and model multi-decadal modulation and observational gaps. The most contentious interval is the ACRIM-gap (1989–1992), which arose from the delay in launching ACRIM-II after the ACRIM-I mission ended—a delay caused by the Challenger disaster in 1986, which disrupted NASA's launch schedule.

Between June 1989 and October 1991, no high-precision, self-calibrating satellite instruments were available to measure TSI. During this ACRIM-gap, data came from less stable instruments: the ERB radiometer aboard Nimbus-7 and the ERBE sensor on ERBS.

The ERB/Nimbus-7 data show a marked upward trend in TSI during this period (see Figure 24.1). However, the PMOD group (*Physikalisch-Meteorologisches Observatorium Davos*) applied substantial corrections to this dataset, along with ACRIM-I, ACRIM-II, and ERBE/ERBS, citing instrumental degradation, thermal drift, and cavity aging.

Using empirical solar models, Claus Fröhlich of PMOD removed the upward trend evident in the ERB/Nimbus-7 data, producing a TSI composite record with a stationary or slightly declining trend (Fröhlich & Lean, 1998; Fröhlich, 2012).

---

[5]Fröhlich, C., Lean, J., 1998. The Sun's total irradiance: Cycles, trends and related climate change uncertainties since 1976. Geophysical Research Letters, 25(23), 4245–4392.
Fröhlich, C., 2012. Total solar irradiance observations. Surveys in Geophysics, 33(3–4), 453–473.
[6]Chapman, G.A., Cookson, A.M., Preminger, D.G., 2013. Modeling Total Solar Irradiance with San Fernando Observatory ground-based photometry. Solar Physics, 283, 295–305.
Dewitte, S., Clerbaux, N., 2017. Measurement of the Earth radiation budget at the top of the atmosphere—a review. Remote Sensing, 9(11), 1143.

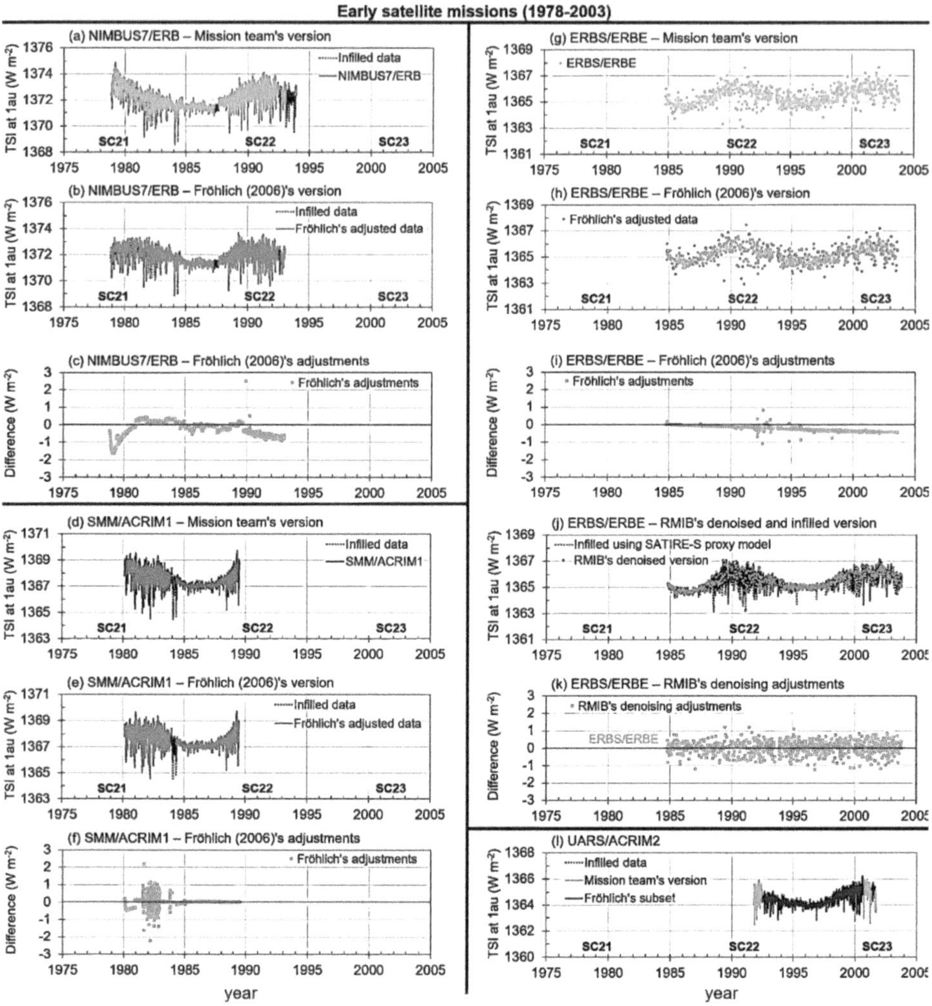

FIGURE 24.2. Monthly satellite TSI data from 1978–2003, shown before and after the application of Fröhlich's (2006) adjustments, RMIB's denoising, and the infilling methods used in this study. Panels (a–c), (d–f), and (g–i) show, respectively, the original and adjusted records for Nimbus 7/ERB, SMM/ACRIM1, and ERBS/ERBE, along with the corresponding F06 corrections. Panel (j) shows RMIB's infilled ERBS/ERBE data; (k) its denoising adjustments. Panel (l) displays UARS/ACRIM2 with Fröhlich's subsetting. All original data gaps were infilled using the GA method, except RMIB's ERBS/ERBE, which used SATIRE-S. Infilled segments are marked with black dashed lines. (Adapted from Connolly et al., 2024).

24.3. Willson and Hoyt's statements on PMOD's TSI adjustments

The PMOD revisions have been strongly contested by Richard C. Willson (Principal Investigator of ACRIM) and Douglas V. Hoyt (Principal Investigator

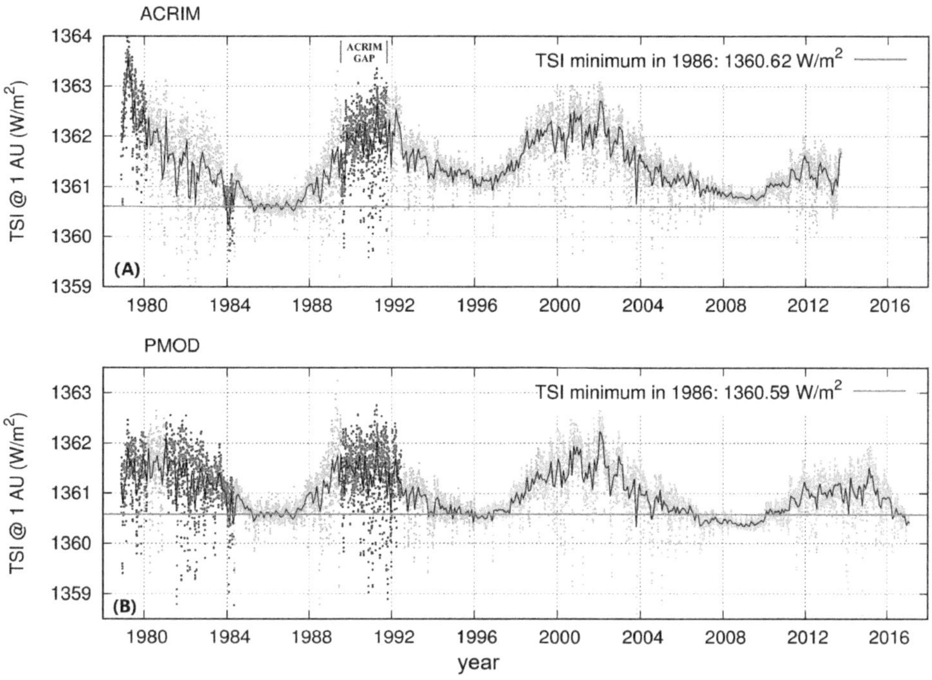

FIGURE 24.3. Comparison of ACRIM composites (using official satellite TSI measurements) and PMOD composites (using modified versions of the original records). (Adapted from Scafetta, 2023c).

of ERB/Nimbus-7). Both researchers argued that the PMOD composite introduced significant adjustments to the published TSI satellite observations without consulting the original science teams. They contended that such modifications were based solely on indirect assumptions, such as solar proxy models, rather than on documented instrumental behavior, resulting in what they describe as arbitrary and unjustified adjustments.

In written statements provided to Nicola Scafetta (see Figure 24.4), Willson and Hoyt emphasized that PMOD's TSI satellite adjustments are inconsistent with the known physical characteristics and calibration history of the instruments involved. They warned that such alterations compromise the integrity of the TSI record and risk distorting long-term solar–climate relationships. These concerns are documented in several peer-reviewed studies (Willson & Mordvinov, 2003; Scafetta & Willson, 2014; Scafetta et al., 2019a),[7] which advocate preserving the original satellite data and caution against overreliance on proxy-driven reinterpretations.

[7]Scafetta, N., Willson, R.C., 2014. ACRIM total solar irradiance satellite composite validation versus TSI proxy models. Astrophysics and Space Science, 350, 421–442.
Scafetta, N., Willson, R.C., Lee, J.N., Wu, D.L., 2019. Modeling quiet solar luminosity variability from TSI satellite measurements and proxy models during 1980–2018. Remote Sensing, 11(21), 2569.

Dr. Richard C. Willson
Principal Investigator
ACRIM Experiments
12 Bahama Bend,
Coronado, CA, 92118
Phone: 619-407-7716
Fax: 619-365-9579
E-mail: rwillson@acrim.com

September 16, 2008

Dear. Dr. Scafetta:

Regarding Frohlich's PMOD TSI composite:

1. Frohlich made unauthorized and incorrect adjustments to the SMM/ACRIM1 and UARS/ACRIM2 TSI results. In the case of ACRIM1 he arbitrarily miss-applied the degradation correction published by the ACRIM1 Science team for the SMM 'spin mode' (1981 – 1984) to the 1980 results. He did this without any detailed knowledge of the ACRIM1 instrument or on-orbit performance, original analysis or consultation with the ACRIM1 team. His intent was clearly to revise the solar cycle 21 TSI to agree with Judith Lean's TSI proxy model.
2. Frohlich chose the ERBS/ERBE database to 'bridge' the ACRIM gap when it was clearly inferior to the Nimbus7/ERB gap data. His justification was based on hypothetical 'upward steps' in the Nimbus7/ERB results ('glitches' in Frohlich's words) that no other researchers, including both the original PI (Hickey) and the final science team (Hoyt and Kyle) believe exist. As with ACRIM1 above, Frohlich had no detailed knowledge of the Nimbus7/ERB instrument and made no original analysis or computations. The only obvious purpose appears to be to obtain a TSI composite that agreed with the predictions of Lean's TSI proxy model.
3. The TSI proxy models, such as Lean's, are not competitive in accuracy or precision with even the worst satellite TSI observations. To 'adjust' satellite data to agree with such models is incompatible with the scientific method.
4. The PMOD TSI composite panders to those who promote anthropogenic causes as the principal component of global warming, despite mounting evidence to the contrary. They cite its lack of significant TSI trending as evidence of relatively insignificant solar climate forcing during the past 30 years.

Sincerely,

Dr. Richard C. Willson

Nicola Scafetta, Ph. D.

From: "Douglas Hoyt" <dhoyt@toast.net>
Date: Tuesday, September 16, 2008 5:59 PM
To: "Nicola Scafetta" <ns2002@duke.edu>
Subject: Re: A question - letter response

September 16, 2008

Dear Dr. Scafetta:

Concerning the supposed increase in N7 sensitivity at the end of September 1989 and other matters as proposed by Frohlich's PMOD TSI composite:

1. There is no known physical change in the electrically calibrated N7 radiometer or its electronics that could have caused it to become more sensitive. At least neither Lee Kyle nor I could never imagine how such a thing could happen and no one else has ever come up with a physical theory for the instrument that could cause it to become more sensitive.

2. The N7 radiometer was calibrated electrically every 12 days. The calibrations before and after the September shutdown gave no indication of any change in the sensitivity of the radiometer. Thus, when Bob Lee of the ERBS team originally claimed there was a change in N7 sensitivity, we examined the issue and concluded there was no internal evidence in the N7 records to warrant the correction that he was proposing. Since the result was a null one, no publication was thought necessary.

3. Thus, Frohlich's PMOD TSI composite is not consistent with the internal data or physics of the N7 cavity radiometer.

4. The correction of the N7 TSI values for 1979-1980 proposed by Frohlich is also puzzling. The raw data was run through the same algorithm for these early years and the subsequent years and there is no justification for Frohlich's adjustment in my opinion.

Sincerely,

Douglas Hoyt

FIGURE 24.4. Statements by Richard C. Willson and Douglas V. Hoyt regarding the modifications made by Claus Fröhlich of PMOD to the TSI series from the ACRIM I and II satellite experiments and ERB/Nimbus-7.

24.4. ACRIM–PMOD divergence and its climatic relevance

Scafetta & Willson (2014)[8] and Scafetta et al. (2019a)[9] examined the modifications to satellite TSI data proposed by Fröhlich and the PMOD group, demonstrating that these revisions contradict multiple empirical findings derived from independent solar records. Figure 24.5 illustrates several key discrepancies.

The principal findings are as follows.

(1) *ACRIM-gap divergence*: During the ACRIM-gap (1989–1991), ERB/Nimbus-7 exhibits an upward TSI trend, whereas ERBE/ERBS shows a slight decline (Figure 24.5A). PMOD interprets this as a sensor sensitivity increase in ERB/Nimbus-7. However, experimental evidence indicates that ERBE/ERBS underwent sensitivity degradation owing to its early exposure to high UV radiation during the solar

[8]Scafetta, N., Willson, R.C., 2014. ACRIM total solar irradiance satellite composite validation versus TSI proxy models. Astrophysics and Space Science, 350, 421–442.

[9]Scafetta, N., Willson, R.C., Lee, J.N., Wu, D.L., 2019a. Modeling quiet solar luminosity variability from TSI satellite measurements and proxy models during 1980–2018. Remote Sensing, 11(21), 2569.

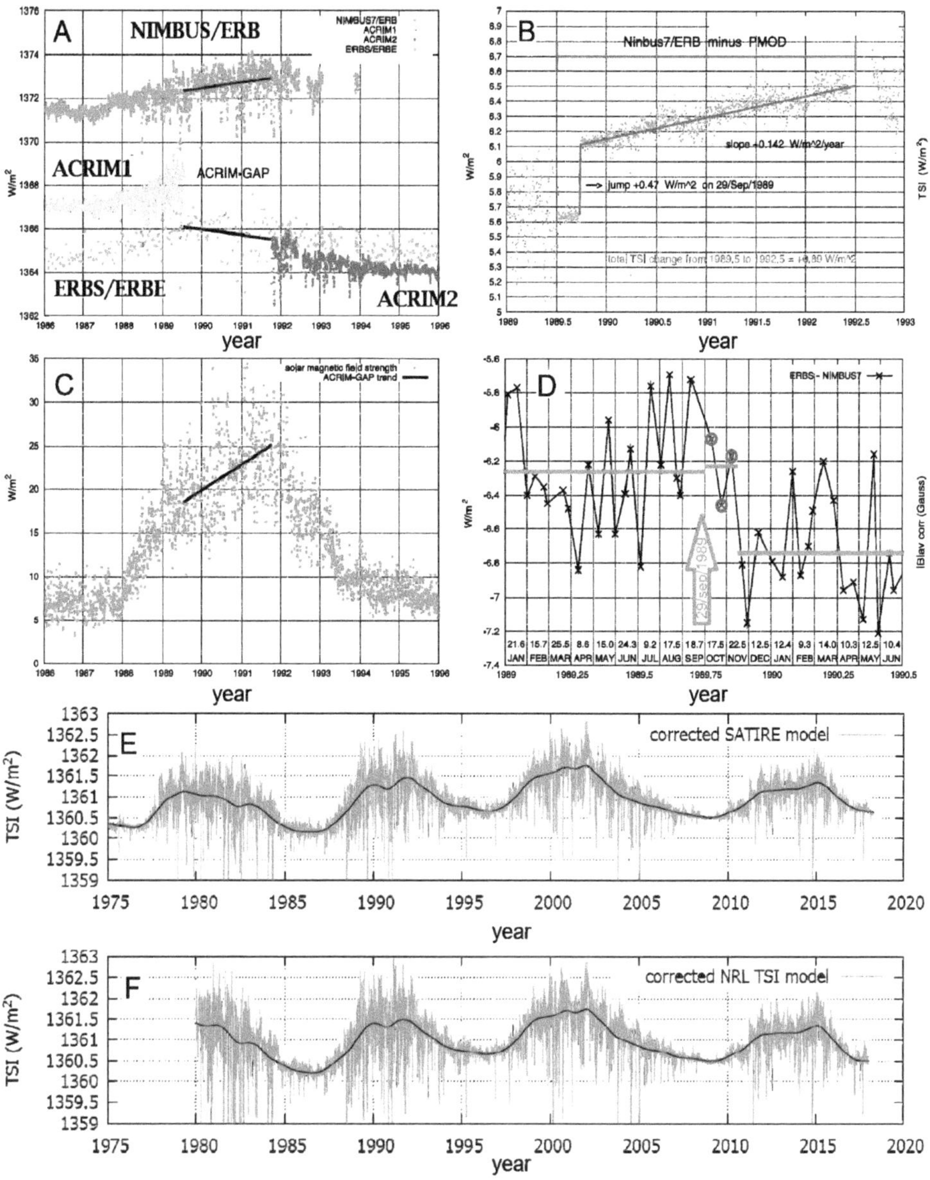

FIGURE 24.5. [A] Original satellite data during the ACRIM-gap period: ACRIM I and II, ERB/Nimbus-7, and ERBE/ERBS. [B] Difference between the ERB/Nimbus-7 TSI record and the modified version proposed by PMOD. [C] Solar magnetic field intensity (*"National Solar Observatory/Kitt Peak Data"*). [D] Difference between the ERB/Nimbus-7 and ERBE/ERBS TSI records around September 29, 1989. [E & F] SATIRE and NRLTSI2 TSI proxy models after being rescaled to accurately reproduce non-controversial satellite TSI data. (Adapted from: Scafetta & Willson, 2014; Scafetta et al., 2019a).

cycle 22 maximum (1989–1992). Even the original ERBE/ERBS data show a modest TSI increase from the 1985 to the 1995 solar minima, which disappears in PMOD's revised record.

(2) *Sensor shutdown anomaly*: PMOD hypothesizes a sudden 0.47 W/m^2 sensitivity shift in ERB/Nimbus-7 on September 29, 1989, following a temporary shutdown. Yet simultaneous ERBE/ERBS measurements confirm the ERB/Nimbus-7 values during that period (Figures 24.5B and D), undermining the revision.

(3) *Linear drift inconsistency*: PMOD estimates a linear sensitivity drift of 0.142 W/m^2/year in ERB/Nimbus-7 during the ACRIM-gap. This figure is incompatible with solar magnetic field evolution and proxy reconstructions, which indicate a rising TSI trend (Figures 24.5B and C).

(4) *Proxy model recalibration*: When TSI proxy models, initially aligned with PMOD, are recalibrated to match satellite data during non-controversial periods, they tend to reproduce the ACRIM composite modulation rather than that of PMOD (Figures 24.5E and F).

These findings reinforce the conclusion that the original TSI data published by the ACRIM and ERB teams are the most reliable, being grounded in direct knowledge of instrument behavior and performance.

Scientific integrity and political overtones. Beyond technical disagreements, some commentators have suggested that ideological motivations may also have influenced the construction and adoption of the PMOD composite. Notably, Judith Lean—coauthor of the first PMOD composite—stated in 2003:

> *"The fact that some people could use Willson's results (the ACRIM TSI composite) as an excuse to do nothing about greenhouse gas emissions was one of the reasons we felt the need to examine the data ourselves"* (Lindsey, 2003).[10]

This statement suggests that the ACRIM TSI composite was perceived in some circles as politically inconvenient.

Climatic implications of divergent TSI composites. Figure 24.3 compares the two principal satellite TSI composites:

- *ACRIM composite*: shows a significant TSI increase from 1980 to 2000, correlating with the rise in global temperatures during that period, followed by a slight decline;
- *PMOD composite*: depicts total solar irradiance (TSI) as stationary or slightly decreasing since 1980, implying a solar dynamic that correlates negatively with the observed warming.

The difference between the two composites exceeds 0.5 W/m^2 during the 1980–2000 interval. The upward trend identified in the ACRIM composite correlates with the observed climate warming of that period and may account for a substantial portion of it (Scafetta & West, 2005, 2008; Scafetta, 2023c).[11]

---

[10] Lindsey, R., 2003. Under a Variable Sun. NASA Earth Observatory. Available at: https://earthobservatory.nasa.gov/features/VariableSun/variable4.php

[11] Scafetta, N., West, B.J., 2005. Estimated solar contribution to the global surface warming using the ACRIM TSI satellite composite. Geophysical Research Letters, 32, L18713.

Two Interpretive Paradigms. The ACRIM–PMOD controversy defines two fundamentally different views of solar–climate interaction:

- *ACRIM composite*: suggests that solar activity varies significantly across Schwabe cycles, with meaningful multi-decadal, secular, and millennial-scale impacts. The observed TSI increase from 1980 to 2000 may have contributed substantially to global warming.
- *PMOD composite*: portrays solar activity as essentially stable or slightly declining since 1980, implying that recent warming is almost entirely anthropogenic.

This divergence is not merely academic; it shapes how we interpret climate history, model future scenarios, and design environmental policy.

24.5. Interpreting the 2000–2014 climate "pause"

Numerous studies have identified a strong correlation between the TSI rise reconstructed in the ACRIM composite and the increase in global surface temperatures between 1980 and 2000 (Scafetta & West, 2005, 2008).[12] After 2000, both series appear to enter a phase of stabilization or slowdown, commonly referred to as the *pause* or *hiatus* in global warming, which persisted until 2014 (see Figure 21.4) (Kaufmann et al., 2011; Medhaug et al., 2017).[13]

In this context, the climate pause may reflect a substantial solar influence: a temporary reduction in TSI could have offset anthropogenic warming, producing a plateau in global temperatures. Between the solar maxima of 2001 (Cycle 23) and 2013–2014 (Cycle 24), the Schwabe cycle entered a phase of low solar activity, with reduced energy output.

By contrast, the CMIP6 GCMs simulate continuous warming over the same interval, assigning the Sun a marginal role in temperature variability (IPCC, 2021).[14] The discrepancy between model projections and observed trends invites a reassessment of the Sun's contribution to recent climate dynamics.

24.6. Possible satellite TSI composites

Figures 24.6 and 24.7 from Connolly et al. (2024) highlight the divergence among various satellite TSI composites. Reconstructions based on the ACRIM records show a rising TSI trend until 2000, while those based on the PMOD-modified records depict a stationary or slightly declining trend.

The principal difference lies in the data adopted:

- ACRIM-based composites use only official satellite data;

Scafetta, N., West, B.J., 2008. Is climate sensitive to solar variability? Physics Today, 61(3), 50–51.
Scafetta, N., 2023c. Empirical assessment of the role of the Sun in climate change using balanced multi-proxy solar records. Geoscience Frontiers, 14(6), 101650.

[12]Scafetta, N., West, B.J., 2005. Estimated solar contribution to the global surface warming using the ACRIM TSI satellite composite. Geophysical Research Letters, 32, L18713.
Scafetta, N., West, B.J., 2008. Is climate sensitive to solar variability? Physics Today, 61(3), 50–51.

[13]Kaufmann, R.K., Kauppi, H., Mann, M.L., Stock, J.H., 2011. Reconciling anthropogenic climate change with observed temperature 1998–2008. Proceedings of the National Academy of Sciences, 108(29), 11790–11793.
Medhaug, I., Stolpe, M.B., Fischer, E.M., Knutti, R., 2017. Reconciling controversies about the global warming hiatus. Nature, 545(7652), 41–47.

[14]IPCC, 2021. Climate Change 2021: The Physical Science Basis. Cambridge University Press.

- PMOD-based composites use empirically adjusted satellite data.

Composites numbered 17 to 21 incorporate data from SORCE/TIM, DIARAD, and TSIS-1 in various ways. Some show a slight increase in TSI even after 2000. When extended with historical ACRIM data (1980–2025), a stronger correlation emerges between solar variability and surface temperature fluctuations.

This evidence suggests that the Sun may have played a more substantial role in recent climate dynamics than estimated by current GCMs, raising questions about the validity of model-based attribution and the completeness of Earth's energy balance assessments.

24.7. The climate paradox of the *"almost"* constant Sun

Beginning with the Fourth Assessment Report (IPCC, 2007),[15] the IPCC adopted the PMOD composite as the sole reference for TSI. This decision had far-reaching implications:

- in the CMIP5 and CMIP6 models, secular solar forcing is represented as nearly constant, effectively minimizing the Sun's role in climate variability over the past century;
- this modeling choice has shaped the architecture of climate simulations and the attribution of warming causes, raising doubts about the adequacy of the underlying assumptions.

The CMIP6 GCMs use the solar model proposed by Matthes et al. (2017),[16] which averages two reconstructions partially validated and calibrated against PMOD:

- SATIRE (Figure 25.3, panel "b"): a semi-empirical model based on solar imagery and magnetic proxies (Krivova et al., 2010; Yeo et al., 2014);[17] two versions exist: SATIRE-S, focused on the satellite era (1974–2013), and SATIRE-M/T, which extends the reconstruction on a secular scale from 1610 to the present.
- NRLTSI2 (Figure 25.3, panel "f"): an empirical model using sunspot numbers as proxies (Kopp et al., 2016; Coddington, 2016);[18]

Both models suggest a slight decline in TSI from 1980 to 2020, consistent with PMOD-based reconstructions. By design, this modeling implies that the Sun has played no significant role in recent global warming.

---

[15]IPCC, 2007. Climate Change 2007: The Physical Science Basis. Cambridge University Press.

[16]Matthes, K., Funke, B., Andersson, M.E., Barnard, L., Beer, J., et al., 2017. Solar forcing for CMIP6 (v3.2). Geoscientific Model Development, 10, 2247–2302.

[17]Krivova, N.A., Vieira, L.E.A., Solanki, S.K., 2010. Reconstruction of solar spectral irradiance since the Maunder minimum. Journal of Geophysical Research: Space Physics, 115, A12112.
Yeo, K.L., Krivova, N.A., Solanki, S.K., Glassmeier, K.H., 2014. Reconstruction of total and spectral solar irradiance from 1974 to 2013 based on KPVT, SoHO/MDI and SDO/HMI observations. Astronomy and Astrophysics, 570, A85.

[18]Kopp, G., Krivova, N., Wu, C.J., Lean, J., 2016. The impact of the revised sunspot record on solar irradiance reconstructions. Solar Physics, 291, 2951–2965.
Coddington, O., Lean, J.L., Pilewskie, P., Snow, M., Lindholm, D., 2016. A solar irradiance climate data record. Bulletin of the American Meteorological Society, 97, 1265.

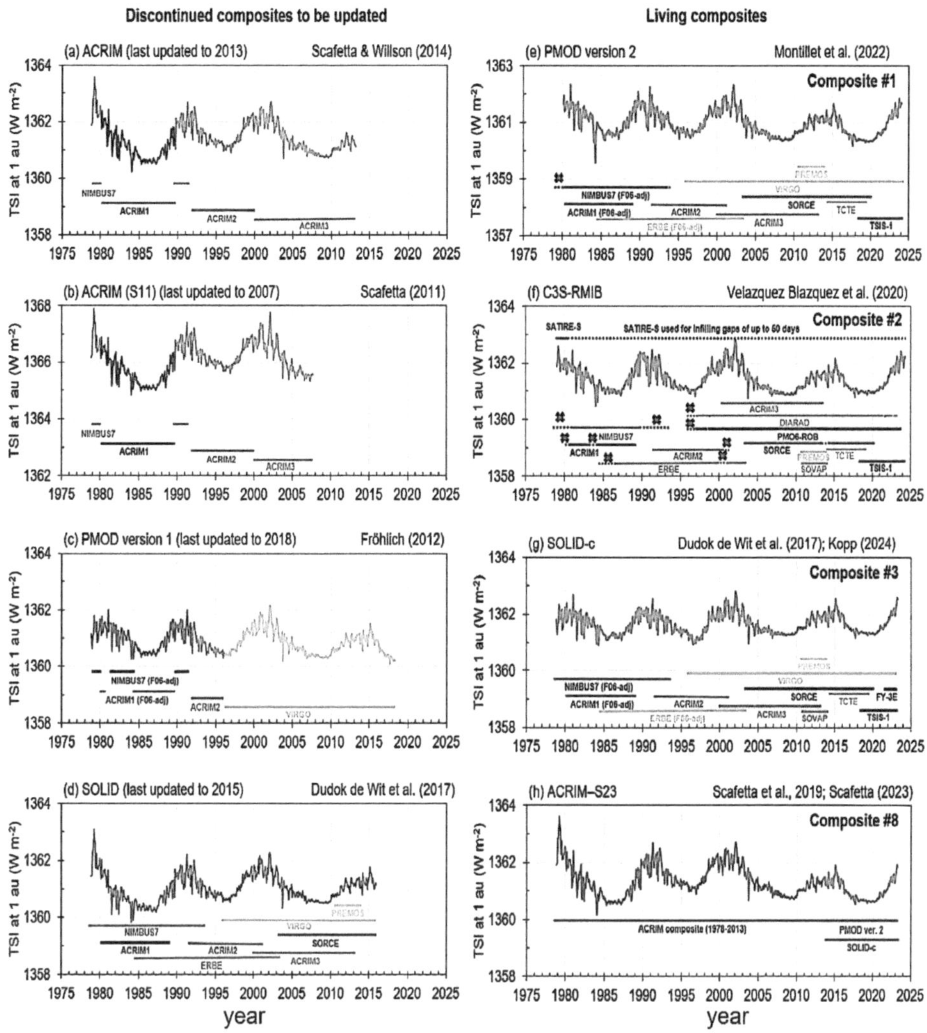

FIGURE 24.6. This figure compares several currently available *"Total Solar Irradiance"* (TSI) composites. The four now-obsolete composites include: (a) original ACRIM; (b) an ACRIM variant proposed by N. Scafetta (2011); (c) PMOD version 1; (d) SOLID. The four currently active living composites include: (e) PMOD version 2; (f) C3S-RMIB; (g) the SOLID-c update by Kopp; (h) a new ACRIM composite based on the methodology described by Scafetta et al. (2019a) and Scafetta (2023c). Colored horizontal lines in the lower panels indicate the satellite missions whose data contribute to each composite. Each panel also notes any modifications made to original TSI records by PMOD, relative to the data published by the satellite mission teams. Dashed lines and X symbols identify periods of satellite data intentionally omitted during composite construction. (Adapted from Connolly et al., 2024).

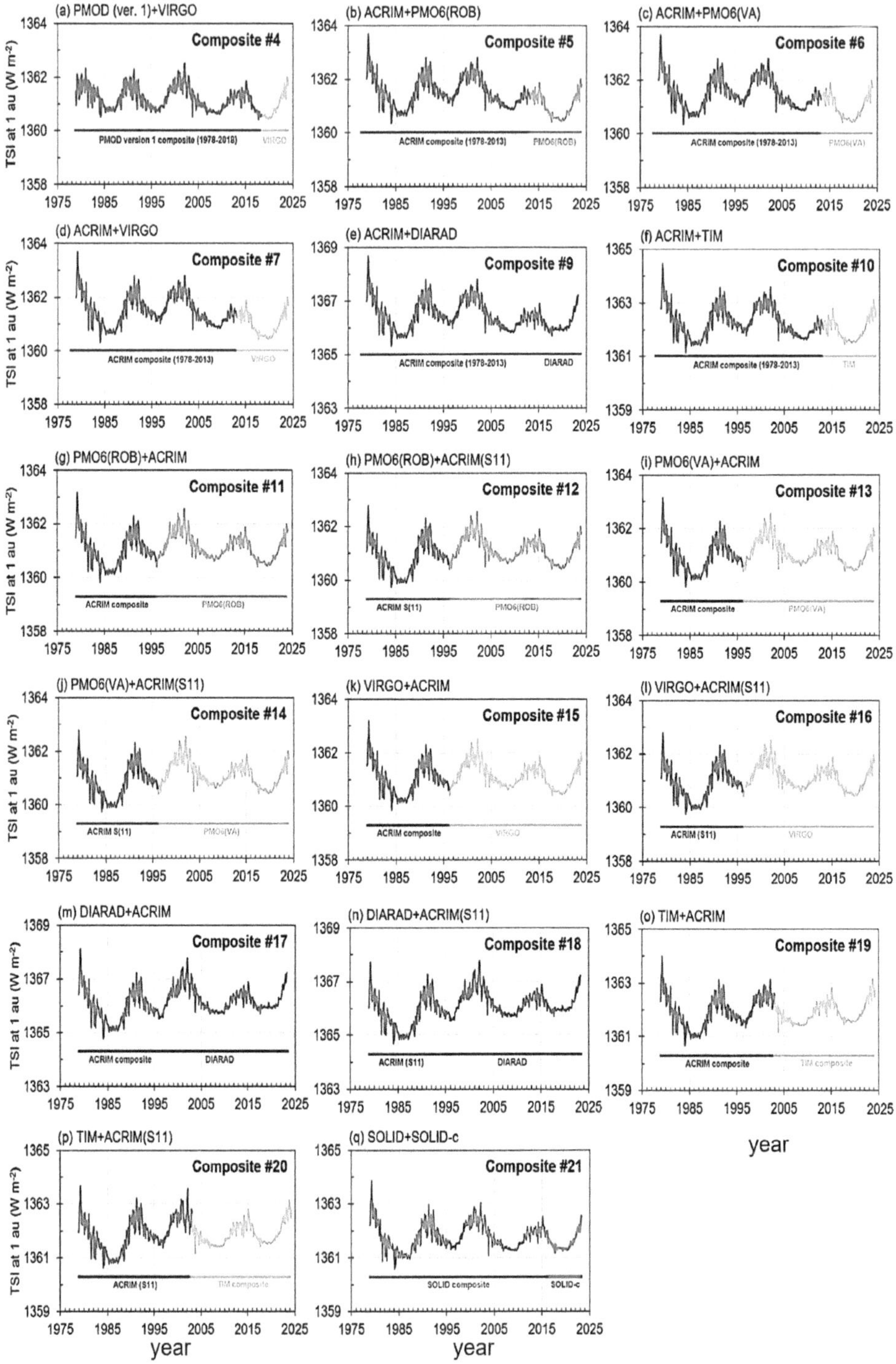

FIGURE 24.7. Various other possible TSI composites. (Adapted from Connolly et al., 2024).

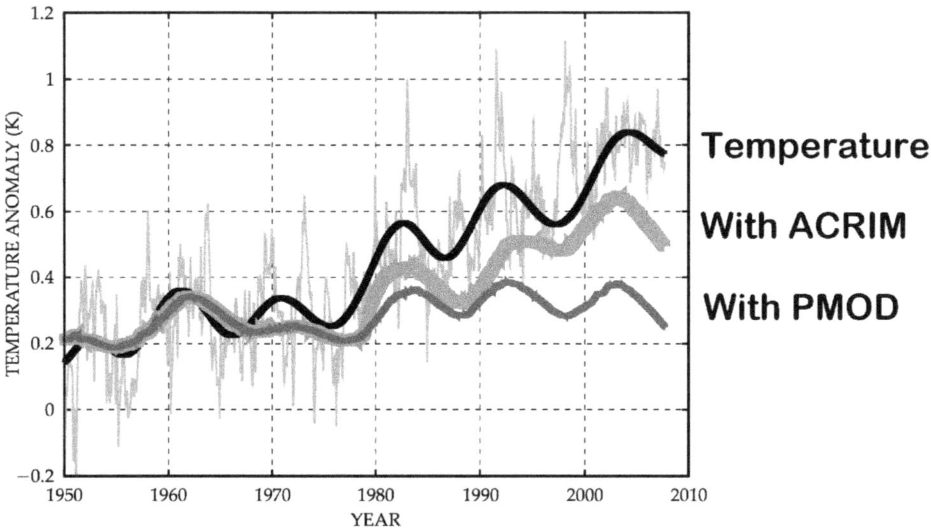

FIGURE 24.8. Global Surface Temperature (GST) anomaly compared with two alternative solar climatic signatures using a TSI proxy reconstruction until 1978 and the ACRIM and PMOD alternative TSI satellite composites after 1978. Note the 11-year solar cycle signature in the temperature record (Adapted from Scafetta & West, 2008).

However, de Wit et al. (2017)[19] pointed out that the SATIRE model is incompatible with the TSI composites, even those reprocessed using PMOD-modified records. This internal inconsistency raises questions about the coherence of SATIRE modeling.

In this context, Connolly et al. (2024)[20] emphasize that the divergence between models and empirical data renders the entire attribution chain in the CMIP6 GCMs controversial, raising fundamental methodological and epistemological concerns. The issue extends beyond calibration; it challenges the core assumptions underlying the solar forcings used in global climate simulations.

24.8. The epistemological paradox of solar and climate models

At the heart of the paradox lies the widespread adoption of TSI composites modified by the PMOD group, which depict solar activity as essentially stable and nearly invariant over time. This portrayal reduces solar variability to a marginal climatic role, despite extensive empirical evidence to the contrary. The differing climatic interpretations derived from ACRIM and PMOD are shown in Figure 24.8: with PMOD, the warming observed since the 1970s cannot be

[19] de Wit, T.D., Kopp, G., Fröhlich, C., Schöll, M., 2017. Methodology to create a new total solar irradiance record: making a composite out of multiple data records. Geophysical Research Letters, 44(3), 1196–1203.

[20] Connolly, R., Soon, W., Connolly, M., Cionco, R.G., Elias, A.G., et al., 2024. Multiple new or updated satellite total solar irradiance (TSI) composites (1978–2023). The Astrophysical Journal, 975, 102.

attributed to solar forcing, whereas with ACRIM a significant portion appears to be solar-induced (Scafetta & West, 2008).[21]

The critical issue is epistemological: a specific theoretical assumption — namely, that solar forcing has remained nearly constant— is elevated to a dominant paradigm. In this process, methodological critiques and alternative reconstructions, including those based on official satellite data already optimized by the original experimental teams, are systematically ignored.

This methodological choice shapes the internal logic of the CMIP climate models, steering them toward exclusive attribution of global warming to anthropogenic factors. As a result, the contribution of solar activity to climate change from 1850–1900 to the present is deemed negligible (see Figure 6.3), despite numerous empirical findings suggesting otherwise.

Recent TSI satellite measurements, including those analyzed by Connolly et al. (2024), document significant multi-decadal variability. These findings indicate that the Sun may have exerted a non-negligible influence on climate change—an influence that current models fail to capture.

In light of this evidence, the adequacy of the solar reconstructions adopted in IPCC-endorsed GCMs appears increasingly questionable. The modeling choices seem oriented toward producing a predetermined outcome rather than reflecting the full complexity of Earth's radiative balance.

24.9. Conclusion: an unresolved epistemological fracture

The debate over satellite measurements of total solar irradiance (TSI) is not a minor technical issue but a fundamental epistemological fracture within climate science. Its implications extend beyond solar physics, reaching the methodological foundations of the discipline.

Because TSI is an astronomical parameter, many climatologists may lack the expertise to critically assess its construction and significance. Consequently, the solar assumptions embedded in the CMIP GCMs are often accepted without rigorous scrutiny or consideration of alternative datasets.

If, as multiple studies suggest, solar variability has contributed substantially to recent warming, then the framework of climate forcings adopted by the IPCC (2021) may be structurally misleading. The exclusive reliance on the PMOD TSI composite in AR6, coupled with the dismissal of alternatives such as ACRIM, reflects a non-neutral epistemic choice that reduces the complexity of the climate system to a single narrative of anthropogenic causation.

Connolly et al. (2024) emphasize that excluding legitimate solar datasets and selectively choosing sources undermines the core principles of empirical science: pluralism of data, transparency of methods, and openness to debate. Until the controversy over total solar irradiance is addressed—rather than ignored—the scientific community will struggle to achieve a robust understanding of modern climate change. Restoring the Sun to the heart of climate discourse requires critical inquiry, methodological rigor, and a renewed appreciation of Earth's systemic complexity.

[21] Scafetta, N., West, B.J., 2008. Is climate sensitive to solar variability? Physics Today, 61(3), 50–51.

CHAPTER 25

TSI proxy models: between uncertainty and revision

Reconstructing solar activity variability across secular and millennial timescales remains one of the most formidable scientific challenges and a major source of uncertainty in both solar physics and climate science. In the absence of direct total solar irradiance (TSI) measurements prior to 1978—the beginning of the satellite era—and given the persistent uncertainties in post-1980 TSI satellite composites, researchers rely on various solar proxies to estimate long-term solar dynamics.

The two primary sources of proxy data are:

- the Wolf sunspot number record, available since 1610 thanks to the telescopic observations initiated by Galileo Galilei;
- cosmogenic isotopes, primarily carbon-14 (^{14}C) from tree rings and biological records, and beryllium-10 (^{10}Be) from polar ice cores, both of which reflect the modulation of galactic cosmic rays by the solar magnetic field.

These proxies allow the identification of periods of high and low solar activity, such as the Maunder Minimum (1645–1715), but they do not provide direct or precise measurements of solar luminosity (TSI). Converting these indirect signals into reliable numerical TSI values remains controversial, rendering empirical validation of proxy-based solar models an exceedingly difficult endeavor.

25.1. Formation of ^{14}C and ^{10}Be in the atmosphere

Carbon-14 (^{14}C). ^{14}C is produced in the upper atmosphere when galactic cosmic rays—primarily high-energy protons—collide with atmospheric atoms, generating secondary particles, including free neutrons. These neutrons interact with nitrogen-14 (^{14}N) through the nuclear reaction:

$$^{14}N + n \rightarrow {}^{14}C + p.$$

The resulting ^{14}C oxidizes to form $^{14}CO_2$, which mixes into the atmosphere and is absorbed by living organisms, making it a key tool for radiocarbon dating. Its production rate varies with solar activity and geomagnetic field strength, which together regulate the incoming flux of cosmic rays.

Beryllium-10 (^{10}Be). ^{10}Be is formed through spallation, when cosmic rays strike atmospheric nuclei (e.g., oxygen, nitrogen, argon), fragmenting them and producing light isotopes. A typical reaction is:

$$p + O \rightarrow ^{10} Be + \text{fragments}.$$

Once formed, ^{10}Be binds to aerosols and is deposited, particularly in polar regions, into ice layers retrievable through core drilling.

Because solar magnetic activity modulates cosmic ray penetration, weaker solar phases yield higher ^{14}C and ^{10}Be concentrations. These records thus serve as valuable paleo-solar proxies, enabling reconstructions of solar variability across centuries and millennia.

25.2. Limitations in the use of solar proxies

Limitations of using the "*sunspot number*" as a TSI proxy.

- *No direct TSI measurement*: sunspots reflect magnetic activity, not total irradiance, which also depends on faculae and quiet solar regions.
- *Nonlinear correlation*: TSI does not scale linearly with sunspot count; bright faculae often offset the dimming effect of sunspots.
- *Historical data uncertainty*: pre-nineteenth-century records are sparse, inconsistent, and subject to retroactive correction.
- *Methodological shifts*: changes in counting criteria (e.g., SN version 1 vs. version 2) introduce discontinuities.
- *Lack of magnetic detail*: sunspot numbers do not capture the magnetic intensity, size, or complexity of active regions.

In short, while useful for tracking general solar trends, the sunspot number is not a reliable proxy for solar luminosity and must be supplemented with other indicators.

General limitations of using cosmogenic records as TSI proxies.

- *Indirect link to TSI*: ^{14}C and ^{10}Be reflect cosmic ray modulation, not solar brightness.
- *Geomagnetic interference*: Earth's magnetic field also affects isotope production, complicating attribution.
- *Environmental filtering*: volcanic, hydrological, and biological processes can alter isotope concentrations.
- *Temporal resolution*: ^{10}Be offers annual resolution, whereas ^{14}C is smoother and more sensitive to biological processes.
- *Model dependence*: translating isotope data into solar indices requires semi-empirical models with uncertain parameters.

Specific limitations for climatological use.

- *Lack of radiometric precision*: proxies cannot provide the accuracy required for energy balance calculations.
- *Divergent reconstructions*: estimates vary widely depending on methodology and calibration.
- *No consensus on pre-satellite TSI*: uncertainty remains too high to draw definitive conclusions about the solar influence on climate change before 1978.

25.3. The "*active Sun*" and the "*quiet Sun*": reconstructing TSI over time

TSI represents the total solar energy received per unit area at 1 AU, integrated across all wavelengths. Although TSI varies with the 11-year Schwabe cycle, the nature and magnitude of its long-term variability, together with its role in climate change, remain subjects of ongoing scientific debate.

To realistically model TSI over time, its variability is often decomposed into two components:

$$\mathrm{TSI}(t) = \mathrm{TSI}_{\mathrm{quiet}}(t) + \Delta\mathrm{TSI}_{\mathrm{active}}(t),$$

where:

- $\mathrm{TSI}_{\mathrm{quiet}}(t)$ represents the background irradiance from magnetically quiet solar regions;
- $\Delta\mathrm{TSI}_{\mathrm{active}}(t)$ captures the brightness variations due to sunspots, faculae, and other active phenomena.

These components are assumed to be independent, although they are physically interdependent. The equation must therefore be interpreted as an approximation.

The TSI active component. Sunspots are cooler, darker regions that reduce TSI, whereas faculae are brighter areas that increase it. Their interplay produces the characteristic oscillations of the solar cycle. Facular brightness typically outweighs sunspot dimming, resulting in a net positive correlation between sunspot number and TSI, although this correlation is not strictly linear.

Models that quantify $\Delta\mathrm{TSI}_{\mathrm{active}}(t)$ include:

- SATIRE ("*Spectral And Total Irradiance REconstruction*"): uses magnetograms to simulate facular and sunspot distributions Krivova et al. (2003);[1]
- NRLTSI2 ("*Naval Research Laboratory Total Solar Irradiance*"): employs empirical proxies such as the sunspot number and the Mg II index (Coddington, 2016).[2]

A simplified formulation is:

$$\Delta\mathrm{TSI}_{\mathrm{active}}(t) = a \cdot R(t) + b \cdot F(t)$$

where $R(t)$ is the relative sunspot number and $F(t)$ is a proxy for faculae.

Hypotheses on modeling the quiet Sun's luminosity. There are two principal approaches.

Constant TSI Models. Some models assume that the quiet Sun's irradiance is constant:

$$\mathrm{TSI}_{\mathrm{quiet}}(t) = TSI_0$$

This assumption underlies models such as NRLTSI2 (Coddington, 2016) and is consistent with the PMOD composite, which attributes TSI variability solely to active regions (Fröhlich, 2009).[3]

Variable TSI Models. Other studies propose that the quiet Sun varies on secular or millennial timescales owing to changes in the solar dynamo. Examples include:

[1] Krivova, N. A., Solanki, S. K., Fligge, M., 2003. Reconstruction of solar total and spectral irradiance variations in cycle 23. Astronomy & Astrophysics, 399, L1–L4.

[2] Coddington, O., Lean, J.L., Pilewskie, P., Snow, M., Lindholm, D., 2016. A solar irradiance climate data record. Bulletin of the American Meteorological Society, 97, 1265.

[3] Fröhlich, C., 2009. Evidence of a long-term trend in total solar irradiance. Astronomy & Astrophysics, 501(3), L27–L30.

- Vieira et al. (2011)[4] and Steinhilber et al. (2009, 2012),[5] who use cosmogenic isotopes (^{10}Be and ^{14}C) to model long-term TSI variability;
- Friis-Christensen & Lassen (1991),[6] who employ solar cycle length (SCL) as a proxy for deep solar dynamo strength. A typical model is:

$$\mathrm{TSI}_{\mathrm{quiet}}(t) = TSI_0 + \alpha\,(\mathrm{SCL}_{\mathrm{ref}} - \mathrm{SCL}(t))$$

  where α is an empirical parameter and $\mathrm{SCL}_{\mathrm{ref}}$ is the reference cycle length, for example 11 years. Shorter cycles imply stronger solar activity, and vice versa. The algorithm used to compute the cycle length $\mathrm{SCL}(t)$ varies across studies.
  - Friis-Christensen & Lassen (1991),[7] later corrected by Lassen & Friis-Christensen (2000),[8] proposed weighted moving averages (e.g., 121 or 12221 schemes);
  - Baliunas & Soon (1991)[9] suggested using two Schwabe cycles to reflect the Hale magnetic cycle. They estimate that TSI may have varied by ~4 W/m² from the Maunder Minimum to the present, on the basis of chromospheric activity in solar-type stars (cf. Judge et al., 2020).[10]
- SATIRE-M: a multi-millennial extension of SATIRE based on proxy data Wu et al. (2018).[11]

Multi-indicator models. Some researchers, such as Hoyt & Schatten (1993),[12] incorporate multiple solar indicators to reconstruct TSI, including:

- *solar cycle length*: regarded as indicative of solar activity intensity, with shorter cycles associated with stronger solar activity;
- *solar cycle decay rate*: normalized to estimate how rapidly solar activity declines after the maximum;

---

[4]Vieira, L.E.A., Solanki, S.K., Krivova, N.A., Usoskin, I., 2011. Evolution of the solar irradiance during the Holocene. Astronomy & Astrophysics, 531, A6.

[5]Steinhilber, F., Beer, J., Fröhlich, C., 2009. Total solar irradiance during the Holocene. Geophysical Research Letters, 36, L19704.

Steinhilber, F., Abreu, J.A., Beer, J., Brunner, I., Christl, M., et al., 2012. 9,400 years of cosmic radiation and solar activity from ice cores and tree rings. Proceedings of the National Academy of Sciences of the United States of America, 109(16), 5967–5971.

[6]Friis-Christensen, E., Lassen, K., 1991. Length of the solar cycle: An indicator of solar activity closely associated with climate. Science, 254, 698–700.

[7]Friis-Christensen, E., Lassen, K., 1991. Length of the solar cycle: An indicator of solar activity closely associated with climate. Science, 254, 698–700.

[8]Lassen, K., Friis-Christensen, E., 2000. Reply to "Solar cycle lengths and climate: A reference revisited." Journal of Geophysical Research: Space Physics, 105(A12), 27493–27495.

The series by Friis-Christensen & Lassen (1991) was subsequently criticized because the model 12221 values proposed after 1970 were speculative owing to a lack of available observables. It nevertheless remains valid for the period 1570–1970.

[9]Baliunas, S.L., Soon, W.H., 1991. Are variations in the length of the activity cycle related to changes in brightness in solar-type stars? The Astrophysical Journal, 374, 463–470.

[10]Judge, P.G., Egeland, R., Henry, G.W., 2020. Sun-like stars shed light on solar climate forcing. The Astrophysical Journal, 891(1), 96.

[11]Wu, C.-J., Krivova, N.A., Solanki, S.K., Usoskin, I.G., 2018. Solar total and spectral irradiance reconstruction over the last 9000 years. Astronomy and Astrophysics, 620, A120.

[12]Hoyt, D.V., Schatten, K.H., 1993. A discussion of plausible solar irradiance variations, 1700–1992. Journal of Geophysical Research: Space Physics, 98(A11), 18895–18906.

- *average solar activity level*: based on aggregated indices such as the sunspot number;
- *sunspot structure*: the ratio between umbral and penumbral area, useful for estimating magnetic intensity;
- *decay rate of individual sunspots*: a proxy for magnetic dynamics and convective energy transport;
- *number of sunspots without an umbra*: an indicator of variations in surface magnetic structure;
- *equatorial rotation rate of the Sun*: linked to internal dynamics and magnetic field generation.

These variables aim to capture the complexity of solar magnetic dynamics and energy transport.

Ongoing controversies. Debates over the proper use of proxies persist. For example, Chatzistergos (2024)[13] challenged the plausibility of long-term TSI variations, prompting responses from Soon et al. (2024)[14] and the CERES team, who defended the validity of multi-proxy reconstructions and identified methodological flaws in the revisionist approaches.

Models predicting strong secular variability. Models such as the PMOD/WRC long-term irradiance model Shapiro et al. (2011)[15] and the CHRONOS model Egorova et al. (2018)[16] predict significant background TSI variability, of up to 6 W/m² between the Maunder Minimum and the minima of modern Schwabe cycles. These reconstructions integrate geomagnetic and cosmogenic isotope proxies to derive coherent estimates of the solar magnetic field and irradiance over millennial scales.

Such models contradict the notion of a "*quiet*" Sun with nearly constant luminosity, suggesting instead that deep solar dynamo processes can generate substantial secular fluctuations in solar output.

Scafetta's hybrid planetary model. Scafetta (2012b, 2014b, 2026)[17] proposes a hybrid model that combines:

- an active component, estimated from the sunspot record;
- a background component, modulated by planetary orbital cycles.

---

[13]Chatzistergos, T., 2024. A discussion of implausible total solar irradiance variations since 1700. Solar Physics, 299, 21.

[14]Soon, W., and CERES team (2024, March 10). The challenges of trying to reconstruct changes in Total Solar Irradiance (TSI) since 1700: A response to Chatzistergos (2024)'s remake of Hoyt & Schatten (1993). CERES-Science.

[15]Shapiro, A. I., Schmutz, W., Rozanov, E., Schoell, M., Haberreiter, M., Shapiro, A.V., Nyeki, S., 2011. A new approach to the long-term reconstruction of the solar irradiance leads to large historical solar forcing. Astronomy & Astrophysics, 529, A67.

[16]Egorova, T., Schmutz, W., Rozanov, E., Shapiro, A.I., Usoskin, I., et al., 2018. Revised historical solar irradiance forcing. Astronomy and Astrophysics, 615, A85.

[17]Scafetta, N., 2012b. Multi-scale harmonic model for solar and climate cyclical variation throughout the Holocene based on Jupiter–Saturn tidal frequencies plus the 11-year solar dynamo cycle. Journal of Atmospheric and Solar–Terrestrial Physics, 80, 296–311.
Scafetta, N., 2014b. Discussion on the spectral coherence between planetary, solar and climate oscillations: a reply to some critiques. Astrophysics and Space Science, 354, 275–299.
Scafetta, N., 2025. Detection, attribution, and modeling of climate change: Key open issues. Gondwana Research, DOI:10.1016/j.gr.2025.05.001.

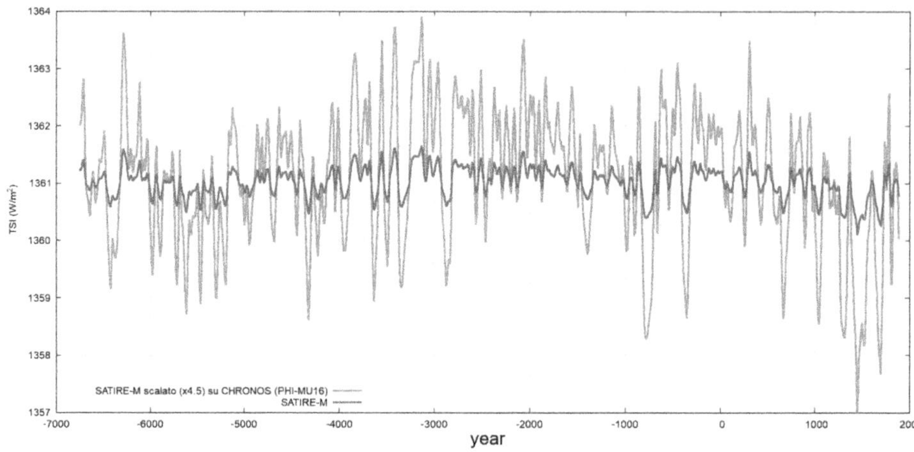

FIGURE 25.1. Comparison between the SATIRE-M TSI model (smaller curve) (Wu et al., 2018) and the same model rescaled by a factor of 4.5 (larger curve) to approximately simulate the average variability of the CHRONOS TSI models by Egorova et al. (2018) shown in Figure 25.4. (See also Chatzistergos et al., 2023).

The model can be expressed as:

$$\text{TSI}(t) = \text{TSI}_{\text{quiet}}(t) + \Delta\text{TSI}_{\text{active}}(t)$$

where:

$$\text{TSI}_{\text{quiet}}(t) = TSI_0 + \sum_i A_i \cos(2\pi f_i t + \phi_i)$$

with f_i representing planetary frequencies derived from the synodic and orbital periods of Venus, Earth, Jupiter, Saturn, Uranus, and Neptune—cycles that appear in solar proxy records.

This approach aligns more closely with multi-secular climate variability and with the ACRIM composite, which exhibits a secular TSI increase from 1980 to 2000. One advantage is that planetary cycles are governed by precise astronomical laws, reducing the uncertainties inherent in isotope-based reconstructions (e.g., ^{14}C and ^{10}Be), which are also influenced by terrestrial factors such as geomagnetic field shifts and atmospheric dynamics (cf. McCracken et al., 2013).[18]

25.4. Comparison between high- and low-variability solar models

Figure 25.1 compares:

- SATIRE-M (Wu et al., 2018): a low-variability model based on cosmogenic proxies;
- a rescaled version of SATIRE-M ($\times$4.5): which simulates the amplitude of secular variability in the CHRONOS model (Egorova et al., 2018).

[18]McCracken, K.G., Beer, J., Steinhilber, F., Abreu, J., 2013. A phenomenological study of the cosmic ray variations over the past 9400 years. Solar Physics, 286, 609.

The discrepancy is substantial. Amplifying secular fluctuations dramatically alters the estimated climatic impact of solar variability, underscoring the importance of model selection in climate attribution studies.

Conclusions. The decomposition of TSI into active and background components offers a useful framework for modeling solar variability. Although simpler models, such as NRLTSI2 (Coddington, 2016), assume a constant quiet Sun, most modern reconstructions (CHRONOS, SATIRE-M, Steinhilber et al., Scafetta) posit that the quiet Sun's luminosity varies over time.

The amplitude of this variability differs significantly across models, however, reflecting divergent assumptions about solar physics and proxy calibration.

Secular and millennial-scale solar variability is likely driven by the deep solar dynamo, modulated by both internal processes and external astronomical forcings. Incorporating these dynamics is essential for a more realistic understanding of the Sun's role in Earth's climate system.

25.5. The challenge of proxy modeling of TSI

In recent decades, various research groups have proposed TSI reconstructions based on combinations of:

- direct solar observations (sunspots, faculae, irradiance);
- cosmogenic isotope data (^{14}C and ^{10}Be);
- magnetic indices;
- physical-statistical modeling frameworks.

The aim is to generate coherent solar forcing datasets for integration into global climate models (GCMs), particularly for simulating preindustrial climate evolution and assessing the Sun's long-term influence on Earth's energy balance.

Notable examples are discussed in Connolly et al. (2021, 2023),[19] which highlight the profound differences among reconstruction approaches and emphasize how methodological choices can shape the attribution of climate change causes.

Figures 25.2, 25.3, 25.4, and 25.5 illustrate various TSI proxy reconstructions. All figures use the same vertical-axis scale to facilitate visual comparison.

What emerges is a striking heterogeneity in simulated secular TSI variability. This divergence is evident not only across different research groups but also between successive versions of the same record, underscoring the persistent methodological and theoretical uncertainty in this field.

Specifically:

(1) Figure 25.3 presents low-variability models, which depict TSI as relatively constant over time;

[19] Connolly, R., Soon, W., Connolly, M., Baliunas, S., Berglund, J., Butler, C. J., Cionco, R. G., Elias, A. G., Fedorov, V. M., Harde, H., Henry, G. W., Hoyt, D. V., Humlum, O., Legates, D. R., Lüning, S., Scafetta, N., Solheim, J.-E., Szarka, L., van Loon, H., Velasco Herrera, V. M., Willson, R. C., Yan, H., & Zhang, W. (2021). How much has the Sun influenced Northern Hemisphere temperature trends? An ongoing debate. Research in Astronomy and Astrophysics, 21(6), 131.
Connolly, R., Soon, W., Connolly, M., Baliunas, S., Berglund, J., Butler, C. J., Cionco, R. G., Elias, A. G., Fedorov, V. M., Harde, H., Henry, G. W., Hoyt, D. V., Humlum, O., Legates, D. R., Scafetta, N., Solheim, J.-E., Szarka, L., Velasco Herrera, V. M., Yan, H., & Zhang, W. (2023). Challenges in the Detection and Attribution of Northern Hemisphere Surface Temperature Trends Since 1850. Research in Astronomy and Astrophysics, 23(10), 105015.

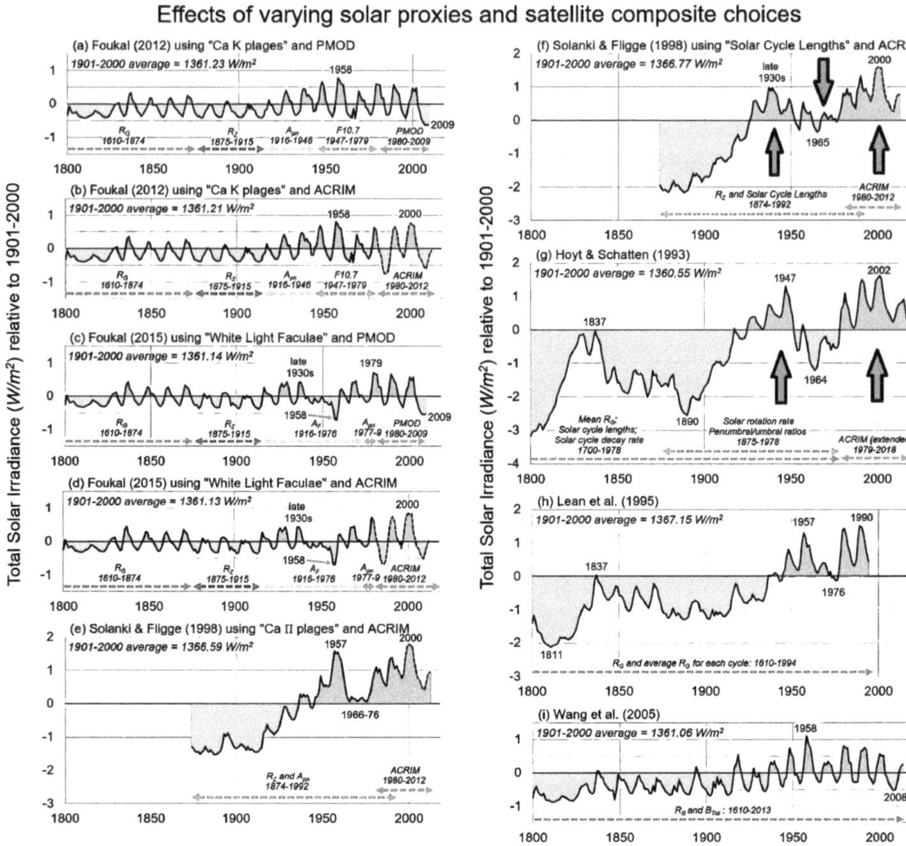

FIGURE 25.2. Examples of TSI reconstructions produced by different authors, varying the choice of solar proxies for the pre-satellite era and the TSI composite used for the satellite era. (Adapted from Connolly et al., 2021).

(2) Figure 25.4 showcases high-variability reconstructions, which attribute a more dynamic energetic role to the Sun on secular scales.

25.6. The TSI model proposed by Matthes et al. (2017)

The most influential TSI reconstruction adopted by the IPCC Sixth Assessment Report (2021) is the model proposed by Matthes et al. (2017),[20] shown in panel "e" of Figure 25.3. This model serves as the official solar forcing input for CMIP6 climate simulations.

The model proposed by Matthes et al. exhibits near stationarity in TSI between 1850 and 2020, with decadal variations of no more than ±0.05%. This feature is pivotal: it implies that CMIP6 models effectively attribute no significant solar contribution to the warming observed between the preindustrial baseline (1850–1900) and the present (see Figures 6.2 and 6.3).

[20]Matthes, K., Funke, B., Andersson, M.E., Barnard, L., Beer, J., et al., 2017. Solar forcing for CMIP6 (v3.2). Geoscientific Model Development, 10, 2247–2302.

For millennial-scale simulations, CMIP6 employs the SATIRE-M (*"Spectral and Total Irradiance Reconstructions for the Millennia"*) model (Wu et al., 2018),[21] which likewise displays an extremely subdued secular modulation of TSI. Together, these models reinforce the view that solar variability has played a secondary or negligible role in long-term climate dynamics.

The adoption of these reconstructions reflects a paradigm in which anthropogenic forcings dominate climate simulations, while solar variability is treated as climatically marginal.

Critical comment. While methodologically consistent with the prevailing framework, this approach raises profound epistemological questions:

- *To what extent do the adopted TSI proxy models truly capture the natural variability of the Sun?*
- *To what extent does the dominant paradigm—minimizing the solar role—constrain a more pluralistic and systemic understanding of Earth's climate evolution?*

These questions are not merely academic. They strike at the core of climate attribution, model architecture, and the integrity of scientific inquiry. A more inclusive and critically reflective approach may be necessary to fully understand the Sun's role in shaping Earth's climate of the past, present, and future.

25.7. Downward revision of solar variability?

Over time, several models that initially attributed significant secular variability to TSI have undergone progressive downward revision, contributing to the consolidation of a minimalist view of solar variability.

A prominent example is the work of Lean et al. (1995),[22] which estimated a TSI reduction of approximately 0.25% (about 3.5 W/m^2) during the Maunder Minimum relative to present-day values (Figure 25.2, panel "h"). However, in subsequent revisions—Lean et al. (2005)[23] and Wang et al. (2005)[24]—this estimate was reduced to about 1.3 W/m^2 (approximately 0.1%), as reflected in the NRLTSI2 model (Coddington, 2016; Kopp et al., 2016).[25]

Similarly, the group led by Sami K. Solanki at the Max Planck Institute for Solar System Research has proposed several TSI models with markedly different characteristics over time. Earlier work (Solanki & Fligge, 1998) indicated notable secular variability (Figure 25.2, panels "e" and "f"), whereas the later SATIRE

[21]Wu, C.-J., Krivova, N.A., Solanki, S.K., Usoskin, I.G., 2018. Solar total and spectral irradiance reconstruction over the last 9000 years. Astronomy and Astrophysics, 620, A120.

[22]Lean, J., Beer, J., Bradley, R., 1995. Reconstruction of solar irradiance since 1610: implications for climate change. Geophysical Research Letters, 22, 3195–3198.

[23]Lean, J., Wang, Y.M., Sheeley, N.R., 2005. SORCE contributions to new understanding of global change and solar variability. Solar Physics, 230, 27–53.

[24]Wang, Y.-M., Lean, J.L., Sheeley, N.R. Jr., 2005. Modeling the Sun's magnetic field and irradiance since 1713. The Astrophysical Journal. 625, 522–538.

[25]Coddington, O., Lean, J.L., Pilewskie, P., Snow, M., Lindholm, D., 2016. A solar irradiance climate data record. Bulletin of the American Meteorological Society, 97, 1265.
Kopp, G., Krivova, N., Wu, C.J., Lean, J., 2016. The impact of the revised sunspot record on solar irradiance reconstructions. Solar Physics, 291, 2951–2965.

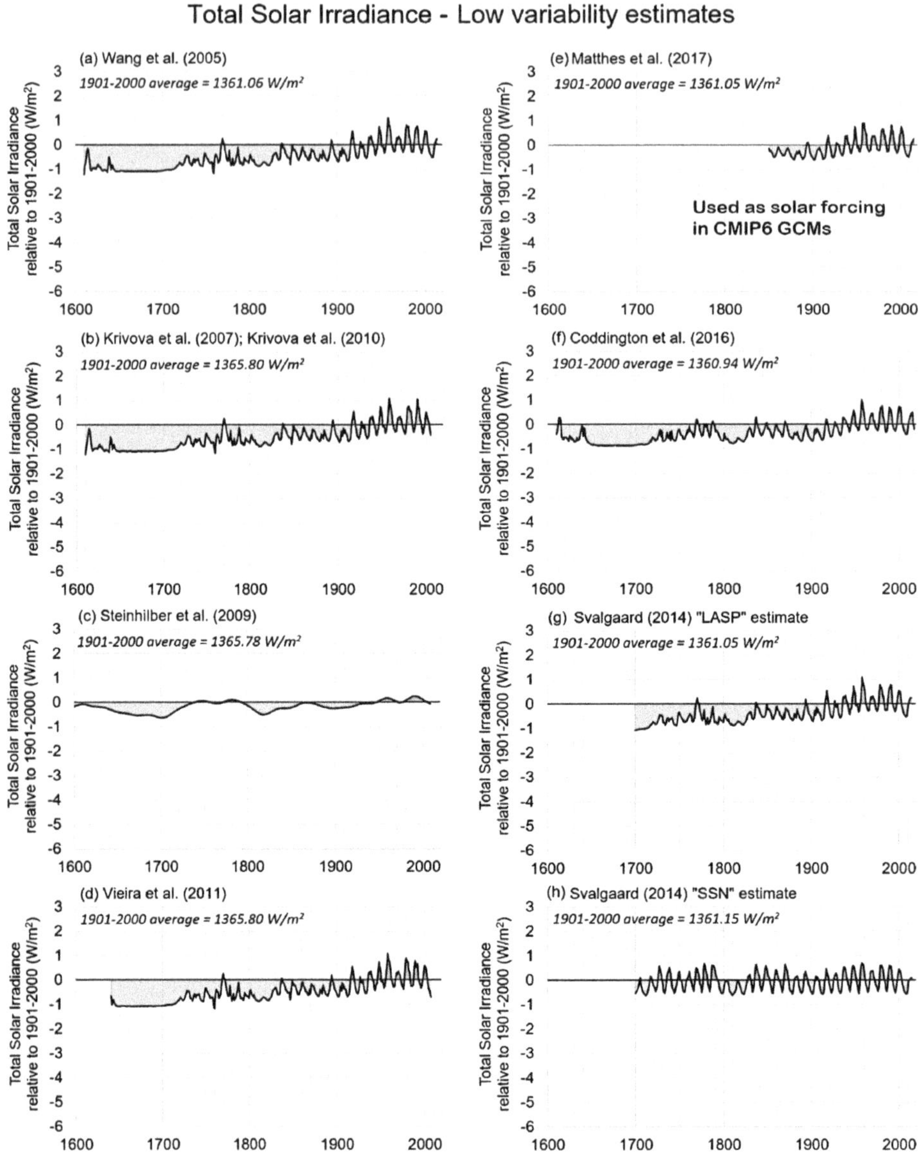

FIGURE 25.3. Eight low-variability estimates of TSI changes relative to the 1901–2000 average. (Adapted from Connolly et al., 2021).

models (Krivova et al., 2010; Vieira et al., 2011; Wu et al., 2018)[26] display markedly reduced variability (Figure 25.3, panels "b" and "d").

---

[26]Krivova, N.A., Vieira, L.E.A., Solanki, S.K., 2010. Reconstruction of solar spectral irradiance since the Maunder minimum. Journal of Geophysical Research: Space Physics, 115, A12112.
Vieira, L.E.A., Solanki, S.K., Krivova, N.A., Usoskin, I., 2011. Evolution of the solar irradiance during the Holocene. Astronomy & Astrophysics, 531, A6.

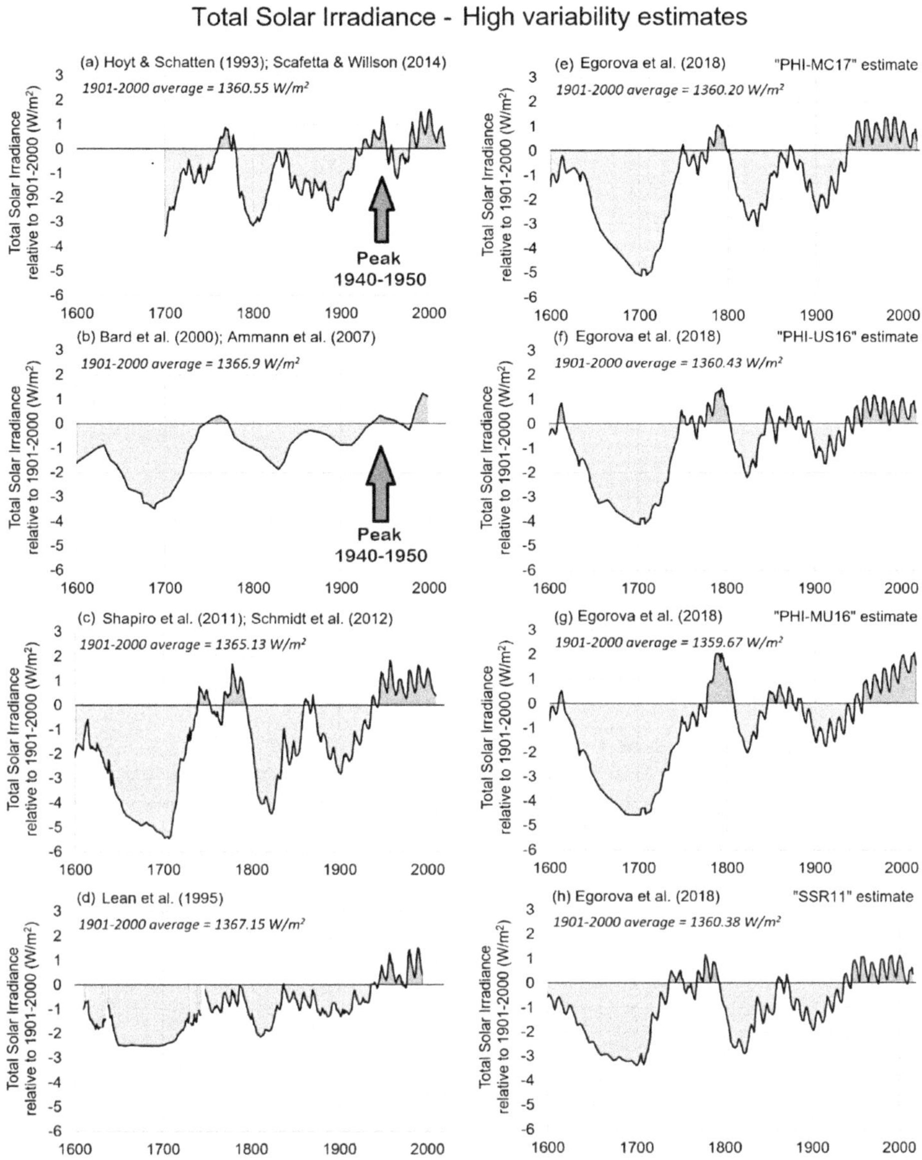

FIGURE 25.4. Eight high-variability estimates of TSI changes relative to the 1901–2000 average. (Adapted from Connolly et al., 2021).

More recently, Yeo et al. (2020)[27] estimated that the dimmest theoretical TSI state the Sun could reach would correspond to a level 2.0 ± 0.7 W/m² lower than the 2019 value (about 1360.4 W/m²). This estimate, supported in part by PMOD-based satellite composites that minimize post-1980 multi-decadal variability,

[27]Yeo, K.L., Solanki, S.K., Krivova, N.A., Rempel, M., Anusha, L.S., et al., 2020. The dimmest state of the Sun. Geophysical Research Letters, 47(19), e2020GL090243.

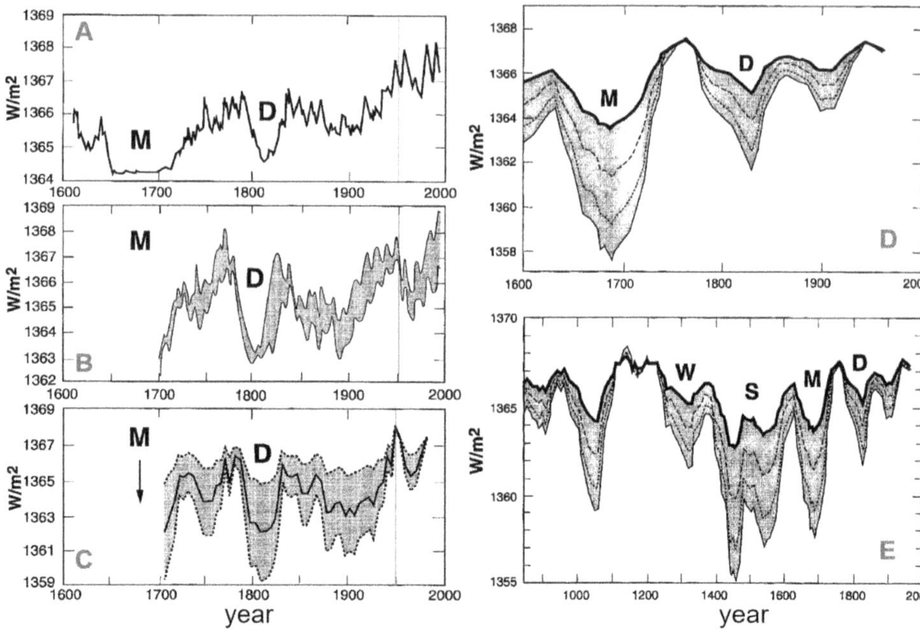

FIGURE 25.5. Comparison of TSI reconstructions. [A] from Lean et al. (1995); [B] from Hoyt & Schatten (1993); [C] from Zhang et al. (1994); [D & E] possible TSI variations inferred from ^{14}C and ^{10}Be production. Letters indicate periods of secular solar minima: "W" = Wolf Minimum; "S" = Spörer Minimum; "M" = Maunder Minimum; "D" = Dalton Minimum. Note that the curves in B, C, and D show a peak occurring before 1950. (Adapted from Bard et al., 2000).

reinforces the hypothesis of low secular solar variability (cf. Chatzistergos et al., 2023).[28] This hypothesis underpins both the SATIRE and NRLTSI2 models, which form the basis of the CMIP6 solar forcing framework (Matthes et al., 2017), used in global climate models and adopted in the IPCC AR6 (2021).

Separately, Rempel (2020)[29] shows that even if, on a secular scale, the quiet-Sun small-scale magnetic field could vary in unobserved ways, high-variability scenarios would be unrealistic, as they would require a drastic reduction of the quiet-Sun field to values far beyond what small-scale dynamo physics permits. His radiative magnetohydrodynamical simulations therefore partially support Yeo et al. (2020), indicating that the quiet-Sun field should vary only weakly and already accounts for nearly all cycle-scale irradiance variability, reinforcing the conclusion that TSI changes arise almost entirely from active-region magnetic fields.

[28]Chatzistergos, T., Krivova, N.A., Yeo, K.L., 2023. Long-term changes in solar activity and irradiance. Journal of Atmospheric and Solar-Terrestrial Physics, 252, 106150.

[29]Rempel, M., 2020. On the contribution of quiet Sun magnetism to solar irradiance variations: Constraints from small-scale dynamo simulations. The Astrophysical Journal, 894(2), 140.

25.8. Alternative TSI models with greater secular variability

The minimalist view of TSI variability adopted by institutional GCMs is not universally accepted. Numerous studies present alternative proxy reconstructions that suggest much more pronounced solar variability over recent centuries (cf. Figure 25.4). These models indicate a significant increase in solar luminosity from the Little Ice Age (LIA) to the present.

Key contributions include:

- Hoyt & Schatten (1993, 1997)[30]—among the first to propose a marked TSI increase between 1700 and 2000;
- Bard et al. (2000)[31]—based on cosmogenic isotopes (^{10}Be and ^{14}C), which identifies secular oscillations consistent with paleoclimatic records;
- Shapiro et al. (2011)[32] and Egorova et al. (2018)[33]—introduce advanced modeling of the internal solar magnetic field, expanding perspectives on long-term solar forcing;
- Penza et al. (2022)[34]—combines solar proxies with regional climate evidence, offering an integrated view of solar signals and climate responses.

Several solar activity reconstructions suggest that TSI may have increased by 2.2 to 5.5 W/m^2 between 1700 (Maunder Minimum) and the solar minimum of 1986 (cf. Chatzistergos et al., 2023; Connolly et al., 2021). This increase correlates significantly with the observed global climate warming from the seventeenth to the twenty-first century, implying a potentially substantial solar contribution to both preindustrial and modern climate change.

Further support comes from studies of the Sun's coronal magnetic field, such as Lockwood et al. (1999),[35] which document a doubling of its intensity between 1900 and 1990. This period coincides with a marked increase in solar magnetic forcing, which may have significantly influenced Earth's energy balance.

25.9. An approximately 60-year cycle in solar activity?

A particularly revealing comparison emerges from the two high-variability TSI reconstructions proposed by Solanki & Fligge (1998) (Figure 25.2):

(1) panel "e": based on the luminous contribution of solar faculae, using the Ca II spectral lines as a proxy; the TSI peak occurs around 1957;

[30]Hoyt, D.V., Schatten, K.H., 1993. A discussion of plausible solar irradiance variations, 1700–1992. Journal of Geophysical Research: Space Physics, 98(A11), 18895–18906.

[31]Bard, E., Raisbeck, G.M., Yiou, F., Jouzel, J., 2000. Solar irradiance during the last 1200 years based on cosmogenic nuclides. Tellus Series B: Chemical and Physical Meteorology, 52, 985–992.

[32]Shapiro, A. I., Schmutz, W., Rozanov, E., Schoell, M., Haberreiter, M., Shapiro, A.V., Nyeki, S., 2011. A new approach to the long-term reconstruction of the solar irradiance leads to large historical solar forcing. Astronomy & Astrophysics, 529, A67.

[33]Egorova, T., Schmutz, W., Rozanov, E., Shapiro, A.I., Usoskin, I., et al., 2018. Revised historical solar irradiance forcing. Astronomy and Astrophysics, 615, A85.

[34]Penza, V., Berrilli, F., Bertello, L., Cantoresi, M., Criscuoli, S., Giobbi, P., 2022. Total solar irradiance during the last five centuries. The Astrophysical Journal, 937(2), 84.

[35]Lockwood, M., Stamper, R., Wild, M., 1999. A doubling of the Sun's coronal magnetic field during the past 100 years. Nature, 399, 437–439.

(2) panel "f": uses solar cycle length as a pcroxy, assuming that shorter cycles correspond to higher solar activity; here, the TSI peak occurs around 1940.

This discrepancy underscores how the choice of proxy can significantly alter the timing and amplitude of reconstructed solar maxima, thereby affecting climate attribution and historical interpretation.

Reconstructions by Hoyt & Schatten (1993) and Bard et al. (2000) also exhibit a TSI peak between 1940 and 1950, consistent with the solar cycle length-based model. Scafetta & Willson (2013a)[36] further corroborated this finding using aurora records from Hungary, which display a peak in the 1940s, suggesting a multi-decadal solar maximum earlier than that indicated by sunspot-based models.

These reconstructions also reveal a decline in solar activity until the 1970s–1980s, followed by a renewed rise into the 2000s. The presence or absence of a solar peak in the 1940s is pivotal for interpreting 60-year climate cycles, such as those observed in the Atlantic Multidecadal Oscillation (AMO) (Wyatt & Curry, 2014),[37] (see Figures 12.2, 12.3, 18.1, and 18.2).

An approximately 60-year modulation has been detected in various solar proxies (e.g., Neff et al., 2001; Ogurtsov et al., 2002)[38] and in auroral records (Scafetta, 2012c; Scafetta & Willson, 2013a),[39] reinforcing the hypothesis of a link between solar cyclicity and multi-decadal climate oscillations. Veretenenko & Ogurtsov (2018)[40] identify an approximately 60-year cycle in Earth's climate and show that the correlation between solar activity and lower atmospheric circulation varies over time, suggesting that solar modulation influences tropospheric dynamics.

If solar activity indeed peaked in the 1940s and again shortly after 2000, this would point to a direct solar contribution to the observed ~60-year climate oscillation. In the absence of supporting astronomical evidence, however, the ~60-year climate oscillation would be attributed primarily to internal, unforced ocean–atmosphere mechanisms.

Comparison between climate models and proxy reconstructions. Figure 25.6, adapted from Soon et al. (2015),[41] reveals a striking contrast:

[36]Scafetta, N., Willson, R.C., 2013a. Planetary harmonics in the historical Hungarian aurora record (1523–1960). Planetary and Space Science, 78, 38–44.

[37]Wyatt, M.G., Curry, J.A., 2014. Role of Eurasian Arctic shelf sea ice in a secularly varying hemispheric climate signal during the 20th century. Climate Dynamics, 42, 2763–2782.

[38]Neff, U., Burns, S.J., Mangini, A., Mudelsee, M., Fleitmann, D., Matter, A., 2001. Strong coherence between solar variability and the monsoon in Oman between 9 and 6 kyr ago. Nature, 411, 290–293. Ogurtsov, M.G., Nagovitsyn, Y.A., Kocharov, G.E., Jungner, H., 2002. Long-period cycles of the Sun's activity recorded in direct solar data and proxies. Solar Physics, 211, 371.

[39]Scafetta, N., 2012c. A shared frequency set between the historical mid-latitude aurora records and the global surface temperature. Journal of Atmospheric and Solar–Terrestrial Physics, 74, 145–163.

[40]Veretenenko, S.V., Ogurtsov, M.G., 2018. 60-year cycle in the Earth's climate and dynamics of correlation links between solar activity and circulation of the lower atmosphere. Geomagnetism and Aeronomy 58, 973–981.

[41]Soon, W., Connolly, R., Connolly, M., 2015. Re-evaluating the role of solar variability on Northern Hemisphere temperature trends since the 19th century. Earth-Science Reviews, 150, 409–452.

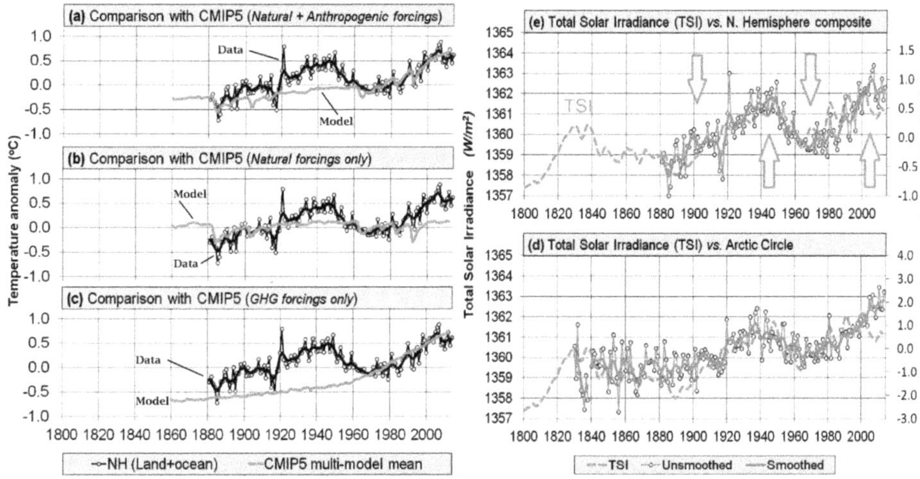

FIGURE 25.6. Comparison between climate observations and solar reconstructions. (Adapted from Soon et al., 2015).
Left: comparison between an estimate of Northern Hemisphere temperatures (black line with dots) and the mean simulations of the CMIP5 GCM climate models (colored curves).
Right: overlay between climate records and the TSI proxy reconstruction proposed by Hoyt & Schatten (1993, 1997), updated with the ACRIM TSI satellite composite by Scafetta & Willson (2014).

- right panel: the CMIP5 GCM simulations, used in IPCC AR5 (2013),[42] fail to reproduce the prominent 1940s warm peak in Northern Hemisphere temperatures;
- left panel: the TSI proxy reconstruction by Hoyt & Schatten (1993, 1997), updated with the ACRIM composite Scafetta & Willson (2014),[43] shows nearly perfect correlation with historical climate oscillations, including the 1940s warm phase.

This finding suggests that the CMIP GCMs, across phases 3, 5, and 6, may have underestimated actual solar variability, thereby compromising their ability to simulate observed climate dynamics. A likely cause is the use of an attenuated solar forcing, which fails to capture decadal-to-centennial solar fluctuations. If a solar maximum occurred in the 1940s rather than in the late 1950s, this temporal mismatch could severely impair model accuracy and reliability.

25.10. Evidence supporting high secular variability in TSI

Despite the contemporary dominance of secular low-variability solar models, multiple lines of indirect evidence challenge their validity:

[42]IPCC, 2013. Climate Change 2013: The Physical Science Basis. Cambridge University Press.
[43]Scafetta, N., Willson, R.C., 2014. ACRIM total solar irradiance satellite composite validation versus TSI proxy models. Astrophysics and Space Science, 350, 421–442.

- Baliunas & Soon (1991)[44] and Judge et al. (2020)[45] observed brightness variations in solar-type stars, suggesting that luminosity changes far greater than those predicted by models such as NRLTSI2 and SATIRE are not only plausible but astrophysically likely.
- Comparative analysis between TSI reconstructions and paleoclimate records (Figures 23.1, 23.2, and 23.3) reveals strong correlations with major climate transitions, such as the shift from the Medieval Warm Period to the Little Ice Age and the subsequent transition toward the contemporary era (Scafetta & West, 2007; Scafetta, 2023c; Soon et al., 2023).[46]

These correlations contradict the hypothesis of a quasi-stationary Sun suggested by the TSI low-variability models (cf. Figure 25.3). The data instead suggest that solar activity has undergone substantial secular variations, sufficient to drive natural climate oscillations on multi-decadal and secular scales.

In this context, Schmutz (2021) offered a pointed critique:

> *"The marked correlation between solar activity reconstructions and climate proxies on millennial scales radically challenges the idea that the Sun has undergone only minimal secular variations."*

This conclusion is supported by Figure 13.2, which shows that climate models employing secular low-variability TSI forcings fail to reproduce the transition from the Medieval Warm Period to the Little Ice Age. This represents a critical limitation that undermines the epistemological validity of the CMIP6 framework.

25.11. Conclusion: the Sun's controversial role in the climate system

The lack of direct TSI measurements prior to 1978, combined with residual uncertainties in post-1980 satellite composites, leaves substantial interpretive margins in assessing the Sun's secular variability and its potential influence on Earth's climate system. As a result, TSI proxy reconstructions diverge widely, reflecting contrasting assumptions about solar magnetic physics and its long-term radiative response. A central difficulty arises from the long-standing discrepancy between the ACRIM and PMOD TSI composites: whereas PMOD applies degradation corrections that yield a nearly flat multidecadal TSI trend, ACRIM preserves a small upward drift between successive solar minima from 1980 to 2000, implying a higher level of secular irradiance variability.

---

[44] Baliunas, S.L., Soon, W.H., 1991. Are variations in the length of the activity cycle related to changes in brightness in solar-type stars? The Astrophysical Journal, 374, 463–470.

[45] Judge, P.G., Egeland, R., Henry, G.W., 2020. Sun-like stars shed light on solar climate forcing. The Astrophysical Journal, 891(1), 96.

[46] Scafetta, N., West, B.J., 2007. Phenomenological reconstructions of the solar signature in the Northern Hemisphere surface temperature records since 1600. Journal of Geophysical Research, 112, D24S03.

Scafetta, N., 2023c. Empirical assessment of the role of the Sun in climate change using balanced multi-proxy solar records. Geoscience Frontiers, 14(6), 101650.

Soon, W., Connolly, R., Connolly, M., et al., 2023. The detection and attribution of Northern Hemisphere land surface warming (1850–2018) in terms of human and natural factors: Challenges of inadequate data. Climate, 11(9), 179.

Low-variability TSI scenarios are supported by results such as those of Yeo et al. (2020), which rely on interpretations of modern high-resolution magnetograms and on radiative MHD modeling of present-day quiet-Sun conditions. These reconstructions are internally consistent and broadly aligned with PMOD-like TSI satellite corrections, which, by construction, tend to minimize long-term instrumental trends. However, present-day quiet-Sun observations may not fully capture the physical state of the quiet Sun during Grand-Minimum epochs such as the Maunder Minimum, when cosmogenic isotope records indicate a substantially weakened heliospheric magnetic field and altered large-scale solar magnetism. Under such conditions, it remains physically plausible that the efficiency of the small-scale dynamo, or its coupling with deeper convective layers, differed from the modern regime.

Similarly, the radiative magnetohydrodynamical simulations of Rempel (2020) provide strong constraints on the radiative response of quiet-Sun magnetism under present-day conditions. However, their local-box geometry and the use of "fixed" lower boundary conditions only about 1000 km beneath the solar surface limit their ability to capture the dynamics of the deep convection zone, where processes capable of modulating small-scale magnetic fields on secular timescales originate. As Rempel (2020) explicitly acknowledges, these simulations do not constitute an *ab initio* model of long-term solar variability. Consequently, their applicability to multi-century or grand-minimum scenarios remains inherently limited.

Independent lines of evidence further suggest caution in extrapolating modern solar constraints to historical epochs: Schmutz (2021) argued that reproducing Maunder-Minimum-type cold climate excursions would require a TSI decrease on the order of 10 W/m^2, a magnitude difficult to reconcile with strictly low secular-variability TSI baselines, while stellar-comparative studies indicate that Sun-like stars with similar fundamental parameters can exhibit a broader range of long-term brightness variability than is observed for the modern Sun alone (Baliunas & Soon, 1991; Judge et al., 2020).

Taken together, these considerations suggest that, while large secular TSI variations may appear unlikely within the framework of current solar modeling, a broader range of long-term irradiance variability —such as that implied by some TSI proxy reconstructions and by ACRIM-consistent trends—cannot yet be definitively excluded on historical timescales.

The fact that CMIP6 GCMs rely on reconstructions exhibiting low secular solar variability—such as those by Matthes et al. (2017) and Wu et al. (2018)—raises the risk of systematically underestimating the solar contribution to global warming. While this approach aligns with dominant narratives, it risks narrowing the interpretive spectrum and obscuring the complexity of solar forcing in the climate system.

If climate science is to remain faithful to Karl Popper's principle of falsifiability, it must embrace a pluralistic methodological stance —one that considers alternative solar models and compares competing hypotheses.

A crucial question therefore remains open: *Did the Sun play, and does it continue to play, a more significant role in climate variability than is currently acknowledged?*

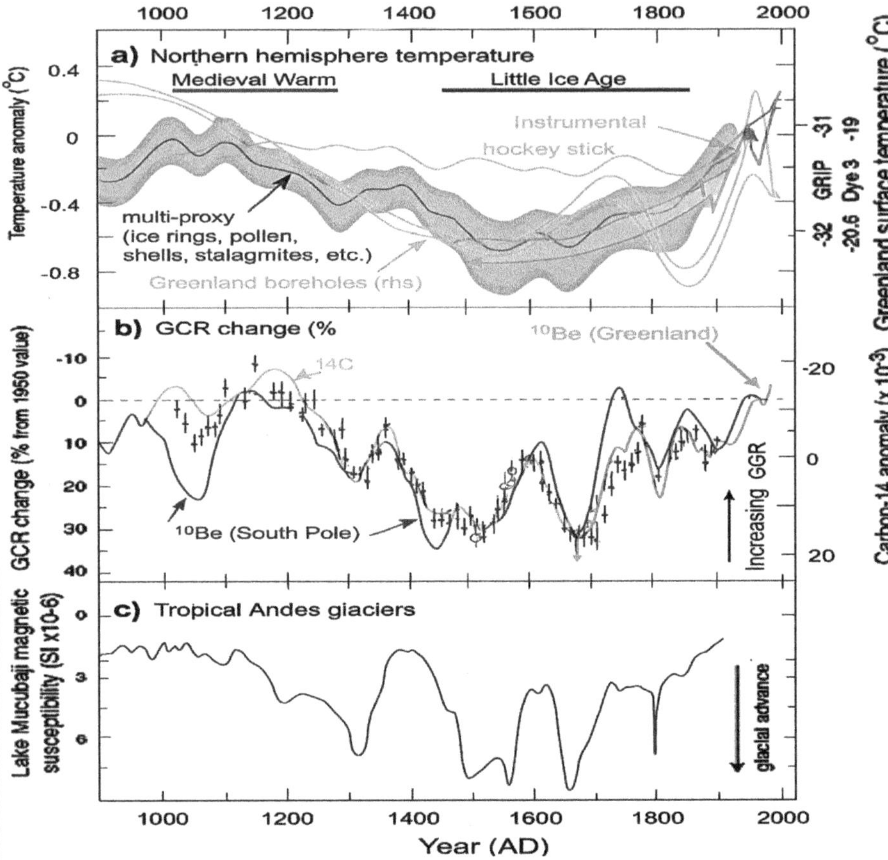

FIGURE 25.7. **Climate and solar variability over the past 2000 years.**
a) Northern Hemisphere temperature anomalies (°C), reconstructed from multiple proxies including tree rings, pollen, stalagmites, and boreholes.
b) Changes in galactic cosmic ray (GCR) intensity, alongside ^{14}C and ^{10}Be isotope records from tree rings and polar ice cores. These proxies reflect solar modulation of cosmic rays, reinforcing the link between solar variability and climate.
c) Advance and retreat of tropical glaciers in the Andes (Lake Mucubaji, Venezuela), illustrating the cryospheric response to temperature fluctuations. Glacial expansion aligns with cooler periods, whereas retreat corresponds to warming phases. (Adapted from Kirkby, 2007).
Together, the panels support the hypothesis that solar variability —tracked through isotopic and cosmic ray proxies—has played a significant role in modulating regional and hemispheric climate patterns prior to the industrial era.

CHAPTER 26

Particle-Driven climate forcing modulated by solar variability

26.1. The Sun does not emit light alone

In the framework adopted by the global climate models (GCMs) of the CMIP6 program, the Sun's influence on Earth's climate is interpreted primarily through its luminous output, particularly through variations in total solar irradiance (TSI). Solar irradiance is treated as both an energetic and a chemical forcing. For example:

- ultraviolet radiation (UV, 200–300 nm) affects stratospheric ozone through the photolysis of O_2 and O_3 (Chapman mechanism), influencing stratospheric temperature and atmospheric circulation;
- visible light (400–700 nm), or photosynthetically active radiation (PAR), drives photosynthesis, regulating the carbon cycle, oxygen production, and biospheric feedbacks;
- near-infrared radiation (750–4000 nm) is partially absorbed by water vapor and CO_2, enhancing the greenhouse effect (see Figure 4.3).

However, this radiative-centric view is increasingly regarded as incomplete. Numerous studies suggest that Earth's climate responds to TSI variations with greater sensitivity than can be explained by the modest energy fluctuations traditionally modeled.

As noted by Soon et al. (2000)[1] and Shaviv (2008),[2] climate data imply a hypersensitivity to solar forcing that cannot be fully accounted for by radiative or chemical mechanisms alone.

A discrepancy in historical climate response. Schmutz (2021),[3] argues that to explain the cooling during the Maunder Minimum (1650–1715) and the subsequent warming, the TSI in the seventeenth century would have needed to be about 10 W/m^2 lower than today. This estimate is:

- more than ten times greater than the TSI variation used in the CMIP6 models;

[1]Soon, W., Posmentier, E., Baliunas, S., 2000. Climate hypersensitivity to solar forcing? Annales Geophysicae, 18(5), 583–588.
[2]Shaviv, N.J., 2008. Using the oceans as a calorimeter to quantify the solar radiative forcing. Journal of Geophysical Research: Space Physics, 113, A11101.
[3]Schmutz, W.K., 2021. Changes in the total solar irradiance and climatic effects. Journal of Space Weather and Space Climate, 11, 40.

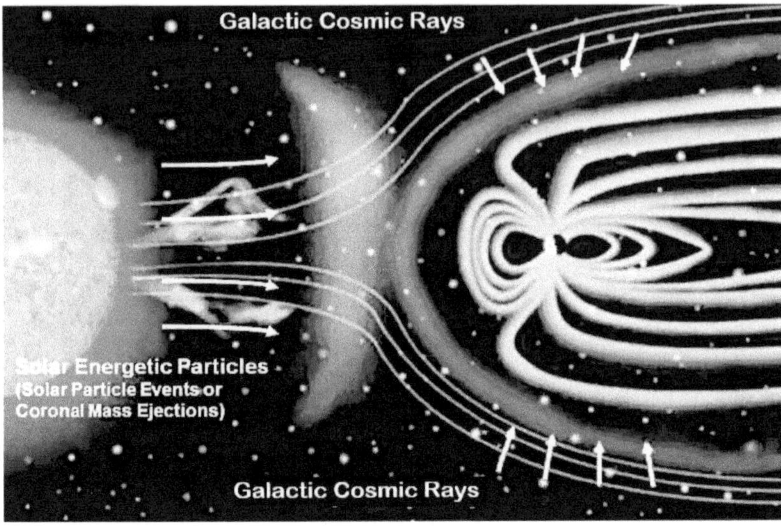

FIGURE 26.1. Schematic representation of the solar wind, Earth's magnetosphere, and cosmic rays.

- still 2–3 times higher than that indicated by the solar high-variability proxy models (e.g., Scafetta, 2023c)[4]; see Figure 25.4.

This mismatch raises a crucial question:

Is it possible that the Sun influences climate through non-radiative mechanisms?

26.2. Beyond light: cosmic rays, cloud formation, and electric circuits

In 1959, Ney[5] hypothesized that the galactic cosmic ray (GCR) flux, modulated by solar activity, could influence atmospheric ionization and cloud formation. This hypothesis launched decades of research into solar–cosmic–climate interactions. For example:

- Svensmark & Friis-Christensen (1997)[6] linked GCR flux to low cloud cover;
- Miyahara et al. (2023)[7] extended this connection to high-altitude clouds in tropical regions.

Another often overlooked factor is the global atmospheric electric circuit, extensively studied by Brian A. Tinsley.[8] This circuit connects Earth's surface, atmosphere, and ionosphere through electric currents modulated by:

[4]Scafetta, N., 2023c. Empirical assessment of the role of the Sun in climate change using balanced multi-proxy solar records. Geoscience Frontiers, 14(6), 101650.

[5]Ney, E.P., 1959. Cosmic Radiation and the Weather. Nature, 183, 451–452.

[6]Svensmark, H., Friis-Christensen, E., 1997. Variation of cosmic ray flux and global cloud coverage—a missing link in solar-climate relationships. Journal of Atmospheric and Solar-Terrestrial Physics, 59(11), 1225–1232.

[7]Miyahara, H., Kusano, K., Kataoka, R., Shima, S.-i., Touber, E., 2023. Response of high-altitude clouds to galactic cosmic ray cycles in tropical regions. Frontiers in Earth Science, 11, 1157753.

[8]Tinsley, B.A., 2008. The global atmospheric electric circuit and its effects on cloud microphysics.

- internal generators (e.g., thunderstorms);
- external drivers (e.g., the solar wind, GCR flux).

Tinsley proposed that variations in solar activity alter atmospheric conductivity, thereby influencing cloud microphysics and the energy balance:

> *"The ionosphere–surface current (Jz) can alter aerosol scavenging processes and droplet nucleation, with measurable effects on cloud cover and atmospheric pressure."* (Tinsley, 2008)

Tinsley & Zhou (2014)[9] showed that small changes in electric current flow can induce significant shifts in tropospheric dynamics, particularly at high latitudes. Although modest on a daily basis, these effects may accumulate over decades. The issue is significant because even sub-1% changes in planetary albedo can substantially affect the global energy balance (cf. van Wijngaarden & Happer, 2025).[10]

Reframing albedo: forcing or feedback? In the CMIP6 GCMs, albedo is treated solely as a passive feedback, responding to radiative forcings that are assumed to be predominantly anthropogenic (Figure 6.2). However, if Earth's albedo is also modulated by astronomical drivers (e.g., the solar wind or cosmic ray flux), it becomes an active forcing, reshaping the climate narrative. This possibility opens the door to a reinterpretation of climate processes, in which part of the observed variability may stem from external mechanisms not included in standard models.

CERES observations: a case for solar–albedo coupling. Figure 26.2 compares CERES-based datasets, including:

- incoming total solar irradiance (TSI) at the top of the atmosphere (TOA);
- Earth's albedo derived from reflected shortwave radiation;
- global surface air temperature (GSAT) from independent observational datasets.

An inverse correlation emerges: when albedo decreases, GSAT increases. According to Loeb et al. (2024),[11] the recent decline in albedo is primarily driven by a reduction in cloud fraction, which increases shortwave absorption at the TOA. Furthermore, Nikolov & Zeller (2024)[12] report that GSAT variations lag changes in absorbed shortwave radiation by 0–9 months, suggesting that surface

---

Reports on Progress in Physics, 71(6), 066801.

[9]Tinsley, B.A., Zhou, L., 2014. Effects of global circuit downward current flow on atmospheric dynamics: modeling of relevant cloud microphysics. Proceedings of the International Conference on Atmospheric Electricity (ICAE 2014), Paper 282. https://www.nssl.noaa.gov/users/mansell/icae2014/preprints/

[10]van Wijngaarden, W.A., Happer, W., 2025. Radiation transport in clouds. Science of Climate Change, 5(1), 1–12.

[11]Loeb, N.G., Ham, S.-H., Allan, R.P., Thorsen, T.J., Meyssignac, B., Kato, S., Johnson, G.C., Lyman, J.M., 2024. Observational Assessment of Changes in Earth's Energy Imbalance Since 2000. Surveys in Geophysics, 45(6), 1757–1783.

[12]Nikolov, N., Zeller, K.F., 2024. Roles of Earth's Albedo Variations and Top-of-the-Atmosphere Energy Imbalance in Recent Warming: New Insights from Satellite and Surface Observations. Geomatics, 4, 311–341.

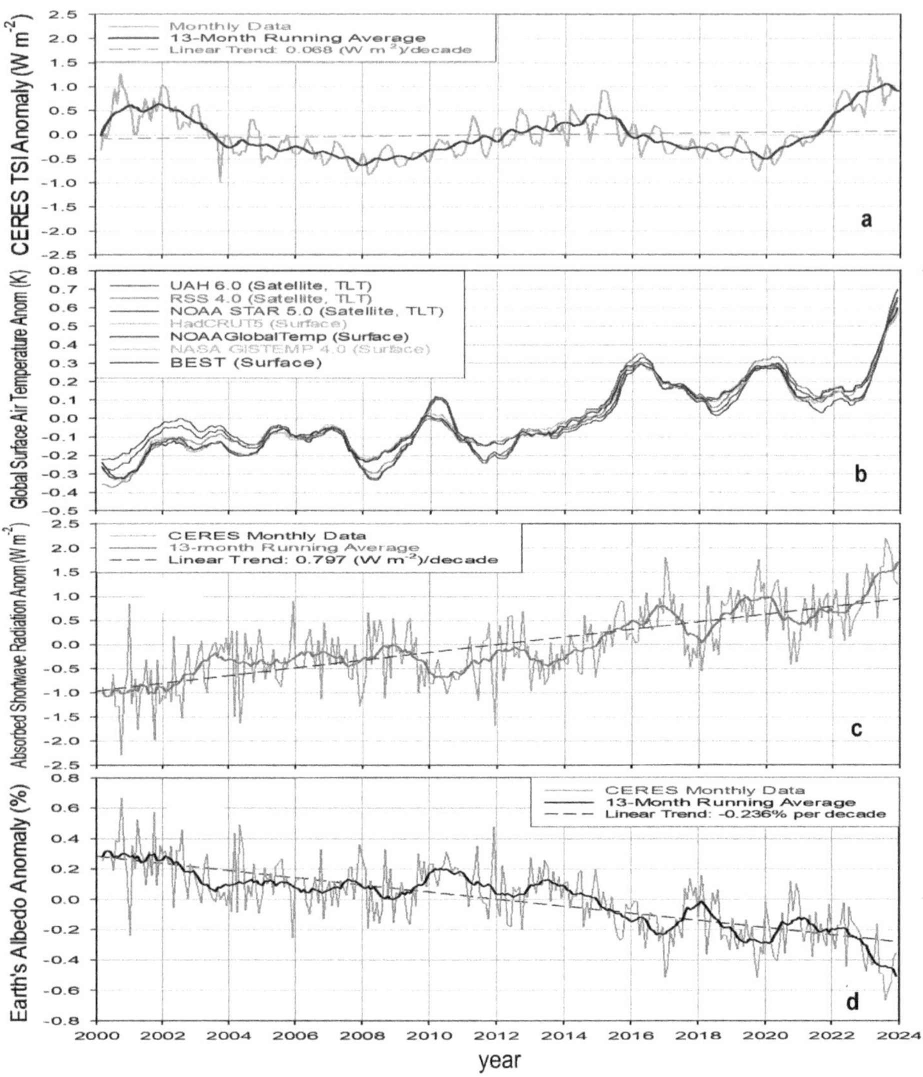

FIGURE 26.2. Comparison between monthly radiative anomalies from the *Clouds and the Earth's Radiant Energy System* (CERES) project and global surface temperatures. (Adapted from Nikolov & Zeller, 2024):
(a) deseasonalized anomalies of total solar irradiance (TSI) at TOA;
(b) 13-month average anomalies of global surface air temperature (GSAT), based on 7 satellite datasets (TLT from MSU) and ground-based thermometers;
(c) absorbed solar flux, obtained by inverting the reflected radiation anomaly;
(d) percentage variations in global terrestrial albedo. The data refer to the average from March 2000 to December 2023.

temperature responds to variations in absorbed shortwave solar energy, not solely to internal climate feedbacks.

In addition, some satellite TSI composites (Figure 24.7) display a slight increase in solar input, partially correlating with the warming observed between 2000 and 2023. In particular, solar maxima in 2000–2004 and 2012–2016 preceded temperature peaks in 2002–2007 and 2014–2020, respectively.

Taken together, these observations point to a significant solar contribution to recent warming and climate modulation—compatible with the solar cycle, yet largely unaccounted for in conventional GCMs.

26.3. The hypothesized physical mechanism

The functioning of particle-based solar forcing can be summarized in two steps.

(1) Solar minima: the weakened solar magnetic field allows more galactic cosmic rays (GCRs) to penetrate Earth's atmosphere, enhancing ionization and promoting the formation of aerosols and cloud condensation nuclei (CCN).
(2) This process increases cloud cover, raises planetary albedo, reflects more solar radiation back into space, and produces climatic cooling.

Conversely, during solar maxima, the strengthened solar magnetic field reduces the influx of GCRs, decreasing CCN formation and cloud cover, thereby resulting in global warming.

This mechanism suggests that solar activity modulates climate not only through radiative output but also through electromagnetic shielding and particle flux, with cascading effects on cloud microphysics and Earth's energy balance.

26.4. Cosmic rays, cosmogenic isotopes, and clouds

The hypothesis that cosmic rays influence climate is supported by the behavior of cosmogenic isotopes such as carbon-14 (^{14}C) and beryllium-10 (^{10}Be), which correlate strongly with climate sequences (cf. Figures 23.1 and 23.2). These isotopes are formed in the atmosphere through GCR interactions, and their abundance reflects cosmic ray intensity.

Solar activity modulates GCR flux as follows:

- high solar activity strengthens magnetic shielding, reducing GCR flux and isotope production;
- low solar activity weakens shielding, increasing GCR flux and isotope production.

Beyond solar cycles, extraordinary astrophysical events—such as supernova explosions—can temporarily produce spikes in GCR flux. Brakenridge (2025)[13] identifies eleven ^{14}C peaks in tree rings over the last 15,000 years that coincide with nearby supernovae. Such events may:

- thin the ozone layer;
- degrade stratospheric methane;
- trigger climate cooling, selective extinctions, and wildfires.

[13]Brakenridge, G.R., 2025. Late Quaternary supernovae in Earth history. Monthly Notices of the Royal Astronomical Society, 539(4), 3201–3219.

Galactic modulation of climate. According to Svensmark & Friis-Christensen (1997) and subsequent studies (Kirkby, 2007; Shaviv & Veizer, 2003),[14] increased atmospheric ionization from GCRs promotes CCN formation and cloud cover.

In particular, Shaviv & Veizer (2003) propose that Phanerozoic climate variability (about 545 million years) was strongly influenced by GCR flux variations linked to the solar system's orbit through the spiral arms of the Milky Way, with a periodicity of about 143 $\pm$ 10 million years. Their analysis suggests that approximately 66% of climate variance may be attributable to GCR modulation.

Other studies identify an approximately 32-million-year climate oscillation, possibly tied to the solar system's oscillating motion perpendicular to the galactic plane (Shaviv et al., 2014; Rampino & Caldeira, 2020).[15]

Quantifying the Impact. Svensmark (2007)[16] estimates that the decrease in GCR flux since the nineteenth century may have reduced low cloud cover, contributing to a radiative forcing increase of approximately 2 W/m^2, comparable to the total anthropogenic forcing over the same period (cf. Figure 6.2).

When the solar system passes through denser regions of interstellar material and supernova remnants (Figure 26.3), Earth is exposed to a higher GCR flux, which enhances atmospheric ionization and cloud formation. These passages appear to leave cyclical imprints in paleoclimate records, suggesting that galactic events modulate climate in ways not yet represented in conventional GCMs.

Supporting evidence.

- Czymzik et al. (2016) document a correlation between cosmogenic proxies and flood frequency in Europe during the Holocene, interpreted as evidence of systematic cloud cover variation driven by GCRs.
- Shaviv (2008)[17] shows that ocean heat content varies with the 11-year solar cycle, but with an amplitude 5–7 times greater than expected from TSI variation alone.
- Soon et al. (2000) and Easterbrook (2019)[18] reach similar conclusions, pointing to an amplifying mechanism absent from conventional models.

A comprehensive treatment of these mechanisms is provided in *Force Majeure: The Sun's Role in Climate Change* by Svensmark (2019), which synthesizes the proposed theory of solar, cosmic, and climatic interactions.

---

[14]Kirkby, J., 2007. Cosmic rays and climate. Surveys in Geophysics, 28(5–6), 333–375.
Shaviv, N.J., Veizer, J., 2003. Celestial driver of Phanerozoic climate? GSA Today, 13(7), 4–10.

[15]Shaviv, N.J., Prokoph, A., Veizer, J., 2014. Is the Solar System's Galactic Motion Imprinted in the Phanerozoic Climate? Scientific Reports, 4, 6150.
Rampino, M.R., Caldeira, K., 2020. A 32-million year cycle detected in sea-level fluctuations over the last 545 Myr. Geoscience Frontiers, 11(6), 2061–2065.

[16]Svensmark, H., 2007. Cosmoclimatology: a new theory emerges. Astronomy & Geophysics, 48(1), 1.18–1.24.

[17]Shaviv, N.J., 2008. Using the oceans as a calorimeter to quantify the solar radiative forcing. Journal of Geophysical Research: Space Physics, 113, A11101.

[18]Easterbrook, D.J., 2019. Solar Magnetic Cause of Climate Changes and Origin of the Ice Ages. Independently Published.

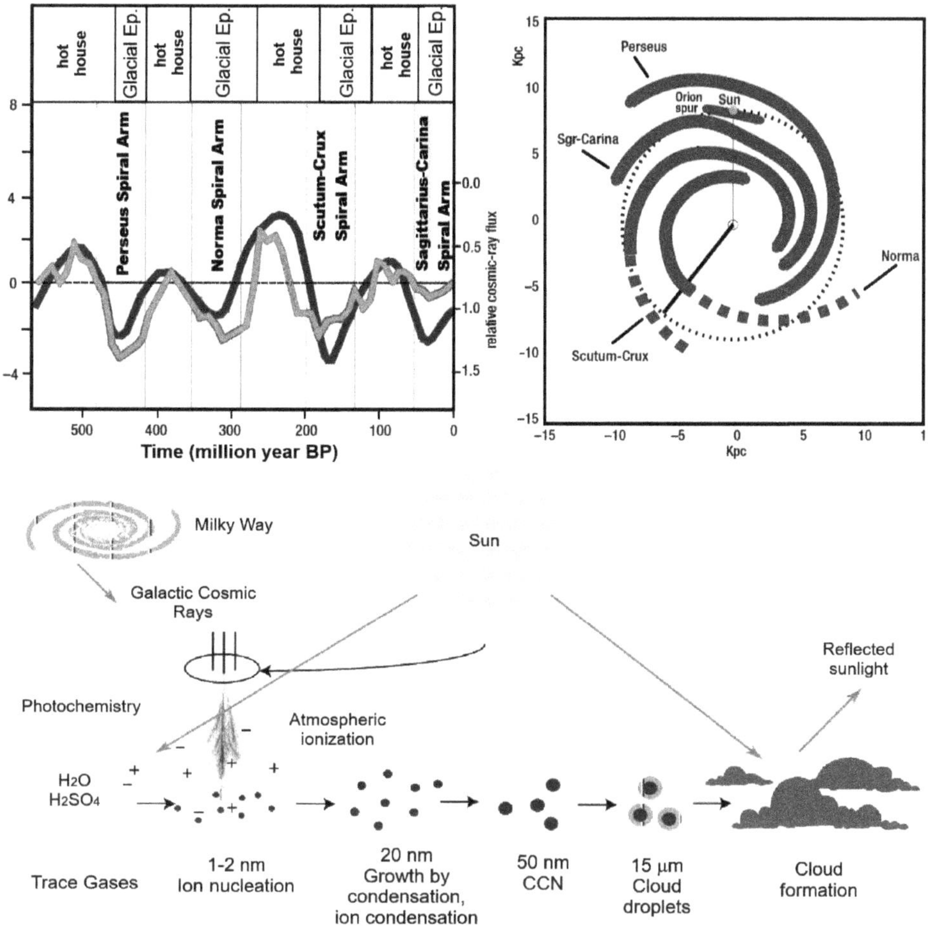

FIGURE 26.3. Left: comparison between galactic cosmic ray (GCR) variations obtained from meteorites and Phanerozoic climate variability (last 545 million years). Right: solar orbit through the spiral arms of the Milky Way. Bottom: hypothesized particle-based mechanisms leading to cloud formation. Adapted from Shaviv & Veizer (2003) and Svensmark (2007, 2019).

26.5. Critiques overcome by strong empirical evidence

The hypothesis that cloud cover is modulated by galactic cosmic ray (GCR) flux remains controversial. The CLOUD experiment at CERN (Kirkby et al., 2011)[19] concluded that, under laboratory-simulated atmospheric conditions, GCR-induced ionization alone was insufficient to generate new cloud condensation nuclei (CCN) in quantities large enough to produce a significant climatic

[19] Kirkby, J., Curtius, J., Almeida, J., Dunne, E., Duplissy, J., et al., 2011. Role of sulphuric acid, ammonia and galactic cosmic rays in atmospheric aerosol nucleation. Nature, 476, 429–433.

impact. This finding initially appeared to undermine the theory of particle-based solar forcing.

However, subsequent empirical studies have revived and strengthened the hypothesis.

- Svensmark et al. (2016)[20] and Matsumoto et al. (2022)[21] demonstrated that Forbush events are followed by a rapid and significant reduction in cloud cover within days. Forbush events are caused by coronal mass ejections (CMEs) and solar flares. These solar storms temporarily deflect GCRs, triggering a rapid reduction (the Forbush decrease) in the intensity of galactic cosmic rays observed at Earth (cf. Figure 26.1). The cloud response is nonlinear and may be amplified by complementary atmospheric factors.
- Miyahara et al. (2023) identified a clear 11-year modulation in tropical high clouds, consistent with the solar cycle.
- Svensmark (2022)[22] extended the theory to a cosmic timescale, showing that supernova rates over the past 3.5 billion years correlate with climatic and oceanic parameters. A higher rate of supernovae implies:
  - a colder climate;
  - stronger winds;
  - greater ocean mixing;
  - higher nutrient availability;
  - increased marine productivity;
  - more buried organic matter.

This chain of effects suggests that supernovae indirectly regulate atmospheric oxygen and may have influenced the evolution of complex life.

The increase in cosmic rays around 1 and 2 billion years ago may be linked to the recurring passages of the Sagittarius Dwarf Galaxy through the Milky Way (Ruiz-Lara et al., 2020),[23] which may have triggered bursts of star formation and supernova activity.

26.6. Alternative particle mechanisms: interplanetary dust

Meteoric dust and climate modulation. Beyond cosmic rays, interplanetary dust falling on Earth—predominantly composed of ionized meteoric particles—may play an active role in modulating cloud cover and Earth's radiative balance.

Studies by Scafetta (2012c)[24] and Scafetta, Milani, and Bianchini (2020),[25] reveal an approximately 60-year cycle in meteorite fall frequency in China (Yu et

[20]Svensmark, J., Enghoff, M.B., Shaviv, N.J., Svensmark, H., 2016. The response of clouds and aerosols to cosmic ray decreases. Journal of Geophysical Research: Space Physics, 121(9), 8152–8181.

[21]Matsumoto, H., Svensmark, H., Enghoff, M.B., 2022. Effects of Forbush decreases on clouds determined from PATMOS-x. Journal of Atmospheric and Solar–Terrestrial Physics, 230, 105845.

[22]Svensmark, H., 2022. Supernova rates and burial of organic matter. Geophysical Research Letters, 49(1), e2021GL096376.

[23]Ruiz-Lara, T., Gallart, C., Bernard, E.J., Cassisi, S., 2020. The recurrent impact of the Sagittarius dwarf on the Milky Way star formation history. Nature Astronomy, 4(10), 965–973.

[24]Scafetta, N., 2012c. A shared frequency set between the historical mid-latitude aurora records and the global surface temperature. Journal of Atmospheric and Solar–Terrestrial Physics, 74, 145–163.

[25]Scafetta, N., Milani, F., Bianchini, A., 2020. A 60-year cycle in the meteorite fall frequency suggests a possible interplanetary dust forcing of the Earth's climate driven by planetary oscillations.

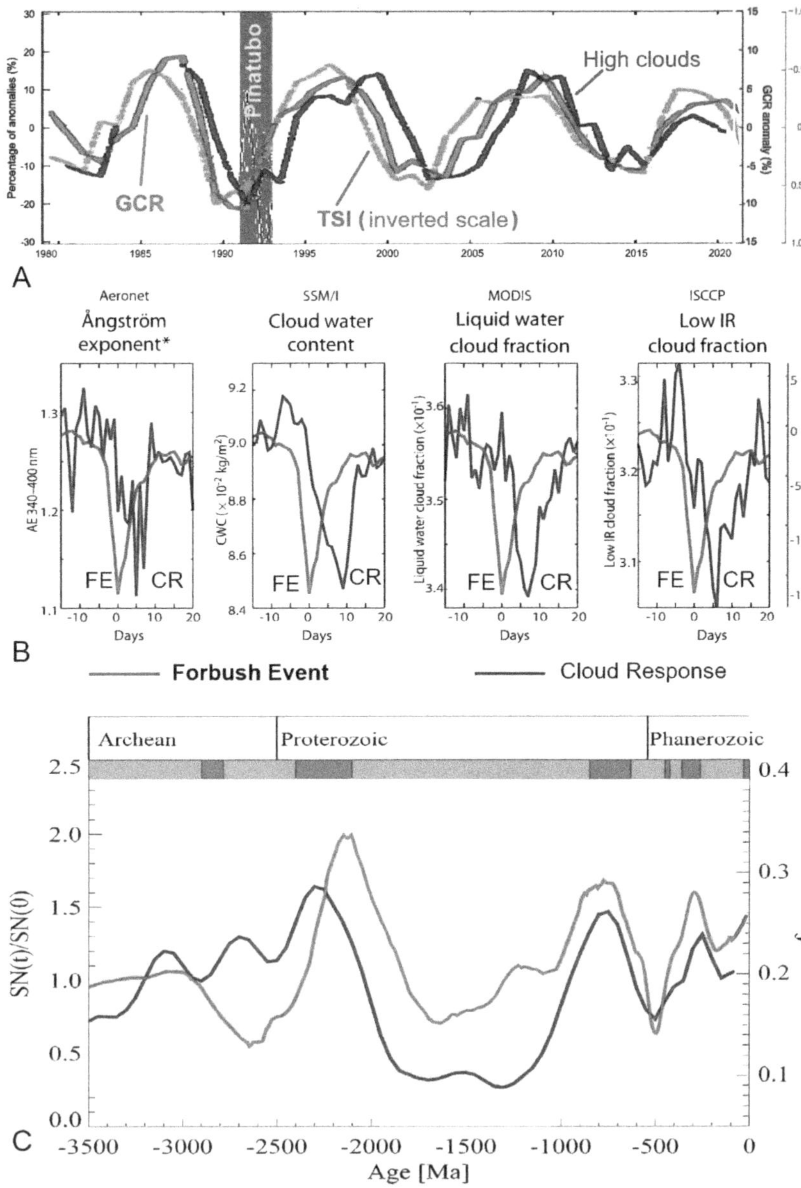

FIGURE 26.4. [A] Comparison between a high-altitude cloud index and the GCR flux anomaly monitored by neutron monitors and TSI. [B] Comparison between the rapid GCR decrease (Forbush event) and the observed cloud cover reduction a few days later. Adapted from Miyahara et al. (2023) and Svensmark (2019). [C] Comparison between a reconstruction of relative cosmic ray intensity (wide band) and the fraction of carbon buried as organic matter in sediments (thin curve) over the last 3.5 billion years. Adapted from Miyahara et al. (2023), Svensmark (2019), and Svensmark (2022).

al., 1983),[26] consistent with multi-decadal climate oscillations (Figure 26.5). The authors hypothesize that this cyclicity is driven by planetary orbital oscillations, particularly the eccentricity of Jupiter modulated by the gravitational pull of Saturn.

In fact, the near 5:2 gravitational resonance between Jupiter and Saturn may exert a dynamical influence on both the spatial configuration and the ionization processes of interplanetary dust within the inner solar system. Dust interacting with solar radiation and magnetic fields may contribute to aerosol formation or modify cloud nucleation, particularly in tropical and subtropical regions.

Figure 26.5 shows that periods of increased meteorite falls—implying higher dust accumulation—coincide with surface temperature decreases, and vice versa. The data suggest that when Jupiter's eccentricity decreases, its gravitational deflection of dust weakens, allowing more material to reach Earth. A reduction in eccentricity would thus diminish Jupiter's ability to deflect interplanetary dust —including meteorites and comets—away from Earth through the "gravitational slingshot" effect.

Meteoric smoke particles (MSPs) and polar clouds. MSPs are now recognized as key agents in polar mesospheric cloud (PMC) formation. Laboratory studies on iron-silicate nanoparticles— analogs of MSPs—confirm their role as heterogeneous ice nuclei at approximately 83 km altitude (Duft et al., 2019).[27]

A new microphysics framework, the charged-MSP nucleation (CMN) scheme, proposes that electrically charged MSPs act as efficient ice nuclei throughout the PMC column, offering a more coherent model than classical sedimentation processes (Zhang et al., 2024b).[28]

In support of this view, researchers at Dome C station in Antarctica analyzed ice cores and identified more than 2,000 extraterrestrial particles, estimating a global deposition of approximately 5,200 tons of cosmic dust per year, primarily from Jupiter-family comets Rojas et al. (2021).[29] This includes 20–100 tons of extraterrestrial carbon, which may enhance heterogeneous nucleation and influence atmospheric chemistry. Although no quantitative climate impact has yet been established, the vaporization and chemical interaction of these particles in the mesosphere may affect cloud microphysics and formation—altering altitude, density, and duration—and open new perspectives on the interplay between interplanetary matter and Earth's climate system.

Geophysical Research Letters, 47(18), e2020GL089954.

[26]Yu, Z., Chang, S., Kumazawa, M., Furumoto, M., Yamamoto, A., 1983. Presence of periodicity in meteorite falls. National Institute of Polar Research Memoirs, Special Issue, 30, 362–366.

[27]Duft, D., Nachbar, M., Eritt, M., Vrtala, A., Tomsche, L., et al., 2019. Unravelling the microphysics of polar mesospheric cloud formation. Atmospheric Chemistry and Physics, 19(4), 2871–2879.

[28]Zhang, J., Xue, W., Megner, L., Plane, J. M. C., Rapp, M., 2024. Dominant role of charged meteoric smoke particles in mesospheric ice nucleation. EGUsphere: DOI: 10.5194/egusphere-2024-1259.

[29]Rojas, J., Duprat, J., Engrand, C., Dartois, E., Delauche, L., et al., 2021. The micrometeorite flux at Dome C (Antarctica), monitoring the accretion of extraterrestrial dust on Earth. Earth and Planetary Science Letters, 560, 116794.

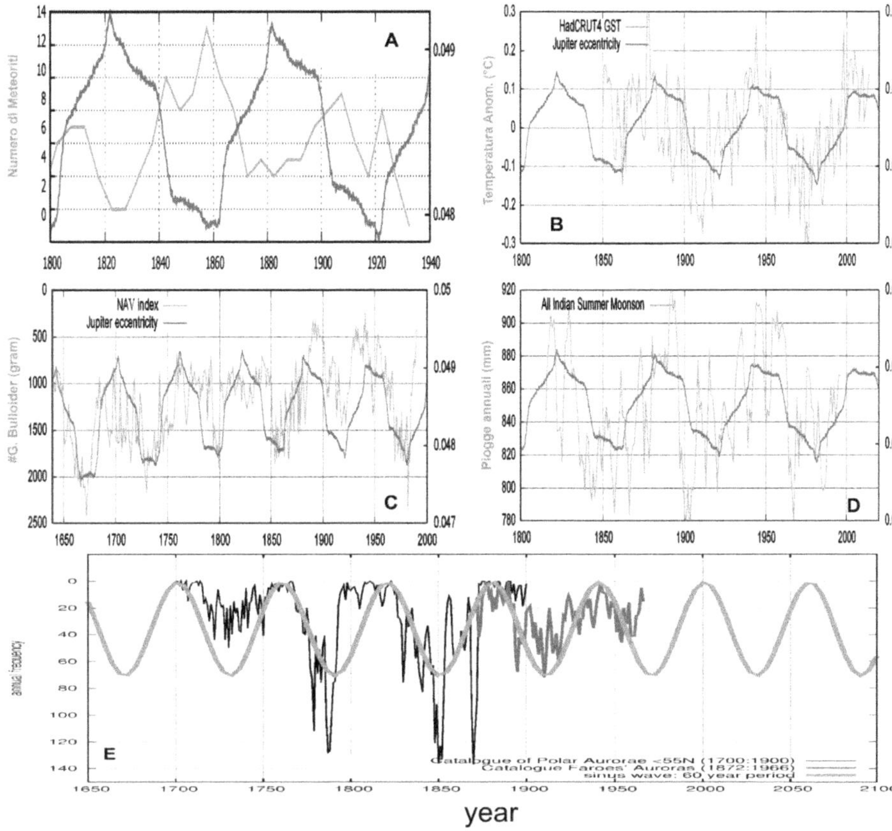

FIGURE 26.5. Comparison between the 60-year modulation of Jupiter's orbital eccentricity (darker curve) and various curves representing: (A) the meteorite fall record in China (Yu et al., 1983); (B) the detrended HadCRUT4 global surface temperature record (Morice et al., 2012); (C) a marine record used as a proxy for North Atlantic temperatures and the *Atlantic Multidecadal Oscillation* (AMO) (Black et al., 1999); (D) the historical record of monsoon rainfall in India (Agnihotri & Dutta, 2003); (F) aurora records from Křivský & Pejml (1988) and from the Faroe Islands from 1872 to 1966. (Adapted from Scafetta et al., 2020; Scafetta, 2012c).

Some researchers (e.g., Firestone et al., 2007)[30] have proposed that the Younger Dryas cooling resulted not solely from changes in the Atlantic Meridional Overturning Circulation (AMOC) caused by meltwater pulses from retreating ice sheets, but may have been triggered by an extraordinary extraterrestrial event. According to the Younger Dryas Impact Hypothesis, a large meteorite or comet fragmented in the atmosphere, producing a massive airburst over

[30]Firestone, R. B., West, A., Kennett, J. P., Becker, L., Bunch, T. E., et al., 2007. Evidence for an extraterrestrial impact 12,900 years ago that contributed to the megafaunal extinctions and the Younger Dryas cooling. Proceedings of the National Academy of Sciences, 104(41), 16016–16021.

North America. The resulting shock waves, high-energy fragments, and widespread wildfires may have destabilized ecosystems, released enormous quantities of soot and aerosols into the atmosphere, and disrupted the climate system. Proponents argue that this chain of events could explain the abrupt return to near-glacial conditions around 12,900 years ago, coinciding with megafaunal extinctions and cultural shifts in Paleoindian populations such as the Clovis culture.

Correlation between temperature and aurora frequency. Auroral activity, driven by the interaction between the solar wind and Earth's magnetosphere, precipitates energetic particles into the upper atmosphere. These particles influence the chemistry and electrical properties of the polar mesosphere and stratosphere, potentially promoting the nucleation of polar mesospheric clouds (PMCs) and noctilucent clouds. Ionization from auroral electrons may enhance ice nucleus formation or generate charged particles that act as precursors to cloud condensation (Mironova et al., 2015).[31]

In two separate studies, Scafetta (2012c) and Scafetta & Willson (2013a)[32] analyzed centuries-old auroral archives, including the detailed Hungarian record (1523–1960), and identified shared frequencies between auroral variability and global surface temperature oscillations. A 60-year cycle was detected (cf. Figure 26.5F), consistent with multi-decadal climate variations and meteorite fall records, and apparently linked to planetary orbital resonances, particularly between Jupiter and Saturn. The presence of an approximately 60-year cycle in aurora and other solar records has been recognized since the nineteenth century (e.g., Olmsted, 1856; Charvátová-Jakubcová et al., 1988; Komitov, 2009).[33]

Auroras thus serve as indirect indicators of solar and interplanetary magnetic activity. Their frequency may reflect deep cycles linked to solar-system dynamics. The fact that these cycles are echoed in climate records strengthens the hypothesis that planetary variability and space weather may act synergistically in modulating Earth's climate response.

This scenario implies that Earth's climate system is sensitive not only to direct TSI forcing but also to indirect astrophysical processes mediated by matter fluxes that propagate throughout the solar system and are modulated by planetary dynamics. It suggests the existence of climate cycles driven by external space weather mechanisms, opening new perspectives for understanding multi-decadal variability and long-term forecasting.

---

[31]Mironova, I.A., Aplin, K.L., Arnold, F., Bazilevskaya, G.A., Beer, J., et al., 2015. Energetic particle influence on the Earth's atmosphere. Space Science Reviews, 194(1–4), 1–96.

[32]Scafetta, N., Willson, R.C., 2013a. Planetary harmonics in the historical Hungarian aurora record (1523–1960). Planetary and Space Science, 78, 38–44.

[33]Olmsted, D., 1856. On the recent secular period of the aurora borealis. Smithsonian Contributions to Knowledge, 8, 1–52. Washington, D.C.: Smithsonian Institution.

Charvátová-Jakubcová, I., Střeštík, J., & Křivský, L., 1988. The periodicity of aurorae in the years 1001–1900. Studia Geophysica et Geodaetica, 32(1), 70–77.

Komitov, B., 2009. The "Sun-climate" relationship. II. The "cosmogenic" beryllium and the middle latitude aurora. Bulgarian Astronomical Journal, 12, 85–93.

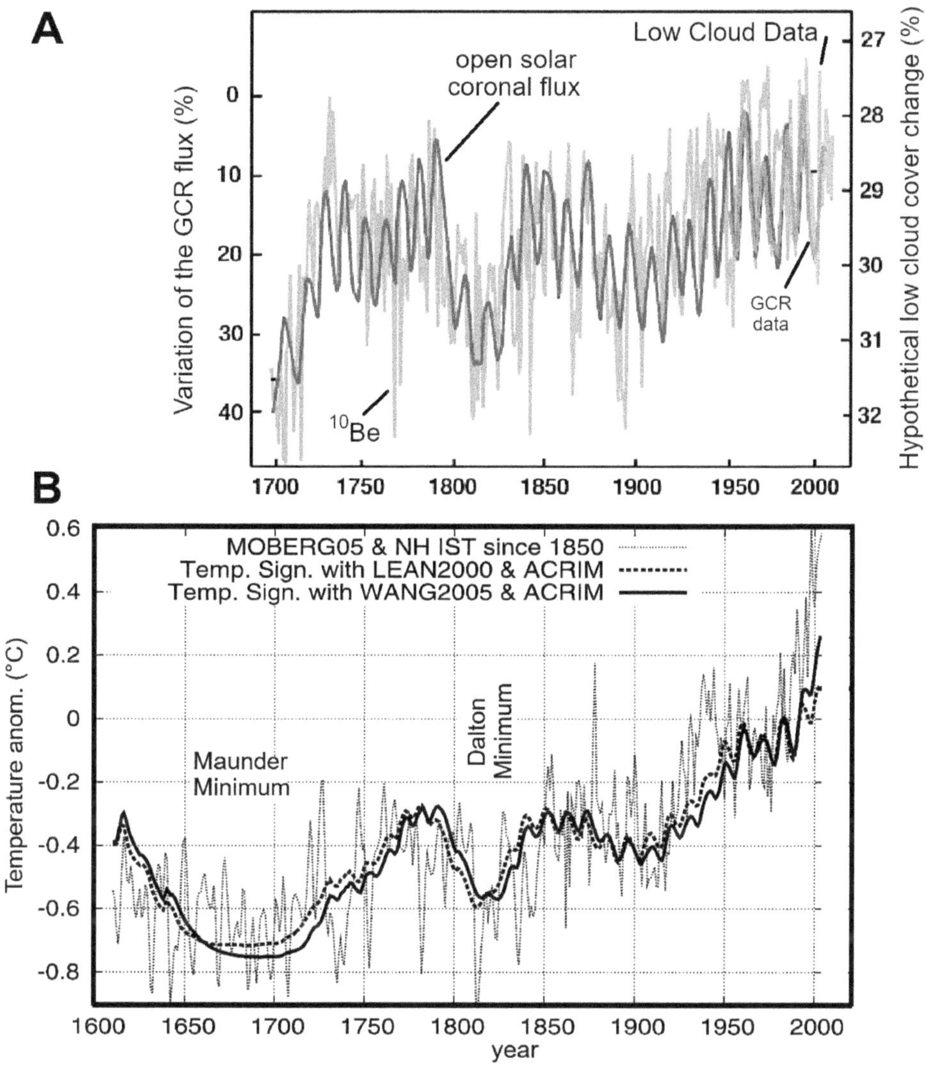

FIGURE 26.6. [A] Hypothetical extension of global low-cloud-cover variation based on ^{10}Be isotope, open solar coronal flux, and Climax cosmic rays for the period 1953–2005 (threshold 3 GeV). (Adapted from Svensmark, 2007).
[B] Comparison between a climate reconstruction from 1600 to 2000 based on the proxy by Moberg et al. (2005) and instrumental surface temperature data for the Northern Hemisphere from 1850 (Brohan et al., 2006), together with an empirical estimate of the solar contribution. Adapted from Scafetta & West (2007).

26.7. Limitations of the GCMs: the cloud cover challenge

One of the most critical limitations of global climate models (GCMs), including those developed under CMIP5 and CMIP6, is their inadequate representation of cloud cover dynamics, particularly in relation to solar activity and extraterrestrial particle forcings. Cloud formation depends on complex microphysical processes and highly variable atmospheric conditions, which are often coarsely parameterized in models. As Scafetta (2013a)[34] noted, this limitation can lead to an underestimation of natural variability and an overestimation of the anthropogenic contribution.

If astronomical particle forcings, such as galactic cosmic rays (GCRs) and interplanetary dust, significantly influence climate, their systematic exclusion from the forcing functions implemented in GCMs constitutes a structural shortcoming in current climate science.

Figure 26.6 compares:

- a reconstruction of cloud cover variation, using GCR data as a proxy;
- estimates of Northern Hemisphere surface temperatures;
- an empirical reconstruction of the solar contribution to climate change Scafetta & West (2007),[35] based on solar proxies (Lean, 2000; Wang et al., 2005),[36] and the ACRIM TSI composite (Willson & Mordvinov, 2003).[37]

A strong correlation is observed across the depicted records, capturing common warming trends, multi-decadal modulations, and the cooling periods during the Maunder (1650–1715) and Dalton (1790–1830) minima. This convergence suggests that solar variability has been the dominant climate driver over the past four centuries, with anthropogenic influence emerging only more recently. Yet this solar contribution is absent from the CMIP6 GCMs used by the IPCC (2021),[38] which attribute a net solar impact of 0 °C from 1850–1900 to the present (cf. Figure 6.3).

26.8. Conclusion

Empirical evidence indicates that the Sun affects Earth's climate not only through changes in luminosity but also through particle forcings such as cosmic rays and interplanetary dust. These mechanisms, although less explored, may account for long-term variability and for correlations with solar records.

If cosmic ray or particle flux modulation proves more influential than currently assumed, the prevailing attribution frameworks may need revision, particularly with regard to CO_2's role over geological timescales. The chapters that follow will examine how solar and planetary rhythms align with climate oscillations.

---

[34] Scafetta, N., 2013a. Discussion on climate oscillations: CMIP5 general circulation models versus a semi-empirical harmonic model based on astronomical cycles. Earth-Science Reviews, 126, 321–357.

[35] Scafetta, N., West, B.J., 2007. Phenomenological reconstructions of the solar signature in the Northern Hemisphere surface temperature records since 1600. Journal of Geophysical Research, 112, D24S03.

[36] Lean, J., 2000. Evolution of the Sun's spectral irradiance since the Maunder Minimum. Geophysical Research Letters, 27, 2425–2428.

Wang, Y.-M., Lean, J.L., Sheeley, N.R. Jr., 2005. Modeling the Sun's magnetic field and irradiance since 1713. The Astrophysical Journal. 625, 522–538.

[37] Willson, R.C., Mordvinov, A.V., 2003. Secular total solar irradiance trend during solar cycles 21–23. Geophysical Research Letters, 30(5), 1199.

[38] IPCC, 2021. Climate Change 2021: The Physical Science Basis. Cambridge University Press.

Part 5

Modeling Climate Change with Empirical Models

CHAPTER 27

From reductionist models to empirical–holistic frameworks

27.1. Starting from data and acknowledging uncertainty

Over the past several decades, climate modeling has been dominated by deterministic numerical approaches rooted in a strongly reductionist paradigm. As discussed in previous chapters, however, numerous theoretical and empirical challenges complicate the identification, attribution, and projection of climate change. Among these are:

- uncertainties in the observational record, particularly in global surface temperature datasets, which are often affected by systematic biases and controversial homogenization practices;
- intrinsic limitations of numerical models, which struggle to accurately simulate key processes such as cloud dynamics, oceanic circulation, and indirect solar influences.

A particularly critical issue concerns the estimation of the actual solar forcing. Although solar variability is widely recognized as a major natural driver of climate change on secular and millennial timescales, both its magnitude and the underlying forcing mechanisms remain highly uncertain. This uncertainty is compounded by our incomplete understanding of how solar activity—both radiative (TSI) and particle-based (cosmic rays, solar wind, and interplanetary dust)—interacts with the atmosphere and influences climate evolution.

In response, some researchers have proposed complementing or partially replacing GCMs with empirical or semi-empirical models, which may be better suited to capturing the actual dynamics of the observed climate system.

The IPCC assessment reports of 2007, 2013, and 2021 have relied heavily on the GCMs developed within the CMIP (*Coupled Model Intercomparison Project*) framework. These models embody the reductionist paradigm, which seeks to understand complex systems by decomposing them into their fundamental components, modeling each in isolation, and then coupling them together.

Although this approach has proven effective in many areas of physics and engineering, it encounters significant limitations when applied to highly nonlinear, interconnected systems such as Earth's climate, in which feedbacks, emergent properties, and unidentified mechanisms play a decisive role.

27.2. Disadvantages of reductionist models

In climate science, reductionism presents several methodological drawbacks:

- *loss of systemic context*—isolating subsystems (atmosphere, ocean, biosphere) risks overlooking dynamic interactions and underestimating feedback loops and emergent behaviors;
- *simplified representations*—despite their numerical sophistication, GCMs still struggle with cloud formation, vertical energy fluxes, and tropical convection, thereby compromising regional projections;
- *inadequate modeling of interconnected systems*—the nonlinear feedbacks among the surface, atmosphere, oceans, and outer space are often poorly captured, leading to simulations that diverge from observations;
- *limited practical applicability*—even the most refined models may fail to yield actionable insights for regional climate adaptation or policy planning.

For these reasons, a growing number of researchers advocate methodological integration, combining reductionist rigor with holistic frameworks that better reflect the complexity of the climate system, even if doing so entails sacrificing some degree of physical detail.

27.3. Advantages of empirical models

Empirical models rely on direct observation and statistical analysis of climate data. Rather than simulating all physical processes from first principles, they seek to identify historical regularities and attribute variations to specific drivers.

Their key advantages include:

- *adherence to real data*—being grounded in measurements rather than theoretical assumptions, they offer analyses anchored in empirical evidence;
- *simplicity and transparency*—unlike GCMs, which are often perceived as "*black boxes*," empirical models are more readable and verifiable;
- *robust statistical structure*—techniques such as multivariate regression, Fourier and wavelet analysis, and ARIMA models enable the direct examination of relationships between climate variables and forcings;
- *effective attribution*—comparing historical time series of temperature, solar activity, oceanic indices, and other variables facilitates the identification of dominant climate drivers;
- *policy relevance*—their simplicity makes them useful tools for flexible, context-specific decision-making, especially at the regional scale.

27.4. Limitations and precautions

Although empirical models offer valuable insights, they are not without critical limitations:

- *dependence on data*—their reliability hinges on the quality, continuity, and coverage of observational datasets;
- *structural simplifications*—they may overlook deep or poorly understood physical mechanisms, thereby risking misinterpretation;
- *difficulty with nonlinearity*—the climate system is chaotic, and linear approaches may miss bifurcations, tipping points, and emergent dynamics;

- *limited geographic validity*—a model calibrated for a specific region or timescale may not generalize to other contexts;
- *uncertain attribution*—although adept at detecting correlations, empirical models often struggle to quantify the precise contribution of individual forcings.

These limitations underscore the need for cautious interpretation and for methodological cross-validation.

27.5. Empirical and semi-empirical approaches in climate modeling

Empirical Climate Models (ECMs). ECMs are streamlined tools for analyzing historical climate trends. They rely on statistical techniques such as:

- multiple linear regression;
- principal component analysis (PCA);
- spectral decomposition (Fourier and wavelet);
- multivariate correlation analysis.

These methods enable the exploration of relationships between temperature evolution and key forcing agents, including:

- *ENSO (El Niño–Southern Oscillation)*—a source of interannual variability;
- *volcanic aerosols*—transient cooling agents;
- *solar activity*—via TSI, galactic cosmic rays, and other channels;
- *greenhouse gases and anthropogenic aerosols*—indicators of human influence.

ECMs provide initial estimates of the relative contributions of these factors. Their robustness, however, depends on variable selection, data integrity, and statistical rigor. Crucially, they describe correlations rather than causation, and must therefore be interpreted with care.

Energy Balance Models (EBMs). EBMs are grounded in the principle that Earth absorbs solar energy and re-emits it as infrared radiation. The net balance determines energy accumulation or loss.

Their configurations include:

- *zero-dimensional (0-D)*—treating Earth as a single point and estimating the global mean temperature;
- *one-dimensional (1-D)*—accounting for latitudinal variations in insolation and albedo;
- *multidimensional or coupled*—incorporating atmospheric and oceanic components while remaining simpler than GCMs.

EBMs are useful for testing climate sensitivity to radiative forcings and for comparing theoretical and empirical estimates. Their simplicity, however, limits their ability to simulate feedbacks (e.g., clouds and hydrology) and transient phenomena.

Statistical Downscaling Models (SDMs). SDMs bridge the gap between coarse-resolution GCM outputs and localized climate projections. They establish statistical relationships between large-scale variables (e.g., geopotential height and 500 hPa temperature) and local observations (e.g., precipitation and temperature extremes).

Once calibrated, these relationships can be applied to GCM outputs to generate regionally refined projections. SDMs are particularly valuable in:

- water resource management;
- agriculture;
- urban planning.

They share, however, the limitations of ECMs: dependence on historical data, sensitivity to methodological choices, and difficulty in extrapolating beyond the observed range of variability.

Machine Learning Models (MLMs). MLMs represent an emerging frontier in climate modeling. They employ algorithms such as:

- neural networks;
- support vector machines;
- random forests;
- generative models and ensemble methods.

These tools can detect nonlinear patterns and latent correlations in large datasets, even in the absence of explicit physical models. Their applications include:

- extreme event prediction;
- satellite image classification;
- greenhouse gas estimation;
- noise reduction in observational data.

Their principal limitation is low interpretability, especially in deep learning architectures. MLMs often function as *"black boxes,"* and their reliability depends heavily on the quality and diversity of the training data.

27.6. Toward an integrated vision

Climate modeling stands to benefit from a synergistic integration of numerical and empirical approaches. Physical models are indispensable for hypothesis testing and counterfactual simulations, whereas empirical models—anchored in observation—are essential for validation and attribution. A robust strategy should transcend the dichotomy between reductionism and holism, and between theory and observation, by integrating physics and statistics within a complementary framework.

The following chapters will critically review empirical models published in recent literature. Various methodologies for estimating the relative contributions of anthropogenic and natural forcings to observed warming will be examined, with particular attention to the role of the Sun, whose climatic influence remains a subject of active debate. Although estimates vary depending on the methods and assumptions adopted, empirical models have played a pivotal role in reopening the discussion on the attribution of contemporary climate change.

CHAPTER 28

Climate attribution using multilinear regression

28.1. Introduction

The attribution of observed climate changes to specific drivers—such as greenhouse gas emissions, solar variability, or oceanic circulation—typically begins with the comparison of time series and their statistical cross-correlations. This preliminary step can reveal potential synchronisms between climatic variations and candidate forcings. Its interpretive power, however, is limited: correlation does not imply causation, and causal relationships may not manifest through simple linear associations.

Indeed, climate-relevant correlations are frequently nonmonotonic and nonlinear. They may emerge only after cross-spectral analysis, particularly when the climate system's response is modulated by cycles, time lags, selective amplification, or nonlinear dynamics. Coherence between time series may be evident only at specific temporal scales—interannual, multidecadal, centennial, or millennial—or during particular dynamical phases of the system. In such cases, advanced analytical tools such as wavelet analysis, harmonic decomposition, and spectral-coherence methods are essential to uncover relationships that elude conventional linear correlation techniques.

Thus, although simple linear correlation can serve as a useful starting point, it is insufficient for robust climate attribution. A more comprehensive analysis is required—one that accounts for signal structure, the dynamic nature of the climate system, and the potential interdependence between causes and effects. Only through sophisticated statistical methodologies, combined with physically coherent interpretation, can the role of each factor in driving observed climate changes be accurately assessed.

28.2. Multilinear regression

Linear regression models, particularly in their multivariate form, are widely employed in climate science to attribute observed changes to various forcings. Their appeal lies in their ability to synthesize the relative influence of multiple predictors—such as radiative forcings—on a dependent variable, typically the global temperature anomaly. This approach assumes an additive and linear climate response, with regression coefficients quantifying the magnitude of each forcing's effect. It enables, for instance, the estimation of the anthropogenic contribution relative to natural ones (e.g., solar or volcanic) over long-term periods, even in the absence of a complete physical understanding of the underlying mechanisms.

This statistical elegance, however, conceals significant epistemological challenges. The climate system is inherently nonlinear, characterized by feedback loops, thresholds, inertia, and complex interactions among its components (atmosphere, ocean, cryosphere, and biosphere). A simple weighted sum of forcings cannot adequately capture these dynamics, especially when the predictors exhibit mutual correlations. For example, both greenhouse gas concentrations and solar variability can influence atmospheric chemistry, creating interdependencies that complicate attribution assessments.

Moreover, regression outcomes are highly sensitive to the choice and nature of the input data, which are often subject to revision and methodological uncertainty. The inclusion of internal-variability indices, such as the El Niño–Southern Oscillation (ENSO) or the Atlantic Multidecadal Oscillation (AMO), as independent predictors introduces conceptual ambiguities. These indices represent internal dynamics rather than external forcings, and treating them as exogenous variables can distort their interpretation. Without rigorous spectral analysis, control of multicollinearity, and a physically grounded framework, regression models risk producing results that are statistically plausible yet physically misleading.

Consequently, multilinear regression should be applied with caution and always complemented by advanced techniques capable of capturing the cyclical and dynamic nature of the climate system.

28.3. A typical application of multilinear regression

A representative example of multilinear regression assessments in climate science is the study by Chylek et al. (2014),[1] which analyzed global warming from 1900 to 2010 with particular emphasis on the post-1975 period. The authors aimed to quantify the relative contributions of known radiative forcings using the following model:

$$T(t) = a_0 + a_1 \cdot F_{GHG}(t) + a_2 \cdot F_{AER}(t) + a_3 \cdot F_{SOL}(t) + a_4 \cdot F_{VOLC}(t) \\ + a_5 \cdot I_{ENSO}(t) + a_6 \cdot I_{AMO}(t).$$

Here, a_i (for $i = 0, ..., 6$) are seven regression coefficients representing the statistical weights of each predictor, and $\varepsilon(t)$ is the stochastic error term.

The study employed radiative forcing data from the GISS ModelE Hansen et al. (2007),[2] and solar irradiance reconstructions from Lean (2000).[3] The predictor functions included:

- well-mixed greenhouse gases (GHG),
- anthropogenic sulfate aerosols (AER),
- solar irradiance variations (SOL),
- volcanic eruptions (VOLC),
- the ENSO index (El Niño–Southern Oscillation),
- the AMO index (Atlantic Multidecadal Oscillation).

---

[1]Chylek, P., Klett, J.D., Lesins, G., Dubey, M.K., Hengartner, N., 2014. The Atlantic Multidecadal Oscillation as a dominant factor of oceanic influence on climate. Geophysical Research Letters, 41, 1689–1697.

[2]Hansen, J., Sato, M., Ruedy, R., Kharecha, P., Lacis, A., et al., 2007. Climate simulations for 1880–2003 with GISS modelE. Climate Dynamics, 29, 661–696.

[3]Lean, J., 2000. Evolution of the Sun's spectral irradiance since the Maunder Minimum. Geophysical Research Letters, 27, 2425–2428.

According to the model, approximately two-thirds of the post-1975 warming is attributable to anthropogenic forcings, with the remaining third linked to the positive AMO phase. Solar variability was deemed negligible in recent decades.

Despite its methodological rigor, the study raises several concerns that warrant a more careful interpretation.

(1) *Sensitivity to the solar dataset*. The model relies on Lean's TSI proxy, supported by the PMOD TSI satellite composite, which has been critiqued. Alternative reconstructions, such as the ACRIM TSI composite (Scafetta & West, 2005, 2008; Scafetta, 2009),[4] suggest increased solar activity since the 1980s, potentially altering the attribution outcomes. As noted by Connolly et al. (2023),[5] the choice of solar dataset significantly influences the regression results, underscoring the impact of the initial assumptions.

(2) *Assumption of linearity and immediate response*. The model presumes a linear and instantaneous climate response to radiative forcings, overlooking the system's thermal inertia, feedback mechanisms, and delayed reactions. For instance, solar variations may exert their effects over extended periods through ocean–atmosphere coupling, rendering static models inadequate.

(3) *Conceptual ambiguity in the use of certain predictors*. The inclusion of ENSO and AMO as independent variables introduces interpretive challenges, as these indices reflect internal variability rather than external drivers. Their potential dependence on other forcings and mutual correlations complicate causal attribution. Without a careful analysis of such interdependencies, the model risks misattributing climate effects to internal dynamics.

These considerations highlight the importance of critically evaluating the assumptions of statistical models, ensuring physical coherence, and assessing sensitivity to methodological choices. Only through such scrutiny can robust empirical relationships be distinguished from statistical artifacts.

28.4. Multicollinearity and attribution ambiguity

A fundamental structural limitation of multilinear regression models lies in the issue of multicollinearity among predictors. When two or more explanatory variables exhibit strong correlation—as in the concurrent rise of greenhouse gases and certain natural forcings—the attribution of individual contributions becomes statistically unstable and conceptually ambiguous. Under such conditions, the estimated coefficients can vary significantly depending on the model configuration, leading to divergent or even contradictory interpretations. This

[4]Scafetta, N., West, B.J., 2005. Estimated solar contribution to the global surface warming using the ACRIM TSI satellite composite. Geophysical Research Letters, 32, L18713. Scafetta, N., West, B.J., 2008. Is climate sensitive to solar variability? Physics Today, 61(3), 50–51.
Scafetta, N., 2009. Empirical analysis of the solar contribution to global mean air surface temperature change. Journal of Atmospheric and Solar–Terrestrial Physics, 71, 17–18.

[5]Connolly, R., Soon, W., Connolly, M., et al., 2023. Challenges in the Detection and Attribution of Northern Hemisphere Surface Temperature Trends Since 1850. Research in Astronomy and Astrophysics, 23(10), 105015.

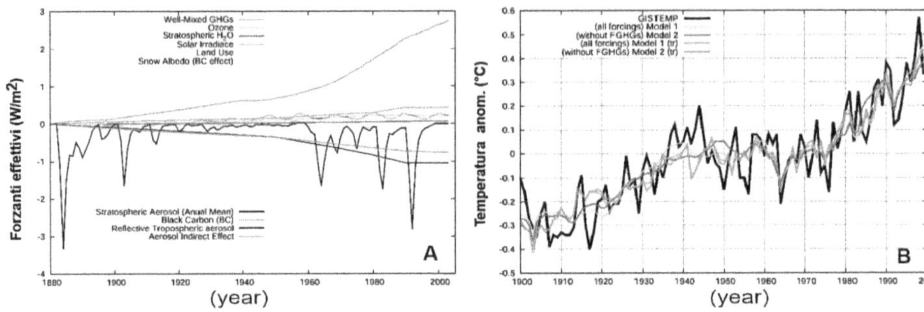

FIGURE 28.1. [A] The effective forcings commonly used in the GCMs, also adopted by Benestad & Schmidt (2009), are employed as predictive variables within a multiple linear regression algorithm. [B] Comparison between four linear regression models applied to the global surface temperature (GISTEMP), constructed by including or excluding the anthropogenic greenhouse gas forcing (GHG), selected among the ten variables shown in [A]. The resulting regression functions are very similar due to the high multicollinearity among predictors. (Adapted from Scafetta, 2013b).

A notable illustration of these methodological challenges is provided by Benestad & Schmidt (2009),[6] who employed a regression model incorporating ten radiative forcing functions derived from the GISS ModelE general circulation model (Hansen et al., 2007)[7] as climate forcings (Figure 28.1A). Their analysis estimated that solar forcing contributed approximately 7% to the global warming between 1900 and 2000. The authors acknowledged, however, that this figure was statistically fragile owing to multicollinearity among the radiative forcings. Subsequently, Scafetta (2013b)[8] demonstrated that the same methodology could yield accurate temperature reconstructions even when well-mixed greenhouse gases (CO_2 and CH_4) were excluded from the model (see Figure 28.1B). This result does not negate the anthropogenic contribution but rather underscores the limitations of multilinear regression when applied without rigorous structural safeguards. High correlation among predictors can cause the algorithm to conflate distinct influences, producing misleading estimates.

Another critical issue concerns the quasi-monotonic nature of the analyzed time series. Between 1850 and 2000, both anthropogenic and solar forcings exhibit upward trends, resulting in strong multicollinearity. In such contexts, the model's reliability depends not on the number of predictors but on their statistical independence. When predictor functions share similar geometries, regression models may misattribute effects, even when the underlying physical processes are distinct. Compounding this is the assumption of linearity between forcings and climate response, which is particularly problematic for solar forcing, whose influence is mediated by nonlinear and temporally distributed mechanisms.

Multicollinearity can be substantially reduced only when the analysis is extended to longer time periods, such as those encompassing paleoclimate

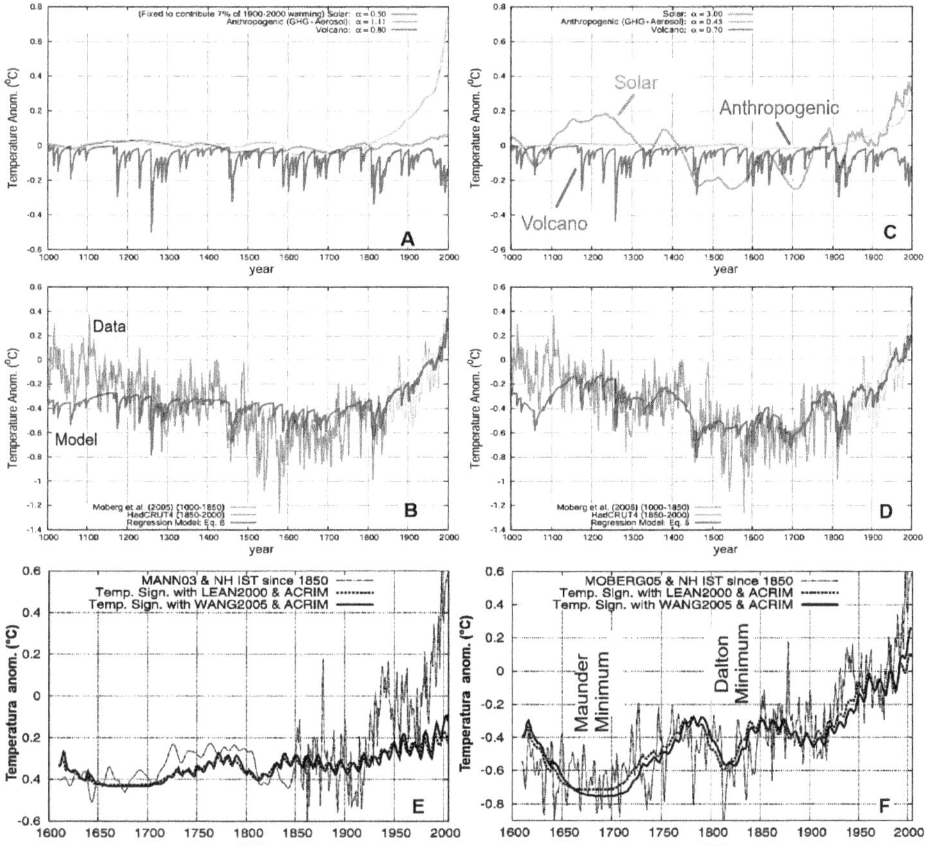

FIGURE 28.2. [A] Temperature components from 1000 to 2000 AD obtained using the regression model assuming the solar contribution estimated by Benestad & Schmidt (2009).
[B] Comparison between the model using the calibrations in [A], the climate record from the proxy by Moberg et al. (2005), and HadCRUT4 temperature (Morice et al., 2012). The model fails to reconstruct climate data from 1000 to around 1750.
[C] Temperature components obtained using the regression model with all coefficients optimized to best reconstruct the climate record.
[D] Comparison between the model using the calibrations in [C] and the climate record, which is well reconstructed thanks to the significant solar contribution estimated by the regression algorithm.
[E] Reconstruction of solar contribution from 1600 to 2000 assuming the paleoclimate reconstruction by Mann et al. (1999).
[F] Reconstruction of solar contribution from 1600 to 2000 assuming the paleoclimate reconstruction by Moberg et al. (2005), extended with HadCRUT4.
(Adapted from Scafetta & West, 2006a, 2007; Scafetta, 2013a).

data from the medieval period. The inclusion of the Medieval Warm Period introduces cyclical signal structures that help discriminate between monotonic forcings (e.g., anthropogenic emissions) and cyclical ones (e.g., solar variability). This distinction is essential for enhancing attribution robustness and mitigating the risk of misinterpretation due to temporal correlation among predictors.

28.5. Toward more sophisticated models

Figure 28.2 exemplifies a more refined methodological approach. Here, linear regression is used not to process the raw forcing functions against the observed temperatures, but to optimize the coefficients that best reproduce a selected paleoclimate reconstruction. The predictors are not direct representations of the external forcings; rather, they are climate responses simulated by separate models—such as energy balance models—each driven by a single forcing component.

This strategy implicitly incorporates aspects of the system's nonlinear dynamics, which may reshape the temporal geometry of the original forcing functions. By using simulated temperature responses as explanatory variables, the regression aims to produce a more physically coherent and statistically robust estimate of each factor's contribution. The optimization of the predictor weights enables a closer fit to the observed paleoclimatic signal, thereby reducing the likelihood of artifacts arising from collinearity.

The model employed is

$$T(t) = \alpha_{T_v} \cdot T_V(t) + \alpha_{T_s} \cdot T_S(t) + \alpha_{T_a} \cdot T_A(t) + c$$

where the α_x are regression coefficients, and the $T_x(t)$ are climate signals associated with the volcanic (V), solar (S), and anthropogenic (A) forcings, derived from the energy balance simulations used by Crowley (2000).[9]

Figure 28.2A shows the contributions of each forcing to the climatic record assuming the solar activity reconstruction by Benestad & Schmidt (2009). Figure 28.2B presents the regression outcome under the constraint that solar forcing accounts for only 7% of the warming from 1900 to 2000. The resulting regression coefficients are:

- $\alpha_{T_a} = 1.1$ (anthropogenic),
- $\alpha_{T_v} = 0.8$ (volcanic),
- $\alpha_{T_s} = 0.5$ (solar).

These values imply that the anthropogenic, volcanic, and solar contributions are approximately 10% higher, 20% lower, and 50% lower, respectively, than Crowley's original estimates. The resulting climate signal aligns with the *"hockey stick"* reconstruction by Mann et al. (1999).[10] The same model, however, fails to

[6]Benestad, R.E., Schmidt, G.A., 2009. Solar trends and global warming. Journal of Geophysical Research: Atmospheres, 114, D14101.

[7]Hansen, J., Sato, M., Ruedy, R., Kharecha, P., Lacis, A., et al., 2007. Climate simulations for 1880–2003 with GISS modelE. Climate Dynamics, 29, 661–696.

[8]Scafetta, N., 2013b. Discussion on common errors in analyzing sea level accelerations, solar trends and global warming. Pattern Recognition in Physics, 1, 37–57.

[9]Crowley, T.J., 2000. Causes of climate change over the past 1000 years. Science, 289, 270–277.

[10]Mann, M.E., Bradley, R.S., Hughes, M.K., 1999. Northern hemisphere temperatures during the past millennium: inferences, uncertainties, and limitations. Geophysical Research Letters, 26, 759–762.

reproduce the Medieval Warm Period (MWP), as captured in the reconstruction by Moberg et al. (2005),[11] which is more consistent with subsequent studies (cf. Esper et al., 2024a).[12]

When, however, the solar regression coefficient α_{T_s} is allowed to vary freely, the regression output successfully reproduces the MWP, the Little Ice Age (1350–1850), and modern warming, reflecting the cyclical common modulation between the solar and the climate records. In this case:

- $\alpha_{T_a} = 0.45$, a significant decrease from the previous value of 1.11;
- $\alpha_{T_s} = 3.0$, a significant increase from the previous value of 0.5.

This represents a sixfold amplification of the solar contribution compared with the results obtained by Benestad & Schmidt (2009), and a threefold increase relative to those by Crowley (2000). If the paleoclimate reconstruction by Moberg et al. (2005) is sufficiently accurate, this implies that solar forcing may account for at least 50% of the warming observed since 1900. This contribution could be even greater if part of the signal attributed to anthropogenic greenhouse gases is in fact indirectly driven by solar-induced feedbacks, such as oceanic CO_2 outgassing (Henry's law).

These findings suggest that the climate system may be more sensitive to solar variability than previously estimated, particularly in studies that suppress preindustrial oscillations or rely on simplified energy balance models. The example underscores the importance of incorporating paleoclimate data extending at least to the medieval period in order to distinguish between monotonic forcings (industrial-era emissions) and cyclical or nonlinear forcings (solar activity), thereby enhancing the credibility and precision of climate attribution analyses.

28.6. An additional proposal: multiscale calibration

An alternative analytical methodology employs multiscale calibration, for example through wavelet decomposition algorithms. Scafetta & West (2006a, 2007)[13] introduced this approach to isolate climate cycles across multiple temporal scales and directly compare them with cyclical forcings, particularly solar variability.

This methodology emphasizes the role of low-frequency solar cycles (e.g., the Gleissberg and Suess cycles) over high-frequency components, which are attenuated by the thermal inertia of the ocean–atmosphere system. The analysis highlights the substantial climatic contribution of the low-frequency component of the solar record.

Figures 28.2E–F summarize the results of this comparative analysis. Panel F demonstrates that the estimated solar contribution is minimal when using the *"hockey stick"* reconstruction by Mann et al. (1999) for the period 1600–1850. By

---

[11]Moberg, A., et al., 2005. Highly variable Northern Hemisphere temperatures reconstructed from proxy data. Nature, 433, 613–617.

[12]Esper, J., et al., 2024. The IPCC's reductive Common Era temperature history. Communications Earth & Environment, 5, 222.

[13]Scafetta, N., West, B.J., 2006a. Phenomenological solar contribution to the 1900–2000 global surface warming. Geophysical Research Letters, 33, L05708.

Scafetta, N., West, B.J., 2007. Phenomenological reconstructions of the solar signature in the Northern Hemisphere surface temperature records since 1600. Journal of Geophysical Research, 112, D24S03.

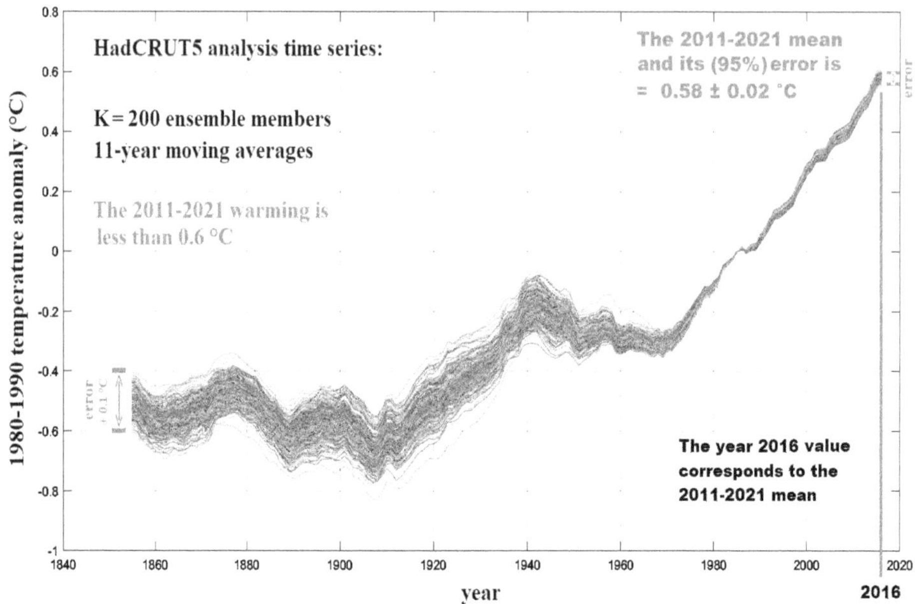

FIGURE 28.3. Ten-year moving averages of the 200 ensemble members of the HadCRUT5 record relative to 1980–1990.

contrast, the reconstruction by Moberg et al. (2005) yields a significantly higher solar contribution, exceeding 50% of the warming from 1900 to the present.

In the latter case, the temporal pattern of the solar imprint from 1600 to 2000 closely aligns with the reconstructed temperature record. Cold episodes such as those during the Maunder (1650–1715) and Dalton (1790–1830) grand solar minima are well reproduced, suggesting a structural correspondence between solar forcing and climate response. This complex correlation supports the empirical attribution and implies that solar variability may have played a more substantial rôle than previously estimated in models based on low-variability reconstructions, such as that of Mann et al. (1999).

The adoption of zero-dimensional (0-D) energy balance models with delayed response times offers an even more realistic representation of climate dynamics. As shown by Scafetta (2009),[14] these models can also approximately simulate the time lag between external forcings and the system's response, producing results consistent with Figure 28.2F, where the analysis suggests that the low-frequency component of the solar forcing could have contributed approximately 50% or more to the post-1900 warming. Such models underscore the need to treat the high- and low-frequency components of the signals separately because of the thermal inertia of the climate system.

28.7. Mastering statistical tools of analysis

The reliability of climate attribution depends not only on the quality of the analyzed data but also on the correct application of analytical methodologies. For

[14]Scafetta, N., 2009. Empirical analysis of the solar contribution to global mean air surface temperature change. Journal of Atmospheric and Solar–Terrestrial Physics, 71, 17–18.

example, Benestad & Schmidt (2009) estimated that the Sun contributed 7% to global warming, which is less than what Scafetta & West (2006a, 2007) estimated. Yet their estimate was still higher than the negligible solar contribution assessed by the IPCC (2021)[15] (see Figure 6.3). It should be noted that Scafetta (2013b) later identified methodological flaws in Benestad & Schmidt (2009)'s use of linear regression and wavelet transforms, which led to an underestimation of the solar attribution to global warming.

In a subsequent debate, Schmidt et al. (2023)[16] again challenged the statistical framework used in Scafetta (2022).[17] Scafetta (2023d)[18] responded by identifying statistical errors in their critique. For example, Schmidt et al. erroneously estimated the uncertainty of the 2011–2021 mean global surface temperature relative to the 1980–1990 baseline as ±0.10 °C, whereas the correct value is ±0.02 °C, as determined through Gaussian error propagation and confirmed by visualizing the full ensemble of HadCRUT5 records (Figure 28.3).

These exchanges underscore the importance of technical rigor and methodological transparency. Without adequate expertise, even sophisticated models may yield misleading results.

28.8. Conclusions

Multilinear regression remains a valuable tool for preliminary climate attribution, providing initial estimates of the relative contributions of various forcings. Its conceptual limitations, however—such as the assumptions of linearity, predictor independence, and instantaneous response—restrict its applicability to a system as complex and nonlinear as Earth's climate.

Structural multicollinearity among the explanatory variables (e.g., between greenhouse gases and solar proxies) compromises the reliability of the coefficient estimates and can lead to interpretive errors. This necessitates a critical and physically grounded approach to regression modeling.

To overcome these limitations, more advanced methodologies are employed, such as spectral decomposition, semi-empirical energy models, and dynamic systems with delayed responses. These techniques better reflect the cyclical, multiscale, and nonlinear nature of climate processes.

The empirical evidence suggests the following.

- Short analysis periods (e.g., 1850–2020) are vulnerable to multicollinearity, which reduces attribution certainty.
- Longer proxy-based reconstructions (extending to at least 1000 AD) reveal significant differences depending on the records used:
  - reconstructions compatible with the *"hockey stick"* by Mann et al. (1999) show minimal preindustrial variability and suggest a

[15] IPCC, 2021. Climate Change 2021: The Physical Science Basis. Cambridge University Press.

[16] Schmidt, G.A., Jones, G.S., Kennedy, J.J., 2023. Comment on "Advanced Testing of Low, Medium, and High ECS CMIP6 GCM Simulations Versus ERA5-T2m" by N. Scafetta (2022). Geophysical Research Letters, 50, e2022GL102530.

[17] Scafetta, N., 2022. Advanced testing of low, medium, and high ECS CMIP6 GCM simulations versus ERA5-T2m. Geophysical Research Letters, 49(6), e2022GL097716.

[18] Scafetta, N., 2023d. Reply to "Comment on 'Advanced Testing of Low, Medium, and High ECS CMIP6 GCM Simulations Versus ERA5-T2m"' by Schmidt et al. (2023). Geophysical Research Letters, 50, e2023GL104960.

negligible solar contribution to post-1900 global warming. Even so, Crowley (2000), using the solar proxy by Bard et al. (2000),[19] estimated a 25% solar contribution to twentieth-century warming;

- reconstructions with greater secular variability, such as Moberg et al. (2005) and other authors (cf. Esper et al., 2024a),[20] indicate a more substantial solar role, with at least a 50% contribution to the warming from 1900 to 2000.

These findings challenge the simulations produced by the CMIP6 GCMs cited by the IPCC (2021), suggesting that observational evidence derived from regression analyses may contradict model-based projections. The difficulty of reproducing the Medieval Warm Period (MWP) in climate models has led to controversial remarks, such as:

> *"We have to get rid of the Medieval Warm Period"*.[21]

Such statements reflect the tension between empirical reconstructions and model narratives. Historical evidence, however, must always be acknowledged.

---

[19]Bard, E., Raisbeck, G.M., Yiou, F., Jouzel, J., 2000. Solar irradiance during the last 1200 years based on cosmogenic nuclides. Tellus Series B: Chemical and Physical Meteorology, 52, 985–992.

[20]Esper, J., Smerdon, J.E., Anchukaitis, K.J., Cook, E.R., D'Arrigo, R., et al., 2024. The IPCC's reductive Common Era temperature history. Communications Earth & Environment, 5, 222.

[21]Deming, D., 2006. U.S. Senate Hearing on Climate Change and the Media (Dec 6, 2006).

CHAPTER 29

Model selection and empirical optimization

29.1. *"Hot"* models and implications for climate projections

Chapter 11 examined the issue of the so-called *"hot"* models, that is, those CMIP6 GCMs that tend to overestimate global surface warming. The IPCC (2021)[1] explicitly acknowledges that several of these models exhibit excessively high equilibrium climate sensitivity (ECS), exceeding 4.5 °C, and therefore excludes them from its core projections. Consequently, the IPCC's climate scenarios and policy recommendations are based only on those GCMs whose ECS values are considered likely.

Independent assessments by Scafetta (2022, 2023b)[2] and Lewis (2023),[3] however, have shown that only models with ECS below 3 °C produce simulations consistent with the global surface temperature record. Furthermore, if the actual warming from 1980 to 2022 is best represented by satellite-based lower troposphere records—such as the UAH-MSU series (Spencer et al., 2017)[4] or the NOAA-STAR series (Zou et al., 2023)[5]—then all CMIP6 GCM simulations are found to overestimate the observed warming, as illustrated in Figures 11.2 and 11.4.

The selection of models directly influences the climate projections for the twenty-first century, depending on the Shared Socioeconomic Pathway (SSP) scenario adopted. Accordingly, climate risk assessments, such as those shown in Figure 6.4, are likewise affected, with significant implications for mitigation strategies and policy design.

Assuming that the actual climate sensitivity is below 3 °C, two corrective strategies can be applied:

(1) adopting only the simulations from the GCMs that are consistent with the observational data;
(2) applying specific scaling algorithms to adjust the outputs of the hot models so as to align them with the observed temperature trend.

---

[1]IPCC, 2021. Climate Change 2021: The Physical Science Basis. Cambridge University Press.

[2]Scafetta, N., 2022. Advanced testing of low, medium, and high ECS CMIP6 GCM simulations versus ERA5-T2m. Geophysical Research Letters, 49(6), e2022GL097716.

Scafetta, N., 2023b. CMIP6 GCM ensemble members versus global surface temperatures. Climate Dynamics, 60, 3091–3120.

[3]Lewis, N., 2023. Objectively combining climate sensitivity evidence. Climate Dynamics, 60, 3139–3165.

[4]Spencer, R.W., Christy, J.R., Braswell, W.D., 2017. UAH version 6 global satellite temperature products: Methodology and results. Asia-Pacific Journal of Atmospheric Sciences, 53(1), 121–130.

[5]Zou, C.-Z., Xu, H., Hao, X., Liu, Q., 2023. Mid-tropospheric layer temperature record derived from satellite microwave sounder observations with backward merging approach. Journal of Geophysical Research: Atmospheres, 128, e2022JD037472.

If the actual global warming trend indicated by satellite lower troposphere records is correct, all CMIP6 GCMs would be inadequate for accurately forecasting twenty-first-century climate evolution. Under such conditions, the actual ECS may be less than half of the value estimated by the low-ECS CMIP6 GCMs, potentially falling within the range of 1–2 °C (Scafetta, 2013a, 2023a; McKitrick & Christy, 2020; Stefani, 2021).[6]

29.2. GCM rescaling: first algorithm

The first rescaling approach assumes that surface-based temperature reconstructions are reliable. Since only the low-ECS models approximately reproduce the observed warming, the proposed rescaling aims to adjust the hot models to better match the surface warming produced by the low-ECS models. The ensemble simulations produced by GCMs with medium ECS ($3.0\,°C < ECS \leq 4.5\,°C$) and high ECS ($4.5\,°C < ECS \leq 6.0\,°C$) are thus scaled down to align them with those produced by the low-ECS models ($1.5\,°C < ECS \leq 3.0\,°C$). The classification of GCMs by ECS is shown in 11.1.

According to Scafetta (2024),[7] the mean warming simulated for the period from 1980–1990 to 2011–2022 is as follows:

- the medium-ECS models simulate an average warming of 0.80 °C, approximately 30% higher than that of the low-ECS models (0.62 °C);
- the high-ECS models simulate 0.97 °C, about 60% higher than that of the low-ECS group.

The ratios $0.80/0.62 = 1.3$ and $0.97/0.62 = 1.6$, respectively, remain remarkably stable throughout the twenty-first century, regardless of the SSP scenario adopted. For the period 2080–2100, the ratios are:

- under SSP1-2.6, approximately 1.4 and 1.7;
- under SSP2-4.5, approximately 1.4 and 1.6;
- under SSP3-7.0, approximately 1.3 and 1.5;
- under SSP5-8.5, approximately 1.4 and 1.6.

This consistency further supports the application of corrective linear scaling factors to the global surface temperature simulations produced by CMIP6 GCMs with medium and high ECS values, in order to generate temperature trends more closely aligned with the observed global surface temperature records. Consequently, the scaled climate projections for the twenty-first century may be considered more credible.

---

[6] Scafetta, N., 2013a. Discussion on climate oscillations: CMIP5 general circulation models versus a semi-empirical harmonic model based on astronomical cycles. Earth-Science Reviews, 126, 321–357.
Scafetta, N., 2023a. CMIP6 GCM validation based on ECS and TCR ranking for 21st-century temperature projections and risk assessment. Atmosphere, 14, 345.
McKitrick, R., Christy, J., 2020. Pervasive warming bias in CMIP6 tropospheric layers. Earth and Space Science, 7, e2020EA001281.
Stefani, F., 2021. Solar and anthropogenic influences on climate: Regression analysis and tentative predictions. Climate, 9, 163.

[7] Scafetta, N., 2024. Impacts and risks of "realistic" global warming projections for the 21st century. Geoscience Frontiers, 15(2), 101774.

29.3. GCM rescaling: second algorithm

The second approach assumes the validity of global temperature reconstructions based either on surface temperature measurements or on satellite observations of the lower troposphere. In both cases, empirical rescaling factors are applied to all GCMs in order to reproduce the reference climate trajectory as accurately as possible.

If the HadCRUT5 record is assumed to represent global surface temperatures, the observed warming from 1980–1990 to 2012–2022 is 0.605 °C. Alternatively, using the UAH-MSU satellite lower-troposphere record, the warming over the same interval is only 0.432 °C. These values are used to derive the scaling factors for each model group in two cases:

- Case #1—HadCRUT5-based scaling:
  - high-ECS models: scale $\approx 0.605/0.97 \approx 0.62$;
  - medium-ECS models: scale $\approx 0.605/0.80 \approx 0.77$;
  - low-ECS models: scale $\approx 0.605/0.62 \approx 0.98$.
- Case #2—UAH-MSU-based scaling (lower troposphere):
  - high-ECS models: scale $\approx 0.432/0.97 \approx 0.45$;
  - medium-ECS models: scale $\approx 0.432/0.80 \approx 0.54$;
  - low-ECS models: scale $\approx 0.432/0.62 \approx 0.70$.

The associated relative uncertainty is $\pm 5\%$.

In Case 1, the correction required for the low-ECS models is minimal, and the resulting projections closely resemble those produced by the first algorithm. In Case 2, however, the adjustments are substantially larger. It should be noted that satellite-based measurements of the lower troposphere may provide a more accurate representation of global surface warming, as they are less influenced by localized surface biases such as those arising from urbanization and heat islands, and may therefore yield a more neutral and consistent estimate of the global warming trend.

29.4. A flexible methodology

The strength of the proposed approach lies in its flexibility. Once rescaled, the GCM simulations can be further refined by integrating empirical simulations of natural cycles—solar, oceanic, and astronomical—which are often underrepresented or oversimplified in the CMIP GCMs. This strategy was adopted by Scafetta (2013a, 2021b)[8] to develop semi-empirical models based on climate harmonics and astronomical cycles, optimizing the correlation between climate simulations and observational data.

[8]Scafetta, N., 2013a. Discussion on climate oscillations: CMIP5 general circulation models versus a semi-empirical harmonic model based on astronomical cycles. Earth-Science Reviews, 126, 321–357. Scafetta, N., 2021b. Reconstruction of the interannual to millennial scale patterns of the global surface temperature. Atmosphere, 12, 147.

29.5. Assessing risks with scaled models

The SSP2-4.5 scenario was adopted as the reference for future projections, as it is regarded as the most realistic SSP by several recent studies, including Hausfather & Peters (2020)[9] (see Figure 2.7) and Pielke et al. (2022).[10]

The climate projections derived from the models rescaled according to the two cases of the second algorithm are presented in Figure 29.1, which illustrates the evolution of the global mean temperature from 1980 to 2100. As shown in Figure 29.1.

- Case #1—based on HadCRUT5:
  - very likely warming between 1.15 °C and 2.21 °C (2040–2060);
  - warming between 1.65 °C and 3.03 °C (2080–2100);
  - median warming in 2100: 2.28 °C.
- Case #2—based on UAH-MSU v.6 (lower troposphere):
  - very likely warming between 0.82 °C and 1.58 °C (2040–2060);
  - warming between 1.18 °C and 2.16 °C (2080–2100);
  - median warming in 2100: 1.63 °C.

In both cases, the projections are significantly lower than those published in IPCC AR6, which are based on models with ECS values between 2.5 °C and 4.0 °C. According to the IPCC estimates for the SSP2-4.5 scenario:

- projected warming: 1.37 °C to 2.61 °C (2040–2060);
- projected warming: 1.98 °C to 3.82 °C (2080–2100).

This discrepancy stems from the substantial influence exerted by both the choice of observational dataset and the assumed ECS values on global surface temperature projections. Models calibrated using climate records that indicate lower warming—such as satellite-based lower troposphere measurements—tend to project less pronounced warming trajectories for the twenty-first century, with significant implications for risk assessment and policy formulation.

29.6. Risk estimates associated with warming

The projections obtained from the rescaled climate models are evaluated against the five "*Reasons for Concern*" (RFC) outlined in IPCC (2022a, 2023),[11] and depicted in the right-hand panels of Figure 29.1. These categories represent distinct domains of climate vulnerability:

- RFC1—unique and threatened systems, such as ecosystems and communities in geographically constrained areas, including coral reefs, the Arctic, mountain glaciers, and biodiversity hotspots.
- RFC2—extreme weather events, such as risks to health, infrastructure, and ecosystems from heatwaves, heavy precipitation, droughts, wildfires, and coastal flooding.

---

[9]Hausfather, Z., Peters, G.P., 2020. Emissions—the 'business as usual' story is misleading. Nature, 577(7792), 618–620.

[10]Pielke, R., Burgess, M.G., Ritchie, J., 2022. Plausible 2005–2050 emissions scenarios project between 2°C and 3°C of warming by 2100. Environmental Research Letters, 17, 024027.

[11]IPCC, 2022. Climate Change 2022: Impacts, Adaptation and Vulnerability. Cambridge University Press.
IPCC, 2023. Climate Change 2023: Synthesis Report. IPCC, Geneva, Switzerland.

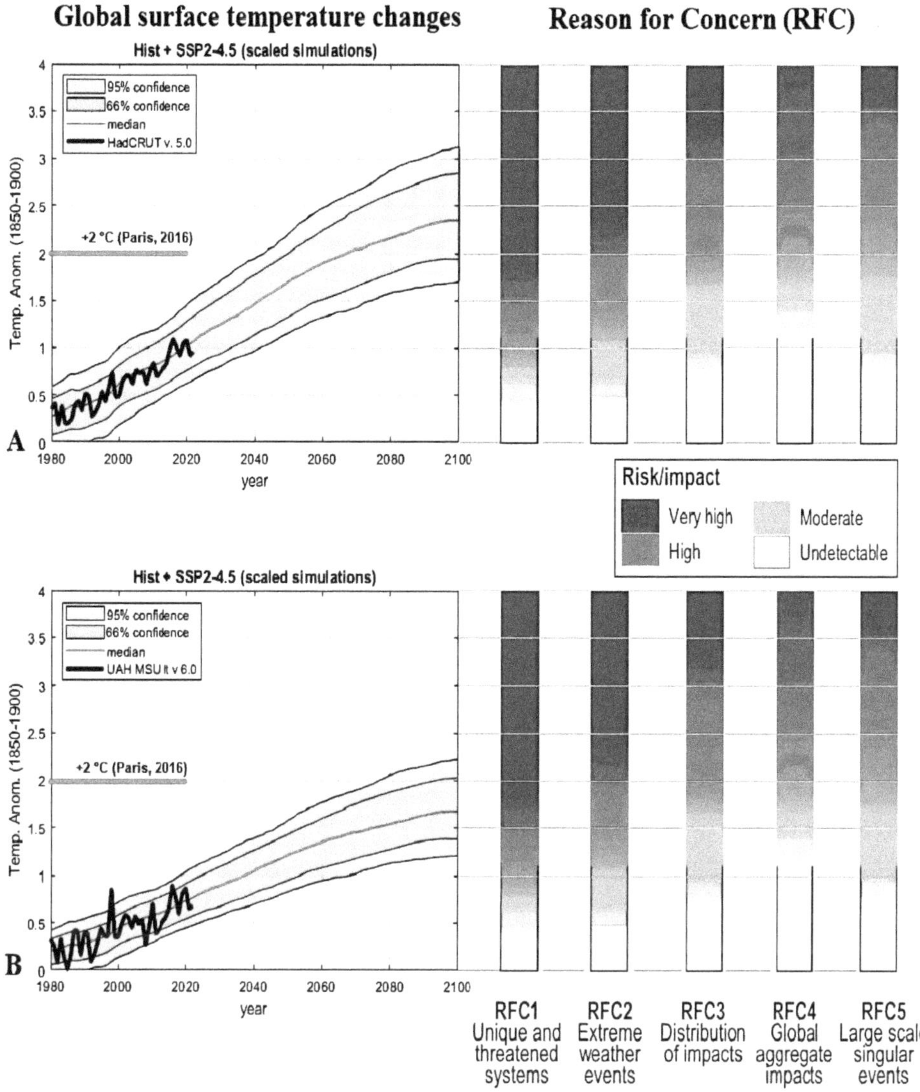

FIGURE 29.1. Temperature reconstructions and risk assessment.
[A] Estimated global temperature changes based on the Hist + SSP2-4.5 scenario, with CMIP6 GCM simulations scaled to optimally reproduce the HadCRUT5 global surface temperature warming between 1980–1990 and 2012–2022.
[B] Same approach applied to the CMIP6 GCM simulations scaled to match the UAH-MSU lower troposphere v6.0 temperature trend over the same period.
[Right] Risk diagrams illustrating the five major reasons for concern (RFC) as a function of global temperature increase, under the assumption of low or no adaptation, as reported by the IPCC (2022a).
(Adapted from Scafetta, 2024).

- RFC3—uneven distribution of impacts, such as disproportionate effects on vulnerable populations due to unequal exposure and adaptive capacity.
- RFC4—global aggregate impacts, such as cumulative damages including economic losses, mortality, species extinction, and ecosystem degradation.
- RFC5—large-scale singular events, such as abrupt and potentially irreversible changes in major climate subsystems, including ice sheet collapse or disruption of the thermohaline circulation.

In Case 1, climate risks remain moderate (yellow/orange zone) until mid-century, with gradual intensification thereafter. In Case 2, the risk profile remains predominantly moderate even by 2100. These findings suggest that, if satellite-based measurements more accurately reflect the actual global warming, the potential impacts of climate change throughout the twenty-first century may be significantly lower than those projected by the IPCC (2021).

The Paris Climate Agreement (2016) stipulates that the global temperature rise should be kept below 2 °C during the twenty-first century. According to the IPCC (AR6, 2021), this target would be achievable only under SSP1-type scenarios, which require aggressive mitigation policies such as the *"Net-Zero 2050"* initiative promoted by the European Union.

The analysis of Figure 29.1, however, reveals that:

- in Case #1, the 2 °C threshold would be slightly exceeded;
- in Case #2, the threshold would not be exceeded at all.

These results suggest that implementing a shared socioeconomic pathway consistent with the SSP2-4.5 scenario, which is a moderate pathway, may be sufficient to meet the targets of the Paris Climate Agreement. This carries significant social implications: policies aligned with SSP2 are considerably less burdensome and more sustainable than those required under SSP1, since SSP2 does not mandate the achievement of net-zero emissions within the twenty-first century.

CHAPTER 30

Modeling climate with astronomical harmonics

30.1. Modeling uncertain forcings

One of the most widely recognized limitations of global climate models (GCMs) is their limited ability to accurately reproduce the natural oscillations that characterize historical climate variability on decadal, centennial, and millennial timescales. This issue has prompted several researchers to explore alternative or complementary approaches, such as the harmonic empirical models developed by Scafetta (2010, 2012b, 2013a, 2021c).[1].

These models are based on the premise that climate variability includes a quasi-cyclic natural component induced by astronomical and solar oscillations that current GCMs fail to simulate. The presence of such oscillations emerges from the spectral analysis of historical climate records, although the underlying physical mechanisms remain poorly understood and are not yet incorporated into conventional climate models. From this perspective, key natural dynamics—still poorly modeled or entirely unmodeled—may influence fundamental processes within the climate system, including cloud formation and distribution, as well as atmospheric and oceanic circulation.

Modern GCMs parameterize complex processes such as cloud formation by treating them as feedbacks to radiative forcings. These processes are calibrated using numerous free parameters (Mauritsen et al., 2019; Mignot et al., 2021).[2] However, these models tend to neglect potential nonradiative forcings, including the influence of galactic cosmic rays and interplanetary dust. These phenomena, modulated by solar magnetic activity and planetary motion, may affect Earth's albedo through mechanisms such as cloud nucleation. This hypothesis was advanced by Shaviv (2002); Shaviv & Veizer (2003)[3] and has been more recently

---

[1]Scafetta, N., 2010. Empirical evidence for a celestial origin of the climate oscillations and its implications. Journal of Atmospheric and Solar–Terrestrial Physics, 72, 951–970.
Scafetta, N., 2012b. Multi-scale harmonic model for solar and climate cyclical variation throughout the Holocene based on Jupiter–Saturn tidal frequencies plus the 11-year solar dynamo cycle. Journal of Atmospheric and Solar–Terrestrial Physics, 80, 296–311.
Scafetta, N., 2013a. Discussion on climate oscillations: CMIP5 general circulation models versus a semi-empirical harmonic model based on astronomical cycles. Earth-Science Reviews, 126, 321–357.
Scafetta, N., 2021c. Testing the CMIP6 GCM simulations versus surface temperature records from 1980–1990 to 2011–2021: High ECS is not supported. Climate, 9, 161.

[2]Mauritsen, T., et al., 2019. Developments in the MPI-M Earth System Model version 1.2 and its response to increasing CO_2. Journal of Advances in Modeling Earth Systems, 11, 998–1038.
Mignot, J., et al., 2021. The tuning strategy of IPSL-CM6A-LR. Journal of Advances in Modeling Earth Systems, 13, e2020MS002340.

[3]Shaviv, N.J., 2002. Cosmic ray diffusion from the galactic spiral arms, iron meteorites, and a possible climatic connection. Physical Review Letters, 89(5), 051102.
Shaviv, N.J., Veizer, J., 2003. Celestial driver of Phanerozoic climate? GSA Today, 13(7), 4–10.

supported by other studies (Svensmark et al., 2016; Svensmark, 2019, 2022; Scafetta et al., 2020; Scafetta, 2023c,a).[4]

It is already acknowledged that GCMs can be tuned to better match observational temperature records by adjusting the available free parameters (Mauritsen & Roeckner, 2020).[5] Such calibration, however, does not guarantee predictive accuracy, nor does it eliminate the possibility that important climate forcings remain unrepresented. In this sense, tuning may improve consistency with the historical data but does not necessarily enhance the physical understanding of the underlying causes of climate change. The systematic omission of other potential influences raises the possibility that GCMs lack crucial physical mechanisms necessary for a more complete and realistic representation of climate dynamics.

30.2. A semi-empirical approach based on astronomical harmonics

The semi-empirical climate model based on natural oscillations is founded on the empirical observation that global temperature time series exhibit multiple cyclicities across spectral scales ranging from interannual to multidecadal. These oscillations recur at characteristic frequencies associated with the natural oscillations of the Solar System generated by the orbital motions of the planets around the Sun and of the Moon around the Earth.

A comprehensive framework was proposed by Scafetta (2010) in the study "*Empirical evidence for a celestial origin of the climate oscillations and its implications,*" and was further developed in subsequent works (Scafetta, 2014b, 2016, 2018).[6] These studies employed robust spectral coherence techniques to validate the hypothesis of a structural link between planetary dynamics and climate variability.

---

[4]Svensmark, J., Enghoff, M.B., Shaviv, N.J., Svensmark, H., 2016. The response of clouds and aerosols to cosmic ray decreases. Journal of Geophysical Research: Space Physics, 121(9), 8152–8181.
Svensmark, H., 2019. Force Majeure: The Sun's Role in Climate Change. The Global Warming Policy Foundation. ISBN 978-0993119095.
Svensmark, H., 2022. Supernova rates and burial of organic matter. Geophysical Research Letters, 49(1), e2021GL096376.
Scafetta, N., Milani, F., Bianchini, A., 2020. A 60-year cycle in the meteorite fall frequency suggests a possible interplanetary dust forcing of the Earth's climate driven by planetary oscillations. Geophysical Research Letters, 47(18), e2020GL089954.Scafetta, N., 2023a. CMIP6 GCM validation based on ECS and TCR ranking for 21st-century temperature projections and risk assessment. Atmosphere, 14, 345.
Scafetta, N., 2023c. Empirical assessment of the role of the Sun in climate change using balanced multi-proxy solar records. Geoscience Frontiers, 14(6), 101650.

[5]Mauritsen, T., Roeckner, E., 2020. Tuning the MPI-ESM1.2 global climate model to improve the match with instrumental record warming. Journal of Advances in Modeling Earth Systems, 12, e2019MS002037.

[6]Scafetta, N., 2014b. Discussion on the spectral coherence between planetary, solar and climate oscillations: a reply to some critiques. Astrophysics and Space Science, 354, 275–299.
Scafetta, N., 2016. High resolution coherence analysis between planetary and climate oscillations. Advances in Space Research, 57, 2121–2135.
Scafetta, N., 2018. Reply on Comment on "High resolution coherence analysis between planetary and climate oscillations" by S. Holm. Advances in Space Research, 62, 334–342.

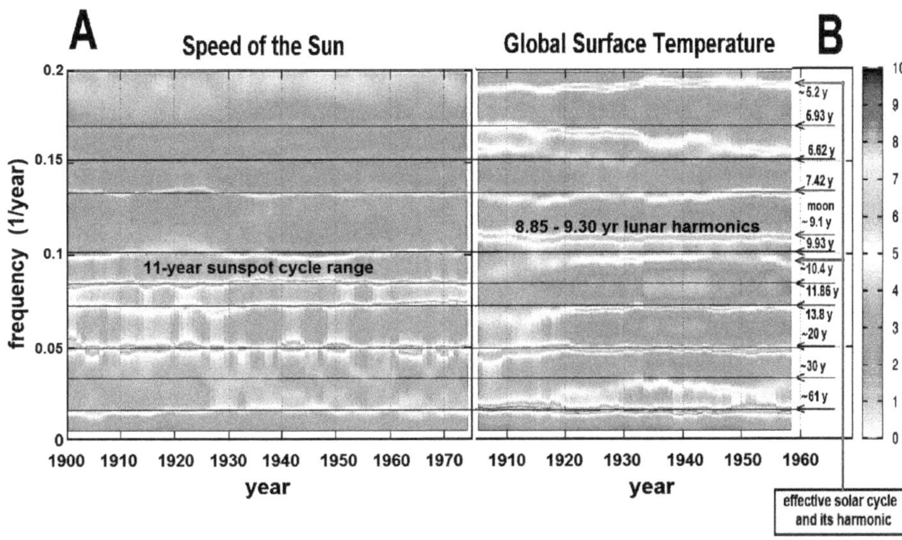

FIGURE 30.1. High-resolution and continuous spectral analysis using a 110-year moving window:
Panel [A] shows the speed of the Sun relative to the barycenter of the Solar System, capturing planetary orbital dynamics;
Panel [B] displays the HadCRUT3 global surface temperature record. Horizontal black lines mark common astronomical spectral frequencies observed in both datasets.
The arrow at 9.1 years highlights the 9.1-year solar–lunar tidal cycle, while the arrow at 10.4 years indicates the solar cycle with a mean periodicity of approximately 10.4 years, which presents a likely harmonic at 5.2 years; these lines are visible only in the climate record. (Adapted from Scafetta, 2014b).

Comparative spectrographic analysis. Figure 30.1, adapted from Scafetta (2014b), presents two panels comparing periodograms calculated using a 110-year moving window. This window length was selected to optimize the resolution of cycles up to approximately 60 years (cf. Scafetta, 2013b, 2014a)

- Panel 1 illustrates the periodogram of the speed of the Sun relative to the barycenter of the Solar System—a dynamic signal encapsulating planetary orbital oscillations;
- Panel 2 displays the periodogram of the HadCRUT3 global surface temperature dataset (Brohan et al., 2006).[7]

The statistical significance of these astronomical oscillations is further demonstrated in Figure 30.2, where panels A1, A2, B1, and B2 show the canonical squared magnitude coherence analysis (MSC-CCA) between the global temperature record and the Sun's barycentric velocity, tested against various statistical significance models.

[7]Brohan, P., et al., 2006. Uncertainty estimates in regional and global observed temperature changes: A new dataset from 1850. Journal of Geophysical Research: Atmospheres, 111, D12106.

In particular, Panels 30.2C1 and 30.2C2 highlight strong spectral coherence at the 20-year and 60-year periodicities. These findings reinforce the hypothesis that astronomical oscillations are not only statistically significant but also structurally correlated with the observed climate variability.

Identified frequencies.

- Spectral analysis of the Sun's barycentric speed reveals distinct periodicities at approximately 5.93, 6.62, 7.42, 9.93, 11.86, 13.8, 20, and 60 years, largely associated with the orbital dynamics of Jupiter and Saturn.
- The global surface temperature record exhibits similar spectral lines: 5.93, 6.62, 7.42, 9.1, 10.4 (variable between 10 and 12 years), 13.8, 20, and 60 years.

This correspondence suggests a structural coherence between planetary motions and climate variability.

Interpretation of apparent anomalies.

- The 9.1-year cycle—as discussed by Scafetta (2012a),[8] and other authors (Cionco & Soon, 2021; Keeling & Whorf, 2000; Le Mouël et al., 2020; Yndestad, 2022)[9]—can be interpreted as a decadal modulation of solar–lunar tidal interactions. This periodicity may arise from the superposition of:
  - the rotation of the lunar apsidal line (8.85 years);
  - the first harmonic of the Saros eclipse cycle (9.0 years);
  - the first harmonic of the solar–lunar nodal cycle (9.3 years).
- The 10.4-year cycle falls within the solar Schwabe cycle band, typically centered around 11 years but known to vary over time. Notably, Figure 30.1B shows that the climatic signature of the solar cycle shortened during the twentieth century, consistent with the observed solar variability.

Development of the semi-empirical model. The observed periodicities formed the foundation of the first semi-empirical climate models based on astronomical harmonics (Scafetta, 2010, 2013a). The model was subsequently refined to include interannual cycles (Scafetta, 2021b)[10] and calibrated using the HadCRUT4 dataset from 1850 to 2014 (Morice et al., 2012).[11]

[8]Scafetta, N., 2012a. Testing an astronomically based decadal-scale empirical harmonic climate model versus the IPCC (2007) general circulation climate models. Journal of Atmospheric and Solar–Terrestrial Physics, 80, 124–137.

[9]Cionco, R.G., et al., 2021. Possible Origin of Some Periodicities Detected in Solar-Terrestrial Studies: Earth's Orbital Movements. Earth and Space Science, 8, e2021EA001805.

Keeling, C.D., Whorf, T.P., 2000. The 1800-year oceanic tidal cycle: A possible cause of rapid climate change. Proceedings of the National Academy of Sciences, 97(8), 3814–3819.

Le Mouël, J.L., et al., 2020. Characteristic time scales of decadal to centennial changes in global surface temperatures over the past 150 years. Earth and Space Science, 7(4), e2019EA000671

Yndestad, H., 2022. Lunar-driven Control of Climate and Barents Sea Ecosystems. Science of Climate Change, 2(1), 74–77.

[10]Scafetta, N., 2021b. Reconstruction of the interannual to millennial scale patterns of the global surface temperature. Atmosphere, 12, 147.

[11]Morice, C.P., et al., 2012. Quantifying uncertainties in global and regional temperature change

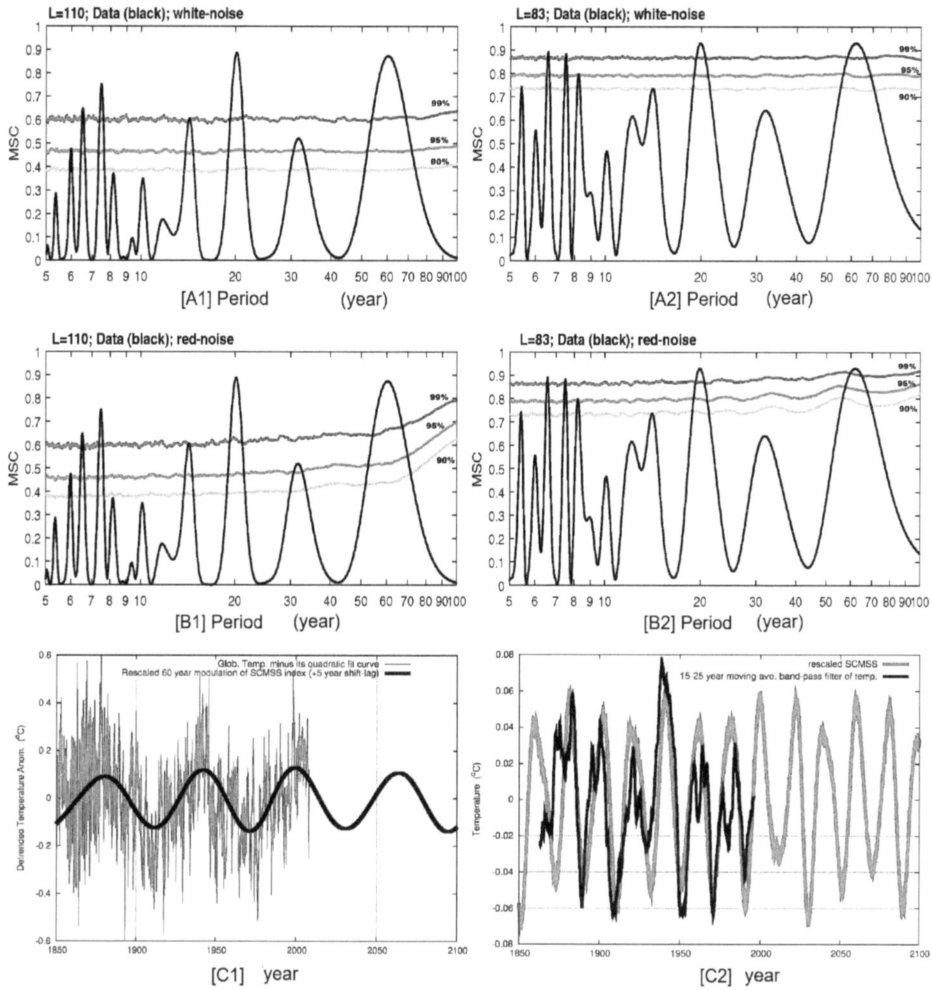

FIGURE 30.2. Panels A1, A2, B1, B2 present the canonical squared magnitude coherence analysis (MSC-CCA) between the global surface temperature record and the Sun's speed relative to the Solar System barycenter, evaluated against various statistical significance models. Several coherence peaks exceed the 95% confidence threshold, notably at the 20-year and 60-year periodicities.
Panel C1 shows the 60-year cycle extracted from the Sun's barycentric velocity (black curve), compared with the global surface temperature record, after removal of the quadratic trend.
Panel C2 illustrates the 20-year cycle derived from the same solar signal, compared with the filtered global temperature record, highlighting the structural coherence between astronomical dynamics and climate variability.
(Adapted from Scafetta, 2010, 2018).

Scafetta's model assumes that global surface temperature datasets (e.g., HadCRUT3 and HadCRUT4) are representative of the actual warming trend. Based on this assumption, the proposed equation models climate change as the sum of two components:

(1) a natural harmonic component, empirically derived from astronomical periodicities (solar, lunar, and planetary);
(2) an anthropogenic and volcanic forcing component, approximated by applying a scaling factor ($\alpha = 0.5$) to the ensemble mean of the CMIP6 GCMs.

This scaling reflects the observation that, once natural harmonics are accounted for, the climate response to anthropogenic forcings appears significantly weaker than in standard GCM simulations.

Scafetta (2010, 2012a, 2013a) demonstrates that this adjustment leads to an effective climate sensitivity (ECS) reduced by approximately half, with estimated values ranging from 0.9 °C to 2.8 °C—consistent with other independent empirical estimates (Lewis, 2023; Spencer & Christy, 2023).[12] Post-2014 forecasts successfully anticipated major climate events, including the strong El Niño episodes of 2015–2016 and 2020.

30.3. Model formulation

The most recent version of the harmonic model, presented in Scafetta (2021b), incorporates 13 astronomical harmonics and is expressed as:

$$T(t) = \sum_{i=1}^{13} A_i \sin\left[2\pi\left(f_i \cdot (t - 2000) + \phi_i\right)\right] + \alpha \cdot GCM(t) + T_0.$$

Here:

- A_i, f_i, and ϕ_i are the amplitudes, frequencies, and phases of the 13 harmonics;
- $\alpha \approx 0.5$ is the scaling factor applied to the CMIP6 ensemble mean;
- $GCM(t)$ represents the CMIP6 simulation output for a given SSP scenario;
- T_0 is the reference temperature anomaly, typically normalized to the 1850–1900 baseline.

The specific harmonic values used in the semi-empirical model are listed in Table 30.3. To represent the quasi-millennial climate cycle, a fictitious frequency corresponding to a 760-year period was introduced, strategically selected to simulate the observed asymmetry in that oscillation. This asymmetry arises from the simultaneous presence of the Bray–Hallstatt cycle, one of the major long-term astronomical oscillations, with an estimated period of approximately 2,318 years, widely documented in the paleoclimate literature.

---

using an ensemble of observational estimates: the HadCRUT4 dataset. Journal of Geophysical Research: Atmospheres, 117, D08101.

[12]Lewis, N., 2023. Objectively combining climate sensitivity evidence. Climate Dynamics, 60, 3139–3165.

Spencer, R.W., Christy, J.R., 2023. Effective climate sensitivity distributions from a 1D model of global ocean and land temperature trends, 1970–2021. Theoretical and Applied Climatology, 155, 299–308.

Cycle	A_i	f_i	T_i	ϕ_i
i	°C	1/Year	Year	1/Year
1	0.3228	0.001316	760	0.171
2	0.0584	0.008696	115	0.424
3	0.0858	0.01639	61	0.152
4	0.0334	0.0500	20	0.148
5	0.0241	0.09615	10.4	0.020
6	0.0265	0.1075	9.3	0.497
7	0.0216	0.1340	7.46	0.711
8	0.0272	0.1666	6.00	0.617
9	0.0260	0.1912	5.23	0.409
10	0.0326	0.2088	4.79	0.931
11	0.0276	0.2754	3.63	0.767
12	0.0254	0.2809	3.56	0.792
13	0.0247	0.3484	2.87	0.975

FIGURE 30.3. Parameters of the harmonic components of the astronomical climate model proposed by Scafetta (2021b).

The astronomical origin of the Bray–Hallstatt cycle was analyzed in detail by Scafetta et al. (2016),[13] and a more in-depth analysis is presented in Chapter 40.

Figure 30.4A illustrates the performance of the proposed semi-empirical climate model, assuming an ECS of approximately 1.5 °C. Figure 30.4B presents the same model extended back to the medieval period. In both cases, the model closely reproduces the observed global surface temperature records, demonstrating strong coherence with historical climate variability.

30.4. Validation and forecasting

Scafetta (2012a) shows that global temperature trends from 1850 to 2011 can be reconstructed with high accuracy even when the calibration of the harmonic model is performed separately for the subperiods 1850–1950 and 1950–2011. A simple visual comparison between the simulated curves and observational

[13]Scafetta, N., Milani, F., Bianchini, A., Ortolani, S., 2016. On the astronomical origin of the Hallstatt oscillation found in radiocarbon and climate records throughout the Holocene. Earth-Science Reviews, 162, 24–43.

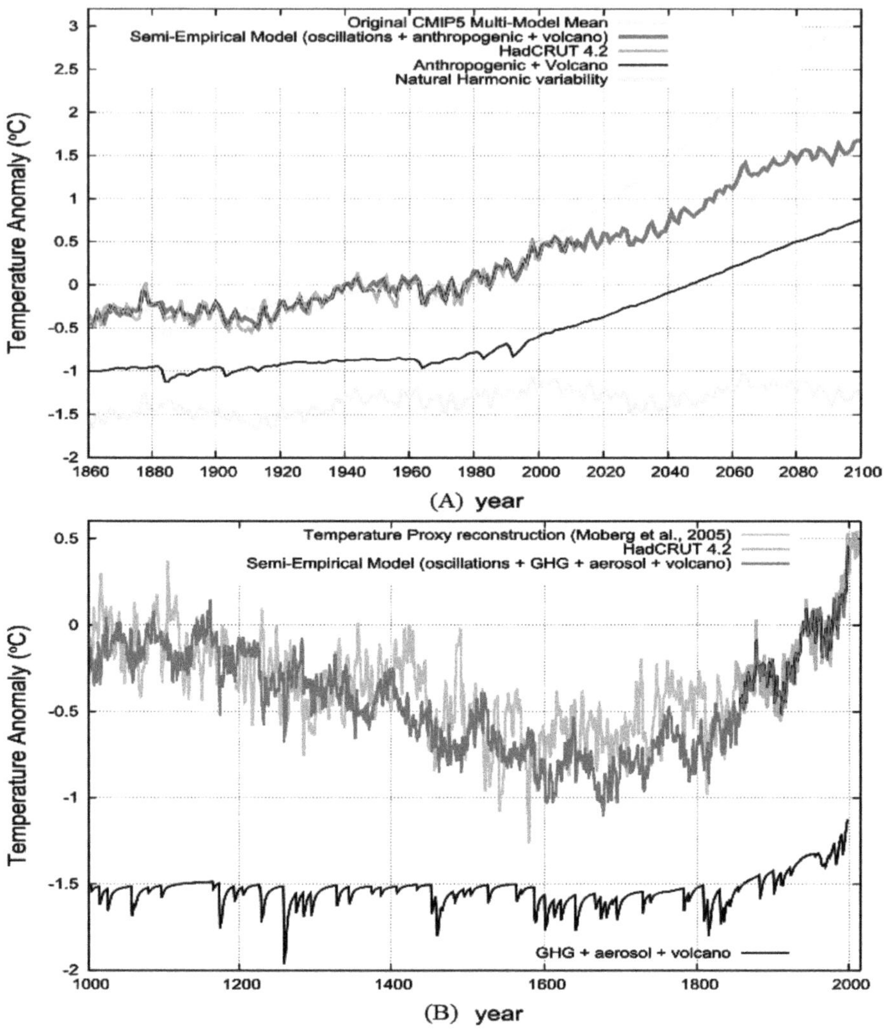

FIGURE 30.4. Astronomically optimized semi-empirical model constructed using the parameters listed in Table 30.3, based on the frequencies and phases of decadal-to-millennial harmonics derived from astronomical considerations. The model is calibrated against both a proxy temperature record (1000–1850) and the instrumental temperature record (1850–2015).
(A) Individual contributions of anthropogenic and volcanic forcings, together with the harmonic natural variability of global surface temperature; these components are combined to produce the model output. For comparison, the ensemble mean of CMIP5 GCM simulations is also shown.
(B) Combined signature of greenhouse gases (GHGs), aerosols, and volcanic forcings—which incorporates the anthropogenic component since 1850—assuming an equilibrium climate sensitivity of approximately 1.5 °C. (Adapted from Scafetta, 2021b).

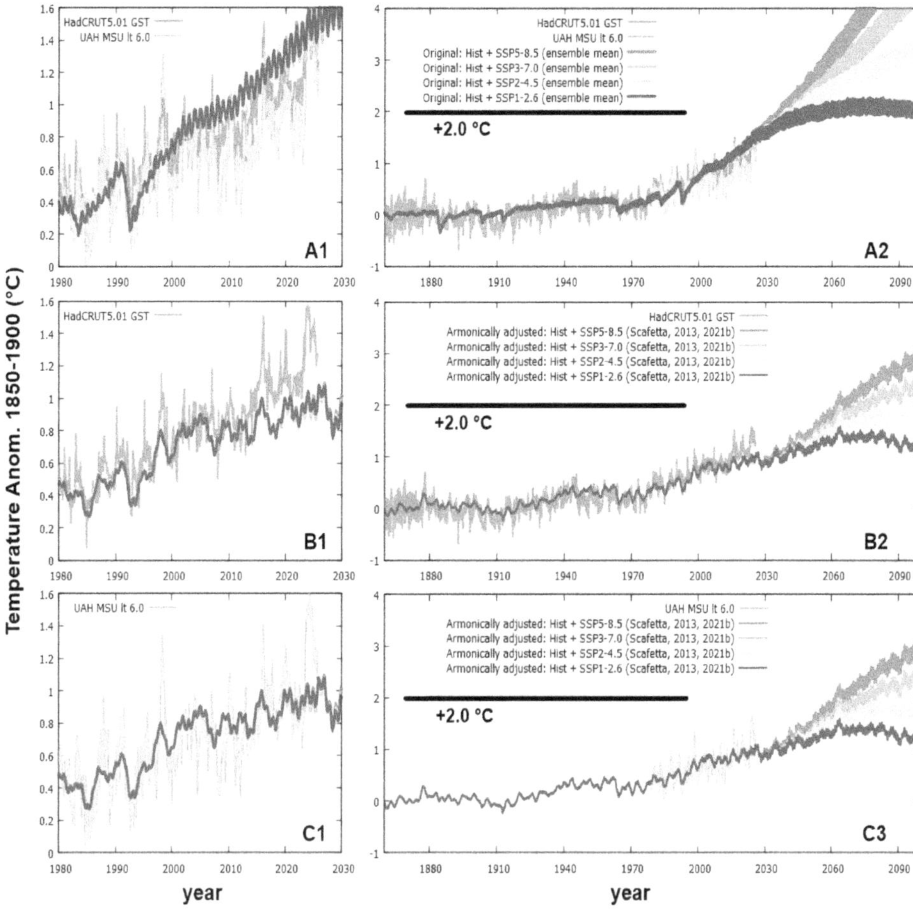

Figure 30.5. Comparison of climate simulations and projections with associated global aggregate impacts and risks obtained using:
[A] CMIP6 GCM ensemble mean simulations;
[B and C] harmonic empirical global climate models proposed by Scafetta (2012a, 2014b, 2021b). Temperature data used include global lower troposphere records from HadCRUT v.5 (Morice et al., 2021) and UAH-MSU lt v.6.0 (Spencer, 2021).

datasets reveals that the model reproduces climate variability with good precision across multiple temporal scales, from interannual to multidecadal. This result confirms the robustness of the model across distinct historical intervals.

Figures 30.5A–C compare the performance of the CMIP6 GCM ensemble mean with that of the harmonic model proposed by Scafetta (2012a, 2014b, 2021b) against the temperature records from HadCRUT5 (Morice et al., 2021) and UAH-MSU v6.0 (Spencer, 2021). Whereas the CMIP6 ensemble tends to overestimate the warming, the harmonic model more faithfully reproduces the observed temperature changes. A strong correlation is found with the satellite lower troposphere temperature record (Figure 30.5C). The discrepancy with HadCRUT5 between 2000 and 2025 shown in Figure 30.5B is likely due to issues with its revisions, which removed the so-called warming *"pause"* between 2000 and 2014, as discussed in Chapter 21.

Figure 30.6 shows the harmonic model under the SSP2-4.5 scenario, which agrees well with the HadCRUT4.6 record (1850–2021) (Morice et al., 2012) in both long-term trends and decadal variability. The right panel presents burning ember diagrams representing climate risk levels as a function of global temperature increase for the five RFCs, as reported by IPCC (2022a).

Importantly, the harmonic model captures the quasi-60-year modulation observed in surface temperature records since 1850, which the GCMs fail to reproduce. Conventional models simulate a nearly monotonic warming interrupted only by major volcanic events.

30.5. Harmonic climate projections to 2100

Figures 30.5A–C extend the climate simulations to the year 2100, based on the four SSP scenarios adopted by the IPCC (2021). In all cases, the harmonic model projects significantly lower warming—approximately half of the CMIP6 estimates. Moreover, the presence of natural oscillations suggests that, after 2060, the descending phase of the 60-year cycle could further mitigate anthropogenic impacts, resulting in more moderate warming.

Specifically, Figure 30.5A shows that only the GCM simulations under SSP1-2.6 are compatible with the Paris Climate Agreement target of limiting warming to below 2 °C relative to preindustrial levels. If the GCM framework is assumed to be correct, this would necessitate Net-Zero policies by mid-century. According to the harmonic model (Figures 30.5B and 30.5C), however, even moderate policies—compatible with SSP2-4.5—could suffice to meet the 2 °C threshold throughout the century. Thus, differences between modeling approaches can have profound social and political implications.

Figure 30.6 shows the harmonic simulation under SSP2-4.5, which is considered the most plausible scenario (cf. Figure 2.7). The harmonic model forecasts:

(1) a moderate warming between 2035 and 2065, driven by the positive phase of the Atlantic Multidecadal Oscillation (AMO);

(2) periods of slowed warming between 2000–2035 and 2065–2100, coinciding with the AMO's negative phases.

According to the harmonic model projections, the global temperature in 2100 will range between 1.15 °C and 2.52 °C above preindustrial levels, with a median of 1.78 °C—consistent with the IPCC estimates for SSP1-2.6. This outcome arises because the harmonic model assumes a lower effective climate sensitivity (ECS). If sufficiently accurate, the Paris Climate Agreement target could be achieved without imposing stringent Net-Zero policies.

A key issue is that the ∼60-year periodicity of the Atlantic Multidecadal Oscillation (AMO) is not reproduced by the CMIP GCMs (Scafetta, 2013a). Yet this oscillation is well documented in numerous climate and paleoclimate reconstructions (Knudsen et al., 2011; Wyatt & Curry, 2014)[14] and may have

[14]Knudsen, M., et al., 2011. Tracking the Atlantic Multidecadal Oscillation over 8,000 years. Nature Communications, 2, 178.

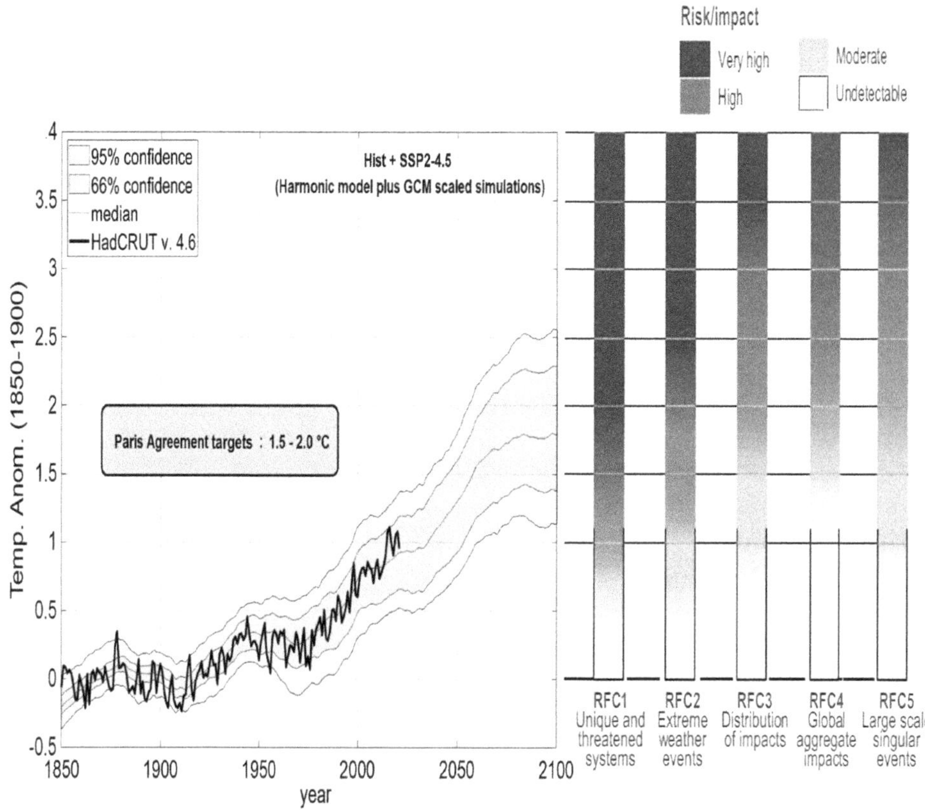

FIGURE 30.6. Performance and projection of the harmonic climate model.
Left panel: The harmonic model is extended from 2014 to 2100 under the SSP2-4.5 scenario and compared with the HadCRUT4.6 temperature record (1850–2021) (Morice et al., 2012), showing strong agreement in both the long-term trend and interannual variability. During the 21st century, the model projects two phases of moderate warming—between 2000 and 2035, and from 2065 to 2100—interrupted by a period of more rapid warming between 2035 and 2065. By the end of the century, the global surface temperature increase is estimated to range between approximately 1.4 °C and 2.4 °C relative to the 1850–1900 preindustrial baseline.
Right panel: "Burning embers" diagrams illustrate climate risk levels as a function of global temperature rise, focusing on the five major Reasons for Concern (RFC) under assumptions of low or no adaptation, as reported by IPCC (2022a). Risk levels are projected to escalate significantly beyond +2 °C of global warming.
(Adapted from Scafetta, 2024).

a solar or astronomical origin (Neff et al., 2001; Scafetta, 2013a; Scafetta et al., 2020).[15] Thus, the proposed empirical model offers a more reliable framework than the current GCM predictions for assessing realistic risks for the twenty-first century.

30.6. Conclusions

The harmonic empirical model offers a methodologically robust and conceptually innovative alternative to traditional GCMs. It provides a more comprehensive interpretive framework that accounts not only for the observed warming but also for the multidecadal and centennial modulations often overlooked or inadequately simulated by conventional models.

Grounded in the hypothesis that the climate system is influenced by astronomical oscillations—solar, lunar, and planetary—the harmonic model integrates these natural dynamics into its predictive structure. The inclusion of such recurring cycles allows for a downscaling of the anthropogenic forcing effect, represented through a simplified scaling factor. This approach suggests that the observed climate change is more compatible with ECS values significantly lower than those estimated by CMIP-class GCMs (cf. Chapter 11).

In summary, the harmonic model:

- more faithfully reproduces the historical observational data, particularly the satellite-based lower troposphere temperature records;
- enables more moderate and realistic climate projections, avoiding extreme scenarios not supported by the empirical trends;
- incorporates the influence of recurring natural oscillations, such as the AMO, which are often not modeled by the GCMs (cf. Figure 12.3).

From both methodological and predictive perspectives, the proposed harmonic model reproduces past climate dynamics with greater fidelity and invites a reassessment of the role of natural forcings in the global energy balance. This has significant implications for climate policy, as it opens the way to more flexible and realistic strategies that emphasize adaptation and moderate mitigation rather than aggressive *Net-Zero* measures. In particular, the projections derived from the harmonic framework suggest that pathways consistent with SSP2-4.5 may suffice to meet international climate targets, offering a more balanced and sustainable approach to risk management.

Wyatt, M.G., Curry, J.A., 2014. Role of Eurasian Arctic shelf sea ice in a secularly varying hemispheric climate signal during the 20th century. Climate Dynamics, 42, 2763–2782.

[15] Neff, U., et al., 2001. Strong coherence between solar variability and the monsoon in Oman between 9 and 6 kyr ago. Nature, 411, 290–293.

Scafetta, N., et al., 2020. A 60-year cycle in the meteorite fall frequency suggests a possible interplanetary dust forcing of the Earth's climate driven by planetary oscillations. Geophysical Research Letters, 47(18), e2020GL089954.

CHAPTER 31

An energy balance–regression climate model

31.1. The Sun, total solar activity, and the challenge to GCMs

The role of solar variability in climate change has long been the subject of intense scientific debate, polarized between two main perspectives. On one side, a line of research attributes a significant influence to solar activity, including in driving recent climate variations. On the other, the prevailing view within the contemporary climatological community—represented by the IPCC—minimizes this contribution, attributing nearly all observed global warming since the preindustrial period (1850–1900) to anthropogenic factors, primarily greenhouse gas emissions and related forcings.

The IPCC's position is grounded in the CMIP6 generation of global climate models (GCMs), which consider only the direct radiative forcing from the Sun—specifically, variations in total solar irradiance (TSI). These models adopt the TSI reconstruction with the lowest long-term variability (Matthes et al., 2017),[1] thereby concluding that the solar influence is limited to minor fluctuations in the Earth's energy balance. This approach neglects potential nonradiative forcings, such as the modulation of galactic cosmic ray fluxes and interplanetary dust flows, which may influence cloud formation, atmospheric circulation, and biospheric dynamics. In practice, this excludes any direct solar impact on the Earth's climate system (cf. Figures 6.2 and 6.3).

By contrast, Scafetta (2023c)[2] proposed a more inclusive methodology by constructing three multi-proxy models of total solar activity (TSA). These models integrate various TSI proxies from the literature, derived from geomagnetic indices, sunspot records, and paleoclimate reconstructions based on cosmogenic isotopes (e.g., carbon-14 and beryllium-10), thereby offering a broader representation of solar variability across secular and millennial timescales.

These TSA models are used as inputs to a zero-dimensional energy balance model (EBM), calibrated through differential multilinear regression using multiple predictors. This approach enables the separation of climate responses to distinct forcings and allows for the estimation of characteristic timescales for energy absorption and dissipation within the climate system.

The key findings include the following:

- When the EBM employs the same radiative forcings used in the CMIP6 GCMs, which include the low-variability TSI record from Matthes et al. (2017) shown in Figure 25.3, the model successfully replicates the

[1]Matthes, K., et al., 2017. Solar forcing for CMIP6 (v3.2). Geoscientific Model Development, 10, 2247–2302.

[2]Scafetta, N., 2023c. Empirical assessment of the role of the Sun in climate change using balanced multi-proxy solar records. Geoscience Frontiers, 14(6), 101650.

behavior of the GCMs. In this configuration, the equilibrium climate sensitivity (ECS) ranges from 1.4 °C to 2.8 °C, with a mean of 2.1 °C, based on calibration with the HadCRUT5 dataset (Morice et al., 2021).[3] This value falls within the lower range of CMIP6 estimates, consistent with the low-ECS models reported by Lewis (2023)[4] and Scafetta (2022, 2023b).[5]

- When the TSI input is replaced, however, with the TSA models proposed by Scafetta, which are characterized by greater secular variability such as those shown in Figure 25.4, and differentiated climate sensitivities are assigned to the solar, anthropogenic, and volcanic forcings, the results shift markedly. The solar contribution becomes substantially larger, and the ECS predicted by the model ranges from 0.9 °C to 1.8 °C, with a mean of around 1.3 °C.

Further analyses using alternative datasets—such as HadSST4 (Kennedy et al., 2019),[6] HadCRUT4 (Morice et al., 2012),[7] and HadSST3 (Brohan et al., 2006)[8]—confirm this behavior, suggesting a reduction in ECS of up to 20% compared with the HadCRUT5-based case. These results imply that a substantial portion of the observed warming since the preindustrial era may be attributable to solar variability, challenging the assumption that anthropogenic emissions have been the sole effective driver since 1850–1900. They also indicate that future climate projections based on high-ECS GCMs are likely to be significantly overestimated.

A particularly important aspect of the study is the demonstration that up to 80% of the solar climate effect may arise from indirect, nonradiative processes. These include solar modulation of cosmic rays—affecting cloud nucleation—variations in energetic particle fluxes, and the influence of the solar magnetic field on atmospheric dynamics. Although the mechanisms remain poorly understood, their existence and importance are supported by extensive empirical evidence and should be incorporated into next-generation climate models.

Ultimately, Scafetta (2023c) advances an interpretive framework that reconciles long-standing empirical evidence of solar influence on Holocene climate variability with the epistemological constraints embedded in current GCMs. The systematic exclusion of alternative solar forcings—including mechanisms

[3]Morice, C.P., et al., 2021. An updated assessment of near-surface temperature change from 1850: the HadCRUT5 data set. Journal of Geophysical Research: Atmospheres, 126, e2019JD032361.

[4]Lewis, N., 2023. Objectively combining climate sensitivity evidence. Climate Dynamics, 60, 3139–3165.

[5]Scafetta, N., 2022. Advanced testing of low, medium, and high ECS CMIP6 GCM simulations versus ERA5-T2m. Geophysical Research Letters, 49(6), e2022GL097716.

Scafetta, N., 2023b. CMIP6 GCM ensemble members versus global surface temperatures. Climate Dynamics, 60, 3091–3120.

[6]Kennedy, J.J., et al., 2019. An ensemble dataset of sea-surface temperature change from 1850. Journal of Geophysical Research: Atmospheres, 124, 7719–7763.

[7]Morice, C.P., et al., 2012. Quantifying uncertainties in global and regional temperature change using an ensemble of observational estimates: the HadCRUT4 dataset. Journal of Geophysical Research: Atmospheres, 117, D08101.

[8]Brohan, P., et al., 2006. Uncertainty estimates in regional and global observed temperature changes: A new dataset from 1850. Journal of Geophysical Research: Atmospheres, 111, D12106.

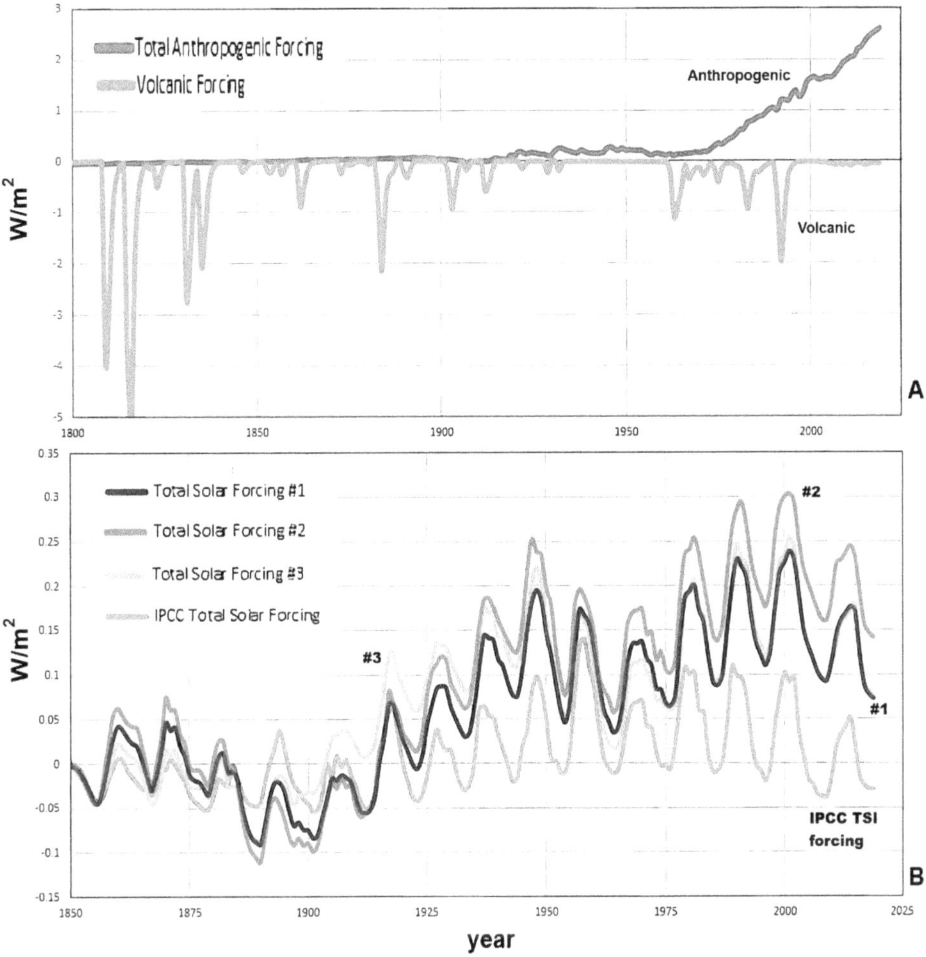

FIGURE 31.1. (A) Effective anthropogenic and volcanic radiative forcing functions adopted by the CMIP6 GCMs (IPCC, 2021). (B) Comparison between the solar radiative forcing function used in CMIP6, based on Matthes et al. (2017), and three alternative TSA-based functions derived from Scafetta (2023c). All records are expressed as anomalies relative to 1850. (Adapted from Scafetta, 2023c).

mediated through cloud-cover modulation, galactic cosmic-ray fluxes, and stratospheric dynamics—compromises the ability of these models to faithfully simulate the Earth's climate system. By contrast, the integration of multi-proxy solar datasets into calibrated energy-balance models (EBMs) yields a more nuanced and physically coherent representation of climate dynamics. Enhancing the reliability of future climate projections thus requires a broader and more inclusive treatment of solar processes within climate-modeling frameworks.

31.2. Development of the EB regression model

Empirical climate models are particularly effective when they combine structural simplicity with scientific rigor, relying on a limited number of physically meaningful predictors selected to minimize multicollinearity. Crucially, such models must avoid rigid linear assumptions, since the Earth's climate system exhibits nonlinear and frequency-dependent responses to different forcings. Low-frequency components, such as solar variability, are processed more slowly because of the high thermal inertia of the ocean–atmosphere system.

Figure 31.1 illustrates the radiative forcing functions used in Scafetta's energy balance model:

- Panel A shows the anthropogenic and volcanic forcings adopted by the CMIP6 GCMs, consistent with the IPCC (2021) assessments. These curves represent the standard inputs for simulations attributing warming primarily to human activities;
- Panel B presents four versions of solar forcing. The lower curve corresponds to the CMIP6 GCM standard, based on the low-variability TSI record by Matthes et al. (2017), which places the solar maximum in the 1950s–1960s.

Three alternative high-variability TSA models developed by Scafetta (2023c) are also shown. Among these, curve #2 (TSI-2) is derived by averaging the reconstructions of Egorova et al. (2018)[9] and the TSI proxy by Hoyt & Schatten (1993),[10] updated with ACRIM satellite data (Scafetta, 2023c). These reconstructions suggest an earlier solar peak in the 1940s, approximately 10–15 years earlier than the Matthes et al. estimate. This temporal discrepancy is highly relevant for climate modeling, as it alters the synchronicity between solar forcing and thermal response. An earlier solar maximum around the 1940s better explains the warming phases observed during the twentieth century, which appear inconsistent with simulations based on other solar proxy records.

Mathematical model. Scafetta (2023c) proposes a simplified empirical model to simulate the evolution of the global surface temperature, expressed as:

$$T(t) = T_A(t) + T_V(t) + T_S(t) + \xi(t).$$

where:

- $T_A(t)$ represents the temperature response to anthropogenic forcing,
- $T_V(t)$ corresponds to the volcanic temperature signature,
- $T_S(t)$ reflects the solar temperature signature,
- $\xi(t)$ is a stochastic component accounting for interannual variability, such as ENSO-related fluctuations.

Each deterministic component is modeled using a zero-dimensional energy balance differential equation, which captures the thermal evolution of the climate

[9] Egorova, T., et al., 2018. Revised historical solar irradiance forcing. Astronomy and Astrophysics, 615, A85.

[10] Hoyt, D.V., Schatten, K.H., 1993. A discussion of plausible solar irradiance variations, 1700–1992. Journal of Geophysical Research: Space Physics, 98(A11), 18895–18906.

system in a compact yet physically coherent form:

$$C\,\frac{dT(t)}{dt} = \frac{1-a}{4}\,S - \sigma\varepsilon\,T(t)^4 + G,$$

where:

- T is the global mean temperature in kelvin,
- C is the effective thermal capacity of the Earth–atmosphere system,
- a is the planetary albedo,
- S is the incoming solar radiation (adjusted for Earth's sphericity),
- ε is the mean emissivity,
- σ is the Stefan–Boltzmann constant,
- G is the radiative effect of greenhouse gases.

The system is linearized in terms of temperature anomalies, $\Delta T(t) = T(t) - T_0$, as follows:

$$\frac{\Delta T_A(t) - \Delta T_A(t-\Delta t)}{\Delta t} = \alpha_A \cdot F_A(t) - \beta \cdot \Delta T_A(t-\Delta t);$$
$$\frac{\Delta T_V(t) - \Delta T_V(t-\Delta t)}{\Delta t} = \alpha_V \cdot F_V(t) - \beta \cdot \Delta T_V(t-\Delta t);$$
$$\frac{\Delta T_S(t) - \Delta T_S(t-\Delta t)}{\Delta t} = \alpha_S \cdot F_S(t) - \beta \cdot \Delta T_S(t-\Delta t).$$

where $X = A, V, S$ denotes the anthropogenic, volcanic, and solar components, respectively, and

- $F_X(t)$ are the radiative forcing functions,
- $\alpha_X = k_X/\tau$ are the response coefficients,
- $\beta = 1/\tau$ is the damping term,
- τ is the characteristic response time of the climate system.

Iterating the equations with a time step of $\Delta t = 1$ year yields:

$$\frac{\Delta T(t) - \Delta T(t-\Delta t)}{\Delta t} = \alpha_A \cdot F_A(t) + \alpha_V \cdot F_V(t) + \alpha_S \cdot F_S(t) - \beta \cdot \Delta T(t-\Delta t).$$

This equation can be solved numerically to reconstruct the deterministic component of the global temperature signal as the sum of the three main contributions:

$$T(t) = T_A(t) + T_V(t) + T_S(t).$$

The coefficients α_A, α_V, α_S, and β are calibrated using multilinear regression against the observational datasets. This approach implicitly accounts for the thermal inertia of the climate system, which introduces time lags and amplitude modulation in the response to external forcings.

31.3. Calculation of ECS and TCR

Under equilibrium conditions, the equilibrium climate sensitivity (ECS)—namely, the warming induced by doubling the atmospheric CO_2 concentration from 280 to 560 ppm—is defined as:

$$ECS = \Delta T_{2\times CO_2} = k_A \cdot \Delta F_{2\times CO_2} = 3.7\,k_A,$$

where $\Delta F_{2\times CO_2} = 3.7\,W/m^2$ is the canonical radiative forcing associated with CO_2 doubling.

The transient climate response (TCR) is a key metric for evaluating the short- to medium-term sensitivity of the climate system to greenhouse gas emissions.

It is defined as the average global temperature increase over the 20-year period centered on the moment when the atmospheric CO_2 concentration doubles, assuming a 1% annual increase, which takes approximately 70 years.

Unlike ECS, which reflects the long-term equilibrium response, TCR captures the transient warming before the system reaches full energy balance. It is particularly relevant for assessing the effectiveness of climate mitigation policies over policy-relevant time horizons.

The TCR is calculated as

$$\mathrm{TCR} = 3.7\,k_A - \frac{3.7}{70}\,k_A\tau\left(1 - e^{-70/\tau}\right) = \mathrm{ECS}\left[1 - \frac{\tau}{70}\left(1 - e^{-70/\tau}\right)\right].$$

Thus, the TCR depends not only on the climate sensitivity k_A, but also on the system's response time τ, which reflects the thermal inertia of the Earth's climate system.

31.4. Formal analysis

Figure 31.2 presents a selection of results obtained by (Scafetta, 2023c) using the proposed empirical model, in which various combinations of climate datasets and radiative forcing functions are tested. The study explores two principal configurations for the climate response coefficients:

- Configuration #1 assumes uniform sensitivity across all forcings, i.e., $\alpha_A = \alpha_V = \alpha_S$, which is consistent with the structure of the CMIP6 GCMs.
- Configuration #2 is subdivided into four variants and assumes $\alpha_A = \alpha_V \neq \alpha_S$, introducing a differentiated response to solar forcing.

The first two simulations shown in Figure 31.2 employ the TSI record from Matthes et al. (2017), characterized by minimal secular variability and adopted in the CMIP6 GCMs. The remaining simulations adopt the high-variability solar proxies proposed by Scafetta (2023c) and illustrated in Figure 31.1, which enable the exploration of scenarios in which solar forcing plays a more prominent role in shaping twentieth-century climate variability.

31.5. Key findings

- The equal sensitivity configuration ($\alpha_A = \alpha_V = \alpha_S$) combined with low-variability TSI yields a negligible solar contribution to global warming between 1850–1900 and 2011–2022. In this setup, the equilibrium climate sensitivity (ECS) is estimated at 2.1 ± 0.7 °C, consistent with the lower range of CMIP6 model outputs and with the findings of Lewis (2023).
- Differentiated sensitivities ($\alpha_A = \alpha_V \neq \alpha_S$) reveal that the climate system is 4 to 6 times more responsive to solar forcing. When using a high-variability TSI composite, solar variations may account for up to 50% of the observed warming, as illustrated in the lower rows of Figure 31.2.
- The most accurate reconstruction is achieved with the TSI-2 solar model, shown in the fourth row of Figure 31.2. In this scenario, the ECS is estimated at 1.1 ± 0.4 °C, suggesting a moderate climate sensitivity to CO_2. This reinforces the hypothesis that solar variability plays a

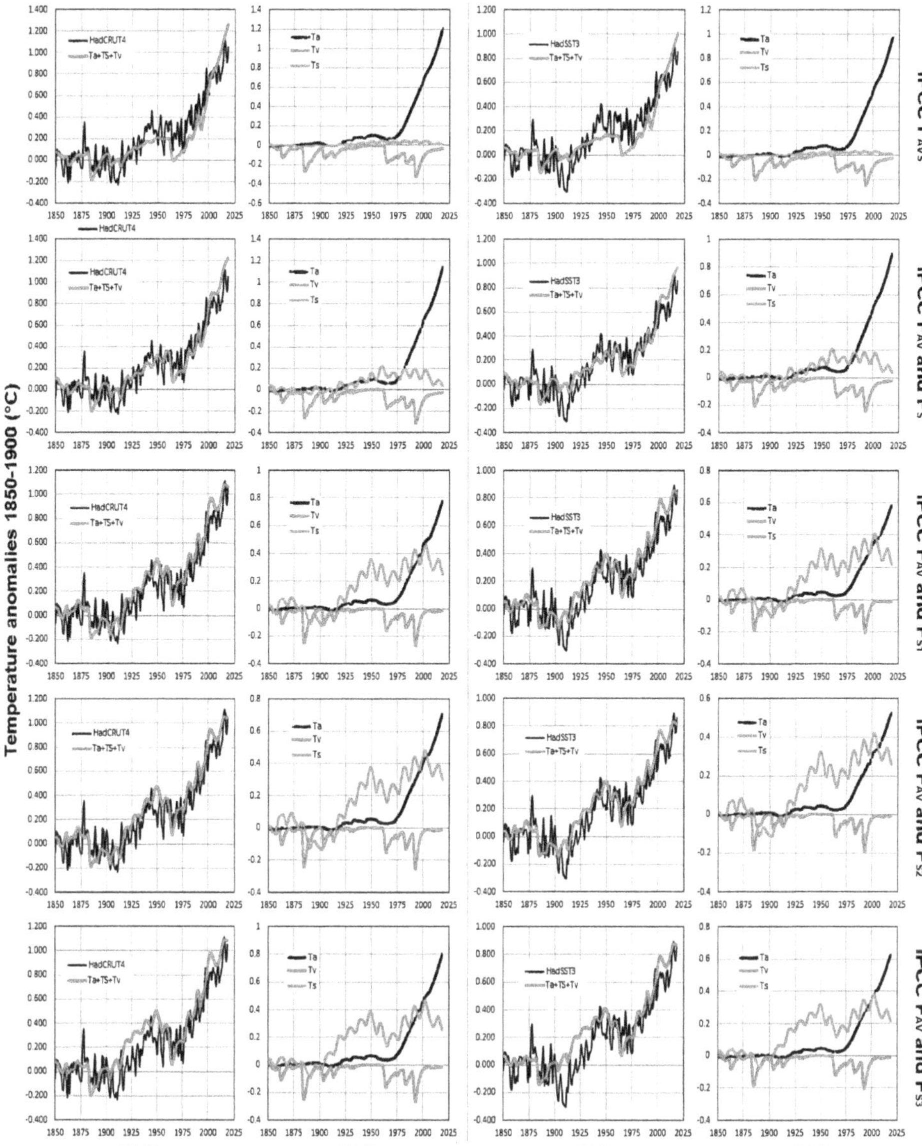

FIGURE 31.2. Results obtained using the proposed EBM with various combinations of datasets and forcing functions. The first row assumes uniform sensitivity ($\alpha_A = \alpha_V = \alpha_S$) and employs the low-variability TSI record from Matthes et al. (2017), consistent with the CMIP6 GCMs. The subsequent rows adopt differentiated sensitivities ($\alpha_A = \alpha_V \neq \alpha_S$) and use alternative high-variability solar proxies developed by (Scafetta, 2023c), as shown in Figure 31.1. The fourth row, using the TSI-2 proxy, yields the most accurate temperature reconstruction. (Adapted from Scafetta, 2023c).

substantial role in climate dynamics, particularly through nonradiative mechanisms.

Model comparison: Figure 31.3.

- Panel A: the CMIP6 ensemble output shows a steadily rising temperature curve dominated by anthropogenic forcing. This simulation fails to reproduce the observed 60-year climate modulation.
- Panel B: the empirical model using TSI-2 successfully reproduces the 60-year oscillation, aligning more closely with the HadCRUT5 trend.

Panels A′ and B′ present the same data detrended using a quadratic function. The correlation coefficients are:

- Panel A′: $r = 0.45$;
- Panel B′: $r = 0.55$.

This comparison indicates that the model in Panel B better captures the multidecadal climate modulation, supporting the hypothesis that incorporating a more variable solar component and differentiated sensitivity improves the predictive skill of climate models.

31.6. Comparing EBMs and GCMs: approaches and performance

The comparison between the proposed semi-empirical energy balance model (ECM) and the CMIP6 global climate models (GCMs) reveals substantial methodological and performance-related divergences:

- The CMIP6 GCMs are highly complex, physically based models incorporating numerous free parameters to simulate known climate processes. Despite their sophistication, many of these models (particularly those with ECS values exceeding 3 °C) tend to overestimate the observed warming and fail to reproduce the multidecadal oscillations evident in historical climate records.
- The proposed ECM, developed by Scafetta (2023c), is a zero-dimensional model lacking spatial resolution. Nevertheless, it demonstrates a superior ability to reproduce multidecadal climate variability and to isolate the contributions of individual climate forcings. An advanced version of the ECM based on the hypothesis $\alpha_A = \alpha_V \neq \alpha_S$ and incorporating more realistic total solar irradiance (TSI) composites could explain a substantial portion of the observed climate variability from 1850 to the present more effectively than the CMIP6 GCMs used by the IPCC (2021).

These findings call into question the idea that solar variability is marginal in modern climate change. By integrating broader solar dynamics and distinguishing responses to different forcings, the ECM enables a reassessment of natural climate drivers.

Rather than replacing the GCMs, the semi-empirical approach complements them, bridging theory and observation and strengthening the interpretive power of climate science.

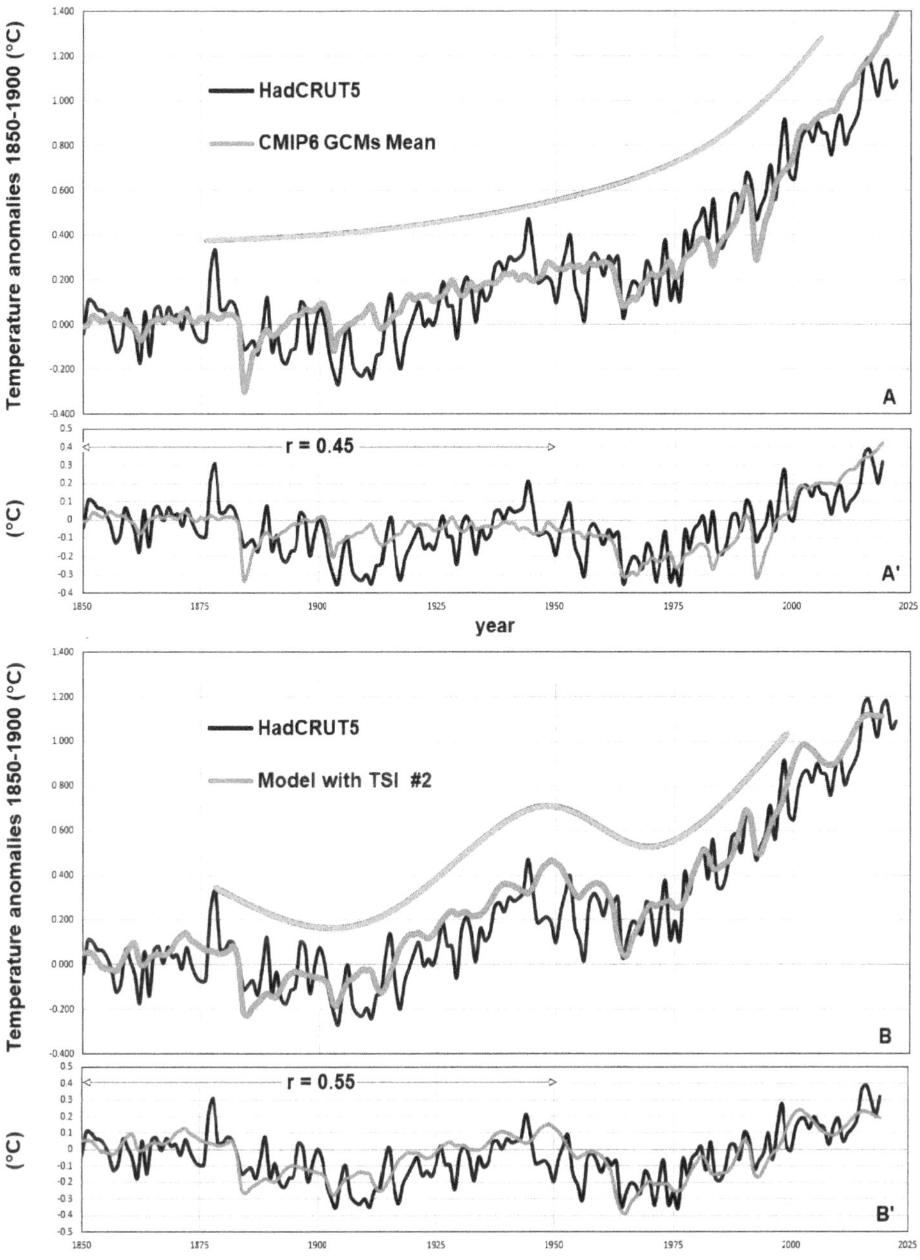

FIGURE 31.3. (A) HadCRUT5 global surface temperature compared with the CMIP6 ensemble mean; (A′) both records detrended using the quadratic function $f(t) = A(x - 1850)^2$. (B) HadCRUT5 compared with the EBM simulation using the TSI-2 proxy; (B′) both records detrended using the same function.
(Adapted from Scafetta, 2023c).

Component	Estimated Contribution	Description
Natural Factors	≥ 50%	Linked to solar activity, astronomical cycles, and oceanic oscillations
Anthropogenic Emissions	~ 30%	Implies a moderate ECS of about 1.1 ± 0.4°C
Local Surface Factors	≤ 20%	Includes the urban heat island effect and artifacts from data homogenization

FIGURE 31.4. Probable attribution of the physical components to global surface warming from 1850–1900 to 2010–2020, as reported in the official global surface temperature records such as HadCRUT5.

31.7. Global warming attribution based on empirical evidence

Based on the analyses presented throughout this chapter, an alternative attribution framework emerges, which differs markedly from the conclusions of the IPCC (2021) summarized in Figure 6.3. Figure 31.4 reports the probable breakdown of the three physical components contributing to the observed warming from 1850 to 2020:

(1) *Natural factors* ($\sim 50\%$), including:
   (a) variations in solar activity, encompassing both irradiance and magnetic effects. The latter likely exerts a dominant influence through corpuscular mechanisms—such as cosmic rays and interplanetary dust—that modulate cloud cover and, consequently, the Earth's albedo. These processes may be linked to astronomical cycles and gravitational resonances.
   (b) multidecadal oceanic oscillations (e.g., AMO and PDO), which redistribute heat and influence regional and global climate variability.

(2) *Anthropogenic emissions* ($\sim$30%), implying an equilibrium climate sensitivity of approximately 1.1 ± 0.4 °C, which is significantly lower than the central estimates of the CMIP6 GCMs, and is effectively aligned with the Planck response ($\sim$1.16 °C; cf. Chapter 6).

(3) *Local non-climatic surface effects* ($\sim$20%), including urban heat island phenomena and methodological artifacts introduced by homogenization techniques in meteorological datasets (e.g., *"blending"*). These factors may generate localized thermal trends that do not represent actual global climate change, yet they remain partially embedded in homogenized datasets, thereby contributing to bias in the published climate records.

Part 6

Spectral Coherence Between Planetary Harmonics and Solar–Climate Cycles

CHAPTER 32

Cosmic timing and Earth's climate cycles

Part VI, the concluding section of this book, presents an innovative theory that offers a physical explanation for the oscillations observed in Earth's climate. It hypothesizes a synchronization with the dynamics of the planetary system, mediated both directly and indirectly by solar activity. This activity manifests itself through a complex hierarchy of oscillations spanning timescales from monthly to multi-millennial.

32.1. Astronomical and climate cycles

The most well-known solar cycle is the approximately eleven-year cycle, first identified around 1850 by Samuel Heinrich Schwabe through systematic observations of sunspot variability (Arlt, 2011).[1] As illustrated in Figure 32.1, these cycles exhibit significant variations in both amplitude and duration, reflecting the intricate and highly nonlinear nature of solar internal dynamics. This variability raises a fundamental question: to what extent is solar activity driven solely by internal processes?

Oscillations in solar activity are crucial not only for understanding the physical mechanisms operating within the Sun, but also for interpreting variability in the Earth's climate system. As discussed in previous chapters, solar and climate records display a notable degree of spectral coherence, suggesting the presence of potential coupling mechanisms between the two systems.

Figure 32.2 presents a reconstruction of global surface temperatures over the past 540 million years (the Phanerozoic), derived from proxy data including oxygen isotopes, chemical elements in ice cores and sediments, and paleoclimate models. The graph is divided into five chronological panels, each reflecting distinct methodological approaches.

- Panel A demonstrates that Earth's climate has consistently undergone significant shifts, with phases much warmer than today, including the four interglacial periods of the past 500,000 years, as well as markedly colder intervals.
- Panel B reveals that this complexity can be described through a series of cyclicities, many of which are of astronomical origin.

Among these, the cycles associated with Earth's orbital variations are particularly prominent. These were first identified by Milutin Milanković between 1920 and 1941, and give rise to the so-called orbital forcings:

- precession of Earth's axis (periodicity of 21,000–26,000 years);

[1] Arlt, R., 2011. The sunspot observations by Samuel Heinrich Schwabe. Astronomische Nachrichten, 332(8), 805–814.

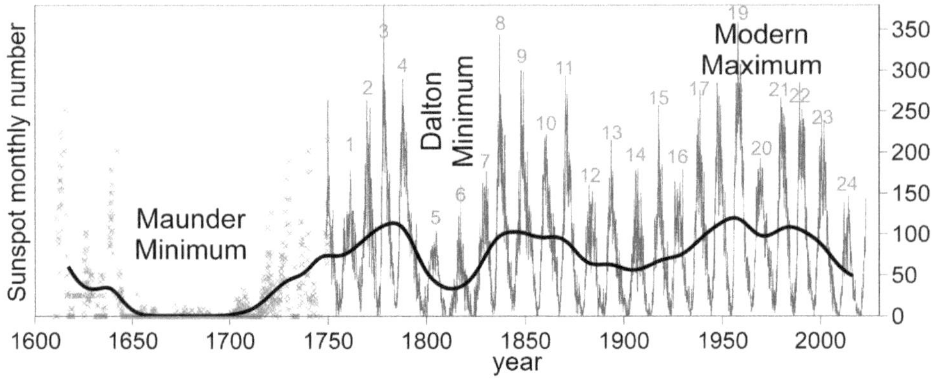

FIGURE 32.1. Sequence of observed sunspot numbers with numbered Schwabe cycles.

- obliquity of Earth's axis (periodicity of 41,000 years);
- eccentricity of Earth's orbit (periodicities of 100,000 and 400,000 years).

These cycles are clearly discernible in paleoclimate data from the past 2–3 million years. Laskar et al. (2004)[2] provided a detailed numerical solution for Earth's orbital evolution spanning ±250 million years. These data are widely used for the astronomical calibration of paleoclimate records and are accessible through the Milankovitch Orbital Data Viewer developed by Colorado State University.[3]

On even longer timescales, cycles spanning tens to hundreds of millions of years emerge, correlated with the motion of the Solar System through the Milky Way galaxy (Shaviv, 2002; Svensmark, 2007).[4] Chapter 26 discusses evidence suggesting that even these distant galactic motions exert a significant influence on Earth's climate system.

Solar–climate coherence in the Holocene. Numerous studies have identified cycles common to both solar and climate records throughout the Holocene on multidecadal, centennial, and millennial timescales (Neff et al., 2001; Kerr, 2001; Ogurtsov et al., 2002; Steinhilber et al., 2012).[5] Scafetta (2010)[6] showed that many major climate oscillations correspond coherently to those predicted

[2] Laskar, J., et al., 2004. A long-term numerical solution for the insolation quantities of the Earth. Astronomy & Astrophysics, 428, 261–285.

[3] Available at: https://biocycle.atmos.colostate.edu/shiny/Milankovitch/

[4] Shaviv, N.J., 2002. Cosmic ray diffusion from the galactic spiral arms, iron meteorites, and a possible climatic connection. Physical Review Letters, 89(5), 051102.
Svensmark, H., 2007. Cosmoclimatology: a new theory emerges. Astronomy & Geophysics, 48(1), 1.18–1.24.

[5] Neff, U., Burns, S.J., Mangini, A., Mudelsee, M., Fleitmann, D., Matter, A., 2001. Strong coherence between solar variability and the monsoon in Oman between 9 and 6 kyr ago. Nature, 411, 290–293.
Kerr, R.A., 2001. A variable sun paces millennial climate. Science, 294(5546), 1431–1433.
Steinhilber, F., Abreu, J.A., Beer, J., Brunner, I., Christl, M., et al., 2012. 9,400 years of cosmic radiation and solar activity from ice cores and tree rings. Proceedings of the National Academy of Sciences, 109(16), 5967–5971.

[6] Scafetta, N., 2010. Empirical evidence for a celestial origin of the climate oscillations and its implications. Journal of Atmospheric and Solar–Terrestrial Physics, 72, 951–970.

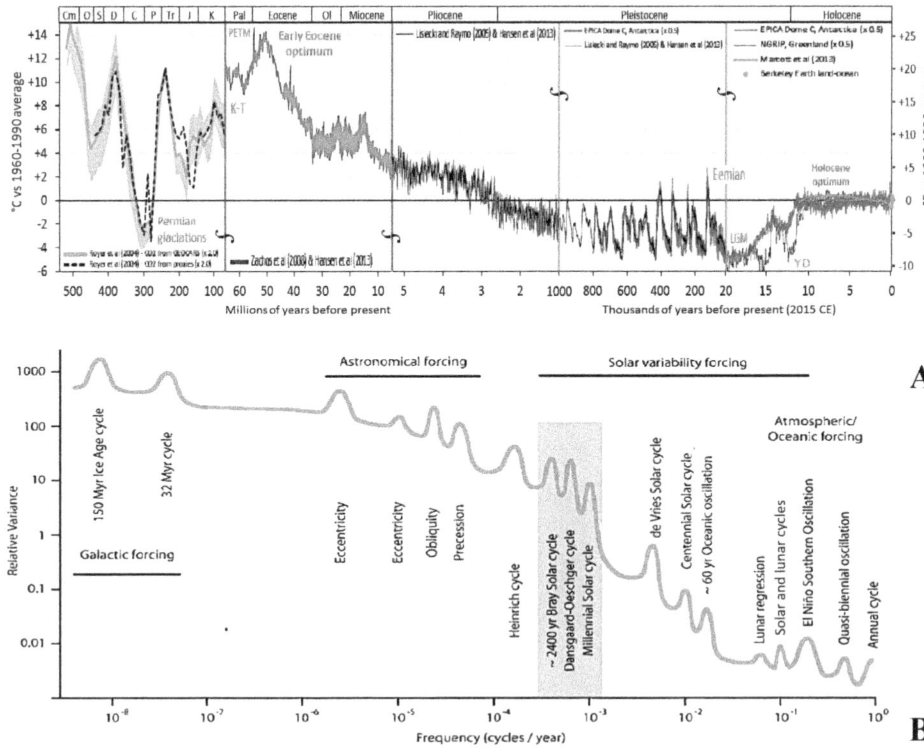

FIGURE 32.2. (A) Composite of global surface temperature records over the past 540 million years (from https://web.archive.org/web/20231210161940/https://gergs.net/all_palaeotemps/).
(B) Schematic periodogram of Earth's climate variance across multiple timescales. (Adapted from Scafetta & Bianchini, 2023).

by the natural dynamics of the planetary system. Scafetta & Bianchini (2022, 2023)[7] review this research by summarizing the most relevant findings.

A central tenet of the proposed theory is that the periodic motion of the planets around the Sun generates gravitational and magnetic forcings varying at specific frequencies that may partially synchronize the solar dynamo. This insight traces back to Rudolf Wolf, the originator of the historical sunspot number series, who in 1859 hypothesized a modulating role for Venus, Earth, Jupiter, and Saturn in generating the Schwabe cycle (Wolf, 1859).[8] The planetary theory ultimately suggests that the Sun and Earth's climate exhibit spectral coherence with the natural oscillations of the planetary system, analogous to ocean tides, which are coherent with the relative motion of the Sun, Moon, and Earth.

[7]Scafetta, N., Bianchini, A., 2022. The planetary theory of solar activity variability: a review. Frontiers in Astronomy and Space Sciences, 9, 937930.
Scafetta, N., Bianchini, A., 2023. Overview of the spectral coherence between planetary resonances and solar and climate oscillations. Climate, 11(4), 77.
[8]Wolf, R., 1859. Extract of a letter from Prof. R. Wolf of Zurich to Mr. Carrington, dated Jan. 12, 1859. Monthly Notices of the Royal Astronomical Society, 19, 85.

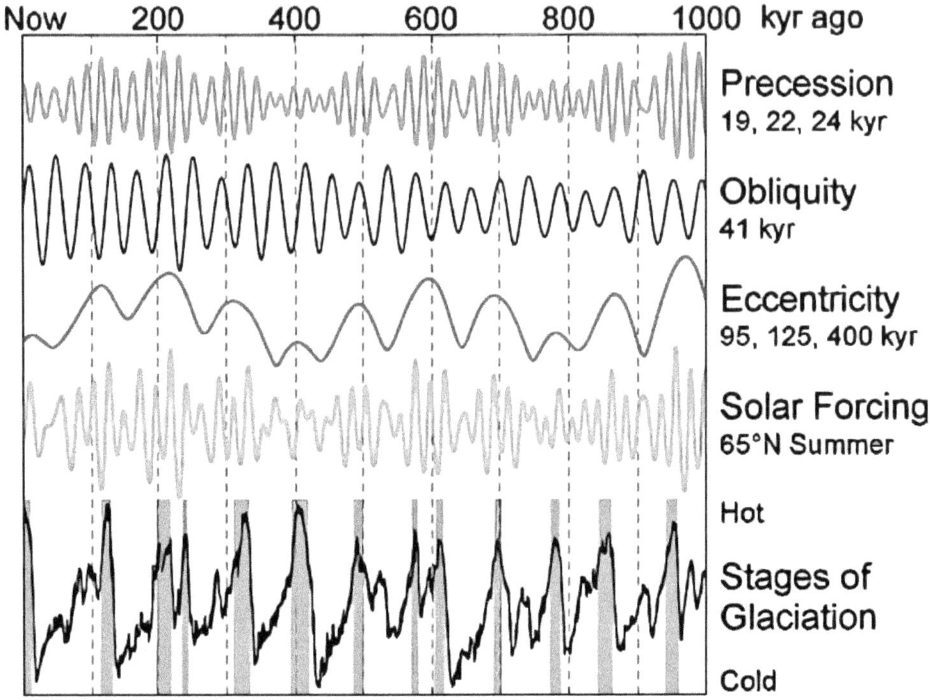

FIGURE 32.3. Milanković orbital cycles. Orbital forcing at 65°N in summer and its correlation with climate oscillations over the past million years.

Rediscovery of the planetary climate theory. Once relegated to the margins of scientific discourse and frequently dismissed as astrology in the pejorative sense, the planetary theory of climate oscillations is now gaining renewed attention. This shift is driven by a growing body of empirical and theoretical evidence that strengthens its credibility and supports its reconsideration within the scientific community. Recent findings encourage a fresh examination of the relationship between astronomical dynamics and Earth's geophysical processes within a more systemic and integrated framework. This perspective moves beyond the traditional divide between space and Earth sciences, and will be explored further in the chapters that follow.

CHAPTER 33

Orbital synchronization and planetary cycles

Since the time of Pythagoras of Samos (circa 570–495 BC) and Johannes Kepler (1571–1630), widely regarded as a founder of modern astronomy, the solar system has been envisioned as a cosmic orchestra: a music of the spheres, a celestial clockwork, and a harmonious dance of planetary bodies.

Contemporary scientific paradigms, however, often downplay the regularities embedded in planetary configurations and motions, favoring reductionist models dominated by stochastic processes and instability. A closer examination, however, reveals an architecture of resonances, proportions, and cycles—a harmonic mechanism that appears to follow an underlying order.

33.1. From astrology to modern science

For centuries, astrology was regarded as the science of the stars par excellence, not merely a divinatory practice but a framework for interpreting cosmic influences on Earth and human affairs. In antiquity, the boundary between astrology and astronomy was fluid; both disciplines were grounded in the belief that celestial bodies, through their cyclical and harmonic motions, exerted measurable effects on the sublunar world.

Claudius Ptolemy (circa 100–168 AD), in his seminal works *Almagest* and *Tetrabiblos*, developed models in which planetary motions not only described the heavens but also corresponded to terrestrial and human phenomena. The influence of lunar tides and the Sun's role in daily and seasonal cycles offered compelling evidence that Earth responds to celestial forces.

Major planetary configurations—especially the conjunctions of Jupiter and Saturn—were historically associated with cycles of approximately 20, 60, and 900–1000 years. These periodicities were linked to the rise and fall of empires, religions, and civilizations, as argued by the Persian scholar Abu Ma'shar al-Balkhi.[1] His treatise *De Magnis Coniunctionibus* (*"On the Great Conjunctions"*) became a cornerstone of medieval astrological thought.

Traditional calendars reflect this cosmological understanding, often being structured around solar, planetary, and lunar cycles. Notably, the Brihaspati cycle in the Hindu-Vedic tradition and the Chinese sexagenary cycle are both based on a 60-year periodicity, corresponding to five revolutions of Jupiter and two of Saturn around the Sun.

The transition to modern science. With the rise of modern science, the astrological paradigm underwent critical reexamination. In the 17th century,

[1]al-Balkhi, Abu Ma'shar (9th century). De Magnis Coniunctionibus, ed. K. Yamamoto, Ch. Burnett, Leiden, 2000, 2 vols. (Arabic and Latin text).

figures such as Kepler—in *De Fundamentis Astrologiae Certioribus* (1601)[2]—and Isaac Newton (1642–1727) contributed to the formal separation of astronomy, understood as the mathematical and physical study of celestial phenomena, from the superstitious aspects of astrology.

In fact, the distinction between science and astrology was not as absolute as is often portrayed. Kepler, who served as imperial astrologer and mathematician from 1601 to 1628 under three Holy Roman Emperors, criticized popular and fatalistic astrology but maintained that celestial configurations could plausibly influence natural phenomena such as weather and climate, and consequently, human societies.

In *Tertius Interveniens* (1610), Kepler wrote:

> *"Astrology should not be completely rejected as mere superstition. That would be like throwing the baby out with the bathwater."*

With this statement, Kepler defended a rational core within astrology, distinguishing its empirical potential from its distorted and superstitious popular forms. He argued that certain astrological principles were compatible with a scientific worldview, and that the error lay in absolutizing or misapplying them without rigorous study.

Newton, while laying the foundations of celestial mechanics, likewise did not dismiss the possibility of planetary forces acting at a distance. On the contrary, his Law of Universal Gravitation was predicated on the premise of invisible yet measurable interactions between bodies across space.

Astrology and science: a shared frontier? Today, certain principles once rooted in astrological tradition have either become scientifically established—such as tidal and seasonal dynamics—or are being reexamined within frontier scientific research, including planetary synchronization and the potential influence of orbital resonances. Contemporary studies investigate whether gravitational and magnetic interactions among planets may modulate solar activity and, consequently, affect Earth's climate variability.

This physical-mathematical recovery of ancient insights reveals that the boundary between science and astrology is historically more nuanced than commonly assumed. Questions once posed within an astrological framework now find rigorous expression in the language of physics, celestial mechanics, and dynamical-systems theory.

33.2. The *"music of the spheres"*: from Pythagoras to Kepler

The concept of orbital resonance has long been interpreted through the philosophical lens of *Musica Universalis*, or the *"harmony of the spheres,"* a notion originating with Pythagoras and his disciples. The Pythagoreans observed that the numerical ratios governing musical intervals—2:1 (octave), 3:2 (perfect fifth), 4:3 (perfect fourth), and 1:1 (unison)—appeared to mirror the proportions underlying the motions of celestial bodies. According to Philolaus (470–385 BC), each celestial body emitted a sound based on its orbital revolution, producing a cosmic harmony audible only to the *"inner ear."*

---

[2]Kepler, J., 1601. On the more certain fundamentals of astrology. In: Brackenridge, J.B., Rossi, M.A. (Eds.), Proceedings of the American Philosophical Society, 123(2), 85–116 (1979).

This vision was revitalized and expanded by Johannes Kepler in foundational works such as *Mysterium Cosmographicum* (1596), *Astronomia Nova* (1609), and especially *Harmonices Mundi* (1619). In the latter, Kepler correlated musical tones with planetary orbital periods, distances from the Sun, and angular velocities. Drawing inspiration from Renaissance polyphony, he conceived of a celestial choir (Mars sang tenor, Saturn and Jupiter the bass, Mercury the soprano, and Venus and Earth the altos), in which the variation in Earth's angular velocity between aphelion and perihelion corresponded to a semitone with a ratio of 16:15.

These insights were instrumental in leading Kepler to formulate his Third Law of planetary motion:

$$a^3 = cT^2 \quad \left[\Longrightarrow a^3 = \frac{4\pi^2}{G(M_\odot + m)}T^2 \approx \frac{4\pi^2}{GM_\odot}T^2\right].$$

The quantities are:

- a: semi-major axis of the planetary orbit;
- c: solar system constant, which simplifies to 1 when a is measured in astronomical units (AU) and T in years, yielding $T \approx a^{3/2}$.
- T: orbital period;
- G: universal gravitational constant;
- $M_\odot$: mass of the Sun;
- m: mass of the planet (negligible compared to $M_\odot$);

The Third Law was later mathematically demonstrated by Isaac Newton through his Law of Universal Gravitation, confirming the profound connection between celestial harmony and physical dynamics.

33.3. Orbits, proportions, and commensurability

Many planetary bodies in the solar system exhibit orbital resonances, in which the ratios between their revolution periods are simple rational numbers. These relationships are not coincidental but arise from gravitational interactions that, in some cases, stabilize orbits over long timescales (Scafetta, 2014c).[3]

The Great Conjunctions of Jupiter and Saturn: 20-year, 60-year, and millennial cycles. Conjunctions between Jupiter and Saturn—the *Great Conjunctions*—occur approximately every 20 years. This cycle has been known since antiquity. In *De Stella Nova* (1609), Kepler showed that these conjunctions repeat in a triangular pattern along the zodiac, forming the so-called *Trigon* (see Figure 33.1). Every three conjunctions (roughly every 60 years), they return near the same zodiacal location, revealing a broader periodicity.

The Trigon itself slowly rotates through the zodiac, marking epochs associated with the four classical elements:

- Fire—Aries, Sagittarius, and Leo;
- Earth—Taurus, Capricorn, and Virgo;
- Air—Gemini, Aquarius, and Libra;
- Water—Cancer, Pisces, and Scorpio.

[3]Scafetta, N., 2014c. The complex planetary synchronization structure of the solar system. Pattern Recognition in Physics, 2, 1–19.

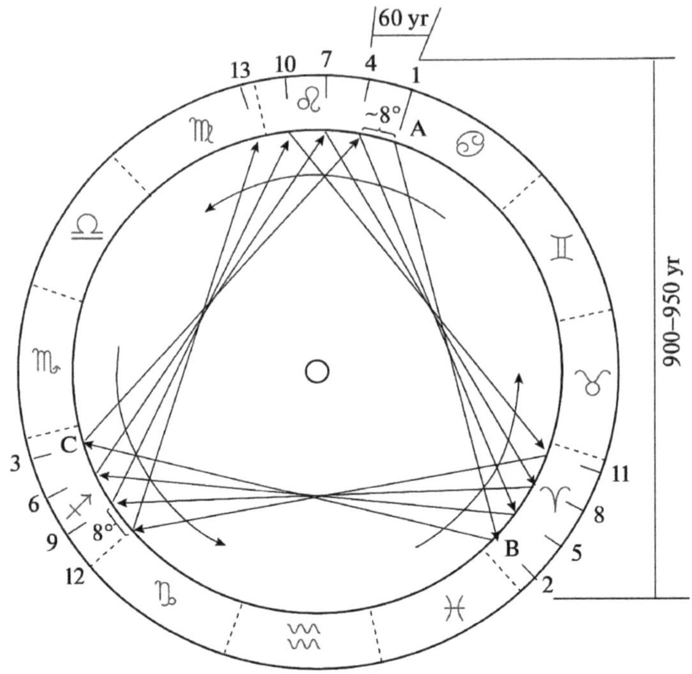

FIGURE 33.1. Schematic representation of 13 consecutive conjunctions between Jupiter and Saturn relative to the 12 zodiac constellations. Each conjunction occurs approximately every 20 years. Every 3 conjunctions—roughly every 60 years—the two planets meet near the same constellation, forming a *Trigon of the Great Conjunctions*. The Trigon slowly rotates by nearly 8° every 60 years, giving rise to the *Great Inequality* cycle of Jupiter and Saturn, which lasts approximately 900–950 years. (Adapted from Scafetta et al., 2019b).

Ancient and medieval astrologers, including Ptolemy and Abu Ma'shar al-Balkhi (IX secolo), estimated a complete Trigon cycle at approximately 960 years, with transitions between elemental trigons occurring every 240 years. The first three elemental triplets were considered auspicious, whereas water signs were often associated with instability and calamity.

The 1345 conjunction in Pisces was interpreted by Jean de Muris and scholars at the University of Paris as a harbinger of misfortune. Shortly thereafter, Europe was ravaged by the *Black Death* (1347–1351), followed by the *Hundred Years' War*, economic collapse, epidemics, and severe winters—events that many contemporaries regarded as confirming the astrological prognosis. This period coincided with the onset of the Little Ice Age, which persisted for centuries.

In *De Stella Nova in Pede Serpentarii*, Kepler (1606)[4] linked the *Great Inequality* between Jupiter and Saturn to quasi-millennial cycles of civilizational rise and decline. He hypothesized that a new phase of development, at least for Europe, could begin between the 17th and 21st centuries. This view was later echoed by

[4]Kepler, J., 1606. De Stella Nova in Pede Serpentarii (On the new star in Ophiuchus's foot).

Newton (1704),[5] who placed the peak of this cycle around 2060, on the basis of biblical chronology.

It was Laplace who ultimately provided the mathematical explanation for the *Great Inequality*, identifying it as a slow oscillation in the motions of Jupiter and Saturn caused by gravitational resonance. This oscillation gives rise to variations in their orbital periods and the progressive rotation of the *Trigon*. The cycle lasts approximately 900–1000 years, and its demonstration constituted one of the earliest successes of Newtonian mechanics applied to planetary systems (Lovett, 1895; Etz, 2000).[6]

Laplace resonance in Jupiter's moons: Io, Europa, and Ganymede. A striking example of resonance is found among Jupiter's Galilean moons. For every orbit completed by Ganymede, Europa completes 2, and Io completes 4. This near-perfect 1:2:4 resonance ensures long-term gravitational stability and regulates phenomena such as internal tidal heating and Io's volcanic activity.

Kirkwood gaps and resonances with Jupiter. In the asteroid belt between Mars and Jupiter, the Kirkwood gaps are regions devoid of asteroids that correspond to destabilizing resonances with Jupiter at ratios such as 4:1, 3:1, 5:2, 7:3, and 2:1. These gaps are produced by gravitational perturbations that eject asteroids from resonant orbits (Moons & Morbidelli, 1995; Moons et al., 1998).[7]

The case of TRAPPIST-1: orbital harmonies in another solar system. Planetary resonances are not unique to our solar system. The TRAPPIST-1 system, composed of a red dwarf and seven terrestrial planets (b–h), exhibits one of the most intricate and stable resonance chains ever observed:

- orbital periods: 1.511, 2.422, 4.049, 6.101, 9.207, 12.352, 18.773 days;
- ratios: 8:5, 5:3, 3:2, 3:2, 4:3, 3:2;
- mean relative error: only 0.6%.

This structure is so elegant that it has even inspired musical interpretations (Agol et al., 2021; Gillon et al., 2017; Luger et al., 2017).[8]

Venus–Earth resonance. One of the most visually and mathematically captivating examples of planetary synchronization is the apparent resonance between Venus and Earth, depicted in Figure 33.2. Every 584 days, an inferior

[5]Newton, I., 1704. Manuscript letter containing prophetic calculations based on the Book of Daniel. Newton Archive, Hebrew University of Jerusalem.

[6]Lovett, E.O., 1895. The Great Inequality of Jupiter and Saturn. The Astronomical Journal, 15, 113–127.
Etz, D.V., 2000. Conjunctions of Jupiter and Saturn. Journal of the Royal Astronomical Society of Canada, 94, 174–178.

[7]Moons, M., Morbidelli, A., 1995. Secular resonances in mean motion commensurabilities: the 4/1, 3/1, 5/2, and 7/3 cases. Icarus, 114, 33–50.
Moons, M., Morbidelli, A., Migliorini, F., 1998. Dynamical structure of the 2/1 commensurability with Jupiter and the origin of the resonant asteroids. Icarus, 135, 458–468.

[8]Agol, E., Dorn, C., Grimm, S.L., Turbet, M., Ducrot, E., et al., 2021. Refining the transit-timing and photometric analysis of TRAPPIST-1. The Planetary Science Journal, 2, 1.
Gillon, M., Triaud, A.H.M.J., Demory, B.-O., Jehin, E., Agol, E., et al., 2017. Seven temperate terrestrial planets around the nearby ultracool dwarf star TRAPPIST-1. Nature, 542(7642), 456–460.
Luger, R., Sestovic, M., Kruse, E., 2017. A Seven-Planet Resonant Chain in TRAPPIST-1. Nature Astronomy, 1, 0129.

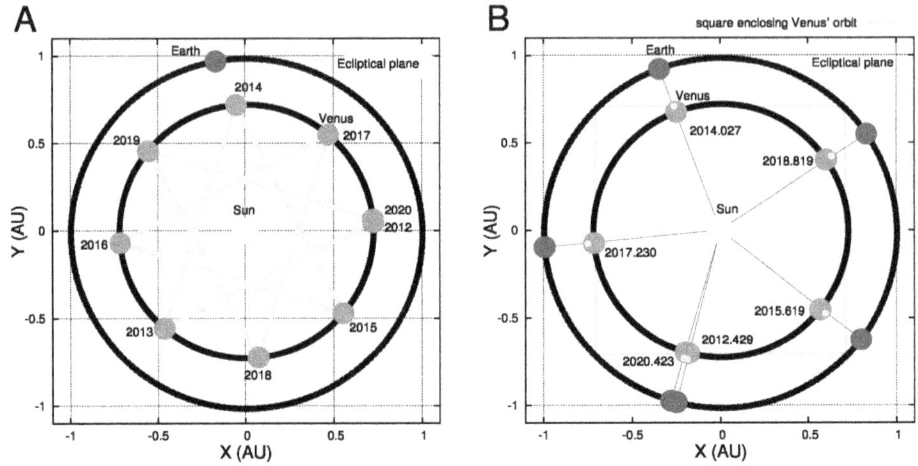

FIGURE 33.2. [A] Orbits of Earth and Venus, together with their positions on January 1st from 2012 to 2020. The Earth–Venus configuration nearly repeats every 8 years. [B] Earth–Venus inferior conjunctions between 2012 and 2020. During each conjunction, the same hemisphere of Venus (indicated by the small dot) faces Earth. (Scafetta, 2014c).

conjunction occurs (closest approach), and after five such conjunctions—roughly every eight years—a regular pentagonal pattern is traced in the sky, known since antiquity as the *Rose of Venus* or *Celestial Pentagram*.

The symmetry extends further: Venus's retrograde rotation appears synchronized with Earth's orbital motion because:

- Sidereal Venusian day: -243 Earth days (retrograde rotation);
- Sidereal Venusian year: 224.7 Earth days.

At each inferior conjunction, the same hemisphere of Venus faces Earth because, in one synodic cycle (584 days):

- Venus completes $584/243 \approx 2.4$ retrograde rotations;
- Earth completes $584/365.25 \approx 1.6$ revolutions;
- together: $2.4 + 1.6 = 4$, an integer, enabling alignment.

Some studies (e.g. Gold & Soter, 1969)[9] hypothesize that Earth's weak tidal potential may have gradually synchronized Venus's rotation over billions of years—a rare example of planet–planet tidal locking, analogous to the Moon's synchronous rotation with Earth, although others regard it as a mere coincidence.

33.4. The Titius–Bode law: coincidence or structure?

In 1766, Johann Daniel Titius[10] proposed an empirical formula to describe the distances of the planets from the Sun:

[9]Gold, T., Soter, S., 1969. Atmospheric tides and the resonant rotation of Venus. Icarus, 11(3), 356–366.

[10]Titius, J.D., 1766. Anmerkungen über die Abstände der Planeten. In C. Bonnet, Betrachtungen

$$a_n = 0.4 + 0.3 \cdot 2^n \qquad \text{with } n = -\infty, 0, 1, 2, 3, \ldots$$

Here, a_n represents the distance in astronomical units (AU), where 1 AU is the mean Earth–Sun distance.

Popularized by Johann Elert Bode in 1772, the formula proved remarkably predictive for the planetary orbits known at the time. It even anticipated the location of a missing planet between Mars and Jupiter, leading to the discovery of Ceres in 1801. Uranus, discovered in 1781, also fit the pattern. However, Neptune (1846) and Pluto (1930) showed significant deviations, prompting many critics to dismiss the law as a numerical coincidence devoid of any physical foundation.

Yet, in 1981, the Italian physicist Giovanni P. Gallavotti (1981)[11] demonstrated that similar sequences can emerge from resonant stability conditions in nonlinear dynamical systems, using tools from chaos theory and Hamiltonian mechanics. Subsequent studies by Pletser (2017)[12] and Scardigli (2003)[13] proposed that Titius–Bode-type relations may reflect emergent quantum-like behaviors in complex gravitational systems. These hypotheses suggest that what once appeared to be a curious numerical artifact may, in fact, conceal a deeper structural principle still awaiting comprehensive theoretical clarification.

33.5. The mirror symmetry rule of the *Asteroid Belt*

Following the discovery of Ceres, astronomers noted that numerous small bodies orbit between Mars and Jupiter, forming what is now known as the Asteroid Belt. No major planet emerged in this region, likely owing to Jupiter's gravitational perturbations, which disrupted accretion processes. Ceres, with a diameter of approximately 1000 km, remains the largest object in the belt and is classified as the only dwarf planet in the inner solar system.

In 1983, Geddes & King-Hele[14] identified a striking mathematical symmetry between the inner planets (Mercury, Venus, Earth, Mars) and the outer planets (Jupiter, Saturn, Uranus, Neptune), arranged as if mirrored across the asteroid belt.

The proposed pairings were:

Mercury ↔ Neptune; Earth ↔ Saturn
Venus ↔ Uranus; Mars ↔ Jupiter

Using the semi-major axes of the planetary orbits, the researchers discovered nearly constant relationships between the products of paired orbits, expressible as ratios between musical notes within an octave. The fundamental harmonic constant employed was:

über die Natur (Übers. J. D. Titius). Leipzig: Weidmanns Erben und Reich.

[11]Gallavotti, G., 1981. Perturbation theory for classical Hamiltonian systems. Springer Lecture Notes in Physics, 93, 32–93.

[12]Pletser, V., 2017. Generalized Titius–Bode relations for the exoplanets of main sequence stars. Astrophysics and Space Science, 362(11), 203.

[13]Scardigli, F., 2003. A quantum-like description of the planetary systems. Foundations of Physics Letters, 16(4), 393–398.

[14]Geddes, A.B., King-Hele, D.G., 1983. Equations for mirror symmetries among the distances of the planets. Quarterly Journal of the Royal Astronomical Society, 24, 10–13.

$$b = 2^{1/8} \approx 1.09$$

This corresponds to the frequency ratio between adjacent notes in an equal-tempered musical scale. Some numerical examples include:

$$\begin{aligned} Me \times Ne &= 1.214 \cdot Ea \times Sa \\ Ve \times Ur &= 1.194 \cdot Me \times Ne \\ Ea \times Sa &= 1.208 \cdot Ma \times Ju \end{aligned}$$

where $b^2 = 2^{2/8} \approx 1.19$. Other relationships involve higher powers of the musical constant:

$$\begin{aligned} Ve \times Ma &= 2.847 \cdot Me \times Ea \\ Sa \times Ne &= 2.881 \cdot Ju \times Ur \end{aligned}$$

where $b^{12} = 2^{12/8} \approx 2.83$.

These relationships exhibit consistent internal symmetries:

$$\frac{Me \times Ne}{Ea \times Sa} \approx \frac{Ve \times Ur}{Me \times Ne} \approx \frac{Ea \times Sa}{Ma \times Ju}$$

and

$$\frac{Me \times Ea}{Ve \times Ma} \approx \frac{Ju \times Ur}{Sa \times Ne}.$$

These analogies suggest the existence of an ordered and symmetric structure within the solar system, with the asteroid belt acting as a reflective boundary between the inner and outer planetary regions.

Geddes & King-Hele (1983) concluded:

> *"The significance of these near-equalities is difficult to assess. Skeptics may dismiss them as mere numerical games; but those with eyes to see and ears to hear may glimpse something more deeply woven into the fact that the average interval between musical notes emerges as the only necessary numerical constant—a result that surely would have delighted Kepler."*

33.6. The approximate matrix of planetary resonances

In a series of pioneering studies, the Russian mathematician Molchanov (1968, 1969)[15] demonstrated that planetary orbital periods can be approximated by a system of linear equations with integer coefficients, revealing quasi-harmonic resonance relationships among planetary bodies.

The system of equations. Molchanov's model is based on the orbital frequency $\omega = T^{-1}$, where T is the orbital period. The system of equations is expressed as:

[15] Molchanov, A.M., 1968. The resonant structure of the solar system: The law of planetary distances. Icarus, 8, 203–215.
Molchanov, A.M., 1969. The reality of resonances in the solar system. Icarus, 11, 104–110.

$$\mathbf{A}\cdot\boldsymbol{\omega} = \begin{pmatrix} 1 & -1 & -2 & -1 & 0 & 0 & 0 & 0 & 0 \\ 0 & 1 & 0 & -3 & 0 & -1 & 0 & 0 & 0 \\ 0 & 0 & 1 & -2 & 1 & -1 & 1 & 0 & 0 \\ 0 & 0 & 0 & 1 & -6 & 0 & -2 & 0 & 0 \\ 0 & 0 & 0 & 0 & 2 & -5 & 0 & 0 & 0 \\ 0 & 0 & 0 & 0 & 1 & 0 & -7 & 0 & 0 \\ 0 & 0 & 0 & 0 & 0 & 0 & 1 & -2 & 0 \\ 0 & 0 & 0 & 0 & 0 & 0 & 1 & 0 & -3 \end{pmatrix} \begin{pmatrix} \omega_{Me} \\ \omega_{Ve} \\ \omega_{Ea} \\ \omega_{Ma} \\ \omega_{Ju} \\ \omega_{Sa} \\ \omega_{Ur} \\ \omega_{Ne} \\ \omega_{Pl} \end{pmatrix} = \begin{pmatrix} 0 \\ 0 \\ 0 \\ 0 \\ 0 \\ 0 \\ 0 \\ 0 \end{pmatrix}$$

where $\mathbf{A}$ is an 8×9 matrix of integer coefficients and $\boldsymbol{\omega}$ is the vector of planetary frequencies from Mercury to Pluto. By convention, Earth's orbital period is set to 1 year ($T_{\text{Ea}} = 1$), yielding the following calculated values:

period				*calculated*	*observed*	*error*
T_{Me}	=	2484/10332	=	0.240	0.241	(0.4%)
T_{Ve}	=	2484/4044	=	0.614	0.615	(0.2%)
T_{Ea}	=	1	=	1.000	1.000	(0.0%)
T_{Ma}	=	2484/1320	=	1.880	1.880	(0.0%)
T_{Ju}	=	2484/210	=	11.83	11.86	(0.3%)
T_{Sa}	=	2484/84	=	29.57	29.46	(0.4%)
T_{Ur}	=	2484/30	=	82.80	84.01	(1.4%)
T_{Ne}	=	2484/15	=	165.6	164.8	(0.5%)
T_{Pl}	=	2484/10	=	248.4	248.1	(0.1%)

Despite its simplicity, the model achieves remarkable accuracy, with all errors below 1.5%. Molchanov interpreted this coherence as evidence of a deep dynamical and harmonic organization within the solar system, suggesting that orbital frequencies are interwoven by a principle of linear resonance, resembling a cosmic symphony governed by ratios of simple integers.

33.7. The "*music of the spheres*" equation by Bank & Scafetta

Building on Geddes & King-Hele (1983), Bank & Scafetta (2022)[16] identified an additional layer of harmonic structure in the solar system, advancing and refining the concept of mirror symmetries among planetary orbital radii.

Bank & Scafetta (2022) demonstrated that this structure can be quantified through a scalar–mirror relationship based on the powers $(a_{planet})^{2/3}$, where a_{planet} is the semi-major axis of each planet or region. Notably, this exponent is the inverse of the one used in Kepler's Third Law, in which the period depends on $(a_{planet})^{3/2}$.

The resulting harmonic equation is the following:

$$\frac{1}{64}\left(\frac{a_{\text{Er}}}{a_{\text{Sz}}}\right)^{2/3} \approx \frac{1}{32}\left(\frac{a_{\text{Pl}}}{a_{\text{Vu}}}\right)^{2/3} \approx \frac{1}{16}\left(\frac{a_{\text{Ne}}}{a_{\text{Me}}}\right)^{2/3} \approx \frac{1}{8}\left(\frac{a_{\text{Ur}}}{a_{\text{Ve}}}\right)^{2/3}$$
$$\approx \frac{1}{4}\left(\frac{a_{\text{Sa}}}{a_{\text{Ea}}}\right)^{2/3} \approx \frac{1}{2}\left(\frac{a_{\text{Ju}}}{a_{\text{Ma}}}\right)^{2/3} \approx \left(\frac{a_{7:3}}{a_{3:1}}\right)^{2/3} \approx \frac{9}{8}$$

[16]Bank, M.J., Scafetta, N., 2022. Scaling, Mirror Symmetries and Musical Consonances Among the

where:

- a_{Sz}: mean radius of the scattered zone within the solar system;
- a_{Vu}: mean radius of the Vulcanoid region, a hypothetical asteroid belt between 0.06 and 0.21 AU;
- $a_{7:3}$ and $a_{3:1}$: semi-major axes associated with the asteroid resonances with Jupiter (7:3 and 3:1, respectively);
- $a_{Er}, a_{Pl}, a_{Ne}, a_{Ur}, a_{Sa}, a_{Ju}, a_{Ma}, a_{Ea}, a_{Ve}, a_{Me}$: semi-major axes of the planets Eris, Pluto, Neptune, Uranus, Saturn, Jupiter, Mars, Earth, Venus, and Mercury.

A hidden harmony. The deviations from the observed values are less than 1%, indicating a high-precision geometric regularity. The numerical sequence on the left (1/64, 1/32, 1/16, ...) forms a binary scale reminiscent of musical octave intervals. The final ratio, 9/8, corresponds to the Pythagorean *epogdoon*, or whole tone (e.g., C to D), a foundational interval in ancient musical theory and a symbol of harmony in Renaissance art, notably illustrated in Raphael's *School of Athens*.

A "fourth law" of planetary motion? The model proposed by Bank & Scafetta (2022), illustrated in Figure 33.3, provides strong empirical support for the harmonic ordering of the solar system and reinforces the analogy with the *Musica Universalis* envisioned by the Pythagoreans. Just as musical notes form a coherent scale, planetary orbits appear to be arranged according to proportional laws, suggesting that the solar system may be governed by underlying mathematical and musical principles.

This model proposes a dual symmetry—mirror and scalar—that links the inner and outer planets through precise harmonic ratios. It adds to the growing body of evidence that planetary dynamics are shaped not only by chaotic perturbations but also by deep geometric constraints, possibly rooted in the formative processes of the protoplanetary disk or in gravitational self-organization.

One might imagine that Kepler, had he intuited this structure, could have proposed it as a Fourth Law of planetary motion, complementing his three canonical laws. Such a law might have enabled predictions of the orbits of Uranus and Neptune from those of Mercury and Venus alone.

33.8. Harmony as a structural physical principle

Harmony, far from being a poetic metaphor, emerges as a genuine structural principle of the solar system. The coherence among planetary orbits, musical intervals, and rotational synchronizations suggests that gravitation acts not only as a force, but also as a rhythmic and geometric organizing principle.

This perspective, situated at the intersection of science and philosophy, offers new interpretive frameworks for understanding the nonlinear behavior of the Sun and its potential influence on Earth's climate. As Pythagoras and Kepler intuited, the universe may indeed resonate.

On a broader scale, synchronization structures emerge from planetary conjunctions and synodic cycles, generating beat patterns that appear to correlate with solar variability and terrestrial climate fluctuations. The hypothesis is that

Distances of the Planets of the Solar System. Frontiers in Astronomy and Space Sciences, 8, 758184.

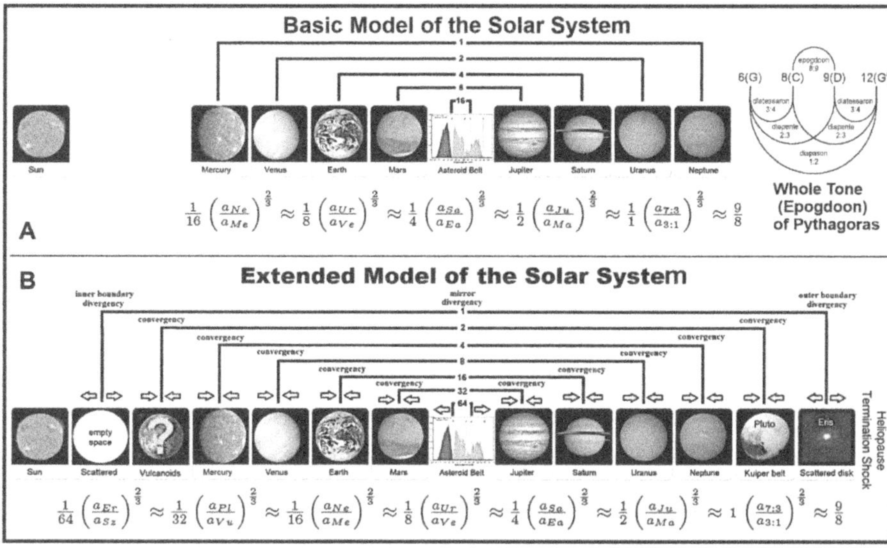

Predictive Accuracy of the Equation

val.	*1.125	*1.125	1.137	1.113	1.128	1.134	1.118	9/8 = 1.125
acc.	(100%)	(100%)	(101%)	(99%)	(100%)	(101%)	(99%)	

FIGURE 33.3. Graphical representation of the *"music of the spheres"* equation, developed by Bank & Scafetta (2022), illustrating the proportions and mirror symmetries among the semi-major axes of the eight planets, including the central region of the asteroid belt. The model builds upon the equations of Geddes–King–Hele and achieves an accuracy of approximately 99%. The ratio sequences derive from the Pythagorean *epogdoon*—the 9/8 interval ratio, which in music corresponds to a whole tone—and its cumulative expression across six octaves. The inset at the top right depicts Pythagoras's musical theory.

interactions among planetary motions, particularly those of the giant planets, may induce tidal variations in the Sun and generate gravitational oscillations. Planetary harmonics are embedded, for example, in the Sun's wobble around the solar system's barycenter. This complex dynamic, commonly referred to as Solar Inertial Motion (SIM), exhibits prominent 20- and 60-year cycles, as shown in Figure 33.4. These cycles are linked to the Trigon pattern formed by the Great Conjunctions of Jupiter and Saturn.

This dynamic could modulate solar activity cycles and influence the propagation of corpuscular currents throughout the heliosphere, ultimately interacting with Earth's magnetosphere and climate system.

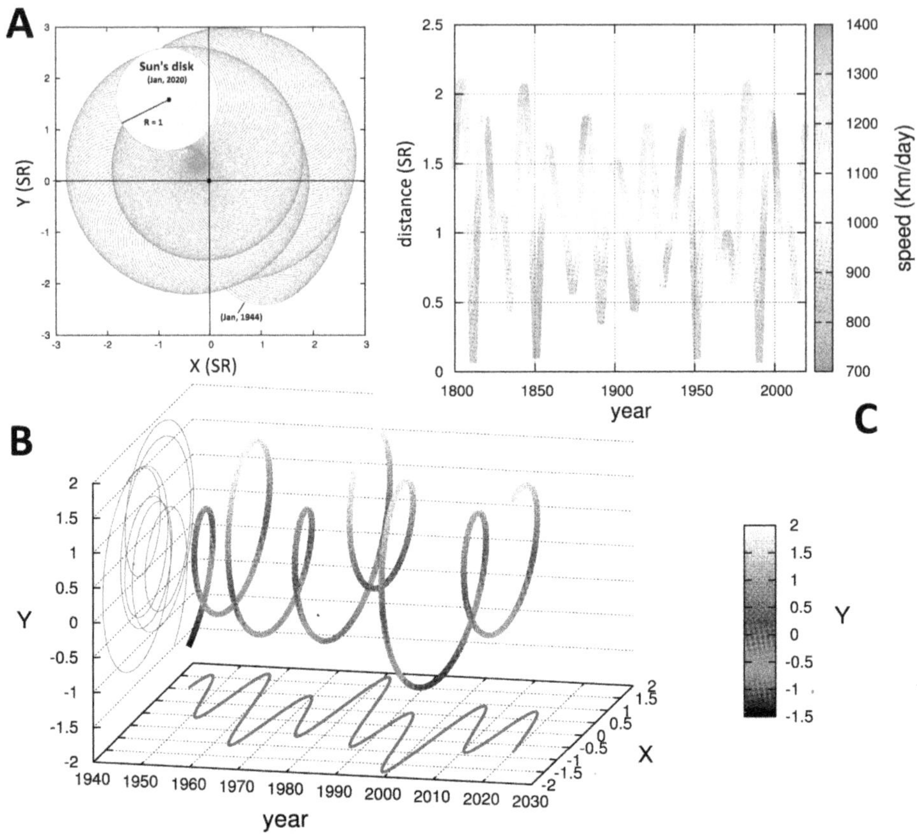

FIGURE 33.4. [A–B] Solar Inertial Motion (SIM) relative to the solar system's barycenter, as viewed along the Z-axis, perpendicular to the ecliptic plane. [C] Temporal variation in the Sun's distance and velocity from 1800 to 2020. A prominent ~20-year oscillation is visible, along with subtler 60-year and 179-year cycles. Solar coordinates were computed using NASA's JPL HORIZONS Ephemeris system, expressed in solar radii (SR). (Adapted from Scafetta, 2014c).

CHAPTER 34

Solar cycles shorter than the Schwabe 11-year cycle

34.1. Introduction to the planetary theory of solar activity variation

Chapters 34–39 are devoted to presenting the empirical and theoretical foundations of the planetary theory of solar activity variations. It will be shown that solar activity exhibits robust signatures of planetary cycles across all timescales, from monthly to multi-millennial. Proposed physical mechanisms are also discussed, together with responses to critiques that have appeared in the literature.

A recurring objection is that the planets are too small and too distant from the Sun to exert any effective influence. This argument, however, neglects the possibility of amplification mechanisms. Moreover, many critiques focus exclusively on the orbital period of Jupiter (11.86 years) and argue that, although this cycle sometimes appears correlated with the sunspot cycle, it is too long to match the $\sim$11-year solar cycle. Such objections can be dismissed because they fail to take into account the complexity of planetary synchronization within the solar system as a whole, which actually points toward a variable $\sim$11-year cycle (cf. Scafetta, 2014b, 2023e).[1]

For example, Okal & Anderson (1975)[2] inferred a mismatch with the solar cycle by adopting a tidal model that excluded Saturn. As a result, they identified only the 11.86-year Jupiter cycle but missed the 9.93-year Jupiter–Saturn tidal period. Nataf (2022)[3] repeated the same oversight and, by further assuming circular planetary orbits, eliminated even the 11.86-year Jupiter tide. Smythe & Eddy (1977)[4] considered both Jupiter and Saturn but failed to recognize that planetary tides can both synchronize with and interfere with the solar dynamo cycle, which is an intrinsic property of the Sun. The observed output is therefore the result of a combination of external forcings and internal dynamics.

Finally, the role of high-frequency tides from the inner terrestrial planets and the potential influence of heliospheric oscillations directly induced by the planets on space weather must also be considered. These topics will be examined in detail in the following chapters.

---

[1]Scafetta, N., 2014b. Discussion on the spectral coherence between planetary, solar and climate oscillations: a reply to some critiques. Astrophysics and Space Science, 354, 275–299.
Scafetta, N., 2023e. Comment on "Tidally Synchronized Solar Dynamo: A Rebuttal" by Nataf (Solar Phys. 297, 107, 2022). Solar Physics, 298, 24.

[2]Okal, E., Anderson, D.L., 1975. On the planetary theory of sunspots. Nature, 253, 511.

[3]Nataf, H.-C., 2022. Tidally synchronized solar dynamo: a rebuttal. Solar Physics, 297, 107.

[4]Smythe, C.M., Eddy, J.A., 1977. Planetary tides during the Maunder Minimum. Nature, 266, 434–435.

34.2. High-frequency variations in solar activity

This chapter explores solar activity variations characterized by periods shorter than the canonical Schwabe cycle, which spans approximately 9 to 13 years. A key reference for this analysis is the study by Scafetta & Willson (2013b),[5] which identified a modulation of total solar irradiance (TSI) linked to the Earth–Jupiter conjunction, with a cycle of approximately 1.09 years.

On sub-decadal timescales, below the Schwabe band, numerous periodic signals have been detected in solar datasets. This spectral domain is particularly significant because only since 1978—following the advent of satellite-based TSI monitoring—have sufficiently precise and continuous observations become available to allow robust assessments of high-frequency solar variability. These datasets enable the detection of monthly and interannual modulations potentially correlated with planetary motions.

Some of these modulations may arise from complex interactions between the Sun's differential rotation and the orbital dynamics of both the inner planets (Mercury, Venus, Earth, Mars) and the outer planets (Jupiter, Saturn, Uranus, Neptune). For instance, Bigg (1967)[6] proposed a possible influence of Mercury on sunspot activity, suggesting that tidal effects from the planet could modulate solar behavior.

Mercury's highly eccentric orbit ($e = 0.206$) generates significant tidal variations on the solar surface, as calculated by Scafetta (2012d)[7]; see Figure 38.3E.

Rapid oscillations in solar activity are studied primarily through satellite-derived TSI data. Since 1978, TSI has been monitored and reconstructed by three major independent scientific teams:

- ACRIM (Willson & Mordvinov, 2003);
- VIRGO aboard the SOHO satellite (Fröhlich, 2006);
- SORCE/TIM (Kopp et al., 2005; Kopp & Lawrence, 2005).

Measurements from 2000 to 2014 indicate a mean TSI of approximately 1361 W/m^2 during the solar minimum of 2008–2009.

34.3. The 22–40 day timescale

Analyzing TSI data from 2003.15 to 2011.00, Scafetta & Willson (2013b) identified significant spectral peaks within the 22–40 day band. Among these, a prominent cycle at 27.3 days corresponds to the well-known Carrington rotation period.

The Carrington period reflects the rotation rate of the Sun at a latitude of approximately 26°, where sunspots are most frequently observed. This value represents the synodic rotation period as seen from Earth, resulting from the

[5]Scafetta, N., Willson, R.C., 2013b. Empirical evidences for a planetary modulation of total solar irradiance and the TSI signature of the 1.09-year Earth–Jupiter conjunction cycle. Astrophysics and Space Science, 348, 25–39.

[6]Bigg, E.K., 1967. Influence of the planet Mercury on sunspots. The Astronomical Journal, 72, 463–466.

[7]Scafetta, N., 2012d. Does the Sun work as a nuclear fusion amplifier of planetary tidal forcing? Journal of Atmospheric and Solar–Terrestrial Physics, 81–82, 27–40.

combination of the Sun's sidereal rotation at 26° latitude (approximately 25.38 days) and Earth's orbital motion (approximately 365.25 days).

In addition to the 27.3-day peak, further spectral signals were detected at: 24.8 days (0.068 years); 34–35 days (0.093–0.096 years); 36–38 days (0.099–0.104 years).

These values fall within the known range of the Sun's differential rotation, which primarily affects the convective zone, a region more dynamic than the stable radiative interior. Solar rotation varies from approximately 24.7 days at the equator (Kotov & Haneychuk, 2020)[8] to approximately 38 days near the poles (Beck, 2000).[9]

Synodic cycles and planetary interactions. The observed spectral peaks can also be interpreted as synodic cycles arising from interactions between the Sun's rotation and the orbital motions of the terrestrial planets. For example, the 24.8-day cycle can be derived by combining Jupiter's orbital period (4332.6 days) with the solar rotation (24.7 days), using the synodic formula:

$$\frac{1}{P_{\text{synodic}}} = \left| \frac{1}{P_1} - \frac{1}{P_2} \right| .$$

Other relevant synodic cycles include: 26.5 days (Earth); 27.75 days (Venus); 34.3 days (Mercury).

Notably, the spectral peak at 34.7 days closely matches the synodic cycle between Mercury and the Sun, suggesting an effective internal solar rotation of approximately 24.89 days.

34.4. The timescale between 0.1 and 1.2 years

The table shown in Figure 34.1 presents the orbital, synodic, and spring tidal cycles between the terrestrial planets (Mercury, Venus, Earth, Mars) and the Jovian planets (Jupiter, Saturn, Uranus, Neptune). These cycles form coherent frequency clusters, revealing structured patterns of planetary interaction. The table also highlights key periodicities, up to 1.6 years, associated with Mercury (Me), Venus (Ve), Earth (Ea), and Jupiter (Ju).

Spectral analyses of TSI data from the ACRIM and PMOD datasets revealed distinct peaks at the following periods: 0.070, 0.097, 0.20, 0.25, 0.30–0.34, 0.39, 0.55, 0.60–0.65, 0.7–0.9, and 1.0–1.2 years (Scafetta & Willson, 2013b). These periods correspond closely to calculated planetary cycles. Notable examples include:

- Mercury–Venus spring tidal cycle: 0.20 years;
- Mercury's orbital period: 0.24 years;
- Venus–Jupiter spring tidal cycle: 0.32 years;
- Venus–Mercury synodic cycle: 0.40 years;
- Venus–Jupiter synodic cycle: 0.65 years;
- Venus–Earth spring tidal cycle: 0.80 years.

[8] Kotov, V.A., Haneychuk, V.I., 2020. Oscillations of solar photosphere: 45 years of observations. Astronomische Nachrichten, 341, 595–599.

[9] Beck, J.G., 2000. A comparison of differential rotation measurements. Solar Physics, 191, 47–70.

Cycle	Type	*P* (day)	*P* (year)
Me	$\frac{1}{2}$ orbital	44 ± 0	0.120 ± 0.000
Me–Ju	spring	45 ± 9	0.123 ± 0.024
Me–Ea	spring	58 ± 10	0.159 ± 0.027
Me–Ve	spring	72 ± 8	0.198 ± 0.021
Me	orbital	88 ± 0	0.241 ± 0.000
Me–Ju	synodic	90 ± 1	0.246 ± 0.002
Ea	$\frac{1}{4}$ orbital	91 ± 3	0.250 ± 0.000
Ve	$\frac{1}{2}$ orbital	112.5 ± 0	0.307 ± 0.000
Me–Ea	synodic	116 ± 9	0.317 ± 0.024
Ve–Ju	spring	118 ± 1	0.324 ± 0.003
Ea	$\frac{1}{3}$ orbital	121 ± 7	0.333 ± 0.000
Me–Ve	synodic	145 ± 12	0.396 ± 0.033
Ea	$\frac{1}{2}$ orbital	182 ± 0	0.500 ± 0.000
Ea–Ju	spring	199 ± 3	0.546 ± 0.010
Ve	orbital	225 ± 0	0.615 ± 0.000
Ve–Ju	synodic	237 ± 1	0.649 ± 0.004
Ve–Ea	spring	292 ± 3	0.799 ± 0.008
Ea	orbital	365.25 ± 0	1.000 ± 0.000
Ea–Ju	synodic	399 ± 3	1.092 ± 0.009
Ea–Ve	synodic	584 ± 6	1.599 ± 0.016

FIGURE 34.1. Main orbital (*"orbital"*), synodic (*"synodic"*), and spring tidal (*"spring"*) periods between Mercury ("Me"), Venus ("Ve"), Earth ("Ea"), and solar Jupiter ("Ju"), with periods up to 1.6 years. (Adapted from Scafetta & Willson, 2013b)

The semiannual cycle (0.50 years) is also evident and may be linked to Earth's passage through the solar equatorial plane, potentially modulating latitudinal patterns in solar irradiance.

Spectral evidence and harmonic modeling. Figure 34.2 summarizes the results of Scafetta & Willson (2013b):

- Panel A: Comparison of TSI records from ACRIMSAT/ACRIM3 (black), SOHO/VIRGO, and SORCE/TIM.
- Panel B: Periodogram of TSI data from 2003.15 to 2011.00, showing monthly-scale solar periodicities.

- Panel C: Spectral analysis of the ACRIM (1992.5–2012.9) and PMOD (1997.75–2004.25) datasets. Yellow bars indicate planetary harmonics from Figure 34.1, revealing a strong correspondence.
- Panel D: Composite TSI during Solar Maximum 23 (1998–2004), with peaks aligned with Earth–Jupiter inferior conjunctions.
- Panel E: High-pass filtered TSI curves compared with the Earth–Jupiter synodic harmonic component (1.092-year period), modulated by the solar cycle.
- Panel F: Long-term comparison of ACRIM and PMOD composites (1978–2013) with the Earth–Jupiter synodic model.
- Panels G–H: Tidal function generated by all the planets and its spectral analysis (Scafetta, 2012d),[10] emphasizing planetary tidal cycles.
- Panels I–J: Velocity of the Sun relative to the solar system barycenter and its spectral analysis, highlighting synodic cycles between planetary pairs.

Panels G and I show that the theoretical planetary curves produce spectral peaks that correspond closely to those observed in satellite TSI data (Panel C), suggesting that both tidal and synodic cycles contribute to solar variability.

The 1.092-year Earth–Jupiter synodic cycle. Among the various cycles identified, the 1.092-year period stands out as the most prominent. This cycle corresponds to the Earth–Jupiter synodic period and is clearly detectable in TSI peaks observed near their conjunctions. Notably, the TSI maximum in 2002 coincided with an exceptionally close Earth–Jupiter alignment (angular separation: 0°13′19″).

To confirm the persistence of this oscillation, Scafetta & Willson (2013b) proposed a harmonic model of TSI modulation based on the Earth–Jupiter synodic cycle, shown in Figure 34.2, panels D–F. The function used is:

$$f(t) = g(t) \cos \left[\frac{2\pi(t - 2002)}{1.09208} \right],$$

where $g(t)$ represents an amplitude modulated by the Schwabe cycle. The phase origin is set to January 1, 2002, the date of the Earth–Jupiter conjunction. This model successfully reproduces the modulation across solar cycles 22, 23, and 24, with stronger effects during solar maxima and diminished amplitude during minima.

Retrospective modeling was performed using:

$$f(t) = S_A(t) + 0.2 \cdot (S_A(t) - 1360.58) \cos \left[\frac{2\pi(t - 2002)}{1.09208} \right]$$

$$f(t) = S_P(t) + 0.2 \cdot (S_P(t) - 1365.3) \cos \left[\frac{2\pi(t - 2002)}{1.09208} \right],$$

where $S_A(t)$ and $S_P(t)$ are biennial moving averages of the ACRIM and PMOD TSI composites, respectively.

TSI peaks near the Earth–Jupiter conjunctions of 1979, 1981, 1984, 1990, 1998, 2011, and 2012 support the validity of the model. Discrepancies in the PMOD

[10]Scafetta, N., 2012d. Does the Sun work as a nuclear fusion amplifier of planetary tidal forcing? Journal of Atmospheric and Solar–Terrestrial Physics, 81–82, 27–40.

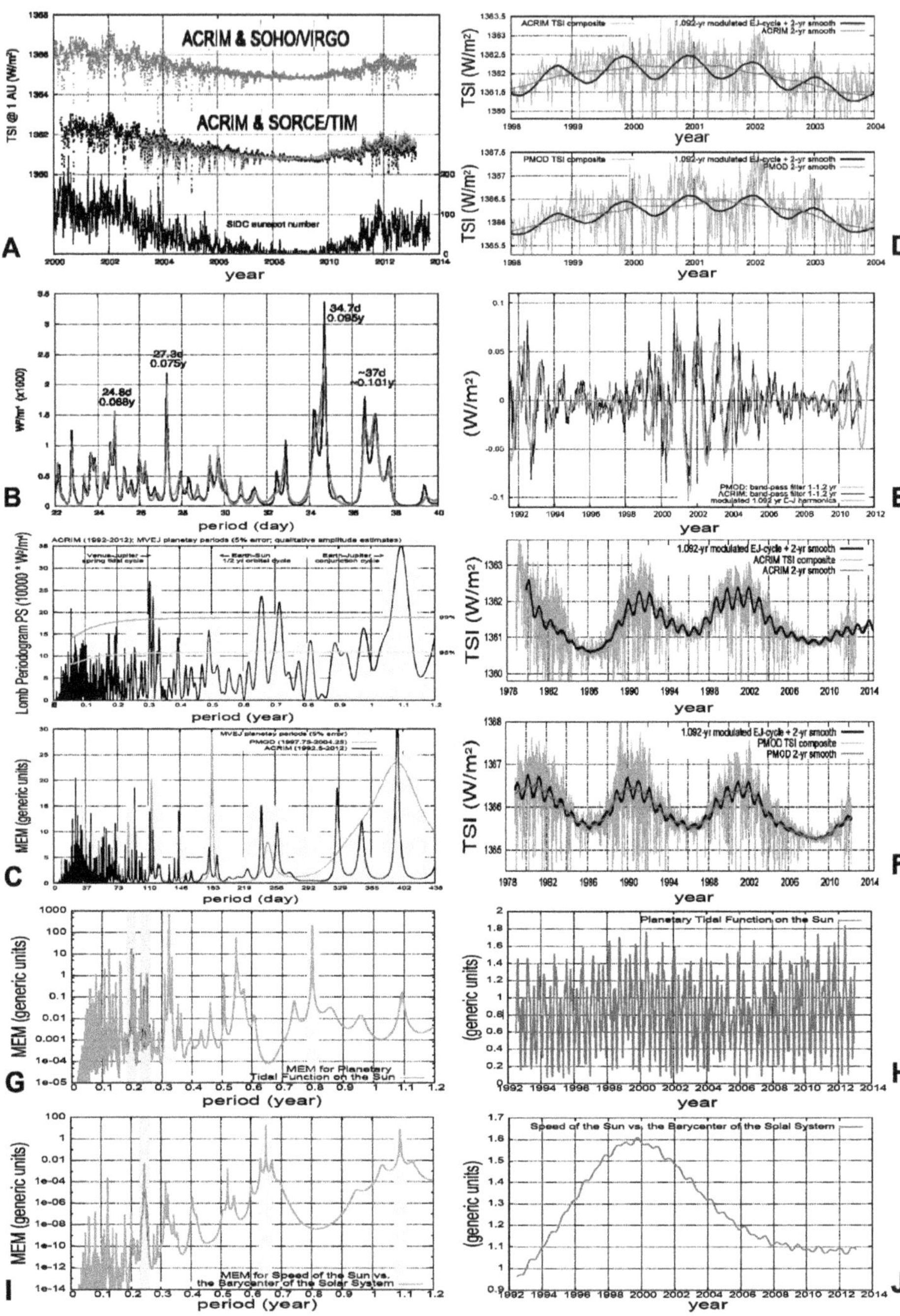

FIGURE 34.2. Spectral analysis of satellite TSI data and interpretation of spectral peaks in light of planetary orbital, tidal, and synodic periods. In particular, the Earth–Jupiter synodic model is highlighted, which reconstructs the 1.092-year oscillations found in the data and satellite TSI composites. (Adapted from Scafetta & Bianchini, 2022).

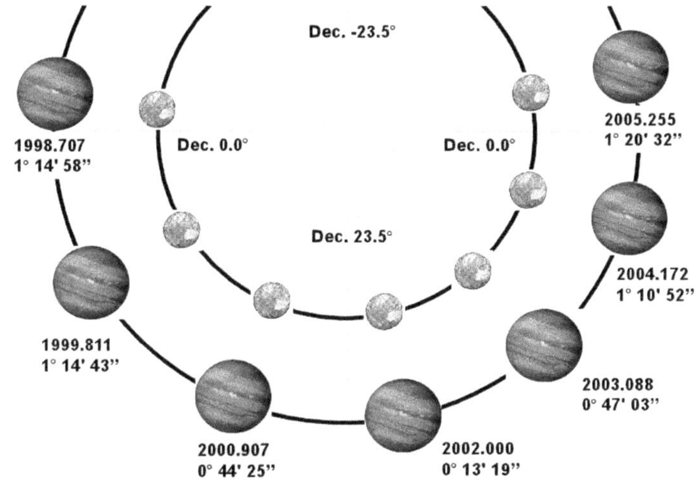

FIGURE 34.3. Earth–Jupiter conjunctions, recurring with a periodicity of 1.09 years during the peak of Solar Cycle 23 (1998–2004), also highlighted in Figures 34.2D and 34.2F. (Adapted from Scafetta & Willson, 2013b)

dataset may stem from revisions to the Nimbus7/ERB data by the Swiss group (Fröhlich, 2006; Scafetta et al., 2019a).[11]

Planet-induced solar brightness modulation? Figure 34.3 depicts the Earth–Jupiter conjunction dates, recurring with a periodicity of 1.09 years during the peak of Solar Cycle 23 (1998–2004). The TSI increase observed at those dates, as highlighted in Figures 34.2D and 34.2F, suggests that the side of the Sun facing Jupiter may be marginally brighter and that, when Earth crosses the Sun–Jupiter conjunction line, it receives slightly greater irradiance.

Similar but more pronounced effects have been observed in stars hosting close-orbiting giant planets—known as Hot Jupiters—which induce chromospheric activity on their host stars (e.g., on HD 179949 and other stars) (Shkolnik et al., 2003, 2005).[12]

Further support comes from Kotov & Haneychuk (2020),[13] who identified two stable pulsation periods in the solar photosphere: $P_1 = 9600.606$ s; $P_2 = 9597.924$ s. The beat between these signals generates a cycle of 397.7 days, nearly identical to the Earth–Jupiter synodic period (398.9 days), suggesting a potential role for these planets in modulating solar pulsations.

[11] Fröhlich, C., 2006. Solar irradiance variability since 1978. Space Science Reviews, 125, 53–65.
Scafetta, N., et al., 2019. Modeling quiet solar luminosity variability from TSI satellite measurements and proxy models during 1980–2018. Remote Sensing, 11(21), 2569.

[12] Shkolnik, E., et al., 2003. Evidence for planet-induced chromospheric activity on HD 179949. The Astrophysical Journal, 556, 296–301.
Shkolnik, E., et al., 2005. Hot Jupiters and hot spots: the short- and long-term chromospheric activity on stars with giant planets. The Astrophysical Journal, 622, 1075–1090.

[13] Kotov, V.A., Haneychuk, V.I., 2020. Oscillations of solar photosphere: 45 years of observations. Astronomische Nachrichten, 341, 595–599.

34.5. Solar cycles between 2 and 9 years

Spectral analyses of sunspot records reveal significant peaks in the 5–6 year band and around 8–8.5 years (Figure 36.1), as discussed by Scafetta (2012b).[14] The first peak may represent a harmonic of the Schwabe cycle (approximately half of 11 years), whereas the second appears to arise from more complex planetary orbital interactions. These periodicities suggest that solar variability on sub-decadal timescales may be modulated by planetary synchronizations.

Orbital synchronizations and beat frequencies. Several planetary configurations yield nearly identical orbital durations, generating beat frequencies that align with observed solar cycles:

- 23 Mercury orbits: $23 \times 0.241 = 5.542$ years;
- 9 Venus orbits: $9 \times 0.615 = 5.535$ years.

These values are nearly indistinguishable and may explain the 5–6 year peak observed in sunspot spectra. Their interaction produces a beat frequency in this range. Moreover, 46 Mercury orbits $\approx$ 18 Venus orbits $\approx$ 11.08 years.

This value closely matches the Schwabe cycle, suggesting that these orbital combinations may act as harmonic generators for the 11-year solar cycle.

The 8-year peak. The frequency peak that occurs near 8 years may be explained by the following synchronizations:

- 8 Earth orbits: $8 \times 1.000 = 8.000$ years;
- 13 Venus orbits: $13 \times 0.615 = 7.995$ years.

The near-perfect coincidence of these periods supports the hypothesis that the 8-year peak may arise from a beat between the Earth and Venus orbital cycles.

Invariant orbital inequalities. More complex harmonic relationships, referred to as *Invariant Orbital Inequalities* (see Chapter 37), involve interactions among Venus, Earth, Jupiter, and Saturn. These relationships generate beat periods that appear in sunspot series periodograms (Figure 36.1):

$$\left(\frac{2}{P_{Ve}} - \frac{3}{P_{Ea}} - \frac{2}{P_{Ju}} + \frac{3}{P_{Sa}}\right)^{-1} = 5.43 \text{ years;}$$

$$2\left(-\frac{1}{P_{Ve}} + \frac{2}{P_{Ea}} - \frac{2}{P_{Ju}} + \frac{1}{P_{Sa}}\right)^{-1} = 8.34 \text{ years.}$$

These equations describe beat frequencies between synodic and tidal cycles among the four planets, strengthening the hypothesis that planetary orbital mechanics contribute to solar variability.

Conclusion. These findings support the hypothesis that high-frequency solar oscillations are induced by planetary synchronizations, particularly tidal and synodic interactions. The coherence between orbital mechanics and solar spectral peaks suggests that solar activity is systematically modulated by gravitational influences rather than being purely stochastic.

[14]Scafetta, N., 2012b. Multi-scale harmonic model for solar and climate cyclical variation throughout the Holocene based on Jupiter–Saturn tidal frequencies plus the 11-year solar dynamo cycle. Journal of Atmospheric and Solar–Terrestrial Physics, 80, 296–311.

CHAPTER 35

The Schwabe solar cycle and the Venus–Earth–Jupiter model

Ilin et al. (2024)[1] provide observational evidence that close-in exoplanets can modulate the activity of their host stars through magnetic and tidal interactions using the high-precision stellar data of Kepler and TESS. This finding is supported by the theoretical insights of Lanza (2018),[2] who suggests that stellar activity may be influenced by star–planet interactions. Taken together, these two studies indicate that stellar variability is not necessarily a purely endogenous phenomenon but can be modulated by the presence and orbital characteristics of planets, opening new perspectives for understanding solar dynamics as well.

What, then, about the Sun and its 11-year solar cycle?

35.1. Historical background

The approximately 11-year solar cycle, known as the Schwabe cycle, was empirically discovered by Heinrich Schwabe in 1843 and later confirmed by Rudolf Wolf, who calculated its mean duration by analyzing over a century of sunspot data from European astronomical observatories.

In 1859, Wolf[3] proposed in a letter to the English astronomer Richard Carrington that this periodicity might be influenced by gravitational forces exerted by the planets, particularly Venus, Earth, Jupiter, and Saturn. Wolf likely recognized that Jupiter, with its 11.86-year orbital period, could not by itself account for the Schwabe cycle. Conversely, the spring tidal cycle of Jupiter and Saturn—approximately 9.93 years—was too short. This led to the hypothesis that only the interaction among multiple planetary cycles could generate the beat frequencies necessary to explain the observed sunspot periodicity.

Despite its conceptual appeal, the idea that planets could modulate solar activity was met with skepticism. The principal objection, discussed in Chapter 38, is that the planets are too small and distant to exert significant tidal forces on the Sun. However, this theoretical critique must be weighed against empirical evidence. If the correlations are robust, then a physical mechanism may exist. Scientific inquiry demands exploration, not premature dismissal.

---

[1] Ilin, E., Poppenhäger, K., Chebly, J., Ilić, N., Alvarado-Gómez, J.D., 2024. Planetary perturbers: Flaring star–planet interactions in Kepler and TESS. Monthly Notices of the Royal Astronomical Society, 527(2), 3395–3417.

[2] Lanza, A.F., 2018. Star–Planet Interactions. In Proceedings of the 18th Cambridge Workshop on Cool Stars, Stellar Systems, and the Sun (Cool Stars 18).

[3] Wolf, R., 1859. Extract of a letter from Prof. R. Wolf of Zurich to Mr. Carrington, dated Jan. 12, 1859. Monthly Notices of the Royal Astronomical Society, 19, 85.

A historical parallel is found in William Thomson (Lord Kelvin), who in 1863 denied the possibility that the solar magnetic field could influence Earth. Despite compelling observational evidence from Wolf, Carrington, and others linking sunspots, solar flares, and auroras, Kelvin maintained his position until 1892. Only with the discovery of the solar wind, which carries the solar magnetic field throughout the solar system, was the direct link between solar activity and geomagnetic disturbances understood (Moldwin, 2008).[4]

Subsequent studies and theoretical developments. Following Wolf's hypothesis, several researchers explored the potential link between planetary dynamics and solar activity:

- Bigg, E.K. (1967)[5] suggested that Mercury's orbital position influences sunspot activity.
- Wood, K.D. (1972)[6] proposed that the orbital configurations of Venus, Earth, and Jupiter generate solar tides capable of modulating solar activity.
- Jose, P.D. (1965)[7] identified a correlation between the Sun's motion around the solar system's barycenter and sunspot cycles, introducing the Jose cycle (~179 years).
- Fairbridge, R.W. & Shirley, J.H. (1987)[8] argued that the Sun's motion relative to the solar system's center of mass could influence its internal dynamics.
- Landscheidt, T. (1988; 1999)[9] introduced the concept of planetary resonance and the role of planetary conjunctions in modulating solar activity.
- Charvátová, I. (1988; 1989; 2009)[10] developed the trefoil model to describe ordered and disordered phases of solar motion, associating them with Grand Solar Minima and the Bray–Hallstatt cycle (~2100–2500 years).

Although the planetary theory of solar variability remains controversial, the idea of a dynamic resonance between planetary configurations and internal solar processes continues to stimulate scientific inquiry in both astrophysics and climate science.

---

[4]Moldwin, M., 2008. An Introduction to Space Weather. Cambridge University Press.

[5]Bigg, E.K., 1967. Influence of the planet Mercury on sunspots. The Astronomical Journal, 72, 463–466.

[6]Wood, K.D., 1972. Physical Sciences: Sunspots and Planets. Nature, 240, 91–93.

[7]Jose, P.D., 1965. Sun's motion and sunspots. The Astronomical Journal, 70, 193.

[8]Fairbridge, R.W., Shirley, J.H., 1987. Prolonged minima and the 179-year cycle of the solar inertial motion. Solar Physics, 110, 191–220.

[9]Landscheidt, T., 1988. Solar rotation, impulses of the torque in the sun's motion, and climatic variation. Climatic Change, 12(3), 265–295.

Landscheidt, T., 1999. Extrema in sunspot cycle linked to Sun's motion. Solar Physics, 189(2), 413–424.

[10]Charvátová, I., 1988. The solar motion and the variability of solar activity. Advances in Space Research, 8(7), 147–150.

Charvátová, I., 1989. On the relation between solar motion and the long-term variability of solar activity. Studia Geophysica et Geodaetica, 33(3), 230–241.

Charvátová, I., 2009. Long-term predictive models based on solar motion. Pattern Recognition in Physics, 1(1), 1–12.

Planet	Period (days)	Period (years)
Mercury	87.969	0.241
Venus	224.701	0.615
Earth	365.256	1
Mars	686.980	1.881
Jupiter	4332.589	11.862
Saturn	10759.22	29.457
Uranus	30685.4	84.011
Neptune	60189.0	164.79

FIGURE 35.1. Sidereal periods of the planets. Note that 1 year is defined by Earth's tropical year, which is 365.242 days. It is recommended to use periods in days and then convert them to Earth tropical years. From the *Planetary Fact Sheet* https://nssdc.gsfc.nasa.gov/planetary/factsheet/ (Scafetta, 2012d).

Toward new models. With the advent of modern computational tools and deeper insights into solar physics and celestial mechanics, the hypothesis of planetary modulation of solar activity has gained renewed momentum. Theoretical models now seek to explain solar cyclicity as the result of planetary interactions. In the following chapters, we will examine two fundamental variants of this perspective, using the sidereal periods of the planets listed in Figure 35.1.

35.2. Planetary tidal forcing and solar variability

The study by Wood (1972), building on Bigg (1967), demonstrated coherence between the solar tides generated by Venus, Earth, and Jupiter and the 11-year Schwabe cycle. More recently, Hung (2007)[11] proposed that solar tides induced by Mercury, Venus, Earth, and Jupiter may influence solar activity, including flares and geomagnetic storms.

Solar tides and flares. Hung (2007) analyzed 38 of the most intense solar flares recorded between 1859 and 2003, finding that 25 occurred during planetary alignments or oppositions relative to the flare's heliographic location. The probability of this correlation arising by chance was estimated at 0.039%, suggesting a possible physical relationship.

According to this hypothesis, the tidal forces exerted by the planets may perturb the equilibrium among the Sun's internal pressure, gravity, and coronal

[11] Hung, C.-C., 2007. Apparent relations between solar activity and solar tides caused by the planets. NASA Technical Memorandum, TM-2007-214817.

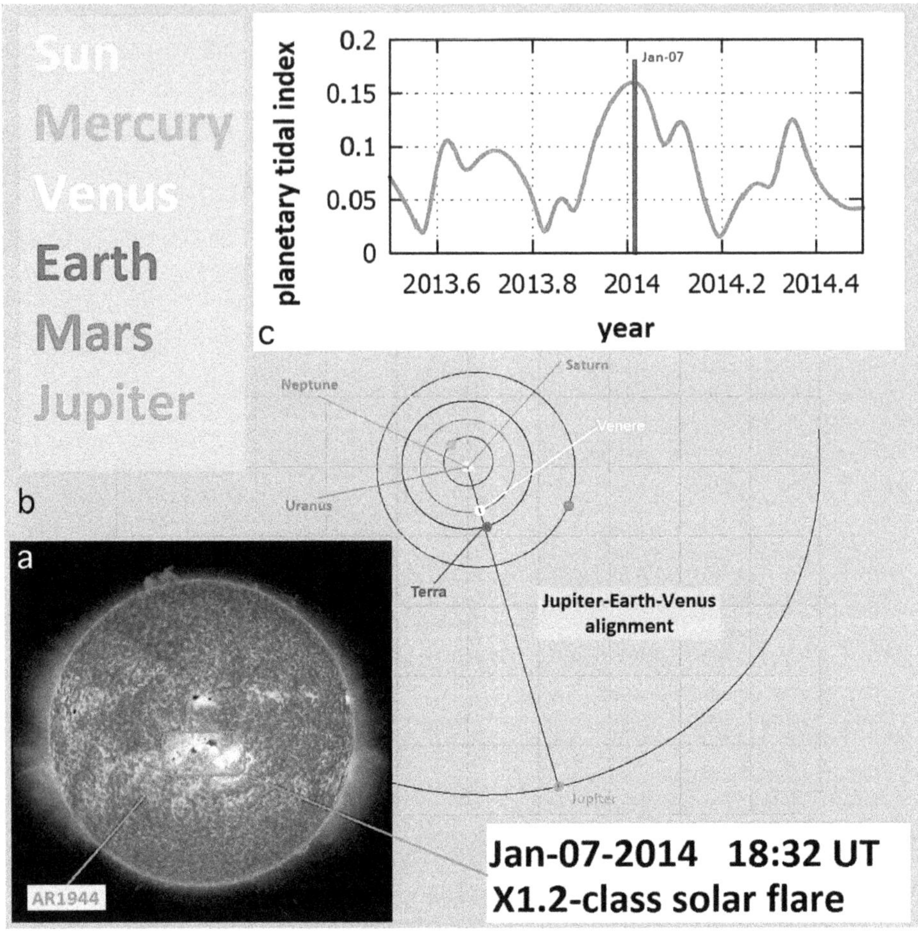

FIGURE 35.2. On January 7, 2014, a major X1.2-class solar flare erupted near the giant sunspot region AR1944 (NASA News, 2014a,b). The flare was directed straight toward Earth. At the same time, Venus, Earth, and Jupiter were aligned in a triple conjunction. The inset shows the peak reached by the planetary tidal index according to the tidal equation of Scafetta (2012d), shown in Figure 34.2H.

magnetic field, potentially triggering magnetic reconnection—the mechanism underlying solar flares.

A notable example is the giant solar flare of January 7, 2014, analyzed by Scafetta & Mörner (2025).[12] As shown in Figure 35.2, the event coincided with a triple alignment of Venus, Earth, and Jupiter, reinforcing the idea that planetary conjunctions may act as catalysts for extreme solar activity.

[12]Scafetta, N., Mörner, N.-A., 2025. The giant solar flare event of January 7, 2014 in light of the planetary theory of solar variability. Science of Climate Change, 5.2, 161–165.

Planetary cycles and the 11-year sunspot cycle. Hung (2007) observed that periodic alignments among Venus, Earth, and Jupiter follow a cycle of approximately 11 years that strikingly matches the Schwabe cycle. This synchronicity suggests a possible resonance between planetary tides and the Sun's internal dynamics, potentially contributing to the modulation of solar activity.

The gravitational forces exerted by these three planets, through their recurring orbital configurations, may influence the Sun's magnetic and thermodynamic equilibrium, acting as a celestial metronome for its cyclic behavior.

Practical implications. If these correlations reflect physical mechanisms, they offer promising prospects for the forecasting of extreme solar events. The predictability of planetary orbits could be exploited to anticipate solar flares and geomagnetic storms, with tangible benefits:

- enhanced protection of satellites and space-based communications;
- increased safety for astronauts during missions;
- safeguarding of terrestrial infrastructure sensitive to space weather, such as power grids and navigation systems.

Ultimately, advancing our understanding of the relationship between planetary dynamics and solar activity is not only a compelling scientific challenge, but also a strategic imperative for improving the resilience of our increasingly interconnected technological systems.

35.3. The Venus–Earth–Jupiter tidal model

The hypothesis that the Schwabe solar cycle may be reconstructed from the recurring orbital configurations of Venus, Earth, and Jupiter was first proposed by Bendandi (1931),[13] as later reported by Battistini (2011).[14]

This idea has since been developed mathematically by several authors, including Bollinger (1952),[15] who identified a 44.77-year Jupiter–Earth–Venus configuration linked to solar–climate cycles; Hung (2007), who explored planetary tidal influences on solar activity; Scafetta (2012d),[16] who proposed a nuclear fusion amplification mechanism for planetary tidal forcing; Tattersall (2013),[17] who examined log-normal distributions and planetary–solar resonance; Wilson (2013),[18] who introduced a spin–orbit coupling model; and Stefani et al. (2016, 2019, 2024),[19] who developed a tidally synchronized solar dynamo model and explored resonance structures in solar variability.

---

[13]Bendandi, R., 1931. Un Principio Fondamentale dell'Universo. Faenza: Osservatorio Bendandi.

[14]Battistini, A., 2011. Il ciclo undecennale del Sole secondo Bendandi. New Ice Age.

[15]Bollinger, C.J., 1952. A 44.77 year Jupiter–Earth–Venus configuration Sun–Tide period in solar–climate cycles. Proceedings of the Oklahoma Academy of Science, 307–311.

[16]Scafetta, N., 2012d. Does the Sun work as a nuclear fusion amplifier of planetary tidal forcing? Journal of Atmospheric and Solar–Terrestrial Physics, 81–82, 27–40.

[17]Tattersall, R., 2013. The Hum: log-normal distribution and planetary–solar resonance. Pattern Recognition in Physics, 1, 185–198.

[18]Wilson, I.R.G., 2013. The Venus–Earth–Jupiter spin–orbit coupling model. Pattern Recognition in Physics, 1, 147–158.

[19]Stefani, F., et al., 2016. Synchronized helicity oscillations: A link between planetary tides and the solar cycle? Solar Physics, 291, 2197–2212.
Stefani, F., et al., 2019. A model of a tidally synchronized solar dynamo. Solar Physics, 294, 60.

Mathematical structure of the Venus–Earth–Jupiter (Ve–Ea–Ju) model. The core rationale of the model lies in the observation that Venus, Earth, and Jupiter exert the strongest tidal forces on the Sun. As noted in Scafetta (2012d), the period of their recurring conjunctions can be expressed as a beat frequency derived from their sidereal orbital periods:

$$\frac{1}{P_{VEJ}} = \left(\frac{3}{P_V} - \frac{5}{P_E} + \frac{2}{P_J}\right),$$

where $P_V = 224.701$ days, $P_E = 365.256$ days, and $P_J = 4332.589$ days are the sidereal orbital periods of Venus, Earth, and Jupiter, respectively.

Solving this yields:

$$P_{VEJ} \approx 22.14 \text{ years.}$$

This period corresponds almost exactly to the Hale magnetic solar cycle (~22 years), which generates the magnetic polarity reversal every 11 years. Since the tidal configurations of Venus–Earth–Jupiter repeat equivalently every half-cycle, the effective beat tidal period becomes:

$$\frac{P_{VEJ}}{2} \approx 11.07 \text{ years.}$$

This value closely matches the historical mean of the Schwabe sunspot cycle, observed consistently since 1750. Both Hung (2007) and Scafetta (2012d) emphasized this correspondence, suggesting a potential resonance mechanism between planetary tidal forces and the solar dynamo.

Further support comes from Vos et al. (2004),[20] who identified a remarkably stable 11.04-year periodicity in Holocene proxy records from Lake Holzmaar (Germany) and GISP2 (Greenland), spanning the interval between 10,000 and 9,000 cal. BP. This finding reinforces the notion of a nearly invariant rhythm in solar behavior, consistent with the predictions of the Ve–Ea–Ju model.

35.4. Orbital synchronicity and vector representation

The planetary alignment index. To quantify the tidal configurations among the planets, Hung (2007) and Scafetta (2012d) introduced an alignment index:

$$I_{ij} = |\cos(\Theta_{ij})|$$

where Θ_{ij} is the angle between the positions of planets i and j relative to the Sun. The index reaches its maximum at conjunction ($\Theta = 0$) or opposition ($\Theta = \pi$), conditions associated with spring tides, and its minimum at quadrature ($\Theta = \pi/2$), corresponding to neap tides.

For a three-planet system (Venus, Earth, Jupiter), the combined alignment index is defined as:

$$I_{VEJ} = \min\left(I_{VE} + I_{VJ},\ I_{EV} + I_{EJ},\ I_{JV} + I_{JE}\right).$$

Stefani, F., et al., 2024. Rieger, Schwabe, Suess–de Vries: the sunny beats of resonance. Solar Physics, 299, 51.

[20] Vos, H., et al., 2004. Phase stability of the solar Schwabe cycle in Lake Holzmaar, Germany, and GISP2, Greenland, between 10,000 and 9,000 cal. BP. In: Fischer, H., et al. (Eds.), The Climate in Historical Times: Towards a Synthesis of Holocene Proxy Data and Climate Models. Springer, Berlin, pp. 293–317.

Figure 35.3A shows that peaks in I_{VEJ} exceeding the 95% confidence level occur every 11.07 years, coinciding with the Schwabe cycle and suggesting a strong synchronizing potential.

Simulation of the conjunction cycle. The orbital inequality defining the P_{VEJ} cycle can be expressed in vector form:

$$(3, -5, 2) = 3 \cdot (1, -1, 0) - 2 \cdot (0, 1, -1).$$

Here:

- $(1, -1, 0)$: synodic frequency vector of the Venus–Earth cycle;
- $(0, 1, -1)$: synodic frequency vector of the Earth–Jupiter cycle.

This harmonic combination generates a beat period of 11.07 years, which may rhythmically influence the solar dynamo, as argued by Stefani et al. (2016, 2019, 2024).

Figure 35.3 illustrates various implementations of the Venus–Earth–Jupiter triple alignment model, all of which successfully reconstruct the Schwabe cycle in both period and phase using purely astronomical parameters.

Sinusoidal modeling of the planetary configuration. Scafetta & Bianchini (2022, 2023)[21] proposed a simplified representation of the Venus–Earth and Earth–Jupiter configurations as the sum of two sinusoids:

$$f(t) = \cos\left(2\pi \cdot 3 \cdot \frac{t - t_{VE}}{P_{VE}}\right) + \cos\left(2\pi \cdot 2 \cdot \frac{t - t_{EJ}}{P_{EJ}}\right),$$

where $P_{VE} = 1.59867$ years, $P_{EJ} = 1.09207$ years, $t_{VE} = 2002.8327$, and $t_{EJ} = 2003.0887$ are the epochs of two recent conjunctions.

The resulting beat pattern aligns closely with the Schwabe sunspot cycle. In fact, Figure 35.3B shows excellent agreement between the modeled signal and observed solar activity, with peaks coinciding with the constructive phases of the tidal beat, when planetary gravitational forces act in synchrony.

Alignment along the Parker spiral. Figure 35.3C presents a variant of the model in panel A, in which planetary alignments are hypothesized to occur not along linear Euclidean conjunctions, but along the Parker spiral, that is, the curved trajectory of the solar wind. This region may facilitate magnetic coupling between the planets, as proposed by (Tattersall, 2013).[22]

35.5. Possible limitations of the model: existence of additional cycles

While the Venus–Earth–Jupiter model reproduces the average periodicity of the Schwabe cycle with high fidelity, it does not account for secular-scale fluctuations in solar activity, such as the Maunder Minimum (1645–1715). These long-term variations likely require modulation by gravitational interactions involving the outer planets.

[21] Scafetta, N., Bianchini, A., 2022. The planetary theory of solar activity variability: a review. Frontiers in Astronomy and Space Sciences, 9, 937930.

[22] Tattersall, R., 2013. The Hum: log-normal distribution and planetary–solar resonance. Pattern Recognition in Physics, 1, 185–198.

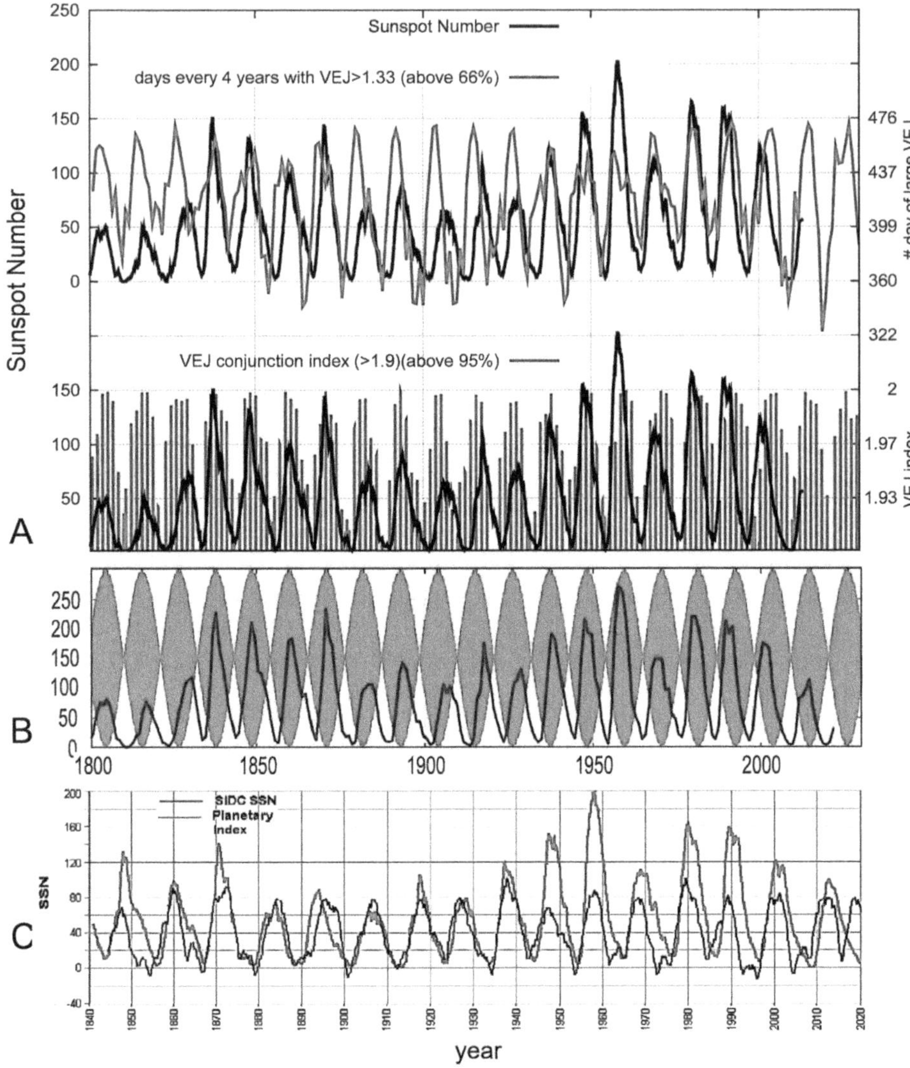

FIGURE 35.3. Reconstruction of the sunspot-number cycle using tidal models based on Venus, Earth, and Jupiter.
[A] (Top) Sunspot number record (black) compared to the alignment index $I_{\mathrm{VEJ}} > 66\%$. (B, bottom) Sunspot number record (black) compared to the number of days with strong alignment ($I_{\mathrm{VEJ}} > 95\%$).
[B] The beat $(3, -5, 2)$ between Venus, Earth, and Jupiter generated by the function $f(t)$ compared to the sunspot number.
[C] Variant of the Venus–Earth–Jupiter model that hypothesizes planetary alignments along the Parker spiral of the solar wind (Tattersall, 2013).
(Adapted from Scafetta & Bianchini, 2022)

In particular, the Jupiter–Saturn cycle (discussed in Chapter 38) and complex resonances with the orbital motions of Uranus and Neptune have been proposed as potential modulating mechanisms.

Recent studies, including Scafetta et al. (2020),[23] suggest that such planetary configurations may influence not only solar activity but also terrestrial phenomena, such as meteorite fall frequency and climate oscillations.

According to Stefani et al. (2016), Stefani (2021), and Stefani et al. (2024), the dominant planetary cycle—particularly that associated with Jupiter and Saturn—may represent a resonance frequency to which the solar dynamo tends to synchronize under conditions of hydromagnetic instability. If this hypothesis is correct, planetary cycles act as external perturbations that entrain the deep rhythm of the Sun's internal dynamics.

[23]Scafetta, N., Milani, F., Bianchini, A., 2020. A 60-year cycle in the meteorite fall frequency suggests a possible interplanetary dust forcing of the Earth's climate driven by planetary oscillations. Geophysical Research Letters, 47(18), e2020GL089954.

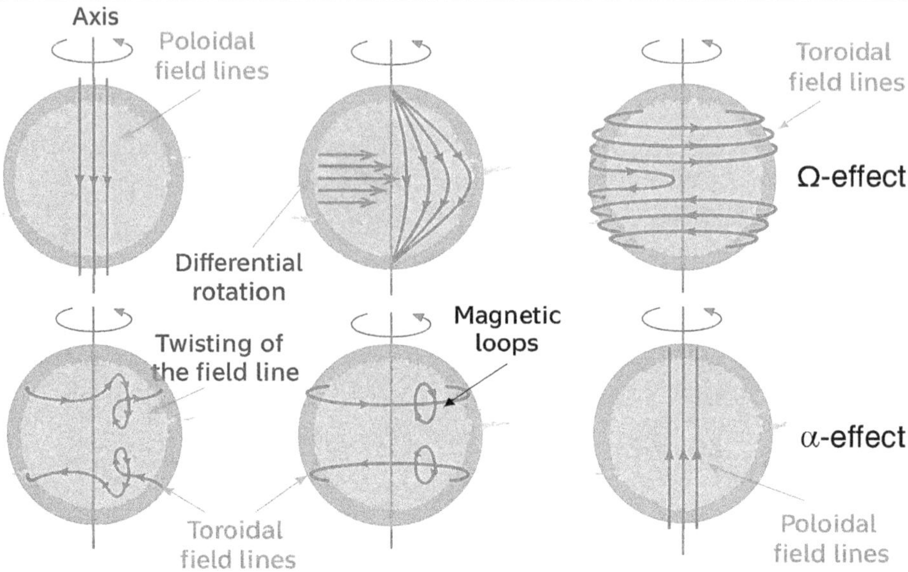

FIGURE 35.4. **The α–Ω solar dynamo model.** The figure illustrates the solar dynamo mechanism through a six-stage sequence capturing the transformation of the Sun's magnetic field over time. In the first panel, poloidal field lines are shown aligned along the Sun's rotational axis, representing the initial configuration of the magnetic field threading from pole to pole. The second panel introduces differential rotation—a key driver of the dynamo—in which the equator rotates faster than the poles. This shearing motion distorts the poloidal lines, setting the stage for the Ω-effect. In the third panel, the Ω-effect is depicted as the conversion of poloidal field lines into toroidal loops encircling the solar equator, amplifying the magnetic field and storing energy in a more compact configuration. The fourth and fifth panels show how these toroidal fields become twisted and buoyant owing to convective turbulence, forming magnetic loops that rise through the solar surface and manifest as sunspots and active regions. Finally, the sixth panel illustrates the α-effect, in which these magnetic loops interact with turbulent flows and helicity in the convection zone, regenerating the poloidal field and closing the cycle. Together, the α- and Ω-effects form the backbone of solar dynamo theory, explaining the periodic reversal of the Sun's magnetic polarity and the emergence of solar phenomena such as flares, prominences, and the 11-year solar cycle. However, the theory remains incomplete: while it successfully describes a generic magnetic cycle, it does not fully account for the precise timing, amplitude modulation, or phase coherence of the observed 11-year solar cycle.

CHAPTER 36

The Jupiter–Saturn model and long-term solar variability

A complementary hypothesis regarding the origin of solar activity variability—particularly the 11-year Schwabe cycle and its secular-to-millennial variability—centers on the tidal influences exerted on the Sun by Jupiter and Saturn, the two most massive planets in the solar system. This approach, known as the *Jupiter–Saturn* tidal model, posits that the observed sunspot cycle variability results from the synchronization and interference of the solar dynamo with the principal tidal oscillations generated by these planets.

Jupiter, with a moderately eccentric orbit ($e = 0.049$), exerts a periodic tidal forcing with a period of approximately 11.86 years, corresponding to its sidereal period. Superimposed on this is the spring tidal oscillation produced by both the synodic alignment (Sun–Jupiter–Saturn) and the opposition alignment (Jupiter–Sun–Saturn). This oscillation has a periodicity of approximately 9.93 years, equal to half the synodic cycle between Jupiter and Saturn, which is about 19.86 years.

These two components form the foundation of a planetary model first suggested by Brown (1900)[1] and other authors, and more recently developed further by Scafetta (2012b).[2] The interaction of their respective tidal forces generates a complex but highly structured behavior, manifesting as an average oscillation of approximately 11 years, in close agreement with the Schwabe cycle, and tends to entrain the solar dynamo cycle through resonance. Importantly, this is not a simple arithmetic mean: the interplay among the three components produces multiple beat frequencies, which modulate solar activity on secular and millennial timescales.

36.1. Observational evidence supporting the model

A growing body of empirical evidence supports the Jupiter–Saturn model. In particular, Scafetta (2012b,d) showed that the duration of solar cycles, defined as the interval between two consecutive solar minima, is not uniformly distributed around the mean value of 11 years, but instead follows a bimodal distribution.

Analyzing 24 solar cycles from February 1755 to December 2019, only the 24th cycle had a duration exactly equal to 11 years. The statistical analysis

[1]Brown, E.W., 1900. A possible explanation of the sunspot period. Monthly Notices of the Royal Astronomical Society, 60, 599–606.

[2]Scafetta, N., 2012b. Multi-scale harmonic model for solar and climate cyclical variation throughout the Holocene based on Jupiter–Saturn tidal frequencies plus the 11-year solar dynamo cycle. Journal of Atmospheric and Solar–Terrestrial Physics, 80, 296–311.

Interval (bin)	Included values (years)	Number of cycles	Count
8.75 - 9.24	9.00	2	1
9.25 - 9.74	9.30, 9.70	3, 8	2
9.75 - 10.24	9.90, 10.10, 10.10, 10.20	22, 15, 16, 18	4
10.25 - 10.74	10.40, 10.50, 10.50, 10.50	17, 7, 19, 21	4
10.75 - 11.24	11.00	24	1
11.25 - 11.74	11.30, 11.30, 11.30, 11.40, 11.50	1, 10, 12, 20, 14	5
11.75 - 12.24	11.80, 11.80	11, 13	2
12.25 - 12.74	12.30, 12.30, 12.40	5, 23, 9	3
12.75 - 13.24	12.80	6	1
13.25 - 13.74	13.60	4	1

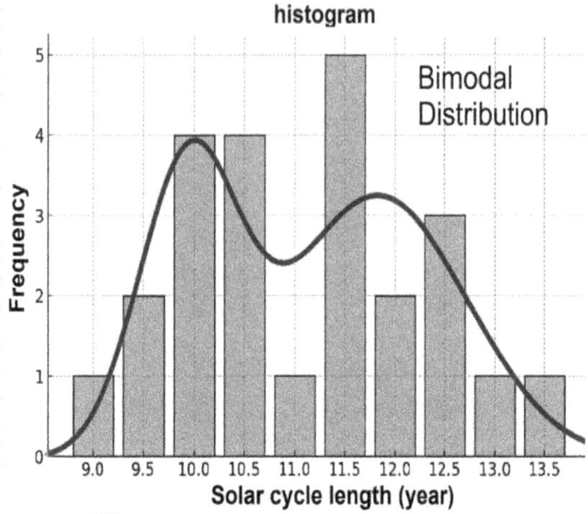

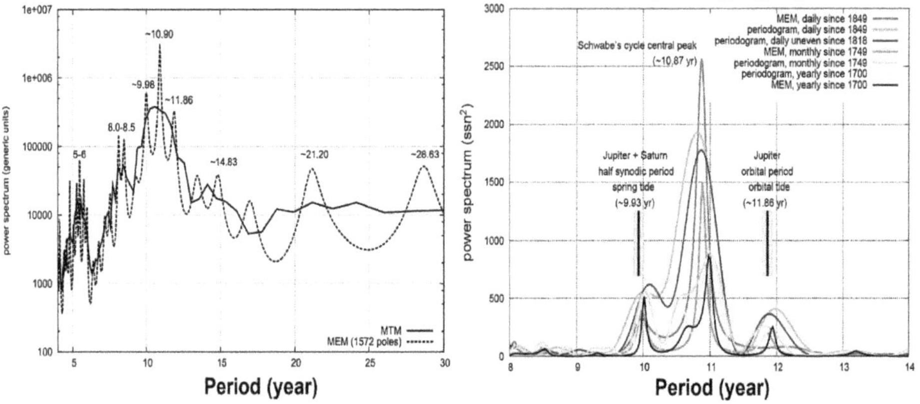

FIGURE 36.1. (Top) Distribution of the lengths of the 24 Schwabe cycles between 1755 and 2019. Note that only one cycle has a duration of exactly 11 years. (Bottom) Periodograms of the sunspot number sequence (Adapted from Scafetta, 2012b).

shown in Figure 36.1 reveals two distinct clusters: one centered around 10 years, and the other near 12 years. This pattern suggests the existence of two natural dynamic attractors, corresponding to the periods of the two main planetary tidal oscillations, 9.93 years and 11.86 years, associated with Jupiter and Saturn.

The fact that the historical record of Schwabe cycles contains only one cycle with a duration precisely close to 11 years reinforces the notion that this value is not an intrinsic property of the Sun, but rather a dynamic outcome of external planetary forcing, since the solar cycle appears to oscillate between two attractor states defined by planetary tidal periodicities. These two states generate the bimodal distribution observed in the histogram of sunspot-number cycle length depicted in Figure 36.1.

These conclusions are further corroborated by the spectral analysis of sunspot time series dating back to 1749. The lower panels of Figure 36.1 display periodograms of the monthly sunspot numbers, revealing a dominant spectral peak centered around 10.87 years. This frequency, also observed in some solar dynamo simulations (cf. Macario-Rojas et al., 2018),[3] represents an intermediate value between the two planetary tidal periods.

Additional support for the Jupiter–Saturn model comes from the presence of two well-defined secondary peaks in the periodograms, with periods of:

- 9.93 years, corresponding to the spring tide between Jupiter and Saturn;
- 11.86 years, matching Jupiter's sidereal orbital period.

This near-perfect alignment between planetary frequencies and observed solar cycle periodicities strongly suggests that the origin of the Schwabe oscillation is astronomical in nature, rather than solely the product of internal solar dynamics.

36.2. Implications for long-term solar variability

The harmonic interaction between the two principal planetary oscillations, 9.93 years (spring tidal cycle between Jupiter and Saturn) and 11.86 years (Jupiter's sidereal period), mediated by a third intermediate frequency at 10.87 years, generates a cascade of beat cycles that extend across decadal, centennial, and millennial timescales. These modulations are mathematically robust and find strong support in historical solar activity records.

Remarkably, the simple superposition of three well-defined sinusoidal waves can reconstruct, with notable precision, the major epochs of solar maxima and minima observed in the past.

Base periodicities. The model is built upon the following three harmonics:

- $P_{S1} = 9.929656$ years,
- $P_{S2} = 10.87$ years,
- $P_{S3} = 11.862242$ years.

Here, P_{S1} and P_{S3} correspond to the Jupiter–Saturn tidal cycle and Jupiter's orbit, respectively. The intermediate frequency P_{S2}, also predicted by solar dynamo models (Macario-Rojas et al., 2018), may represent an internal resonance frequency of the Sun. Alternatively, it may be interpreted as a sub-harmonic derived from an invariant orbital inequality involving Venus, Earth, Jupiter, and Saturn:

$$2\left(\frac{2}{P_{Ve}} - \frac{3}{P_{Ea}} - \frac{2}{P_{Ju}} + \frac{3}{P_{Sa}}\right)^{-1} \approx 2 \cdot 5.43 \text{ years} \approx 10.87 \text{ years},$$

as derived from the sidereal periods in days listed in Figure 35.1.

Beat frequencies. These three harmonics generate four principal beat cycles.

- Between P_{S1} and P_{S3}:

$$P_{13} = \left|\frac{1}{P_{S1}} - \frac{1}{P_{S3}}\right|^{-1} \approx 60.95 \text{ years}$$

- Between P_{S1} and P_{S2}:

$$P_{12} = \left|\frac{1}{P_{S1}} - \frac{1}{P_{S2}}\right|^{-1} \approx 114.78 \text{ years}$$

[3] Macario-Rojas, A., et al., 2018. Solar activity simulation and forecast with a flux-transport dynamo. Monthly Notices of the Royal Astronomical Society, 479, 3791–3803.

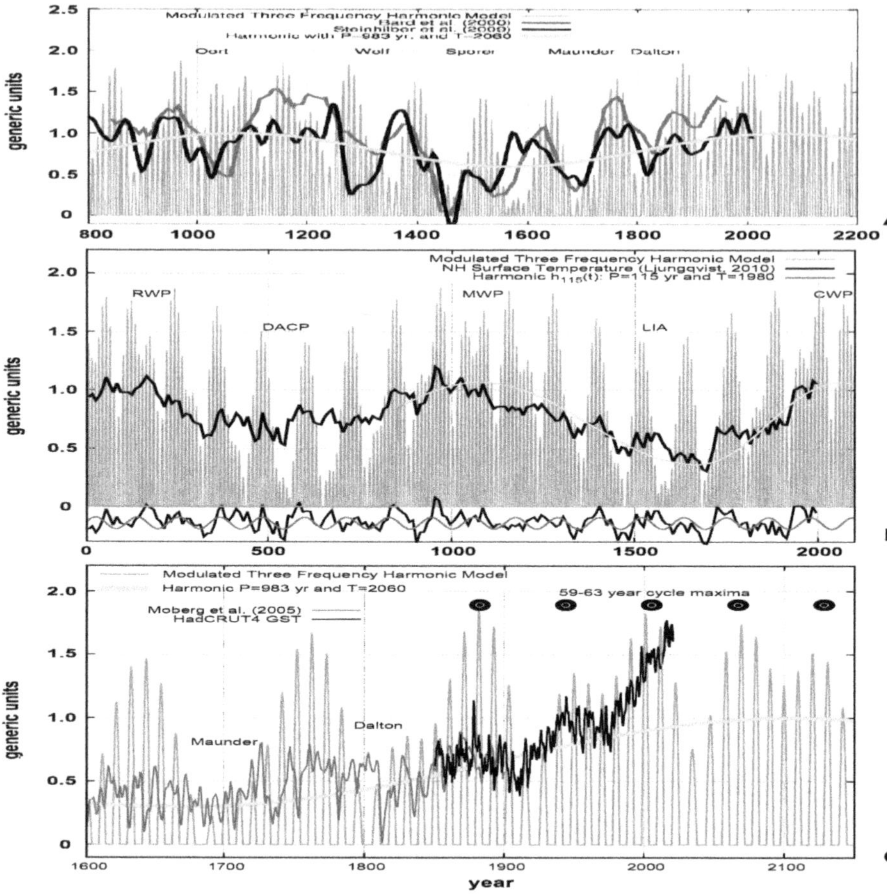

FIGURE 36.2. Comparison between the Jupiter–Saturn model function—$f_{123}(t)$ (harmonic curve)—and solar proxies from the last century (Bard et al., 2000; Steinhilber et al., 2009), the reconstruction of Northern Hemisphere extratropical temperatures by Ljungqvist (2010), and the temperature reconstruction by Moberg et al. (2005) extended with the HadCRUT4 instrumental record (Morice et al., 2012). (Adapted from Scafetta, 2012b).

- Between P_{S2} and P_{S3}:

$$P_{23} = \left| \frac{1}{P_{S2}} - \frac{1}{P_{S3}} \right|^{-1} \approx 129.95 \text{ years}$$

Using the invariant orbital inequality formalism introduced in Chapter 37, a fourth beat with a quasi-millennial periodicity is obtained from the interference between P_{12} and P_{23}. This combination is expressed by the equation:

$$(1, -1, 0) - (0, 1, -1) = (1, -2, 1).$$

The resulting period is:

$$P_{123} = \left| \frac{1}{P_{S2}} - \frac{2}{P_{S2}} + \frac{1}{P_{S3}} \right|^{-1} \approx 983 \text{ years.}$$

The above calculations require high-precision orbital periods owing to their sensitivity to numerical truncation.

Harmonic model construction. To test the effectiveness of the model, Scafetta (2012b) constructed three constituent harmonic functions with amplitudes proportional to the spectral power of the sunspot periodogram peaks, normalized to $A_{S2} = 1$, as follows:

- $A_{S1} = 0.83$;
- $A_{S2} = 1.00$;
- $A_{S3} = 0.55$.

Temporal phases were derived from astronomical events:

- $t_{S1} = 2000.475$: Jupiter–Saturn conjunction (June 23, 2000);
- $t_{S2} = 2002.364$: estimated via regression on sunspot data;
- $t_{S3} = 1999.381$: Jupiter's perihelion (May 20, 1999).

Beat phases were calculated using:

$$t_{ij} = \frac{P_j t_i - P_i t_j}{P_j - P_i}.$$

This yields:

$$t_{S12} = 2095.311, \quad t_{S13} = 2067.044, \quad t_{S23} = 2035.043, \quad t_{S123} = 2059.686$$

The full harmonic model is:

$$h_{123}(t) = \sum_{i=1}^{3} h_i(t) = \sum_{i=1}^{3} A_{Si} \cos\left(2\pi \frac{t - t_{Si}}{P_{Si}}\right).$$

With individual terms:

$$h_1(t) = 0.83 \cos\left(2\pi \frac{t - 2000.475}{9.929656}\right);$$

$$h_2(t) = 1.0 \cos\left(2\pi \frac{t - 2002.364}{10.87}\right);$$

$$h_3(t) = 0.55 \cos\left(2\pi \frac{t - 1999.381}{11.862242}\right).$$

To isolate constructive interference, a threshold function is defined as:

$$f_{123}(t) = \begin{cases} h_{123}(t) & \text{if } h_{123}(t) \geq 0 \\ 0 & \text{if } h_{123}(t) < 0 \end{cases}$$

Comparison with proxy data. Figure 36.2 compares $f_{123}(t)$ with several solar and climate proxies:

- cosmogenic radionuclides (^{14}C, ^{10}Be) reflecting solar irradiance over the last 1200 years (Bard et al., 2000; Steinhilber et al., 2009);
- a two-millennial Northern Hemisphere temperature reconstruction (Ljungqvist, 2010);

- a secular temperature reconstruction by Moberg et al. (2005) extended with the HadCRUT4 instrumental data (Morice et al., 2012).

The model successfully reproduces:

- the quasi-millennial cycle, evident in both solar and climate proxies;
- centennial oscillations between 100 and 150 years;
- the Schwabe cycle with reasonable fidelity;
- the 60-year cycle, especially prominent in temperature records since 1850.

Five black dots in the lower panel mark simulated maxima near the years 1940 and 2000, corresponding to the constructive phases of the 60-year beat cycle. In particular:

- a local maximum around 1880, preceding the early 20th-century warming;
- the second near 1940, coinciding with a well-documented solar and climatic peak according to several records;
- the third around 2000, aligning with satellite-era TSI maxima and the global temperature rise;
- a projected fourth maximum near 2060, in phase with the quasi-millennial cycle peak.

This result is particularly relevant for a proper interpretation of solar physics and for assessing the impact of solar activity variability on climate. In particular, it appears to confirm the hypothesis that solar activity exhibited local maxima around 1880, in the 1940s, and in the early 2000s, as suggested by some solar models (see Figure 25.2) and by the ACRIM satellite TSI composite (Scafetta & Willson, 2009; Scafetta et al., 2019a),[4] and as reflected in various climate reconstructions.

The coherence found between the function $f_{123}(t)$ and solar–climate proxy records confirms the validity of the Jupiter–Saturn harmonic model. It offers a robust astronomical and predictive framework for understanding the evolution of solar activity and its long-term connection to terrestrial climate variability.

36.3. Further implications for long-term solar variability

Harmonic structures and beat cycles. The harmonic interaction between the two principal planetary oscillations, 9.93 years (spring tidal cycle between Jupiter and Saturn) and 11.86 years (Jupiter's sidereal period), mediated by a third intermediate frequency at 10.87 years, generates a cascade of beat cycles that extend across decadal, centennial, and millennial timescales. These modulations are precisely defined mathematically and find strong support in historical solar activity records.

The base periodicities used in the model are:

- $P_{S1} = 9.929656$ years;
- $P_{S2} = 10.87$ years;
- $P_{S3} = 11.862242$ years.

[4]Scafetta, N., Willson, R.C., 2009. ACRIM-gap and TSI trend issue resolved using a surface magnetic flux TSI proxy model. Geophysical Research Letters, 36, L05701.
Scafetta, N., Willson, R.C., Lee, J.N., Wu, D.L., 2019a. Modeling quiet solar luminosity variability from TSI satellite measurements and proxy models during 1980–2018. Remote Sensing, 11(21), 2569.

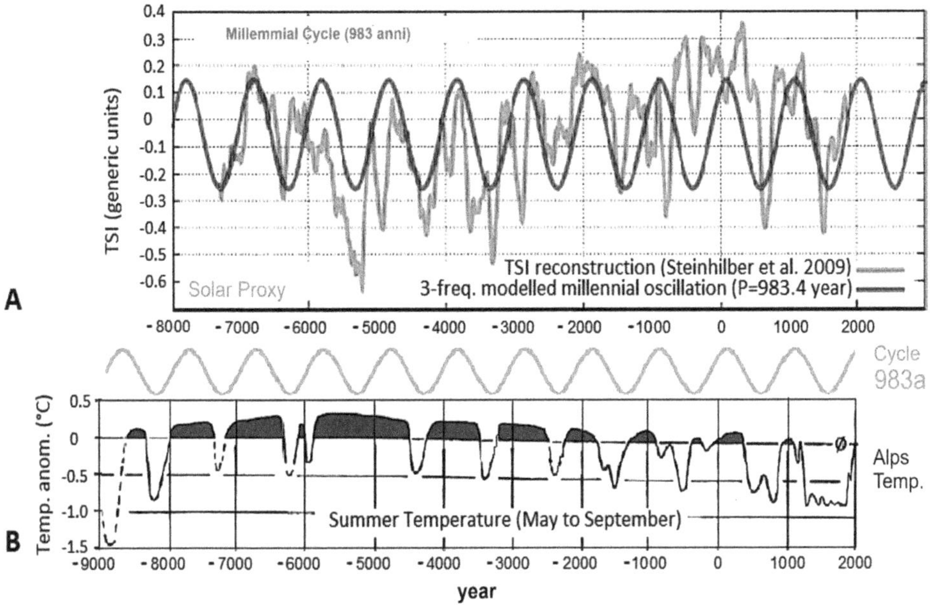

FIGURE 36.3. Comparison of the 983-year cycle predicted by the Jupiter–Saturn model with [A] the solar proxy by Steinhilber et al. (2009) and [B] Alpine temperature records north of Italy by Kutschera et al. (2020). (Adapted from Scafetta, 2012b; Scafetta & Bianchini, 2023).

The intermediate frequency P_{S2}, also predicted by solar dynamo models (Macario-Rojas et al., 2018), may represent a natural resonance of the Sun. Alternatively, it can be derived from an invariant orbital inequality involving Venus, Earth, Jupiter, and Saturn:

$$2\left(\frac{2}{P_{Ve}} - \frac{3}{P_{Ea}} - \frac{2}{P_{Ju}} + \frac{3}{P_{Sa}}\right)^{-1} \approx 10.87 \text{ years.}$$

These three harmonics generate four principal beat frequencies:

- $P_{13} \approx 60.95$ years;
- $P_{12} \approx 114.78$ years;
- $P_{23} \approx 129.95$ years;
- $P_{123} \approx 983.4$ years (quasi-millennial).

The envelope functions for these beats are:

$$b_{12}(t) = 0.60\cos\left(2\pi\frac{t-1980.528}{114.783}\right);$$

$$b_{13}(t) = 0.40\cos\left(2\pi\frac{t-2067.044}{60.9484}\right);$$

$$b_{23}(t) = 0.45\cos\left(2\pi\frac{t-2035.043}{129.951}\right).$$

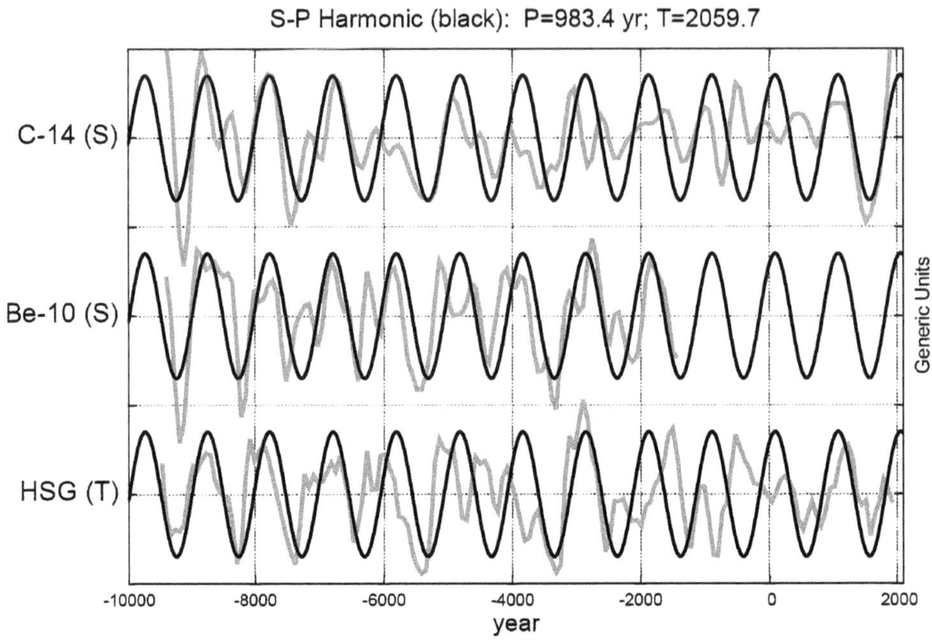

FIGURE 36.4. Comparison of the 983-year cycle predicted by the Jupiter–Saturn model (black harmonic curve) with millennial oscillations found in ^{14}C and ^{10}Be records used as proxies for solar activity, and a composite (HSG MC52-VM29-191) of deep-sea sediment drift ice index records used as a proxy for subpolar North Atlantic surface temperature during the Holocene (Bond et al., 2001). (Adapted from Scafetta, 2012b; Scafetta & Bianchini, 2023).

The millennial-scale function is:

$$g_m(t) = \cos\left(2\pi \frac{t - 2059.686}{983.401}\right).$$

Figures 36.3 and 36.4 show that $g_m(t)$, derived from the Jupiter–Saturn model, correlates well with the Eddy cycle, a millennial solar oscillation detected in radionuclide data (^{14}C and ^{10}Be) (Kerr, 2001; Steinhilber et al., 2012; Scafetta, 2012d, 2014b),[5] and reflected in paleoclimate records such as Alpine glacier

[5]Kerr, R.A., 2001. A variable sun paces millennial climate. Science, 294(5546), 1431–1433.
Steinhilber, F., Abreu, J.A., Beer, J., Brunner, I., Christl, M., et al., 2012. 9,400 years of cosmic radiation and solar activity from ice cores and tree rings. Proceedings of the National Academy of Sciences, 109(16), 5967–5971.
Scafetta, N., 2012d. Does the Sun work as a nuclear fusion amplifier of planetary tidal forcing? Journal of Atmospheric and Solar–Terrestrial Physics, 81–82, 27–40.
Scafetta, N., 2014b. Discussion on the spectral coherence between planetary, solar and climate oscillations: a reply to some critiques. Astrophysics and Space Science, 354, 275–299.

movements (Kutschera et al., 2020)[6] and North Atlantic temperature proxies (Bond et al., 2001).[7]

The model predicts a peak around 2060, aligning with Holocene climate oscillations. This supports the hypothesis that part of the warming observed since 1850 may be attributable to a natural solar modulation.

Historical resonance and symbolic convergence. This prediction curiously echoes a letter written by Isaac Newton,[8] preserved at the Hebrew University of Jerusalem, in which he speculated that 2060 could mark the end of the current world order, on the basis of symbolic calculations from the Book of Daniel. Although prophetic in nature, this date coincides with the peak of the quasi-millennial cycle anticipated by the Jupiter–Saturn model, as previously suggested by al-Balkhi (IX secolo) and Kepler (1606).[9]

This convergence between ancient intuition and modern astrophysical modeling offers a compelling narrative: that long-term solar variability may be governed by deep harmonic structures rooted in planetary dynamics.

Reconstruction of solar grand minima and climate cycles. The harmonic model successfully reconstructs the major secular solar minima of the past millennium, including the Oort, Wolf, Spörer, Maunder, and Dalton minima. These events coincide with destructive interference phases among the three harmonic components, while constructive interference corresponds to periods of maximum solar activity.

Between 1850 and 2150, the model predicts five cycles of approximately 59–63 years, aligning with observed climate maxima around 1880, 1940, and 2000. A solar minimum is projected around 2030, nestled between two maxima: one that already occurred around 2000 and another expected near 2060.

This dynamic is consistent with the upward trend in TSI from 1980 to 2000 shown by the ACRIM composite (Willson & Mordvinov, 2003),[10] which reveals significant secular variability during solar cycles 21–23 and contradicts the PMOD composite (Fröhlich, 2006),[11] which suggests a continuous decline in TSI since 1980. See also Scafetta et al. (2019a)[12] and Connolly et al. (2024).[13]

Anomalous cycles and emerging harmonics. Scafetta (2012b) showed that the model can also account for an unusually long Schwabe cycle of approximately

[6]Kutschera, W., et al., 2020. The movements of Alpine glaciers throughout the last 10,000 years as sensitive proxies of temperature and climate changes. EPJ Web of Conferences, 232, 02002.

[7]Bond, G., et al., 2001. Persistent solar influence on North Atlantic climate during the Holocene. Science, 294(5549), 2130–2136.

[8]Newton, I., 1704. Manuscript letter containing prophetic calculations based on the Book of Daniel. Newton Archive, Hebrew University of Jerusalem.

[9]Kepler, J., 1606. De Stella Nova in Pede Serpentarii (On the new star in Ophiuchus's foot).

[10]Willson, R.C., Mordvinov, A.V., 2003. Secular total solar irradiance trend during solar cycles 21–23. Geophysical Research Letters, 30(5), 1199.

[11]Fröhlich, C., 2006. Solar irradiance variability since 1978. Space Science Reviews, 125, 53–65.

[12]Scafetta, N., et al., 2019. Modeling quiet solar luminosity variability from TSI satellite measurements and proxy models during 1980–2018. Remote Sensing, 11(21), 2569.

[13]Connolly, R., et al., 2024. Multiple new or updated satellite total solar irradiance (TSI) composites (1978–2023). The Astrophysical Journal, 975, 102.

15 years between 1680 and 1700, observed in isotope proxies (^{18}O and ^{14}C) (Yamaguchi et al., 2010).[14]

To refine the model, Scafetta (2014b)[15] proposed that planetary forcing is nonlinearly modulated by the Sun's internal dynamics, generating additional harmonics with periods of approximately 57, 61, 65, 103, 115, 130, and 150 years. These frequencies appear in solar power spectra (Ogurtsov et al., 2002)[16] and in long-term solar proxies (^{10}Be, ^{14}C) dating back over 325,000 years (Cauquoin et al., 2014).[17]

Robustness of the harmonic model: astronomical foundations and observational data. The strength of the harmonic model lies in its physical foundations, not in mere empirical fitting. The frequencies and phases are derived from astronomical parameters and observational data, rather than being optimized to match recent trends.

In particular, the model integrates planetary cycles and orbital configurations with the historical sunspot record from 1750 to 2010, offering enhanced predictive capability and avoiding the distortions typical of models calibrated solely on modern datasets.

36.4. The Suess–de Vries cycle (190–230 years)

The Suess–de Vries cycle has a periodicity between 190 and 230 years. It is one of the principal secular oscillations of solar variability, well documented in paleoclimate records, particularly in radiocarbon data (de Vries, 1958; Suess, 1965).[18] It has also been recognized as a significant modulator of solar activity in numerous recent studies (Neff et al., 2001; Abreu et al., 2012; McCracken et al., 2013; Beer et al., 2018).[19] The best estimates place its length between 200 and 215 years, although some analyses extend the range from 170 to 240 years.

[14] Yamaguchi, Y.T., et al., 2010. Synchronized Northern Hemisphere climate change and solar magnetic cycles during the Maunder Minimum. Proceedings of the National Academy of Sciences, 107, 20697–20702.

[15] Scafetta, N., 2014b. Discussion on the spectral coherence between planetary, solar and climate oscillations: a reply to some critiques. Astrophysics and Space Science, 354, 275–299.

[16] Ogurtsov, M.G., et al., 2002. Long-period cycles of the Sun's activity recorded in direct solar data and proxies. Solar Physics, 211, 371.

[17] Cauquoin, A., et al., 2014. No evidence for planetary influence on solar activity 330,000 years ago. Astronomy and Astrophysics, 561, A132.

[18] de Vries, H., 1958. Variations in concentration of radiocarbon with time and location on Earth. Proceedings of the Royal Netherlands Academy of Arts and Sciences B, 61, 94–102.
Suess, H.E., 1965. Secular variations of the cosmic-ray-produced carbon-14 in the atmosphere and their interpretations. Journal of Geophysical Research, 70, 5937–5952.

[19] Neff, U., et al., 2001. Strong coherence between solar variability and the monsoon in Oman between 9 and 6 kyr ago. Nature, 411, 290–293.
Abreu, J.A., et al., 2012. Is there a planetary influence on solar activity? Astronomy and Astrophysics, 548, A88.
McCracken, K.G., et al., 2013. A phenomenological study of the cosmic ray variations over the past 9400 years. Solar Physics, 286, 609.
Beer, J., Tobias, S.M., Weiss, N.O., 2018. On long-term modulation of the Sun's magnetic cycle. Monthly Notices of the Royal Astronomical Society, 473, 1596–1602.

Theoretical origins and astronomical synchronizations. Stefani et al. (2024)[20] proposed that the Suess–de Vries cycle, along with the Hale cycle (∼22 years) and the Gleissberg cycle (∼80–100 years), may arise from synchronization phenomena between two major astronomical forcings:

(1) the 11.07-year tidal oscillation generated by the interaction among Venus, Earth, and Jupiter;
(2) the 19.86-year gravitational modulation caused by the Sun's motion around the solar system's barycenter, primarily influenced by Jupiter and Saturn.

This interaction is hypothesized to produce a harmonic cycle of ∼193 years, closely matching the Suess–de Vries periodicity.

More specifically, the model proposes that magneto-Rossby waves in the Sun are excited by planetary tidal forces. The Schwabe cycle (11.07 years) is synchronized by Venus–Earth–Jupiter tides, while the Suess–de Vries cycle emerges from the beat between the Hale cycle and the rosette-like motion of the Sun around the barycenter, dominated by Jupiter–Saturn conjunctions.

Figure 36.5 illustrates this mechanism, showing strong spectral coherence between the 193-year periodicity derived from the harmonic planetary model and climate proxy data from Lake Lisan varved sediments (Prasad et al., 2004).[21] The coherence extends to other bands, including the Gleissberg cycle (∼90 years), suggesting that solar cycles may be induced by planetary orbits through a stable external cadence, modulating terrestrial climate on secular timescales.

Orbital invariant inequalities and harmonic construction. The ∼193-year cycle can also be derived from orbital invariant inequalities, based on linear combinations of planetary synodic frequencies. Specifically, the vector $(-3, 5, -1, -1)$ is obtained as the difference between:

- $(0, 0, 1, -1)$: Jupiter–Saturn synodic cycle (19.86 years);
- $(3, -5, 2, 0)$: Venus–Earth–Jupiter alignment (22.14 years, Hale cycle).

Thus:

$$(0, 0, 1, -1) - (3, -5, 2, 0) = (-3, 5, -1, -1) \Rightarrow \text{Period} \approx 193 \text{ years}$$

The sum of these vectors yields another harmonic:

$$(0, 0, 1, -1) + (3, -5, 2, 0) = (3, -5, 3, -1) \Rightarrow \text{Period} \approx 10.47 \text{ years}$$

This 10.47-year cycle is well documented in the solar and climate literature and may represent a secondary harmonic influencing the Schwabe cycle (Scafetta, 2014b, 2020).[22]

Alternative: planetary beats and cyclical overlaps. Spectral analyses of ^{14}C and ^{10}Be suggest a Suess–de Vries periodicity closer to 205–210 years (Abreu

---

[20] Stefani, F., Horstmann, G.M., Klevs, M., Mamatsashvili, G., Weier, T., 2024. Rieger, Schwabe, Suess-de Vries: the sunny beats of resonance. Solar Physics, 299, 51.

[21] Prasad, S., et al., 2004. Evidence from Lake Lisan of solar influence on decadal- to centennial-scale climate variability during marine oxygen isotope stage 2. Geology, 32, 581.

[22] Scafetta, N., 2014b. Discussion on the spectral coherence between planetary, solar and climate oscillations: a reply to some critiques. Astrophysics and Space Science, 354, 275–299.
Scafetta, N., 2020. Solar oscillations and the orbital invariant inequalities of the solar system. Solar Physics, 295(2), 33.

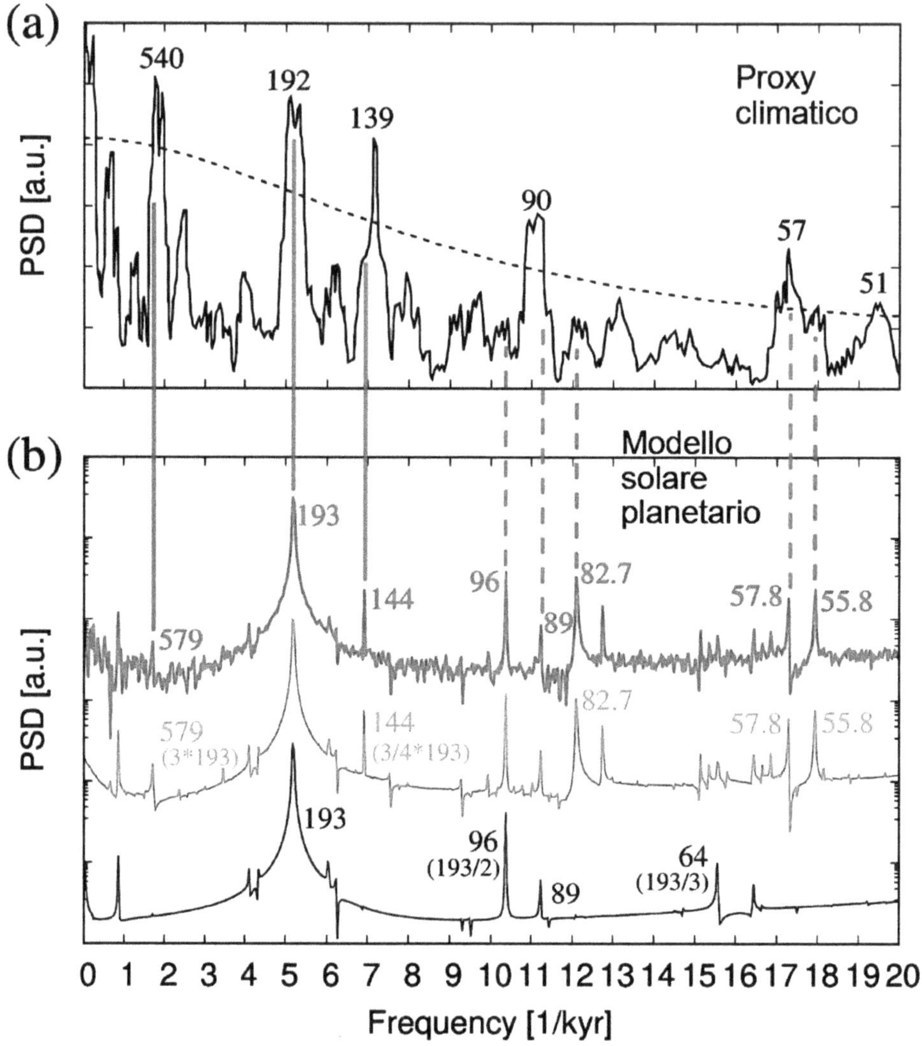

FIGURE 36.5. Comparison between climate proxy spectra and solar dynamo simulations.
(A) Power spectrum of varved lake sediment data from Lake Lisan, adapted from Prasad et al. (2004). The dashed black line marks the 95% confidence threshold; significant periodicities are annotated in years.
(B) Spectral output from a 1D α–Ω solar dynamo model. The bottom curve shows the model response when forced by Jupiter and Saturn via orbital angular momentum; the middle curve includes tidal contributions from all the planets; the top curve incorporates stochastic noise in the α-term to simulate internal solar variability.
(Adapted from Stefani et al., 2024).

et al., 2012; McCracken et al., 2013).[23] The model of orbital invariant inequalities offers a complementary explanation, involving interactions among Jupiter, Saturn, Uranus, and Neptune.

For example, a beat between the 60-year Jupiter–Saturn cycle and Uranus's 84-year orbit yields:

$$\frac{1}{60} - \frac{1}{84} = \frac{1}{210} \Rightarrow \text{Period} \approx 210 \text{ years}$$

Similarly, the *Trigon of Great Conjunctions* ($\sim$59.6 years) produces:

$$\frac{1}{59.6} - \frac{1}{84} = \frac{1}{205} \Rightarrow \text{Period} \approx 205 \text{ years}$$

These estimates align with the radiocarbon data of Wagner et al. (2001),[24] and converge near the commonly cited estimate of 208 years.

36.5. Conclusion: classification of planetary forcings

Planetary cycles that may influence solar activity can be broadly classified into two categories.

(1) *Periodicities from orbital invariant inequalities*: linear combinations of synodic frequencies between planetary pairs, assuming circular orbits.
(2) *Periodicities from orbital eccentricity and precession*: effects arising from elliptical orbital dynamics and long-term secular variations.

Both mechanisms appear to interact and synchronize the Sun's internal dynamics, generating complex but physically plausible modulations. These mechanisms offer a coherent and predictive framework capable of explaining the full spectrum of solar activity oscillations observed in paleoclimate proxies.

[23] Abreu, J.A., et al., 2012. Is there a planetary influence on solar activity? Astronomy and Astrophysics, 548, A88.
McCracken, K.G., et al., 2013. A phenomenological study of the cosmic ray variations over the past 9400 years. Solar Physics, 286, 609.
[24] Wagner, G., et al., 2001. Presence of the solar de Vries cycle (205 years) during the last ice age. Geophysical Research Letters, 28(15), 303–306.

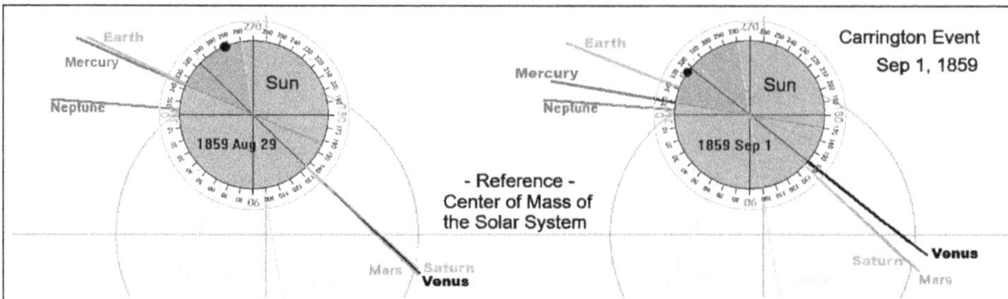

TABLE I.—PLANET POSITIONS DURING FLARES OF THE NINE LARGEST SOLAR FLARE EVENTS

Event period number	Event position[a]		Flare intensity[b]	Event onset time[c]		Heliocentric ecliptic longitude,[d,e] deg					Event position relative to nearest high-tide position	
				Date	UT	M	V	E	J	F	Deviation,[f] deg	Planet[e]
1	20N	15E	-------	1859 Sept. 01	11:20	352.4	143.3	340.4	101.2	326.1	3	V
2	7N	4E	3	1942 Feb. 28	12:42	215.4	176.0	160.1	84.3	155.8	–4	E
3	20N	15E	3+	1946 July 25	15:04	285.4	232.6	302.8	211.1	289.6	4	M
4	2S	70W	3+	1949 Nov. 19	10:30	233.0	14.2	57.5	308.5	127.5	–1	J
5	22N	90E	3?	1956 Feb. 10	21:10	192.4	58.1	141.7	146.7	49.1	–9	V
	20N	4W	3	1956 Feb. 17	11:00	214.5	68.7	148.4	147.2	151.5	3, 4	E, J
	23N	80W	[g]3	1956 Feb. 23	3:31	231.6	77.9	154.1	147.7	237.1	5	M
6	20N	60E	3+	1959 July 10	2:06	235.5	254.5	287.7	241.8	230.0	–5	M
	17N	4E	[g]3+	1959 July 14	3:25	246.8	260.9	291.5	242.1	288.9	–3	E
	16N	31W	3+	1959 July 16	21:14	254.4	265.3	294.2	242.3	325.5	31	E
7	28N	28E	3	1960 Nov. 10	10:09	62.6	317.7	48.6	281.1	18.9	–30	E
	27N	4W	[g]3+	1960 Nov. 12	13:15	75.9	321.0	50.7	281.2	51.7	1	E
	25N	35W	3	1960 Nov. 15	2:07	91.9	325.1	53.3	281.4	85.0	–7	M
8	13N	27E	2B	1972 Aug. 02	19:58	299.2	340.0	311.0	277.0	285.0	8	J
	14N	8E	3B	1972 Aug. 04	6:20	303.8	342.3	312.3	277.1	305.1	1, –7	M, E
	14N	36W	[g]3B	1972 Aug. 07	14:43	315.0	347.6	315.6	277.4	351.2	4	V
9	16S	8E	X17.2	2003 Oct. 28	8:32	220.7	258.9	34.5	154.1	28.3	–6	E
	15S	2W	X10.0	2003 Oct. 29	20:25	225.2	261.3	36.0	154.2	40.0	4, –5	E, M
	14S	56W	X8.3	2003 Nov. 02	16:41	236.3	267.4	39.8	154.5	97.5	10	V
	19S	83W	[g]X28.0	2003 Nov. 04	19:08	242.2	270.7	41.9	154.6	126.5	–28	J

FIGURE 36.6. **Solar flares and solar tides caused by the planets.** The Carrington Event was the most powerful geomagnetic storm ever recorded. It was triggered by a massive solar flare and coronal mass ejection on September 1, 1859, and one day later caused auroras to be visible as far south as the Caribbean and disrupted telegraph systems worldwide. The astronomical configurations preceding the event suggest a possible planetary influence: on August 29, 1859, a triple conjunction occurred among Saturn, Venus, and Mars, while Earth was aligned with Mercury. The event developed near the peak of the tidal forces exerted by these planets, partially overlapping with Jupiter's tidal influence. In the accompanying figure, the black dot in the left panel marks the location on the solar surface where the flare originated, while in the right panel it indicates the corresponding heliographic position of the event taking solar rotation into account. The Table lists the positions of the planets during the nine largest solar flare events (From Hung, 2007).

CHAPTER 37

The stable harmonics of the solar system

37.1. Solar activity cycles

Beyond the well-established Schwabe sunspot cycle (~11 years) and the Hale magnetic cycle (~22 years), solar activity exhibits a rich spectrum of recurring oscillations spanning multiple timescales. These include:

- millennial cycles:
  - Bray–Hallstatt cycle: 2100–2500 years,
  - Eddy cycle: 800–1200 years;
- secular cycles:
  - Suess–de Vries cycle: 200–250 years,
  - Jose cycle: 150–190 years,
  - Gleissberg cycle: 80–100 years;
- multidecadal bands:
  - periodicities between 40 and 65 years.

It is hypothesized that these frequencies are closely linked to specific harmonics of the planetary system, particularly the orbital dynamics of the giant gas planets Jupiter, Saturn, Uranus, and Neptune. Their gravitational and magnetic fields may influence the Sun's magnetic structure and coronal activity, potentially entraining solar variability to planetary periodicities.

Orbital invariant inequalities and planetary harmonics. In this context, Scafetta (2020)[1] introduced the concept of "*orbital invariant inequalities*", a system of mathematically derived periods based on linear combinations of planetary synodic frequencies.

These planetary frequencies aim to identify the spectral architecture capable of synchronizing a differentially rotating system such as the Sun. The approach provides a theoretical framework for evaluating the plausibility of solar–planetary synchronization across a wide range of timescales, from short-term oscillations to millennial-scale cycles.

Although this interpretation remains under active scientific scrutiny, it offers a coherent and physically grounded foundation for integrating the observed cyclic behavior of the Sun with long-term astronomical dynamics. By bridging solar physics and celestial mechanics, the orbital invariant inequality framework contributes to a deeper understanding of the stable harmonic structure underlying solar variability.

---

[1]Scafetta, N., 2020. Solar oscillations and the orbital invariant inequalities of the solar system. Solar Physics, 295(2), 33.

37.2. Historical background and challenges

The discovery of the 11-year solar cycle led Rudolf Wolf (1859)[2] to question the physical mechanism underlying such periodicity. He hypothesized that it could result from planetary influences, particularly those exerted by Venus, Earth, Jupiter, and Saturn, suggesting that periodic forcings linked to orbital motions, such as gravitational tides, might synchronize the Sun's internal dynamics and modulate its activity at specific frequencies.

This theory was initially met with skepticism, since the tidal forces exerted by the planets on the Sun are extremely weak, on the order of a few millimeters (Scafetta, 2012d),[3] and generally considered negligible. Yet, despite decades of research, no complete explanation has been provided for the origin of the Schwabe cycle, nor for other observed solar periodicities.

Expanding the solar paradigm. Over the past fifty years, the science of solar activity has expanded considerably. It is now evident that the 11-year cycle is merely the most prominent among many cyclicities. It varies over time and, in certain historical periods, disappears entirely, as during the Maunder Minimum (1645–1715), which coincided with a period of global climate cooling (Eddy, 1976).[4]

Beyond the Schwabe (~11 years) and Hale (~22 years) cycles, numerous longer-period oscillations have been identified, extending to multi-millennial scales. These cycles mirror patterns found in climate data, suggesting a direct link between solar variability and Earth's climate (Neff et al., 2001; Kerr, 2001; Ogurtsov et al., 2002; Steinhilber et al., 2012, and other studies).[5]

This connection elevates the study of solar dynamics from a purely astrophysical pursuit to a critical component in understanding and forecasting climate change.

Limitations of traditional solar dynamo models. The most advanced solar dynamo theories attribute solar cycles exclusively to internal mechanisms within the Sun (de Jager & Versteegh, 2005).[6] However, these models fail to reproduce the period and phase of the observed solar cycles from first principles (Tobias,

---

[2]Wolf, R., 1859. Extract of a letter from Prof. R. Wolf of Zurich to Mr. Carrington, dated Jan. 12, 1859. Monthly Notices of the Royal Astronomical Society, 19, 85.

[3]Scafetta, N., 2012d. Does the Sun work as a nuclear fusion amplifier of planetary tidal forcing? Journal of Atmospheric and Solar–Terrestrial Physics, 81–82, 27–40.

[4]Eddy, J.A., 1976. The Maunder Minimum. Science, 192, 1189–1202.

[5]Neff, U., Burns, S.J., Mangini, A., Mudelsee, M., Fleitmann, D., Matter, A., 2001. Strong coherence between solar variability and the monsoon in Oman between 9 and 6 kyr ago. Nature, 411, 290–293.
Kerr, R.A., 2001. A variable sun paces millennial climate. Science, 294(5546), 1431–1433.
Ogurtsov, M.G., et al., 2002. Long-period cycles of the Sun's activity recorded in direct solar data and proxies. Solar Physics, 211, 371.
Steinhilber, F., Abreu, J.A., Beer, J., Brunner, I., Christl, M., et al., 2012. 9,400 years of cosmic radiation and solar activity from ice cores and tree rings. Proceedings of the National Academy of Sciences, 109(16), 5967–5971.

[6]de Jager, C., Versteegh, G.J.M., 2005. Do planetary motions drive solar variability? Solar Physics, 229, 175–179.

2002).[7] Theoretical limitations arise because α–Ω dynamo models require *ad hoc* calibration of free parameters to generate the 11-year cycle (Jiang et al., 2007).[8]

This has led to some scientific debate. For example, Callebaut et al. (2012)[9] criticized the idea of planetary synchronization using physical arguments similar to those proposed by de Jager & Versteegh (2005), asserting in addition that no study had successfully correlated the five main solar periodicities—Schwabe, Hale, Gleissberg, Suess–de Vries, and Bray–Hallstatt—with planetary cycles:

> *"It should be remarked in this connection that virtually none of the papers on planetary influences on solar variability succeeded in identifying these five periodicities in the planetary attractions."*

However, following a rebuttal by Scafetta et al. (2013),[10] Callebaut et al. (2013)[11] conceded:

> *"It is well-known that there are some periodicities that are common to solar activity and planetary motions."*

This shift underscores that criticism of planetary theories often stems more from philosophical bias than from empirical refutation. The fact remains: the Sun exhibits quasi-regular oscillations that align with planetary motions. Since this remains an unresolved enigma, it merits open-minded scientific inquiry.

Re-emergence of Wolf's hypothesis. Owing to the limitations of traditional models, the original hypothesis of Wolf (1859) that planetary motions are the only forcings capable of generating coherent harmonic oscillations has re-emerged in updated form over the past two decades.

Experimental evidence includes temporal coincidences between solar flares and specific planetary alignments (Hung, 2007; Bertolucci et al., 2017; Scafetta & Mörner, 2025),[12] and the spectral coherence between solar oscillations and orbital motions, particularly the Sun's motion relative to the solar system barycenter, used as a dynamic proxy (Fairbridge & Shirley, 1987; Abreu et al., 2012; Scafetta & Willson, 2013b; Scafetta et al., 2016).[13]

---

[7]Tobias, S.M., 2002. The solar dynamo. Philosophical Transactions of the Royal Society A: Mathematical, Physical and Engineering Sciences, 360(1801), 2741–2756.

[8]Jiang, J., et al., 2007. Solar activity forecast with a dynamo model. Monthly Notices of the Royal Astronomical Society, 381, 1527–1542.

[9]Callebaut, D.K., de Jager, C., Duhau, S., 2012. The influence of planetary attractions on the solar tachocline. Journal of Atmospheric and Solar-Terrestrial Physics, 80, 73–78.

[10]Scafetta, N., Humlum, O., Solheim, J.-E., & Stordahl, K., 2013. Comment on "The influence of planetary attractions on the solar tachocline" by Callebaut, de Jager and Duhau. Journal of Atmospheric and Solar-Terrestrial Physics, 102, 368–371.

[11]Callebaut, D.K., de Jager, C., Duhau, S., 2013. Reply to "The influence of planetary attractions on the solar tachocline" by N. Scafetta, O. Humlum, J.E. Solheim, K. Stordahl. Journal of Atmospheric and Solar–Terrestrial Physics, 102, 372–373.

[12]Hung, C.-C., 2007. Apparent relations between solar activity and solar tides caused by the planets. NASA Technical Memorandum, TM-2007-214817.

Scafetta, N., Mörner, N.-A., 2025. The giant solar flare event of January 7, 2014 in light of the planetary theory of solar variability. Science of Climate Change, 5.2, 161–165.

Bertolucci, S., et al., 2017. The Sun and its planets as detectors for invisible matter. Physics of the Dark Universe, 17, 13.

[13]Fairbridge, R.W., Shirley, J.H., 1987. Prolonged minima and the 179-yr cycle of the solar inertial motion. Solar Physics, 110, 191.

Abreu, J.A., et al., 2012. Is there a planetary influence on solar activity? Astronomy and Astrophysics,

Addressing the *"cherry-picking"* objection. A common objection is that the large number of planetary frequencies allows for arbitrary correlations with solar or climate cycles, raising concerns of statistical coincidence or "cherry-picking," akin to criticisms of astrology. To overcome this critique, researchers focus on the major planetary harmonics, those generated by the most massive or closest planets to the Sun, which may have a plausible physical role, even if not yet fully understood.

More decisively, Scafetta (2020) and Scafetta & Milani (2025)[14] developed a planetary model that identifies the entire harmonic structure of the solar system and compares it with solar and climate cyclicities. This approach resolves the cherry-picking problem at its root, as it does not rely on the selection of isolated configurations, but rather on the unique and complete harmonic spectrum of the system.

Toward a unified harmonic framework. This chapter has described the planetary structure derived from the orbits of the giant gas planets Jupiter, Saturn, Uranus, and Neptune. A simplified version of this model, extended to the terrestrial planets, was used in earlier chapters.

A key property of these harmonics is their spectral invariance with respect to all rotating reference frames, including the Sun and the heliosphere. This invariance is physically necessary for the activation of synchronization mechanisms between weak harmonic forcings and oscillating dynamical systems.

The general mechanism recalls the phenomenon observed by Huygens in the 17th century, when two pendulums suspended from the same structure synchronized their oscillations. The Huygens principle was later formalized in the study of coupled oscillators (Strogatz, 2009).[15]

37.3. Orbital invariant inequalities generated by the Jovian planets

The concept of orbital invariant inequalities, introduced by Scafetta (2020), provides a powerful framework for identifying stable harmonic structures within the solar system. These inequalities describe quasi-commensurable relationships among planetary orbital periods and offer a theoretical basis for understanding how planetary motions may synchronize solar activity.

In celestial mechanics, a resonance between two harmonics with periods T_1 and T_2 and integers a_1 and a_2 occurs when:

$$\frac{T_1}{T_2} = \frac{a_1}{a_2}.$$

A classic example is the Laplace resonance among Jupiter's Galilean moons—Io, Europa, and Ganymede—with orbital ratios of 1:2:4.

548, A88.

Scafetta, N., Willson, R.C., 2013b. Empirical evidences for a planetary modulation of total solar irradiance and the TSI signature of the 1.09-year Earth–Jupiter conjunction cycle. Astrophysics and Space Science, 348, 25–39.

Scafetta, N., et al., 2016. On the astronomical origin of the Hallstatt oscillation found in radiocarbon and climate records throughout the Holocene. Earth-Science Reviews, 162, 24–43.

[14] Scafetta, N., Milani, F., 2025. Spectral Structure of the Solar Inertial Motion from 12999 BC to 16998 AD. Publications of the Astronomical Society of the Pacific, 137, 054402.

[15] Strogatz, S.H., 2009. Exploring complex networks. Nature, 410, 268.

	Orb. Inv. Ineq.	Period (yr)	Julian Date	Long.
Jup–Sat	$(1, -1, 0, 0)$	19.8593	2451718.4	52° 01′
Jup–Ura	$(1, 0, -1, 0)$	13.8125	2450535.8	305° 22′
Jup–Nep	$(1, 0, 0, -1)$	12.7823	2450442.1	297° 21′
Sat–Ura	$(0, 1, -1, 0)$	45.3636	2447322.1	269° 05′
Sat–Nep	$(0, 1, 0, -1)$	35.8697	2447725.6	281° 14′
Ura–Nep	$(0, 0, 1, -1)$	171.393	2449098.1	289° 22′

FIGURE 37.1. Mean synodic periods of conjunctions among the Jovian planets, with date (in Julian days near the year 2000 AD) and heliocentric longitude using the synodic periods from Figure 35.1 (Adapted from Scafetta, 2020).

Quasi-commensurability and synodic periods. In most cases, exact resonance does not occur, and a *"quasi-commensurability"* or *"orbital inequality"* with frequency f or period T can be defined as:

$$T = \frac{1}{f} = \left| \frac{a_1}{T_1} - \frac{a_2}{T_2} \right|^{-1}.$$

Perfect resonance yields $T = \infty$ and $f = 0$. The simplest case of orbital inequality is the synodic period between two planets (i.e., the interval between successive conjunctions), which is calculated as:

$$T_{12} = \frac{1}{f_{12}} = \left| \frac{1}{T_1} - \frac{1}{T_2} \right|^{-1}.$$

This corresponds to $a_1 = 1$ and $a_2 = -1$, and satisfies the condition $a_1 + a_2 = 0$.

Figure 37.1 presents the synodic periods between the four gas giants—Jupiter, Saturn, Uranus, and Neptune—along with their mean temporal phases near the year 2000. Figure 37.2 illustrates gravitational couplings in a hypothetical solar system composed of four equal-mass planets orbiting a central body ten times more massive. Although not to scale, the image conveys how planetary motion may induce vibrational modes in the solar system's gravitational field.

Generalization to multiple orbital inequalities. The synodic equation can be generalized to a system of n harmonics:

$$f = \frac{1}{T} = \left| \sum_{i=1}^{n} \frac{a_i}{T_i} \right|,$$

where a_i are integers and T_i are the orbital periods of the planets $i = 1, ..., n$. A restricted subset of these frequencies is defined by the condition:

$$\sum_{i=1}^{n} a_i = 0,$$

which ensures rotational invariance of the resulting harmonics.

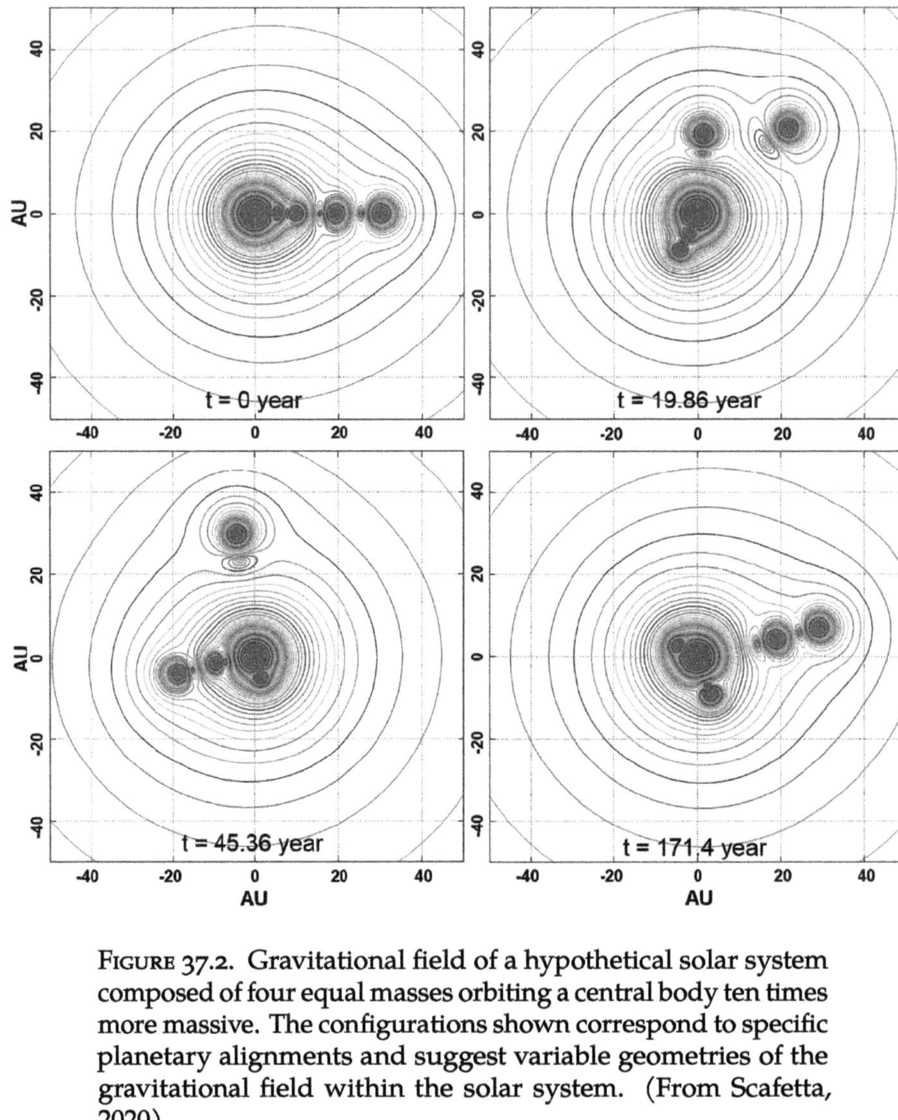

FIGURE 37.2. Gravitational field of a hypothetical solar system composed of four equal masses orbiting a central body ten times more massive. The configurations shown correspond to specific planetary alignments and suggest variable geometries of the gravitational field within the solar system. (From Scafetta, 2020).

In a rotating reference frame centered on the Sun with period P, the observed orbital frequencies become:

$$f'_i = \frac{1}{T'_i} = \frac{1}{T_i} - \frac{1}{P}.$$

Thus, in the rotating system the orbital inequality transforms as:

$$f' = \frac{1}{T'} = \left| \sum_{i=1}^{n} \frac{a_i}{T'_i} \right| = \left| \sum_{i=1}^{n} \frac{a_i}{T_i} - \frac{\sum_{i=1}^{n} a_i}{P} \right| = \left| \sum_{i=1}^{n} \frac{a_i}{T_i} \right| = \frac{1}{T} = f.$$

Given $\sum a_i = 0$, we have $f' = f$ and $T' = T$, meaning that the periods are invariant under rotation of the reference system. This implies that planetary synodic cycles and their beats are observable from any rotating frame centered on the Sun. Since the Sun exhibits differential rotation, each solar latitude experiences the same planetary beats. This property is necessary in order to render harmonic forcings effective synchronizing agents of the solar dynamo.

Classification and structure of planetary harmonics. Figure 37.4 lists the main orbital invariant inequalities generated by the gas giants, using the formalism:

$$P_{(a_1,a_2,a_3,a_4)} = \left| \sum_{i=1}^{n} \frac{a_i}{T_i} \right|^{-1} \equiv (a_1, a_2, a_3, a_4),$$

where a_1, a_2, a_3, and a_4 are integers and $a_1 + a_2 + a_3 + a_4 = 0$. Each index refers to a gas giant in the order: (1) Jupiter; (2) Saturn; (3) Uranus; and (4) Neptune.

The harmonics are classified using two indices:

(1) M: the maximum absolute value among the coefficients $|a_i|$;
(2) $K = \frac{1}{2} \sum |a_i|$: representing the number of beating planetary pairs.

Harmonic clustering and the Jose cycle. It is observed that planetary harmonics are not randomly distributed but rather clustered in groups that correspond to known solar oscillations. Figure 37.3 visually displays these harmonics, revealing their organization within a structure governed by a base period of approximately 179 years, which corresponds to the cycle of Jose (1965),[16] discussed in Section 37.6.

This result demonstrates that the natural frequencies of the solar system exhibit a precise harmonic organization, which has nearly synchronized the planetary orbits and may facilitate other synchronization mechanisms within the solar–planetary system.

37.4. Solar activity, orbital forcing, and climate coherence

The relevance of planetary harmonics for both solar physics and climate science is illustrated in Figure 37.5, which compares two independent reconstructions:

- a record of solar variability based on carbon-14 (^{14}C) concentrations;
- a paleoclimate reconstruction based on oxygen-18 (^{18}O) data, covering the interval between 9500 and 6000 years ago.

These datasets are strongly correlated; they reveal the presence of multiple shared frequencies corresponding to well-known solar activity cycles:

- Eddy cycle: 800–1200 years;
- Suess–de Vries cycle: 200–250 years;
- Jose cycle: 155–185 years;
- Gleissberg cycle: 80–100 years;
- a cluster around 55–65 years;
- a cluster around 40–50 years, among others.

In Figure 37.5B, the spectral peaks common to both records are compared with the clusters of orbital invariant inequalities (bars) shown in Figure 37.3. The comparison demonstrates that the orbital model predicts frequency clusters that

[16] Jose, P.D., 1965. Sun's motion and sunspots. The Astronomical Journal, 70, 193.

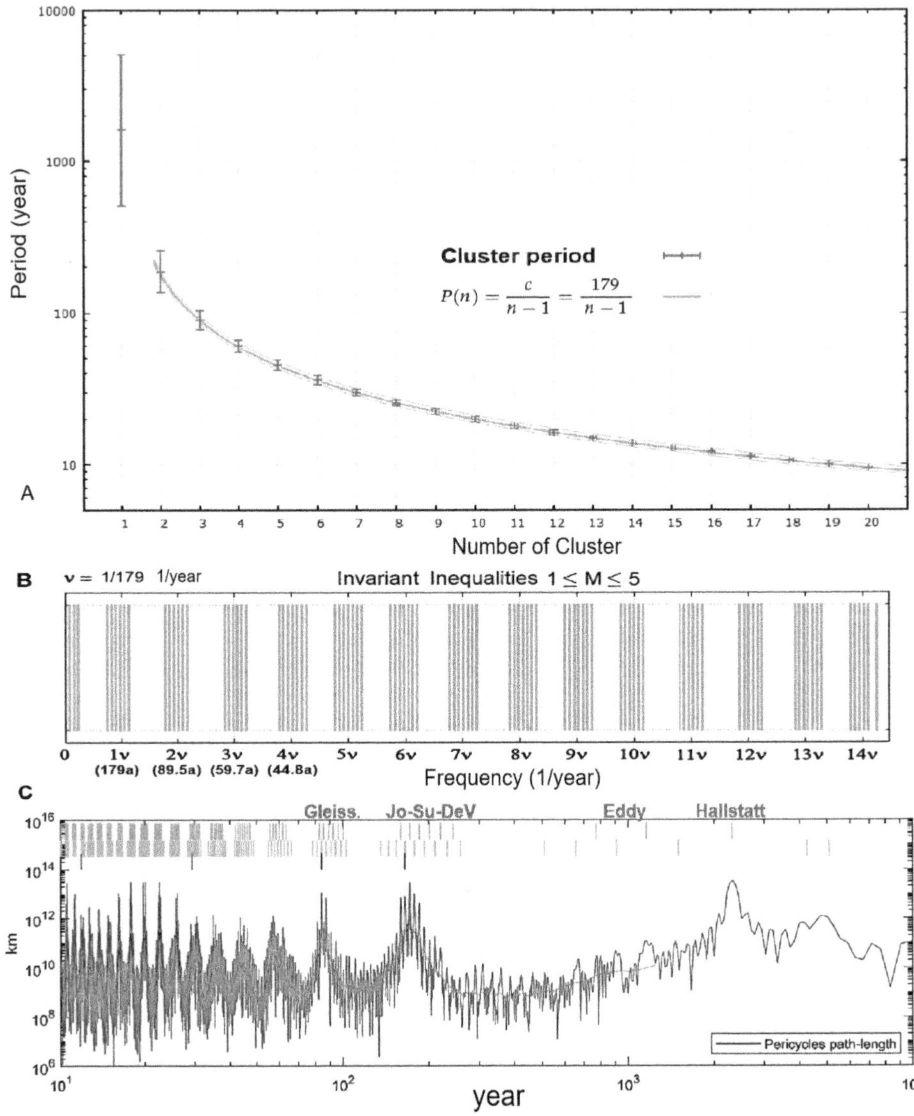

FIGURE 37.3. [A] Distribution law of planetary frequency clusters up to harmonic #20 of the base period proposed by Jose (1965), found by Scafetta & Milani (2025).
[B] Clustering of invariant inequalities up to harmonic #14 of the base period of Jose (1965).
[C] Periodogram of the lengths of the solar inertial motion pericycles, highlighting harmonic periodicity clusters. Colored bars represent orbital periods derived from the extended model of "*orbital invariant inequalities*," showing harmonic clustering consistent with solar activity cycles.

Orbital Invariant Inequalities: Period $P \geqslant 40$ yr and $M \leqslant 5$					Their Beats with Jupiter		
(Jup, Sat, Ura, Nep)	(M, K)	P_{model} (year)	$P_{\text{Periodogram}}$	cluster	(Jup -1, Sat, Ura, Nep)	P_{model} (year)	$P_{\text{Periodogram}}$
(1, −3, 5, −3)	(5, 6)	42.069 ± 0.002	42.08 ± 0.06		(0–1, 3, −2, −1)	81.072 ± 0.004	81.2 ± 0.25
(0, 0, 4, −4)	(4, 4)	42.847 ± 0.002	42.86 ± 0.06		(2–1, −3, 4, −3)	84.265 ± 0.007	84.1 ± 0.25
(2, −5, 1, 2)	(5, 5)	43.653 ± 0.002	43.65 ± 0.06		(2–1, −3, 0, 1)	87.171 ± 0.004	87.3 ± 0.25
(1, −3, −3, 5)	(5, 6)	43.654 ± 0.002	43.65 ± 0.06	∼45 yr	(1–1, 0, 3, −4)	87.445 ± 0.005	87.3 ± 0.25
(1, −2, 0, 1)	(2, 2)	44.591 ± 0.001	44.58 ± 0.07		(0–1, 2, 3, −5)	90.575 ± 0.006	90.8 ± 0.3
(0, 1, −1, 0)	(1, 1)	45.362 ± 0.001	45.37 ± 0.07		(3–1, −5, 0, 2)	90.870 ± 0.007	90.8 ± 0.3
(1, −4, 2, 1)	(4, 4)	46.268 ± 0.002	46.27 ± 0.07		(0–1, 3, 2, −5)	90.873 ± 0.006	90.8 ± 0.3
(1, −1, −5, 5)	(5, 6)	47.209 ± 0.002	47.21 ± 0.08		(1–1, −1, 4, −4)	94.259 ± 0.007	94.5 ± 0.3
(1, −3, 4, −2)	(4, 5)	55.754 ± 0.003	55.71 ± 0.1		(2–1, −2, −1, 1)	94.579 ± 0.004	94.5 ± 0.3
(0, 0, 3, −3)	(3, 3)	57.129 ± 0.003	57.14 ± 0.1		(2–1, −4, 5, −3)	98.25 ± 0.01	98.5 ± 0.4
(2, −5, 0, 3)	(5, 5)	58.571 ± 0.003	58.55 ± 0.1		(1–1, 1, −2, 0)	98.603 ± 0.004	98.5 ± 0.4
(1, −3, −2, 4)	(4, 5)	58.573 ± 0.003	58.55 ± 0.1	∼60 yr	(0–1, 4, −3, −1)	102.98 ± 0.01	103.1 ± 0.4
(1, −2, −1, 2)	(2, 3)	60.090 ± 0.003	60.1 ± 0.15		(1–1, 1, −5, 3)	135.82 ± 0.02	135.3 ± 0.6
(0, 1, −2, 1)	(2, 2)	61.690 ± 0.003	61.7 ± 0.15		(2–1, −2, −4, 4)	144.28 ± 0.02	144.5 ± 0.7
(1, −4, 3, 0)	(4, 4)	63.377 ± 0.004	63.3 ± 0.15		(0–1, 3, −1, −2)	153.85 ± 0.02	154.1 ± 0.8
(1, −3, 3, −1)	(3, 4)	82.638 ± 0.005	82.6 ± 0.25		(3–1, −5, −3, 5)	153.86 ± 0.03	154.1 ± 0.8
(0, 0, 2, −2)	(2, 2)	85.693 ± 0.005	85.7 ± 0.25		(2–1, −3, 3, −2)	165.77 ± 0.02	165.3 ± 0.9
(2, −5, −1, 4)	(5, 6)	88.981 ± 0.007	89.0 ± 0.3		(2–1, −3, 1, 0)	177.40 ± 0.02	177.6 ± 1.0
(1, −3, − 1, 3)	(3, 4)	88.984 ± 0.005	89.0 ± 0.3	Gleissberg	(1–1, 0, 2, −3)	178.54 ± 0.02	177.6 ± 1.0
(1, −2, −2, 3)	(3, 4)	92.533 ± 0.005	92.5 ± 0.3		(3–1, −5, −1, 3)	193.43 ± 0.04	192.7 ± 1.3
(0, 1, −3, 2)	(3, 3)	96.382 ± 0.006	96.4 ± 0.4		(0–1, 3, 1, −4)	193.44 ± 0.02	192.7 ± 1.3
(1, −4, 4, −1)	(4, 5)	100.56 ± 0.01	100.7 ± 0.5		(1–1, −1, 5, −5)	209.45 ± 0.05	209.6 ± 1.5
(1, −3, 2, 0)	(3, 3)	159.585 ± 0.02	159.7 ± 1.0		(2–1, −2, −2, 2)	211.04 ± 0.02	212 ± 1.5
(0, 0, 1, −1)	(1, 1)	171.39 ± 0.01	171.5 ± 1.0	Jose	(1–1, 1, −3, 1)	232.18 ± 0.04	231 ± 2
(2, −5, −2, 5)	(5, 7)	185.06 ± 0.03	185 ± 1.2		(0–1, 4, −4, 0)	258.04 ± 0.07	258 ± 2
(1, −3, 0, 2)	(3, 3)	185.07 ± 0.02	185 ± 1.2		(0–1, 4, −5, 1)	510.4 ± 0.3	513 ± 10
(1, −2, −3, 4)	(4, 5)	201.12 ± 0.03	201 ± 1.5		(1 −1, 1, −4, 2)	654.5 ± 0.4	671 ± 15
(0, 1, −4, 3)	(4, 4)	220.23 ± 0.04	220 ± 2.0	Suess de Vries	(2 −1, −2, −3, 3)	912.1 ± 0.6	925 ± 30
(1, −4, 5, −2)	(5, 6)	243.36 ± 0.07	243 ± 2.0		(0 −1, 3, 0, −3)	1503 ± 2	1470 ± 80
(0, 1, −5, 4)	(5, 5)	772.7 ± 0.6	781 ± 20	Eddy	(3 −1, −5, −2, 4)	1504 ± 2	1470 ± 80
(1, −2, −4, 5)	(5, 6)	1159 ± 2	1149 ± 44		(1 −1, 0, 1, −2)	4279 ± 4	4167 ± 700
(1, −3, 1, 1)	(3, 3)	2318 ± 3	2326 ± 180	Bray Hallstatt	(2 −1, −3, 2, −1)	5056 ± 16	5000 ± 700

FIGURE 37.4. (Left) Complete list of orbital invariant inequalities with periods greater than 40 years. The frequencies are divided into 7 clusters, each consistent with a specific solar cycle. (Right) Periods of selected beats between orbital invariant inequalities and Jupiter's period. (Adapted from Scafetta, 2020; Scafetta & Milani, 2025).

accurately reproduce all major periodicities observed in solar and paleoclimate data on millennial scales.

This spectral coherence supports the hypothesis that planetary harmonics, especially those derived from the Jovian planets, play a fundamental role in modulating both solar activity and terrestrial climate variability.

37.5. The Bray–Hallstatt cycle (2100–2500 years)

The longest orbital invariant inequality analyzed by Scafetta (2020) is defined by the invariant inequality:

$$(1, -3, 1, 1).$$

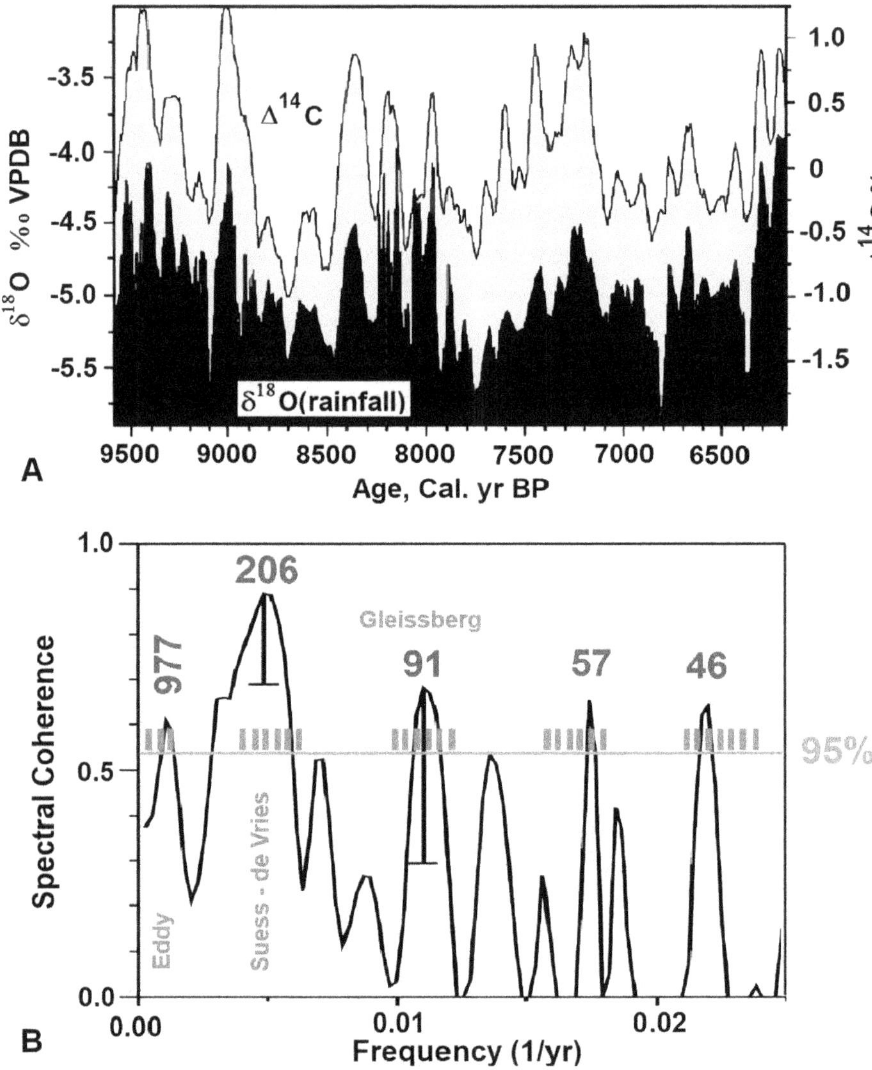

FIGURE 37.5. [A] Visual correlation between a solar activity proxy based on ^{14}C analysis and a precipitation proxy based on ^{18}O analysis; (adapted from Neff et al., 2001).
[B] Comparison of spectral coherence analysis of the two records in A against the orbital invariant inequalities of the solar system listed in the table in Figure 37.4A (bars). Spectral peaks are labeled with their period (in years). Error bars indicate 95% lower confidence limits.
Adapted from Scafetta (2020) using the paleoclimate analyses by Neff et al. (2001).

Its period is calculated as:

$$T_{(1,-3,1,1)} = \left(\frac{1}{T_{Ju}} - \frac{3}{T_{Sa}} + \frac{1}{T_{Ur}} + \frac{1}{T_{Ne}} \right)^{-1} = 2318 \text{ years.}$$

This periodicity corresponds to the Bray–Hallstatt cycle, a long-term solar oscillation identified in multiple paleoclimate records. The harmonic has an index $K = 3$; therefore, it can be decomposed into three synodic components:

$$(1, -3, 1, 1) = (1, -1, 0, 0) - (0, 1, -1, 0) - (0, 1, 0, -1).$$

This decomposition represents a beat formed by the combination of synodic cycles between Jupiter and Saturn, Saturn and Uranus, and Saturn and Neptune.

The Bray–Hallstatt cycle is generated by the beat envelope of

$$f(t) = \sin\left(2\pi \cdot \frac{t - t_{JS}}{T_{JS}}\right) + \sin\left(2\pi \cdot \frac{t - t_{SU}}{T_{SU}}\right) + \sin\left(2\pi \cdot \frac{t - t_{SN}}{T_{SN}}\right),$$

where t_{ij} and T_{ij} are the epochs and periods of conjunctions between planets i and j, reported in Figure 37.1 (cf. Scafetta et al., 2016).[17]

Empirical validation. Figure 37.6 shows that the Bray–Hallstatt cycle is clearly identifiable in the ^{10}Be and ^{14}C data series, which are commonly used to reconstruct solar activity over the Holocene (McCracken et al., 2013).[18]

Using the formalism described in Scafetta (2020), the orbital reconstruction of the Bray–Hallstatt cycle, derived from the inequality $(1, -3, 1, 1)$, exhibits remarkable coherence in both period and phase with the empirical reconstruction extracted from the sequence.

This result is particularly significant because both the period and phase of the theoretical oscillation are derived entirely from planetary orbital parameters, without any empirical calibration. It reinforces the idea that long-term solar variability may be governed by stable harmonic structures rooted in the architecture of the solar system.

37.6. A further development of the orbital model

A significant advancement in the orbital harmonic framework was proposed by Scafetta & Milani (2025). Using high-precision astrometric data from NASA's latest ephemerides (DE440 and DE441), the study analyzes the Solar Inertial Motion (SIM)—the trajectory of the Sun around the solar system's barycenter—over an extended interval from 12999 BC to 16998 AD.

The primary objective was to estimate accurately the temporal sequences and durations of SIM orbital cycles, with the aim of identifying recurring patterns in solar dynamics. As shown in Figures 39.1 and 39.2, spectral analysis reveals that the SIM orbits generate regular beats, the most prominent of which corresponds to the Bray–Hallstatt cycle, with a period of approximately 2318 years, in perfect agreement with the oscillation predicted by the harmonic model.

Extension of the orbital inequality formalism. Spectral analysis of the SIM cycle durations reveals the presence of thousands of harmonic peaks, necessitating an extension of the original formalism based on orbital invariant inequalities. To address this, Scafetta & Milani (2025) complemented the model of Scafetta

[17] Scafetta, N., Milani, F., Bianchini, A., Ortolani, S., 2016. On the astronomical origin of the Hallstatt oscillation found in radiocarbon and climate records throughout the Holocene. Earth-Science Reviews, 162, 24–43.

[18] McCracken, K.G., et al., 2013. A phenomenological study of the cosmic ray variations over the past 9400 years. Solar Physics, 286, 609.

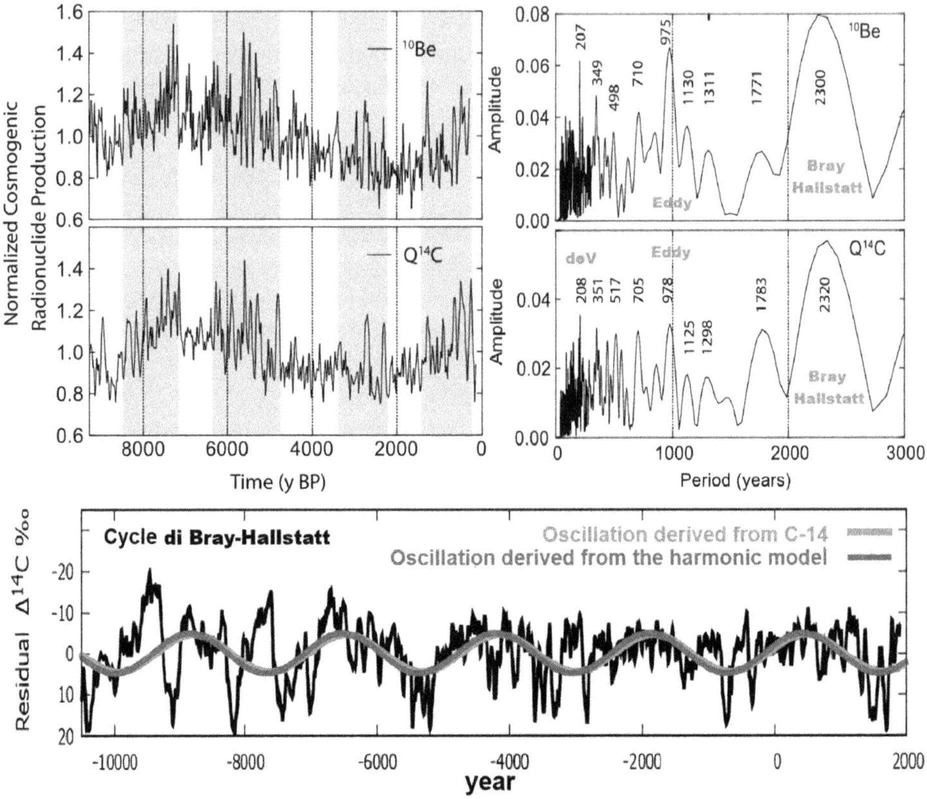

FIGURE 37.6. [Top left] Sequences of ^{10}Be and ^{14}C isotopes used as proxies for solar activity; [right] their respective periodograms showing spectral peaks corresponding to the Suess–de Vries, Eddy, and Bray–Hallstatt cycles (adapted from McCracken et al., 2013).
[B] Comparison between the Bray–Hallstatt cycle extracted from the data and the oscillation derived from the orbital invariant inequality $(1, -3, 1, 1)$. (Adapted from Scafetta et al., 2016; Scafetta, 2020).

(2020) by incorporating the frequency of a single planetary orbit—for example, Jupiter—using the generalized formula:

$$T_{(a_1,\ldots,a_j+b_j,\ldots,a_n)} = \frac{1}{f_{(a_1,\ldots,a_j+b_j,\ldots,a_n)}} = \left| \left(\sum_{i=1}^{n} \frac{a_i}{T_i} \right) + \frac{b_j}{T_j} \right|^{-1}.$$

This extension allows for the modeling of compound harmonics, combining synodic beats with individual planetary frequencies.

Some of these periodicities are listed in the right-hand table of Figure 37.4A. The model proves highly effective in reproducing the full spectrum of frequencies generated by all possible combinations of planetary orbits.

Harmonic clustering and the 179-year base cycle. An important feature of these frequencies is that they are not randomly distributed but aggregate into harmonic clusters whose periods P_n follow the empirical law (Figure 37.3A):

$$P_n \approx \frac{179}{n-1} \text{ years} \qquad \text{for } n = 2, 3, 4, \ldots$$

The cluster corresponding to $n = 1$ includes the longest harmonics, such as:

- the Bray–Hallstatt cycle (~2318 years);
- three quasi-millennial Eddy cycles:
  - 772.7 years from the inequality $(0, 1, -5, 4)$,
  - 912 years from the beat of $(1, -2, -3, 3)$ with Jupiter,
  - 1159 years from the inequality $(1, -2, -4, 5)$.

The base cycle of approximately 179 years, identified with the cycle of Jose (1965), constitutes the fundamental frequency around which the entire harmonic structure of the solar system is organized.

Figure 37.3C presents the periodogram of SIM pericycle lengths, demonstrating that all spectral peaks of the Sun's inertial motion derive from the extended orbital inequality model. These spectral clusters correspond to typical periodicities observed in solar activity.

37.7. Conclusions and climatic implications

Observations of solar variability and climate data over the past 11,000 years reveal a striking spectral coherence, characterized by recurring periodicities. As discussed in earlier sections, these frequencies are consistently predicted by a class of orbital harmonics known as orbital invariant inequalities—beats derived from the synodic cycles of the gas giants (Jupiter, Saturn, Uranus, Neptune), further refined through interactions with Jupiter and individual planetary dynamics.

A key property of these inequalities is their invariance under rotating reference frames, which makes them robust candidates for synchronizing the solar dynamo. This invariance ensures that planetary beats are coherently experienced across solar latitudes, potentially modulating solar activity through stable harmonic patterns.

Revisiting Wolf's hypothesis. Scafetta (2020) builds on a long research trajectory, including earlier models developed to reproduce the 11-year solar cycle using the orbital harmonics of Venus, Earth, Jupiter, and Saturn. These studies revisit Wolf's original intuition in 1859, proposing that planetary configurations modulate solar activity through synchronized interactions (Scafetta, 2012b,d, 2014b).[19]

Criticism has often focused on the absence of a fully articulated physical mechanism linking planetary motion to solar dynamics. Yet the lack of a complete explanation does not invalidate the observed coherence. To date, no solar

[19] Scafetta, N., 2012b. Multi-scale harmonic model for solar and climate cyclical variation throughout the Holocene based on Jupiter–Saturn tidal frequencies plus the 11-year solar dynamo cycle. Journal of Atmospheric and Solar–Terrestrial Physics, 80, 296–311.
Scafetta, N., 2014b. Discussion on the spectral coherence between planetary, solar and climate oscillations: a reply to some critiques. Astrophysics and Space Science, 354, 275–299.

model excluding planetary influences has succeeded in replicating the full spectral structure of solar activity.

Toward a unified solar–planetary theory. In summary, the growing body of evidence supports the idea that solar activity is modulated by planetary harmonics, the same harmonics that appear to shape millennial-scale climate variability. Advancing this understanding requires an in-depth analysis of the Sun's internal dynamics and an exploration of solar system physics, with particular focus on the interaction mechanisms that regulate matter circulation and energy balance within the heliosphere.

CHAPTER 38

How the planets could influence the Sun

The hypothesis that the planets may exert a physical influence on solar activity, and by extension on Earth's climate, has ancient origins. From the Pythagorean concept of the *"Music of the Spheres"* to contemporary models involving gravitational resonances, tidal interactions, angular momentum transfer, and internal solar dynamics, this idea has evolved into a multifaceted scientific inquiry.

In recent decades, the hypothesis has gained renewed momentum thanks to the identification of striking synchronizations between planetary and solar cycles, as discussed in previous chapters. Nonetheless, the physical plausibility of such connections remains a subject of active debate, dividing the scientific community between interdisciplinary openness and methodological skepticism.

Emerging evidence suggests a connection: Sun-like stars hosting planets tend to rotate more slowly than similar stars without planets, indicating a possible link among planetary formation, stellar angular momentum evolution, and tidal interactions (Sibony et al., 2022).[1]

This chapter critically examines the principal mechanisms proposed to explain planetary influence on solar variability, explores their astrophysical and climatic implications, and evaluates the objections raised, with the aim of constructing the most rigorous and up-to-date framework possible for one of the most intriguing and controversial questions in solar system science.

38.1. The challenge of planetary forcing

The primary objection to the idea that planets can influence solar activity is based on scale: the planets are too small and too distant to exert any significant force on the Sun (de Jager & Versteegh, 2005; Charbonneau, 2022).[2] For example, the gravitational accelerations exerted by Jupiter or Saturn on the solar center are less than one billionth of the Sun's surface gravity. Similarly, the tidal elongations induced by these planets are on the order of fractions of a millimeter (Scafetta, 2012d).[3]

However, in the physics of dynamical systems, the weakness of a forcing does not necessarily imply the absence of an effect. Even minimal interactions

---

[1]Sibony, Y., Helled, R., Feldmann, R., 2022. The rotation of planet-hosting stars. Monthly Notices of the Royal Astronomical Society, 513(2), 2057–2075.

[2]de Jager, C., Versteegh, G.J.M., 2005. Do planetary motions drive solar variability? Solar Physics, 229, 175–179.

Charbonneau, P., 2022. External Forcing of the Solar Dynamo. Frontiers in Astronomy and Space Sciences, 9, 853676.

[3]Scafetta, N., 2012d. Does the Sun work as a nuclear fusion amplifier of planetary tidal forcing? Journal of Atmospheric and Solar–Terrestrial Physics, 81–82, 27–40.

can produce amplified responses if they act in resonance with the system's natural oscillatory modes.

A historically illuminating analogy is Huygens synchronization. In 1665, the Dutch physicist Christiaan Huygens observed that two pendulum clocks mounted on the same wall spontaneously synchronized their swings. He attributed this to subtle mechanical interactions transmitted through the shared support.

Today, this phenomenon is recognized as a classic example of weak coupling between oscillators, with applications in biology (e.g., heart rhythms, firefly flashing), electronics, and celestial mechanics (Kuramoto, 1984; Strogatz, 2009).[4]

Huygens synchronization offers a compelling analogy for understanding how weak but persistent interactions between celestial bodies can induce coherent cyclic behavior, even in the absence of dominant forces. This opens the possibility of synchronization between planetary cycles and solar activity.

Yet a fundamental challenge remains. For such coupling to be effective over long timescales, the forcing must maintain sufficient coherence and persistence to avoid being overwhelmed by the Sun's intrinsic stochastic fluctuations.

The focus thus shifts to identifying a physical mechanism within the Sun capable of amplifying weak planetary forcing and transforming it into a driver of solar dynamo synchronization. Several hypotheses in this direction have emerged in the recent literature and are explored in the following sections.

38.2. Coupling through angular momentum

The study by Abreu et al. (2012). Abreu et al. (2012)[5] investigated the potential influence of planetary orbits on millennial-scale solar variability. Using cosmogenic isotopes (^{10}Be and ^{14}C) from ice cores and tree rings as proxies for past solar activity, they compared the spectral content of these records with the barycentric torque exerted by the planets on the solar tachocline—the transition region between the Sun's radiative and convective zones. The results are shown in Figure 38.1.

The analysis revealed a strong correspondence between the dominant periodicities in solar proxies (88, 104, 150, 208, 506, and 1000 years) and those found in planetary torque oscillations, particularly from Jupiter, Saturn, Uranus, and Neptune. Cycles of $\sim$104 and $\sim$150 years, along with peaks around 115 and 130 years, were visible in both datasets.

Notably, cycles of approximately 104 and 150 years, along with additional peaks around 115 and 130 years, appear in both solar records and planetary torque spectra. Scafetta (2014b)[6] demonstrated that these cycles can be explained by a planetary model of the Schwabe cycle, based on Jupiter–Saturn tidal interactions with the solar dynamo.

The authors hypothesized that planetary orbital dynamics may modulate solar activity through gravitational or resonant mechanisms, acting as a *"cosmic*

[4]Kuramoto, Y., 1984. Chemical Oscillations, Waves, and Turbulence. Springer, Berlin.
Strogatz, S.H., 2009. Exploring complex networks. Nature, 410, 268.

[5]Abreu, J.A., Beer, J., Ferriz-Mas, A., McCracken, K.G., Steinhilber, F., 2012. Is there a planetary influence on solar activity? Astronomy and Astrophysics, 548, A88.

[6]Scafetta, N., 2014b. Discussion on the spectral coherence between planetary, solar and climate oscillations: a reply to some critiques. Astrophysics and Space Science, 354, 275–299.

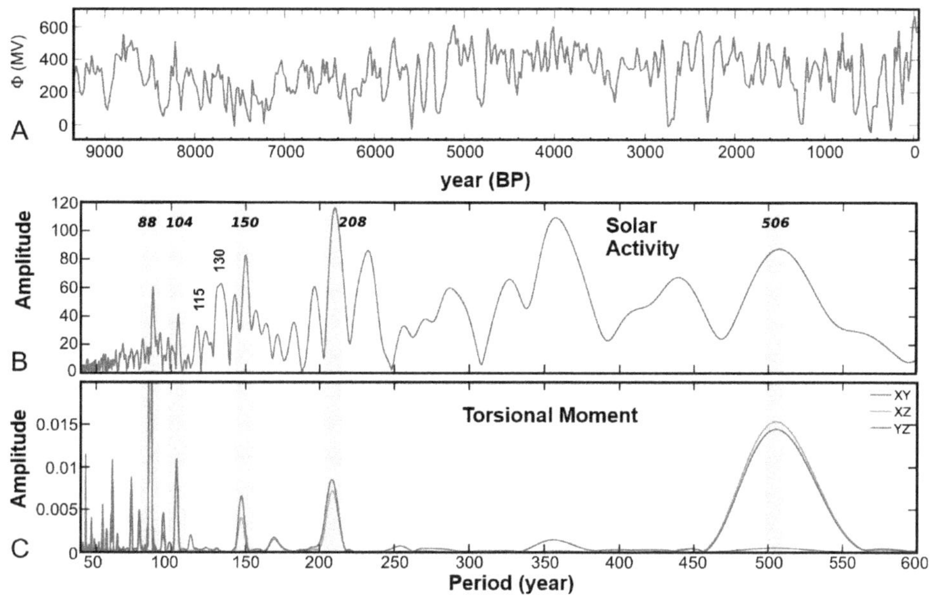

FIGURE 38.1. Comparison between solar activity and planetary torque in the frequency domain.
[A] Solar activity reconstructed from cosmogenic radionuclides;
[B] Fourier spectrum of solar activity quantified via the solar modulation potential;
[C] Fourier spectrum of the annual mean torque modulus induced by the planets on the solar tachocline.
(Adapted from Abreu et al., 2012).

metronome" capable of synchronizing the solar dynamo. Although the physical mechanism remains unclear, the study significantly advanced the debate on planetary–solar coupling.

The study by Wilson et al. (2008). Wilson et al. (2008)[7] proposed a mechanism involving spin–orbit coupling between the Sun and the Jovian planets, particularly Jupiter and Saturn. The Sun's motion around the solar system's barycenter, which shifts with planetary configurations, was suggested to influence its differential rotation, a key component of the magnetic dynamo.

According to the model, variations in the Sun's orbital angular momentum relative to the barycenter could be transferred to solar rotation, analogously to the spin–orbit coupling in binary stars or planetary satellites. This transfer would be facilitated by asymmetries in the Sun's moment of inertia, potentially induced by convective currents or internal mass anisotropies.

The 11-year solar cycle, along with longer-term modulations, would thus be interpreted as an internal resonant response to barycentric dynamics, primarily driven by Jupiter and Saturn. Although hypothetical, this study provided a

[7]Wilson, I.R.G., Carter, B.D., Waite, I.A., 2008. Does a spin–orbit coupling between the Sun and the Jovian planets govern the solar cycle? Publications of the Astronomical Society of Australia, 25, 85–93.

conceptual foundation for linking planetary orbital mechanics to solar internal dynamics, challenging the view of the solar dynamo as an isolated system.

38.3. Wolff and Patrone's internal solar amplification

Wolff & Patrone (2010)[8] proposed an internal mechanism by which planetary motions could influence solar activity. Their model suggests that the Sun's motion around the barycenter induces regular fluctuations in gravitational potential energy within its deep interior, particularly in the core and radiative zones.

Periodic variations in the Sun's barycentric position, caused by the giant planets, generate subtle perturbations in plasma equilibrium. If coupled to a sensitive internal structure such as a system of oscillators in the solar core, these perturbations could trigger enhanced nuclear fusion by inducing localized compression, expansion, and mixing of plasma.

As the authors concluded:

> *"This would cause stars like the Sun with appropriate planetary systems to burn somewhat more brightly and have shorter lifetimes than identical stars without planets."*

This model suggests that small but persistent gravitational fluctuations in the solar core may be amplified by internal solar processes, resulting in measurable increases in solar luminosity.

38.4. The tidal model by Stefani and collaborators

Stefani et al. (2016)—synchronized oscillations of turbulent helicity. Stefani et al. (2016)[9] proposed a novel mechanism by which planetary tides, particularly those generated by Venus, Earth, and Jupiter, could influence the solar dynamo through the generation of turbulent currents in the solar tachocline.

The core of the model is parametric resonance: the authors demonstrate that even extremely weak tidal motions ($\sim$1 mm/s), if applied with coherent periodicity, can harmonically synchronize turbulence in the solar plasma. This synchronization directly affects the α-effect in the α–Ω dynamo model, which governs the regeneration of the Sun's magnetic field.

In this framework:

- the α-effect arises from turbulent and vortical motions in the convective zone, regenerating the poloidal field from the toroidal one;
- the Ω-effect occurs in the tachocline, the transition zone between the radiative and convective layers, where the poloidal magnetic field is sheared into a toroidal field.

This reciprocal conversion sustains the solar magnetic field and drives its 22-year Hale magnetic cycle, which manifests as the 11-year Schwabe sunspot cycle.

Numerical simulations show that a periodic forcing of 11.07 years, which corresponds to half the Venus–Earth–Jupiter tidal cycle represented by the harmonic vector (3, –5, 2) discussed in Chapter 35 (Scafetta, 2012d), can lock

[8] Wolff, C.L., Patrone, P.N., 2010. A new way that planets can affect the Sun. Solar Physics, 266, 227–246.

[9] Stefani, F., Giesecke, A., Weber, N., Weier, T., 2016. Synchronized helicity oscillations: A link between planetary tides and the solar cycle? Solar Physics, 291, 2197–2212.

the phase and frequency of the solar dynamo. This mechanism may explain the near-regularity of the Schwabe cycle.

The study concludes that, despite the extreme weakness of the forces involved, their harmonic persistence may activate internal amplification mechanisms capable of producing macroscopic effects on solar magnetic variability.

Stefani et al. (2019)—synchronization by planetary forcings. In their 2019 follow-up study, Stefani et al.[10] extended their hypothesis by modeling the solar dynamo as a nonlinear oscillator subject to periodic tidal forcing.

Using a simplified α–Ω model, the authors cyclically modulated the α-effect with planetary tides. Simulations revealed that even very weak but temporally coherent forcing can induce phase locking, whereby the dynamo synchronizes with the external periodicity or its harmonics (e.g., 11, 22, 44 years).

This behavior is characteristic of parametric resonance in nonlinear systems, analogous to forced pendulum dynamics. The model supports the physical plausibility that planetary tides could generate regular solar cycles and even modulate longer-term phenomena such as the Maunder Minimum.

Stefani et al. (2024)—resonances of secular solar cycles. In their most recent work, Stefani et al.[11] explored the resonance dynamics underlying key solar cycles, including:

- the Rieger cycle ($\sim$150–160 days);
- the Schwabe cycle ($\sim$11 years);
- the Suess–de Vries cycle ($\sim$200 years).

The solar dynamo is modeled as a system of coupled nonlinear oscillators, each with its own characteristic frequency. Interactions among these frequencies generate beats resulting from the superposition of nearby oscillations. This mechanism explains the multi-scale variability of solar activity.

The model successfully simulates not only the Schwabe cycle but also other quasi-periodic oscillations of medium and long duration, offering a unified framework that integrates both internal solar dynamics and external planetary influences.

Overall conclusion of Stefani's work. The body of research by Stefani and collaborators outlines an advanced dynamical model linking weak planetary tidal forcings, especially those acting on the solar tachocline, to the modulation of the convective zone dynamo.

Grounded in robust physical principles such as:

- parametric resonance;
- nonlinear synchronization;
- coupled oscillator theory,

the model demonstrates how planetary orbital periodicities could synchronize solar activity, generating a spectrum of quasi-regular oscillations consistent with observed solar cycles and planetary harmonics.

---

[10] Stefani, F., Giesecke, A., Weier, T., 2019. A model of a tidally synchronized solar dynamo. Solar Physics, 294, 60.

[11] Stefani, F., Horstmann, G.M., Klevs, M., Mamatsashvili, G., Weier, T., 2024. Rieger, Schwabe, Suess-de Vries: the sunny beats of resonance. Solar Physics, 299, 51.

This work represents a significant step toward reconciling the subtle gravitational influences of planetary motion with the complex internal processes of the Sun, offering new insights into the mechanisms driving solar variability.

38.5. Scafetta's thermonuclear amplification of planetary tides

One of the most frequent criticisms of the hypothesis that the planets can significantly influence solar activity concerns the presumed energetic insignificance of the tidal forces they exert on the Sun. Skeptical authors such as Charbonneau (2022)[12] have described these forces as "*homeopathic*," that is, too weak to produce measurable effects on the Sun's internal dynamics.

However, ten years earlier, Scafetta (2012d)[13] had already proposed an original theoretical model that reframes the problem: even an extremely weak energy input, such as that from planetary tides, could be strongly amplified in the solar core, leading to observable variations in solar luminosity and magnetic activity, synchronized with planetary orbital frequencies.

Critiques of the planetary theory based on poor understanding or outright ignorance of the relevant literature, which has already addressed well-known specific issues, only reveal persistent ideological bias without adding anything substantive to the scientific debate.

The Sun as a "*nuclear amplifier*" of tidal forcing. The mechanism relies on the mass–luminosity relation of main-sequence stars, which implies that perturbations in a star's internal structure can result in changes to its energy output. Scafetta (2012d) suggests that planetary tides, although weak, could induce gravitational potential energy oscillations in the solar core. These oscillations, if resonant with internal fusion dynamics, could modulate the rate of thermonuclear fusion, thereby influencing solar cyclicity.

This concept envisions the Sun not as a passive recipient of external forces, but as an active amplifier—analogous to a loudspeaker that transforms a weak electrical signal into a powerful acoustic wave.

38.6. Mathematical framework

The mass–luminosity relation for main-sequence stars can be approximated as:

$$\frac{L}{L_\odot} \approx \left(\frac{M}{M_\odot}\right)^4 \approx 1 + \frac{4\Delta M}{M_\odot},$$

where L is the stellar luminosity, M is its mass, and the symbol "$\odot$" denotes solar reference values.

This implies that small variations in solar mass—or, equivalently, in internal gravitational energy—can produce proportional changes in luminosity.

Scafetta models the solar luminosity as:

$$L(t) \approx L_\odot + A \cdot \dot{\Omega}_{\text{tidal}}(t),$$

where:

---

[12]Charbonneau, P., 2022. External Forcing of the Solar Dynamo. Frontiers in Astronomy and Space Sciences, 9, 853676.

[13]Scafetta, N., 2012d. Does the Sun work as a nuclear fusion amplifier of planetary tidal forcing? Journal of Atmospheric and Solar–Terrestrial Physics, 81–82, 27–40.

- $L_\odot$ is the baseline solar luminosity;
- $\dot{\Omega}_{\text{tidal}}(t)$ is the gravitational tidal power dissipated in the core;
- A is the nuclear amplification factor, which converts tidal power into luminosity via hydrogen nuclear fusion (4H $\to$ He).

Deriving the amplification factor. Using Einstein's equation $E = mc^2$, the luminosity generated in a spherical shell between radius r and $r + dr$ is related to the mass loss due to fusion:

$$\frac{d\dot{m}(r)}{dr} = \frac{1}{c^2} \cdot \frac{dL(r)}{dr},$$

where $c = 2.998 \cdot 10^8 \sim \text{m/s}$ is the speed of light and $L(r)$ is the luminosity profile derived from the standard solar model (Bahcall et al., 2001, 2005).[14]

The gravitational power associated with mass loss is:

$$\begin{aligned}\dot{U}_\odot &= \frac{1}{2} G \int_0^{R_S} m_\odot(r) \cdot \frac{d\dot{m}(r)}{dr} \cdot \frac{1}{r}\, dr \\ &= \frac{1}{2} \cdot \frac{G}{c^2} \int_0^{R_S} m_\odot(r) \cdot \frac{dL(r)}{dr} \cdot \frac{1}{r}\, dr = 3.6 \cdot 10^{20} \sim \text{W},\end{aligned}$$

where $m_\odot(r)$ is the mass enclosed within radius r and R_S is the solar radius. The factor 1/2 derives from the virial theorem.

This quantity should also represent the gravitational power $\dot{\Omega}_\odot$ that the Sun must generate to sustain the equilibrium of nuclear fusion—for example, through the precipitation of generated helium toward the center of the star and the inward transport of additional hydrogen into the core. Thus, it is assumed that $\dot{U}_\odot \approx \dot{\Omega}_\odot$.

Without such gravitational compensation, the fusion process would be hindered by the loss of pressure and temperature due to the continuous reduction in the number of particles, since each reaction replaces four hydrogen atoms with one helium atom:

$$4\,{}^1\text{H} \to {}^4\text{He} + \text{energy}.$$

During this process, approximately 0.7% of the initial mass is converted into energy.

Planetary tides are assumed to act as additive sources of gravitational energy dissipated in the solar core:

$$\dot{\Omega}_\odot(t) = \dot{\Omega}_\odot + \Delta\dot{\Omega}_\odot(t) = \dot{\Omega}_\odot + \dot{\Omega}_{\text{tidal}}(t).$$

For small perturbations, the following equivalence is assumed:

$$\frac{\Delta M}{M_\odot} \equiv \frac{\dot{\Omega}_{\text{tidal}}}{\dot{\Omega}_\odot}.$$

Applying this relation to the mass–luminosity formula yields the key equation:

$$L(t) \approx L_\odot + \frac{4L_\odot}{\dot{\Omega}_\odot} \cdot \dot{\Omega}_{\text{tidal}}(t) = L_\odot + A \cdot \dot{\Omega}_{\text{tidal}}(t),$$

[14] Bahcall, J.N., Pinsonneault, M.H., Basu, S., 2001. Solar Models: Current Epoch and Time Dependences, Neutrinos, and Helioseismological Properties. The Astrophysical Journal, 555, 990–1012. Bahcall, J.N., Serenelli, A.M., Basu, S., 2005. New Solar Opacities, Abundances, Helioseismology, and Neutrino Fluxes. The Astrophysical Journal, 621, L85–L88.

	m_P (kilogram)	Sidereal orbital period T_P (day)	Semi-major axis d_P (m)	Aphelion d_{un}(m)	Aphelion t_{un}	Angular Sep. $\alpha_{J,2000}$
Mercury	3.30100×10^23	87.969	5.790900 × 10^10	6.981800 × 10^10	2000.0047	143.6°
Venus	4.86730×10^24	224.701	1.082100 × 10^11	1.089410 × 10^11	2000.2219	145.5°
Earth	5.97220×10^24	365.256	1.495980 × 10^11	1.521000 × 10^11	2000.5087	63.6°
Mars	6.41690×10^23	686.98	2.279560 × 10^11	2.492610 × 10^11	2000.8405	37.1°
Jupiter	1.89813×10^27	4332.589	7.784790 × 10^11	8.163630 × 10^11	2005.2841	0.0°
Saturn	5.68320×10^26	10759.22	1.432041 × 10^12	1.506527 × 10^12	1988.6954	9.5°
Uranus	8.68110×10^25	30685.4	2.867043 × 10^12	3.001390 × 10^12	2009.1554	79.8°
Neptune	1.02409×10^26	60189	4.514953 × 10^12	4.558857 × 10^12	1959.5619	92.3°

FIGURE 38.2. Planetary orbital data. Source: https://nssdc.gsfc.nasa.gov/planetary/factsheet, accessed on March 24, 2023.

where the nuclear amplification factor is:

$$A = 4 \cdot \frac{L_\odot}{\dot{\Omega}_\odot} \approx 4 \cdot \frac{L_\odot}{\dot{U}_\odot} \approx 4.25 \cdot 10^6.$$

This result implies that each watt of tidal power dissipated in the solar core could generate approximately 4.25 million watts of additional luminosity owing to the Sun's intrinsic sensitivity to fusion dynamics.

Conceptual implications. With this amplification, sub-millimetric planetary tides are effectively magnified to a kilometer scale, becoming potentially effective in modulating the solar dynamo. The model provides a physically grounded solution to the longstanding objection regarding the weakness of planetary forcing, reframing the Sun as a resonant energy system capable of responding to periodic external signals.

This hypothesis opens new avenues for understanding how planetary harmonics might modulate solar activity—not through sheer force, but through resonant amplification within the Sun's fusion engine.

38.7. The tidal equation of the planets on the Sun

To quantify the influence of planetary tides on solar dynamics, two complementary equations are proposed. The first provides a simplified estimate of tidal elongations and their periodicities; the second models the conversion of tidal power dissipated in the solar core into fluctuations in solar luminosity.

(1) Tidal elongation.
The first equation calculates the combined tidal elongation induced by all planets at a reference point on the Sun—typically along the

Sun–Jupiter axis, where tidal effects are strongest. The function is:

$$Tide(t) \propto \sum_P \frac{m_P}{\left[d_P + (d_{Pa} - d_P)\cos\left(2\pi \frac{t - t_{Pa}}{T_P/365.25}\right)\right]^3} \cdot \left[\cos^2\left(2\pi \frac{t-2000}{T_S/365.25} - 2\pi \frac{t-2000}{T_P/365.25} - 2\pi \frac{\alpha_{PJ,\,2000}}{360°}\right) - \frac{1}{3}\right],$$

where d_P is the mean distance of planet P from the Sun, m_P its mass, $T_S = 27$ days is the solar rotation period, and T_P is the orbital period of planet P. In addition, d_{Pa} and t_{Pa} are the aphelion distance and one of its occurrence epochs, while $\alpha_{PJ,\,2000}$ is the angular distance of planet P from Jupiter on January 1, 2000, at 00:00 UTC. Planetary orbital data are listed in Figure 38.2.

(2) Tidal power dissipation.
Scafetta (2012d) emphasizes that what physically matters is the tidal power dissipated in the solar interior. This is approximated by the rate of vertical displacement:

$$f(t) = \left|\frac{d\,Tide(t)}{dt}\right| \approx \left|\frac{Tide(t) - Tide(t - 1\,\text{day})}{1\,\text{day}}\right|.$$

It is hypothesized that the fraction of tidal power dissipated in the solar core triggers an increase and modulation of the Sun's luminous output. Thus, the full equation for the luminosity fluctuation induced by planetary tides is:

$$I_P(t) = \frac{3GR_S^5}{2Q\Delta t}\int_0^1 K(\chi)\chi^4\rho(\chi)\,d\chi \cdot \int_{\theta=0}^{\pi}\int_{\phi=0}^{2\pi}\left|\sum_P m_P\left[\frac{\cos^2(\alpha_{P,t}) - \frac{1}{3}}{R_{SP}^3(t)} - \frac{\cos^2(\alpha_{P,t-\Delta t}) - \frac{1}{3}}{R_{SP}^3(t-\Delta t)}\right]\right|\sin(\theta)\,d\theta\,d\phi.$$

Where:

- $R_{SP}(t)$—distance of planet P from the Sun;
- m_P—mass of planet P;
- $\alpha_{P,t}$—angles denoting the position of planet P relative to angular position ϕ on the Sun;
- Q^{-1}—effective tidal dissipation factor;
- R_S—solar radius;
- Δt—integration time interval, set to 1 day;
- $K(\chi)$—function to convert gravitational power into TSI at 1 AU from the Sun;
- $\rho(\chi)$—solar density function;
- $\chi = r/R_S$—normalized distance from the solar center;
- G—universal gravitational constant.

This function integrates tidal effects from the solar core to the surface, incorporating real-time ephemeris data for all eight planets.

Analysis of tidal functions on the Sun. Figure 38.3 presents the outputs of both equations:

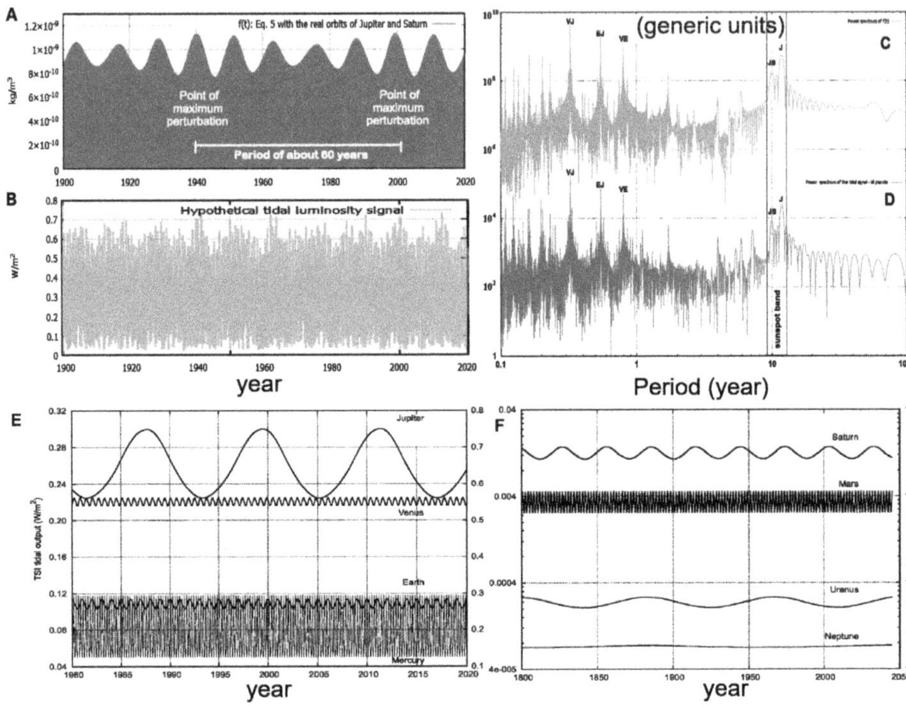

FIGURE 38.3. [A] Tidal signal generated solely by Jupiter and Saturn on the Sun (basic equation), showing a beat of approximately 60 years.
[B] Luminosity signal (TSI, in W/m²) generated by the combination of all planetary tides (complete equation).
[C] Periodogram of the tidal function generated by all planets (basic equation).
[D] Periodogram of the luminosity function generated by planetary tides (complete equation).
[E,F] Luminosity signal (TSI, in W/m²) generated by the tide of each individual planet.
(Adapted from Scafetta, 2012d; Scafetta & Bianchini, 2023).

- Panel A: Tidal signal from Jupiter and Saturn, showing a ~60-year beat. Peaks around 1940 and 2000 appear in some TSI reconstructions (Figure 25.2F–G) and precede the observed temperature maxima (see Chapter 12), suggesting a link between planetary tides and climate oscillations.
- Panel B: Luminosity signal (TSI in W/m²) from the complete tidal model, $I_P(t)$, showing amplitudes compatible with solar dynamo synchronization.
- Panels C & D: Periodograms of the tidal and luminosity functions, revealing strong power in the Schwabe band (9–13 years). Peaks at 9.93 years (Jupiter–Saturn syzygy) and 11.86 years (Jupiter's orbit) match the sunspot cycle band. Additional signals emerge from minor syzygy

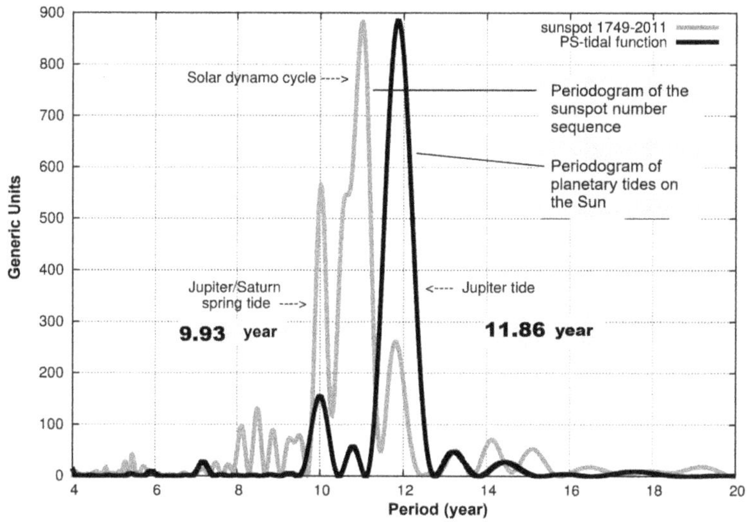

FIGURE 38.4. Comparison of the periodogram of the sunspot number sequence from 1749 to 2011 with the periodogram of the theoretical function of total solar irradiance variation generated by all tides on the Sun (black curve). Note the presence of two peaks at periods of approximately 9.93 years (Jupiter–Saturn syzygy tide) and 11.86 years (Jupiter's orbital tide), common to both sequences. (Adapted from Scafetta, 2012d; Scafetta & Bianchini, 2023).

tides such as Venus–Jupiter (0.324 years), Earth–Jupiter (0.546 years), and Venus–Earth (0.649 years). Beats among these tides generate the orbital invariant inequality $(3, -5, 2)$, whose first harmonic produces a tidal component with a period of approximately 11.07 years, in strong agreement with the average Schwabe solar cycle, as shown in Figure 35.3. Thus, the five strongest planetary tides can be related to the Schwabe cycle band, reinforcing the hypothesis that the solar dynamo is synchronized by planetary tides generated by Venus, Earth, Jupiter, and Saturn, as originally proposed by Wolf (1859).

- Panels E & F: TSI contributions from individual planets, highlighting their relative influence.

Figure 38.4 compares the sunspot number periodogram (1749–2011) with the theoretical TSI variation. Both exhibit peaks at 9.93 and 11.86 years, confirming the correspondence between planetary tides and solar cycles.

Applications and predictions. The model predicts the following luminous impacts on TSI:

- average: 0.3 to 0.8 W/m^2, depending on the Love number used;
- decadal variations: 0.08 to 0.20 W/m^2, mainly from Jupiter and Saturn;
- theoretical peaks: up to 0.6 to 1.6 W/m^2, assuming minimal internal damping.

Although modest in amplitude, these variations are coherent with planetary periodicities and may be sufficient to synchronize the solar tachocline, which is the zone believed to generate the solar dynamo.

Conclusion. The tidal model offers a physically grounded explanation for the origin of many observed solar periodicities, including:

- the Schwabe cycle;
- TSI fluctuations;
- oscillations in terrestrial climate proxies.

By linking planetary tides to solar luminosity modulation, the model reinforces the hypothesis, first proposed by Wolf (1859), that planetary dynamics play a central role in shaping solar activity.

38.8. Toward a unified physical model

The quest for a physically coherent model capable of explaining the observed relationship between planetary configurations and solar activity has led to the development of three complementary theoretical frameworks:

- *harmonic models*, based on planetary orbital periods, which describe solar activity as the result of cyclical interference patterns among planetary motions;
- *gravitational torque models*, which analyze the dynamic influence of planetary-induced angular momentum variations on the Sun's internal structure;
- *tidal nuclear amplification models*, particularly that proposed by Scafetta (2012d), which hypothesize that planetary tides modulate hydrogen fusion rates in the solar core.

Each model addresses distinct facets of the planetary–solar interaction. A conceptual synthesis of these approaches may offer a unified and plausible physical framework capable of integrating the various dynamics involved.

Orbital harmonics as temporal structure. Harmonic models—especially those based on beat frequencies generated by the Venus–Earth–Jupiter (V–E–J) and Jupiter–Saturn (J–S) systems—describe the temporal architecture of solar variability. The principal solar cycles, including:

- Schwabe (approximately 11 years),
- Hale (approximately 22 years),
- Gleissberg (approximately 60–100 years), and
- Eddy (approximately 1000 years),

coincide with harmonic frequencies derivable from planetary orbital configurations.

In this context, planetary alignments act as cosmic metronomes, imposing regular rhythms that may synchronize internal solar oscillators such as the magnetic dynamo, or modulate the energy balance in the solar core. Although these models do not specify the physical transmission mechanism, they accurately predict the periodicity and phase structure of solar cycles.

Gravitational torques. The second mechanism involves gravitational torques generated by the Sun's motion around the solar system barycenter, primarily influenced by Jupiter and Saturn. These torques act on the Sun's internal layers, especially the tachocline, which is the critical zone for magnetic field generation.

Such torques may:

- influence differential rotation;
- alter angular momentum transport;
- modulate the α–Ω dynamo.

This model provides a mechanical pathway for transmitting planetary signals into the Sun's internal dynamics, offering a plausible link between orbital motion and magnetic variability.

Stochastic resonance: amplification within chaotic noise. A third interpretive mechanism is stochastic resonance, a phenomenon in nonlinear systems in which weak periodic forcings are amplified in the presence of internal noise, such as that generated by solar convection.

Planetary oscillations that approximate the natural frequencies of the solar dynamo may be selectively amplified by chaotic fluctuations, producing tangible modulations in magnetic activity. This represents an indirect amplification channel, bridging external periodicities with internal solar responses.

Harmonic resonant systems and quantized structures. The Solar System has long inspired analogies with quantized harmonic systems, from Kepler's geometries to modern orbital mechanics. Planetary orbits exhibit near-perfect ratios (e.g., 5:2 between Jupiter and Saturn, 13:8 between Venus and Earth), suggesting a deep geometric order.

If the Sun is sensitive to these harmonies, for example via coupling between internal rotation and orbital resonances, it may behave as a synchronized oscillator, generating cycles that reflect the "*fundamental notes*" of the planetary ensemble.

Tides and nuclear amplification: the energetic pathway. The third pillar of the integrated model is tidal amplification in the solar core, as proposed by Scafetta (2012d). The central hypothesis is that planetary gravitational tides, although weak, can modulate the rate of nuclear fusion at the Sun's center.

According to the mass–luminosity relation, even minimal perturbations in the conditions of the solar core can lead to significant variations in luminosity. The proposed mechanism operates as follows:

(1) gravitational tides from Venus, Earth, and Jupiter induce cyclical compressions within the solar core;
(2) these compressions, along with the associated fuel mixing, modulate the rate of nuclear fusion;
(3) this process amplifies the gravitational tidal signal by a factor $A \approx 4 \times 10^6$, and the resulting energy fluctuations could become sufficiently effective because they are approximately comparable in magnitude to the Total Solar Irradiance (TSI) fluctuations observed during the Schwabe cycle.

This model directly addresses the classical objection concerning the energetic insignificance of planetary forces. By introducing a plausible amplification mechanism rooted in core dynamics, it offers a physical and quantitative basis for considering planetary tides as modulators of solar activity.

The integrated model: a celestial symphony. By combining these three approaches, a multi-level model emerges in which each component plays a distinct role in the causal chain linking planetary motion to solar activity:

- harmonic frequencies (V–E–J, J–S...) $\Rightarrow$ external temporal forcings with defined periodicity;
- gravitational torques $\Rightarrow$ mechanical coupling with solar structure;
- tidal nuclear amplification $\Rightarrow$ energetic conversion of gravitational forcings.

This synthesis resolves the limitations of individual models:

- harmonic models predict timing, but not mechanism;
- torque models explain mechanical coupling but not energy amplification;
- nuclear models provide amplification, but require coherent rhythms.

Conclusion. The convergence of orbital harmonies, gravitational interactions, and internal amplification offers a compelling theoretical framework for understanding solar variability across decadal, secular, and millennial timescales.

Although still under investigation, this integrated model represents a promising alternative to purely endogenous theories, suggesting that the symphony of the Solar System may indeed resonate deep within the heart of the Sun.

38.9. Critiques, alternatives, and prospects

While the hypotheses explored in this chapter are grounded in robust astrophysical principles and supported by intriguing empirical correlations, none has yet achieved broad consensus within the scientific community. Standard solar models continue to systematically exclude any direct planetary influence, treating it as negligible.

The principal critiques include:

- the absence of a complete and predictive numerical formulation of the proposed mechanisms;
- technical limitations in modeling the Sun's internal dynamics, particularly within the core and above the tachocline;
- the risk of post hoc fallacy, i.e., attributing causality to correlations between planetary periodicities and solar cycles without a proven physical link.

Nevertheless, an empirical enigma persists: the striking synchronizations between planetary orbital cycles and solar activity, which remain unexplained by current dynamo theories (Charbonneau, 2022). Even skeptical researchers such as Callebaut et al. (2013),[15] have acknowledged that solar variability exhibits cycles traceable to planetary motions, even if the underlying mechanism remains elusive.

[15] Callebaut, D.K., de Jager, C., Duhau, S., 2013. Reply to "The influence of planetary attractions on the solar tachocline" by N. Scafetta, O. Humlum, J.E. Solheim, K. Stordahl. Journal of Atmospheric and Solar–Terrestrial Physics, 102, 372–373.

Works such as Weiss & Tobias (2016)[16] provide theoretical frameworks for understanding deep solar variability through internal dynamo dynamics and symmetry changes. However, explaining long-term solar oscillations such as the Eddy and Hallstatt periods likely requires models that incorporate external astronomical mechanisms beyond the Sun itself.

The photon diffusion objection and the solar response times. A classic objection to planetary influence on the solar core concerns the long timescale required for the energy generated in the core to reach the surface. According to traditional radiative models (Bahcall et al., 2001, 2005), this process occurs via random photon diffusion, with estimated timescales ranging from 100,000 to 170,000 years (Mitalas & Sills, 1992).[17]

However, this long timescale refers to the transport of the total photon energy, not to local dynamic modulations arising from seismic-like activity, which can propagate much more rapidly through acoustic (p-mode) and gravity (g-mode) waves. Simulations by Hartlep & Mansour (2005)[18] indicate that pressure waves can traverse the Sun in less than ten days. This suggests that seismic oscillations, spectrally coherent with tidal perturbations in the core, may rapidly reach the surface while also carrying part of the thermal energy produced in the core. Consequently, the tachocline could receive a signal from the core within a relatively short timescale, potentially synchronizing the dynamo mechanisms and generating oscillations consistent with the observed solar cycles.

Wolff & Patrone (2010) explained: *"an event deep in the Sun that affects the nuclear burning rate will change the amount of energy going into the g-mode oscillations. Some information of this is transported rather promptly by g modes to the base of the Sun's convective envelope (CE). Once these waves deposit energy there, it is carried to the surface in a few months by extra convection, which should increase solar activity. ... This upward transport of luminosity by waves was also advocated by Wolff & Mayr (2004) to explain the east–west reversing flows detected by Howe et al. (2000) and Komm et al. (2003) with characteristic time scales of one to three years."*

Further theoretical work (Ahuir et al., 2021; Goodman & Dickson, 1998; Barker & Ogilvie, 2010)[19] has explored how internal waves generated by planetary tidal forcings can propagate, break, and dissipate in the stellar interior, producing nonlinear responses in the deep plasma.

On long timescales, the solar luminosity would be slightly larger than if the Sun had no planets; Figure 38.3B suggests that the difference would be only about 0.3 W/m^2.

These studies converge on a fundamental insight: stars are not passive linear systems, but highly dynamic entities, capable of responding resonantly to even

[16]Weiss, N.O., Tobias, S.M., 2016. Supermodulation of the Sun's magnetic activity: the effects of symmetry changes. Monthly Notices of the Royal Astronomical Society, 456, 2654–2661.

[17]Mitalas, R., Sills, K.R., 1992. On the photon diffusion time scale for the Sun. The Astrophysical Journal, 401, 759–760.

[18]Hartlep, T., Mansour, N.N., 2005. Acoustic wave propagation in the Sun. Center for Turbulence Research Annual Research Briefs, Stanford University, 357–365.

[19]Ahuir, J., Mathis, S., Amard, L., 2021. Dynamical Tide in Stellar Radiative Zones. Astronomy and Astrophysics, 651, A3.

Barker, A.J., Ogilvie, G.I., 2010. On Internal Wave Breaking and Tidal Dissipation Near the Centre of a Solar-type Star. Monthly Notices of the Royal Astronomical Society, 404, 1849–1868.

weak periodic stimuli—provided that the forcing frequency is well tuned to the system's natural modes.

38.10. Conclusion

If the planets do influence the Sun, they likely do so through nonlinear, resonant mechanisms embedded deep within its internal structure. Planetary forces, although minuscule in magnitude, are remarkably persistent and harmonically coherent, operating over astronomical timescales with rhythmic precision. When coupled with internal solar processes such as nuclear fusion, these forces may act as cosmic metronomes, subtly synchronizing the heartbeat of the Sun with the orbital rhythms of the planetary system.

One plausible mechanism involves gravitational tidal forces from planets such as Venus, Earth, Jupiter, and Saturn, which may induce cyclical compressions in the solar core. These tidal-compression cycles could modulate the fusion rate, leading to measurable variations in solar output (Scafetta, 2012d). There may also be exotic mechanisms, such as the interaction between planetary alignments and flows of invisible matter or dark energy within the heliosphere (Bertolucci et al., 2017).[20] Thus, the Sun and its solar system might function as detectors or amplifiers of both internal and deeper cosmic phenomena.

[20] Bertolucci, S., Zioutas, K., Hofmann, S., Maroudas, M., 2017. The Sun and its planets as detectors for invisible matter. Physics of the Dark Universe, 17, 13.

CHAPTER 39

Heliospheric oscillations in climate records

This chapter reviews key themes explored in previous sections, focusing on the potential role of planetary theory in explaining Earth's climate variability. The hypothesis that planetary motions may leave detectable imprints on climatic and geophysical data has gained increasing attention, especially following the recognition of shared harmonic regularities among solar cycles, climate oscillations, and planetary orbital dynamics.

Over the past two decades, a research trajectory largely advanced by the author of this volume has proposed that several major climate cycles may be interpreted as manifestations of planetary resonances, whose signatures are documented across a wide range of natural archives, including ice cores, sediment layers, and cosmogenic isotopes.

This interdisciplinary perspective opens new horizons in our understanding of the Sun–Earth–planet system, suggesting that Earth's climate is not governed solely by internal processes, but may be modulated by the periodic dynamics of the entire planetary system.

39.1. Global temperatures and planetary harmonics

Numerous studies have demonstrated that global temperature variability can be described with notable precision using harmonic models derived from the principal periodicities of the solar system.

In particular, Scafetta (2010)[1] employed the velocity of the Sun relative to the solar system's barycenter as a proxy for planetary orbital influence. Using spectral analysis techniques such as the Lomb–Scargle method and the Maximum Entropy Method, Figure 30.1 reveals shared frequencies between climate data and solar system harmonics, with significant peaks around 5.2, 5.93, 6.62, 7.42, 9.1, 10.4, 20, and 60 years.

Many of these periodicities are attributable to the orbital dynamics of Jupiter and Saturn, as well as to a 9.1-year solar–lunar tidal cycle (Scafetta, 2014b).[2]

A semi-empirical model composed of six sinusoids centered around 9.1, 10.4, 20, 30, 60, and 115 years successfully reproduced the climate variability observed in the HadCRUT4 dataset (1850–2010) (Scafetta, 2012a,c).[3] The result suggests that Earth's climate possesses an internally harmonic structure, surprisingly spectrally coherent with planetary periodicities known from orbital mechanics.

---

[1]Scafetta, N., 2010. Empirical evidence for a celestial origin of the climate oscillations and its implications. Journal of Atmospheric and Solar–Terrestrial Physics, 72, 951–970.

[2]Scafetta, N., 2014b. Discussion on the spectral coherence between planetary, solar and climate oscillations: a reply to some critiques. Astrophysics and Space Science, 354, 275–299.

[3]Scafetta, N., 2012a. Testing an astronomically based decadal-scale empirical harmonic climate

Oceanic oscillations and planetary coherence. These periodicities are not confined to global temperature records. They also emerge in oceanic indices such as:

- the Atlantic Multidecadal Oscillation (AMO);
- the Pacific Decadal Oscillation (PDO).

Both exhibit cycles of approximately 60 years (cf. Wyatt & Curry, 2014),[4] which significantly influence:

- drought patterns;
- monsoon variability;
- Atlantic hurricane frequency.

The temporal coherence between oceanic dynamics and planetary harmonics reinforces the hypothesis of synchronized modulation between the planetary system and Earth's climate response.

Evidence of an approximately 60-year climate cycle, strongly correlated with the Trigon resonance between Jupiter and Saturn, has also been observed in:

- commercial fisheries (Klyashtorin, 2001; Klyashtorin et al., 2009);[5]
- Indian monsoon rainfall variability Agnihotri & Dutta (2003).[6]

The statistical robustness of this periodicity, combined with its presence in independent and geographically diverse datasets, makes it unlikely to be a mere coincidence or analysis artifact.

39.2. Planetary harmonics in temperature and aurora records

Auroras serve as indirect indicators of solar and geomagnetic activity. These atmospheric phenomena are closely tied to geomagnetic field variability, which in turn reflects solar dynamics.

Scafetta (2012c) analyzed historical records of mid-latitude auroras from 1700 to 1966, comparing them with global temperature data. Spectral analysis revealed shared frequencies around 9.1, 10.4, 20, and 60 years. These match planetary harmonics associated with Jupiter and Saturn, as well as the 9.1-year solar–lunar cycle component.

Further confirmation came from Scafetta & Willson (2013a),[7] who examined the Hungarian aurora record (1523–1960), one of the most extensive pre-instrumental archives of geomagnetic activity. Harmonic analysis revealed strong coherence with planetary orbital cycles, including periodicities around

---

model versus the IPCC (2007) general circulation climate models. Journal of Atmospheric and Solar–Terrestrial Physics, 80, 124–137.

Scafetta, N., 2012c. A shared frequency set between the historical mid-latitude aurora records and the global surface temperature. Journal of Atmospheric and Solar–Terrestrial Physics, 74, 145–163.

[4]Wyatt, M.G., Curry, J.A., 2014. Role of Eurasian Arctic shelf sea ice in a secularly varying hemispheric climate signal during the 20th century. Climate Dynamics, 42, 2763–2782.

[5]Klyashtorin, L.B., 2001. Climate change and long-term fluctuations of commercial catches: the possibility of forecasting. FAO Fisheries Technical Paper, No. 410, Rome, FAO, 86p.

Klyashtorin, L.B., Borisov, V., Lyubushin, A.A., 2009. Cyclic changes of climate and major commercial stocks of the Barents Sea. Marine Biology Research, 5(1), 4–17.

[6]Agnihotri, R., Dutta, K., 2003. Centennial scale variations in monsoonal rainfall: manifestations of solar variability. Current Science, 85, 459–463.

[7]Scafetta, N., Willson, R.C., 2013a. Planetary harmonics in the historical Hungarian aurora record (1523–1960). Planetary and Space Science, 78, 38–44.

43, 57–61, 85–90, 108–115, and 171 years. Many of these derive from multiple synods of Jupiter and Saturn, or from beats involving Uranus and Neptune.

Geomagnetic resonance and planetary modulation. The study suggests that Earth's geomagnetic activity, mediated by solar interactions, may be significantly modulated by planetary configurations. This provides independent confirmation of the same harmonic structure observed in climate data.

In particular, the 20- and 60-year periodicities—not reproduced by conventional climate models—emerge as dominant frequencies in secular auroral records. These correspond to:

- the Jupiter–Saturn synodic cycle ($\sim$19.86 years);
- its third harmonic ($\sim$60 years).

The consistent appearance of these frequencies in both climate and geomagnetic datasets strengthens the hypothesis of a shared astronomical origin.

Historical echoes of planetary influence. These empirical findings breathe new life into ancient intuitions—already formulated by al-Balkhi (IX secolo)[8] and Kepler (1601)[9]—that planetary motions exert a relevant influence on terrestrial phenomena. Rather than being dismissed as archaic speculation, these insights now find support in modern data analysis, suggesting that the harmonic architecture of the solar system may leave a measurable imprint on both the Sun's behavior and Earth's climate system.

39.3. Meteors and interplanetary dust: an alternative forcing?

An intriguing addition to planetary theory emerges from recent research that moves beyond traditional solar and climatic indicators, exploring the potential influence of planetary dynamics on the frequency of meteorite falls on Earth.

Scafetta, Milani, and Bianchini (2020)[10] analyzed two extensive historical archives:

- a chronicle-based dataset from Chinese records (619–1943 AD);
- a global meteorite fall database spanning 1800–1974 (Chang & Yu, 1981; Yu et al., 1983).[11]

These meteorite fall records present a clear oscillation with a periodicity of approximately 60 years, remarkably consistent in phase and duration with the well-known 60-year climate cycle observed in global temperature and oceanic indices.

[8]al-Balkhi, Abu Ma'shar (9th century). De Magnis Coniunctionibus, ed. K. Yamamoto, Ch. Burnett, Leiden, 2000, 2 vols. (Arabic and Latin text).

[9]Kepler, J., 1601. On the more certain fundamentals of astrology. In: Brackenridge, J.B., Rossi, M.A. (Eds.), Proceedings of the American Philosophical Society, 123(2), 85–116 (1979).

[10]Scafetta, N., Milani, F., Bianchini, A., 2020. A 60-year cycle in the meteorite fall frequency suggests a possible interplanetary dust forcing of the Earth's climate driven by planetary oscillations. Geophysical Research Letters, 47(18), e2020GL089954.

[11]Chang, S., Yu, Z., 1981. Historical records of meteorite falls in China and their time series analysis. Memoirs of the National Institute of Polar Research (Special Issue), 20, 276–284.
Yu, Z., Chang, S., Kumazawa, M., Furumoto, M., Yamamoto, A., 1983. Presence of periodicity in meteorite falls. National Institute of Polar Research Memoirs, Special Issue, 30, 362–366.

Planetary modulation of interplanetary dust. The authors hypothesize that this periodicity may reflect a modulation of interplanetary dust density along Earth's orbit, driven by planetary cycles—particularly the gravitational interaction between Jupiter and Saturn. Jupiter's orbital eccentricity, which varies on an approximately 60-year cycle owing to its Trigon with Saturn, appears to be synchronized with oscillations observed in both meteorite-fall records and climate variability.

With a mass over 300 times that of Earth, Jupiter exerts a dominant gravitational influence on the solar system. It plays a dual role:

- as a *cosmic vacuum cleaner*, deflecting or capturing asteroids and comets;
- as a gravitational slingshot, redirecting minor bodies inward toward Earth.

This dynamic is especially relevant for short-period comets and main-belt asteroids, whose trajectories are shaped by orbital resonances with Jupiter.

Resonant structures and dust flux amplification. The Kirkwood gaps in the asteroid belt are caused by mean-motion resonances with Jupiter, which destabilize orbits and can deflect bodies inward. Similarly, Jupiter can cyclically perturb:

- the interplanetary dust cloud;
- populations of Centaurs and Trojan asteroids, and other minor bodies.

These perturbations may amplify the meteoroid flux toward Earth in synchrony with planetary harmonic cycles, such as the 60-year Jupiter–Saturn trigon. This suggests that Jupiter may act as a gravitational regulator, cyclically enhancing or attenuating the influx of cosmic debris, with potential climatic repercussions.

Climatic implications of dust deposition. The deposition of interplanetary dust on Earth—often ionized by the solar wind—could influence climate through:

- direct radiative mechanisms: scattering of sunlight, altering Earth's energy balance;
- indirect mechanisms: increasing cloud condensation nuclei, modifying the atmospheric electric circuit.

This hypothesis introduces a novel variable into the debate on natural climate forcings: the interplanetary environment as a mediator between planetary dynamics and Earth's climate system.

Although still underexplored, the idea that interplanetary dust may exert climatic effects is supported by studies linking meteorite events to episodes of global cooling, such as the Cretaceous–Paleogene extinction 66 million years ago.

The innovation lies in the proposal that not only catastrophic impacts, but also periodic variations in the flux of meteoroids and cosmic debris—modulated by planetary cycles—may act as sufficiently strong climate forcings on multi-decadal scales.

39.4. Astronomical origin of the Bray–Hallstatt cycle

Among the most enigmatic and persistent paleoclimatic oscillations of the late Holocene is the Bray–Hallstatt cycle, a millennial-scale fluctuation with an average periodicity between 2100 and 2500 years, observed in numerous geophysical and cosmogenic proxies, particularly in radiocarbon (^{14}C) and beryllium-10 (^{10}Be) records (McCracken et al., 2013; Scafetta et al., 2016).[12]

Originally identified by Bray (1968, 1971),[13] this cycle spans the entire Holocene and shows temporal correspondence with major climatic and historical transitions, including:

- the Holocene Thermal Maximum;
- the Neolithic revolution;
- Indo-European migrations;
- the Roman expansion;
- the Little Ice Age.

The name *"Hallstatt"* derives from the Austrian town of Hallstatt, a site of archaeological and climatic significance. The cycle coincides with cold epochs that profoundly shaped Central European culture, particularly the Hallstatt culture (ca. 750–400 BC) (see Figure 40.2).

Paleoclimate confirmation and global signatures. More recent studies have confirmed Bray's findings. For example, Mayewski et al. (2004)[14] analyzed over 50 high-resolution paleoclimate records and identified six periods of rapid climate change during the Holocene, including cycles with periodicities between 2000 and 2800 years, consistent with the Bray–Hallstatt cycle.

Additional evidence from marine sediments, ice cores, stalagmites, and paleoclimate assimilation models further confirms the presence of this cycle, even independently of solar proxies (Bond et al., 2001; Erb et al., 2022; Wanner et al., 2015).[15]

Astronomical interpretation: Solar Inertial Motion (SIM) and harmonic modulation. A detailed astronomical analysis proposed by Scafetta et al. (2016) identified a remarkable correspondence between the Bray–Hallstatt cycle and long-term modulations in the Solar Inertial Motion (SIM), the trajectory of the Sun around the solar system barycenter.

The authors suggest that the Bray–Hallstatt cycle emerges as a beat frequency from faster oscillations, particularly the Jose cycle ($\sim$179 years), which reflects specific cyclical planetary configurations governing the Sun's barycentric motion.

Figure 39.1 illustrates:

---

[12] McCracken, K.G., et al., 2013. A phenomenological study of the cosmic ray variations over the past 9400 years. Solar Physics, 286, 609.
Scafetta, N., et al., 2016. On the astronomical origin of the Hallstatt oscillation found in radiocarbon and climate records throughout the Holocene. Earth-Science Reviews, 162, 24–43.

[13] Bray, J.R., 1968. Glaciation and solar activity since the fifth century BC and the solar cycle. Nature, 220, 672–674.
Bray, J.R., 1971. Solar–climate relationships in the post-Pleistocene. Science, 171(3972), 1242–1243.

[14] Mayewski, P.A., et al., 2004. Holocene climate variability. Quaternary Research, 62(3), 243–255.

[15] Bond, G., et al., 2001. Persistent solar influence on North Atlantic climate during the Holocene. Science, 294(5549), 2130–2136.
Erb, M.P., et al., 2022. Reconstructing Holocene temperatures in time and space using paleoclimate data assimilation. Climate of the Past, 18(12), 2599–2629.
Wanner, H., et al., 2015. Holocene climate variability and change: a data-based review. Journal of the Geological Society, 172(2), 254–263.

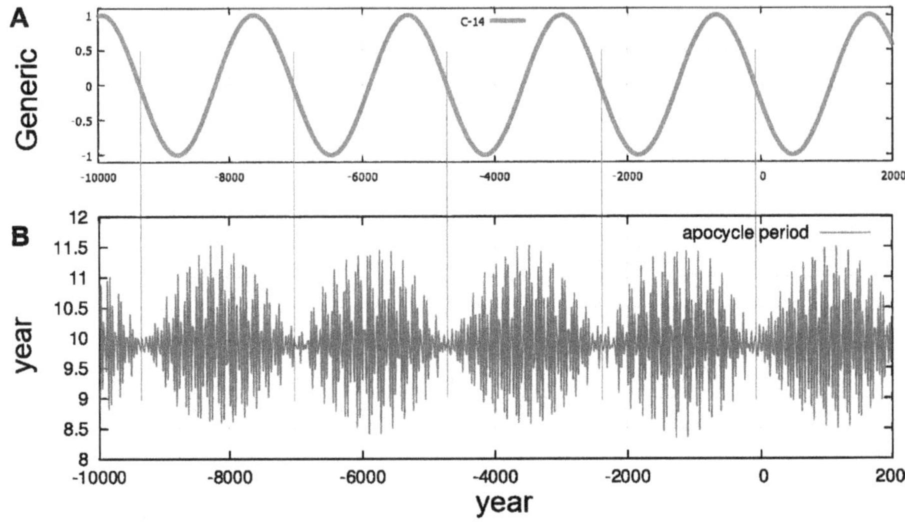

FIGURE 39.1. [A] The Bray–Hallstatt oscillation extracted from the $\Delta^{14}C$ (‰) record, as illustrated in Figure 37.6. Note that the maxima of the ^{14}C curve correspond to solar minima, and vice versa.
[B] The period of the apocycles of the Solar Inertial Motion (SIM). Note the major beat with a periodicity of 2318 years, and the numerous faster Jose beats with periods around 179 years. (cf. Scafetta & Milani, 2025).
(Adapted from Scafetta et al., 2016).

- [A] the Bray–Hallstatt oscillation extracted from the $\Delta^{14}C$ record, derived from the IntCal04 dataset (Reimer et al., 2004),[16] showing that maxima in ^{14}C correspond to solar minima;
- [B] the duration of the SIM apocycles, revealing a beat at 2318 years, alongside faster Jose-type oscillations (~179 years) (cf. Scafetta & Milani, 2025).[17]

The analysis also reveals oscillations with periods of 171.4 and 185 years, both well documented in the scientific literature on solar dynamics and associated with the Jose (1965)[18] band around 179 years.

Harmonic decomposition and planetary coupling. The Bray–Hallstatt orbital invariant inequality:

$$(1, -3, 1, 1) = (0, 0, 1, -1) + (1, -3, 0, 2)$$

[16] Reimer, P.J., et al., 2004. IntCal04 Terrestrial Radiocarbon Age Calibration, 0–26 Cal Kyr BP. Radiocarbon, 46, 1029–1058.

[17] Scafetta, N., Milani, F., 2025. Spectral Structure of the Solar Inertial Motion from 12999 BC to 16998 AD. Publications of the Astronomical Society of the Pacific, 137, 054402.

[18] Jose, P.D., 1965. Sun's motion and sunspots. The Astronomical Journal, 70, 193.

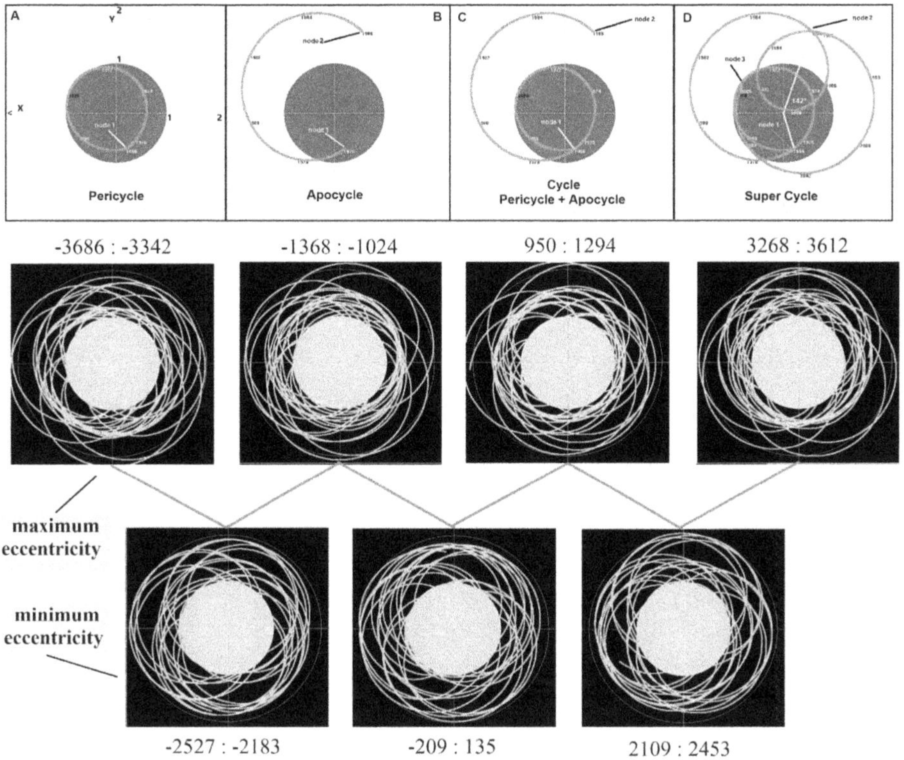

FIGURE 39.2. Motion of the SIM (in solar radii): (top) definition of pericycle, apocycle, cycle (pericycle + apocycle), and supercycle (2 cycles) of a SIM orbit; (middle) periods dominated by disordered cycles (maximum eccentricity); (bottom) periods dominated by ordered cycles (minimum eccentricity). The alternation between these long periods corresponds to the Bray–Hallstatt cycle of 2318 years. (Adapted from Scafetta et al., 2016).

can be decomposed into two harmonics with respective periods of 171.4 and 185 years. This interpretation positions the Jose cycle as part of a hierarchical harmonic structure, modulated over millennia by the Bray–Hallstatt beat.

During phases of favorable planetary alignment, the Sun's motion becomes more regular, and the SIM exhibits greater symmetry. This symmetry may influence the heliosphere, causing it to expand or contract cyclically, as shown in Figure 39.2, thereby affecting the efficiency with which galactic cosmic rays could penetrate the inner solar system.

Heliospheric dynamics and climate implications. Scafetta et al. (2016) propose that millennial-scale SIM dynamics may influence the structure of the heliosphere, including the termination shock, which is the boundary of the solar system at which the solar wind slows abruptly.

During SIM phases of disordered mobility, the heliosphere may offer reduced protection against cosmic rays, resulting in increased flux toward Earth.

This would lead to enhanced production of cosmogenic isotopes (^{14}C and ^{10}Be), with potential climatic effects mediated by low cloud nucleation, as hypothesized by Svensmark & Friis-Christensen (1997).[19]

In this scenario, the Bray–Hallstatt cycle arises from beats among planetary harmonic forcings, producing systemic effects on both solar dynamics and Earth's climate.

Theoretical support and broader implications. There is now solid theoretical and numerical support for an astronomical origin of the Bray–Hallstatt cycle, based on harmonic interactions between planetary orbits and the Sun's barycentric motion.

The implications are profound:

- other millennial or secular climate cycles may have similar astronomical origins;
- Earth's climate may respond more deeply than expected to cyclic external forcings, often overlooked by climate models focused mostly on internal radiative mechanisms.

These findings reinforce the validity of empirical approaches based on spectral analysis and planetary harmonic models, highlighting the key role of Sun–planet interactions in modulating Earth's climate system over long timescales.

39.5. Final considerations

The body of evidence examined across Chapters 32 to 39 ranges from solar proxies and global temperature reconstructions to meteorite-fall frequencies and historical auroral records. Taken together, these datasets outline a coherent and multidisciplinary framework in which planetary orbital dynamics appear to modulate, directly or indirectly, a wide spectrum of terrestrial geophysical processes.

The recurrence of specific temporal harmonics—most notably 9.1, 10.4, 20, 30, 60, 115, and 170 years—across diverse natural archives constitutes a statistically robust and geographically widespread signal. The cross-domain coherence of these periodicities makes it increasingly difficult to dismiss them as coincidental. Instead, they point toward a shared astronomical origin, likely mediated by solar mechanisms and complementary phenomena such as interplanetary-dust flux modulation, gravitational resonance, and cosmic-ray variability.

Integrating planetary harmonics into climate science. Although the precise physical transmission mechanisms remain partially unresolved and subject to ongoing debate, the amplitude, persistence, and spectral clarity of the observed signals reinforce the hypothesis that planetary system dynamics play an active and non-negligible role in Earth's climate variability, particularly on decadal to centennial timescales.

In this light, the harmonic–planetary approach should not be seen as a rival to conventional climate models, but as a complementary and empirically grounded extension. It offers the potential to capture cyclical structures and

[19] Svensmark, H., Friis-Christensen, E., 1997. Variation of cosmic ray flux and global cloud coverage—a missing link in solar–climate relationships. Journal of Atmospheric and Solar–Terrestrial Physics, 59(11), 1225–1232.

long-term modulations that General Circulation Models (GCMs) often fail to reproduce, especially when constrained by short-term calibration windows and limited representation of external forcings.

As Scafetta (2012c) aptly states:

> *"These results suggest that the origin of a significant portion of climate variability is not entirely internal to Earth, but reflects the rhythmic influence of the celestial environment."*

Embracing this perspective calls for a more integrative climate science—one that is open to interdisciplinary insights, respectful of empirical complexity, and committed to exploring the full architecture of planetary–solar–terrestrial interactions. Only through such openness can climate science evolve into a truly holistic discipline, capable of deciphering the deep rhythms that shape our planet's long-term evolution.

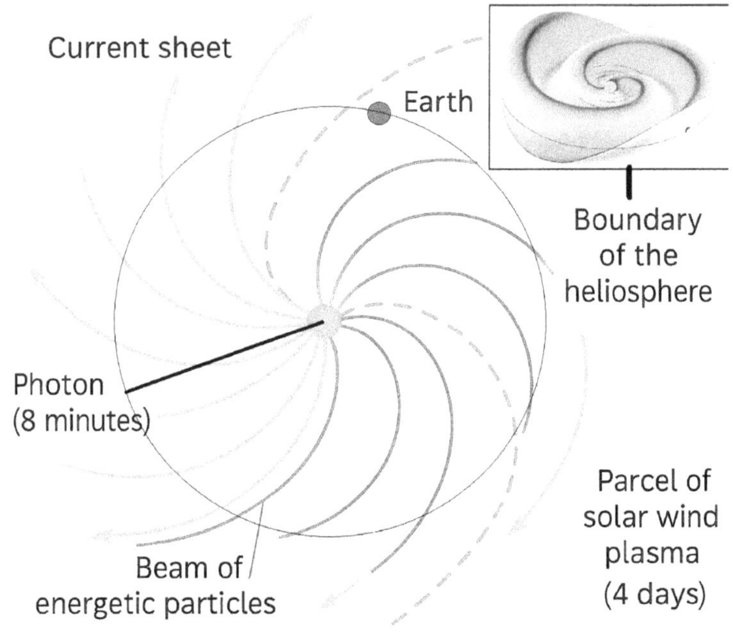

FIGURE 39.3. **Parker spiral of the heliospheric magnetic field.** The figure illustrates the propagation of solar emissions from the Sun to Earth, highlighting the distinct paths and travel times of photons, energetic particles, and solar wind plasma. At the center, the Sun emits light (photons, shown as a black line) that reaches Earth in approximately 8 minutes along a straight trajectory. Energetic particles, by contrast, follow curved magnetic field lines and can arrive more rapidly than the bulk solar wind. The solar wind itself travels outward along the Parker spiral, whose helical structure is generated by the combination of the Sun's radial outflow and its rotation. These spiraling magnetic field lines define the large-scale architecture of the heliosphere, the vast plasma-filled bubble that surrounds the solar system. The circular path in the diagram represents Earth's orbit, which intersects both the Parker spiral and the heliospheric current sheet, where the Sun's magnetic polarity reverses. The Parker spiral not only governs the transport of charged particles but also shapes the geometry of the heliosphere, influencing space weather and modulating the shielding of planetary environments from galactic cosmic rays. In principle, certain low-frequency signals or disturbances (such as Alfvén waves or magnetic perturbations) could propagate along these spiral field lines back toward the Sun, although such reverse transport is limited by the outward flow of the solar wind and the complex topology of the interplanetary magnetic field.

CHAPTER 40

A two-millennial climate reconstruction based on astronomical rhythms

The final chapter of Part VI presents a climate-modeling exercise proposed by Scafetta (2026)[1], which—despite its structural simplicity—demonstrates a remarkable ability to reproduce the principal features of global climate variability over the past two millennia.

The model employed is a synthetic empirical climate model, calibrated for the period 1–2017 AD, and constructed using a coherent formulation of the three primary radiative forcings:

- anthropogenic forcing;
- volcanic forcing;
- solar forcing.

The aim is to evaluate whether a model based on observed natural periodicities can explain long-term climate variability, offering a credible alternative to the General Circulation Models (GCMs) adopted in the CMIP6 project and used in IPCC assessments.

40.1. Structure of the empirical model and forcing components

Figure 40.1 illustrates the model's input data and outputs, with components analyzed as follows.

a) Anthropogenic radiative forcing. Anthropogenic forcing is assumed to be zero prior to 1749, consistent with the hypothesis that human emissions had a negligible climatic impact before the industrial era. From 1750 onward, the model adopts CMIP6 anthropogenic forcing values, ensuring direct comparability with official simulations while preserving the autonomy of the natural components.

b) Volcanic forcing. Volcanic forcing is modeled using CMIP6 estimates from 1750 onward. For the preindustrial period (1–1749 AD), values are extended using the reconstruction by Toohey & Sigl (2017),[2] based on geological and glaciological data. This enables the model to capture the climatic impact of major eruptions, including:

[1]Scafetta, N., 2025. Detection, attribution, and modeling of climate change: Key open issues. Gondwana Research, DOI: 10.1016/j.gr.2025.05.001.

[2]Toohey, M., Sigl, M., 2017. Volcanic stratospheric sulfur injections and aerosol optical depth from 500 BCE to 1900 CE. Earth System Science Data, 9, 809–831.

- Tambora (1815): VEI 7; the most powerful eruption of the past 10,000 years, which triggered the *"year without a summer"* in 1816;[3]
- Samalas (1257):[4] ultra-Plinian eruption with VEI 7; formed the Segara Anak caldera and may have initiated the Little Ice Age;[5]
- Ilopango (431 AD $\pm$ 2 years): Plinian eruption with VEI 6; devastated the region around San Salvador and likely affected the Maya civilization.[6]

c) **Solar forcing: a synthetic TSI model.** The most innovative component is the solar forcing, derived from a synthetic model of Total Solar Irradiance (TSI) based on the planetary harmonics discussed in earlier chapters. Figure 40.1B shows the TSI function, designed to reproduce features observed in solar proxies (^{14}C and ^{10}Be) and integrate solar activity across multiple timescales.

The synthetic TSI model comprises five primary harmonic waves:

- $P_1 = 60.95$ years—multidecadal solar cycle;
- $P_2 = 114.78$ years—secular solar cycle;
- $P_3 = 129.95$ years—secular solar cycle;
- $P_4 = 983.40$ years—Eddy cycle;
- $P_5 = 2318$ years—Bray–Hallstatt cycle.

Harmonics P_1–P_3 simulate the secular solar cycles responsible for grand solar minima such as the Maunder and Dalton events, as reconstructed in Chapter 36 (Scafetta, 2012b).[7]

Harmonics P_4–P_5 represent the millennial cycles documented in the paleoclimate literature, including the Hallstatt oscillation and long-term cosmic ray modulation (McCracken et al., 2014; Scafetta et al., 2016).[8]

An additional component simulates the 11-year solar cycle, with amplitude reduced during solar grand minima. Harmonic phases are fixed on the basis of astronomical considerations, and amplitudes are scaled to match proxy TSI reconstructions (Chapter 36 and Chapter 37). The amplitudes of the harmonics are chosen to ensure compatibility with proxy-based TSI reconstructions, allowing for high variability in secular modulation (see Figure 25.4).

To ensure metric consistency with CMIP6 GCMs, the TSI function is converted into radiative forcing by subtracting a mean value of 1361 W/m² and dividing by 8, following standard methodology.

---

[3]The most powerful eruption of the past 10,000 years, with a VEI of 7. It caused the "year without a summer" in 1816.

[4]An ultra-Plinian eruption with VEI 7. It formed the Segara Anak caldera and may have triggered the onset of the Little Ice Age.

[5]An ultra-Plinian eruption with VEI 7. It formed the Segara Anak caldera and may have triggered the onset of the Little Ice Age.

[6]A Plinian eruption with VEI 6. It devastated the area around San Salvador and likely affected the Maya civilization.

[7]Scafetta, N., 2012b. Multi-scale harmonic model for solar and climate cyclical variation throughout the Holocene based on Jupiter–Saturn tidal frequencies plus the 11-year solar dynamo cycle. Journal of Atmospheric and Solar–Terrestrial Physics, 80, 296–311.

[8]McCracken, K.G., et al., 2014. Evidence for planetary forcing of the cosmic ray intensity and solar activity. Solar Physics, 289, 3207–3229.

Scafetta, N., et al., 2016. On the astronomical origin of the Hallstatt oscillation found in radiocarbon and climate records throughout the Holocene. Earth-Science Reviews, 162, 24–43.

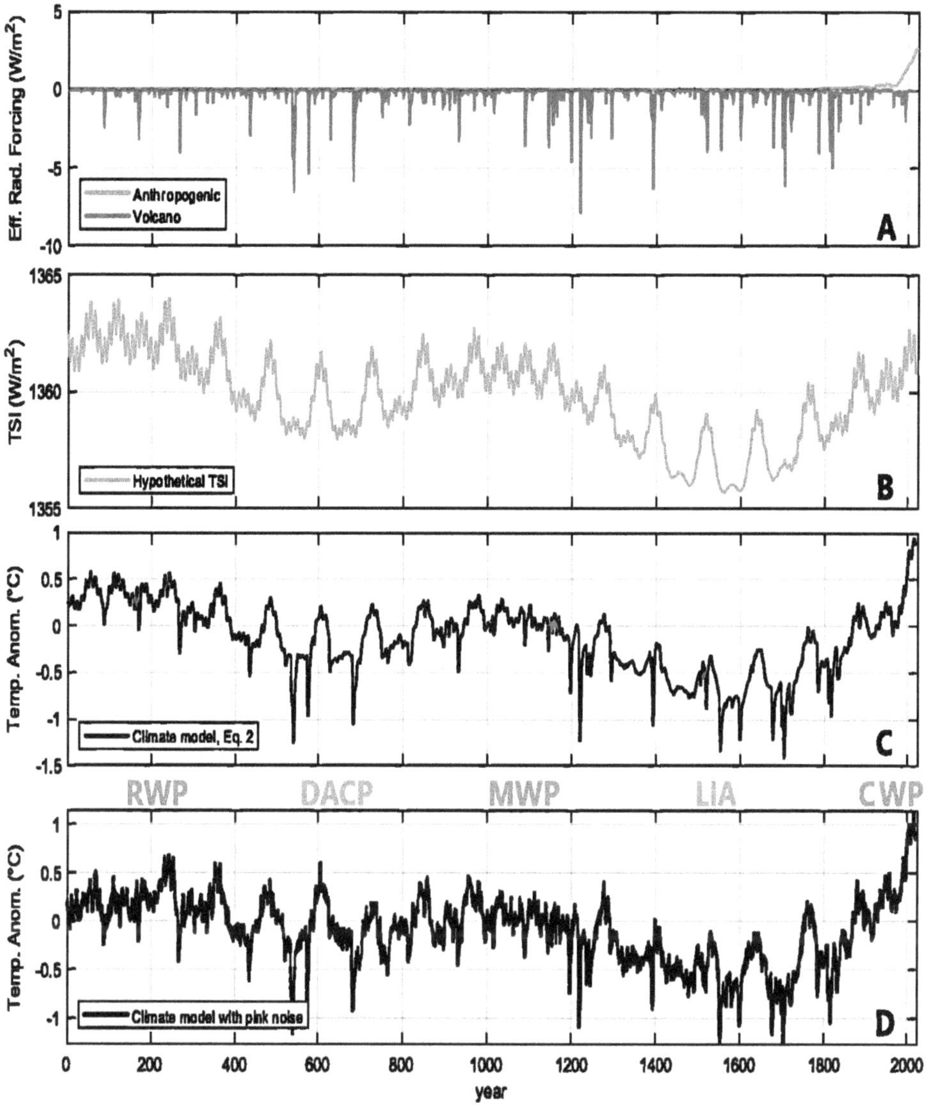

FIGURE 40.1. [A] Effective anthropogenic and volcanic radiative forcings, retroactively extended to the period 1–1750 AD and consistent with CMIP6 GCM values.
[B] Synthetic model of total solar irradiance (TSI), constructed using climate harmonics derived from stable astronomical cycles.
[C] Climate simulation obtained using the Empirical Climate Model (ECM), based on the equations discussed in Chapter 31 and calibrated using the TSI-2 and HadSST4 datasets.
[D] Output as in [C], but with added pink noise to represent internal climate system variability
(Adapted from Scafetta, 2026).

40.2. Parameterization and climate response

The climate simulation is generated by the Empirical Climate Model (ECM) discussed in Chapter 31, forced with the TSI-2 record and calibrated with the HadSST4 dataset (Scafetta, 2023c).[9]

Three climate response coefficients are used:

- $\alpha_{ant} = \alpha_V = 0.083\,°\text{C}\cdot\text{m}^2/\text{y}\cdot\text{W}$ (response to anthropogenic and volcanic forcings);
- $\alpha_S = 0.427\,°\text{C}\cdot\text{m}^2/\text{y}\cdot\text{W}$ (response to solar forcing);
- $\beta = 0.303$ 1/year (system dissipation factor).

These parameters yield an equilibrium climate sensitivity (ECS) of approximately 1 ± 0.3 °C, in stark contrast with CMIP6 GCM estimates, which range from 1.8 to 5.7 °C (IPCC, 2021).

Moreover, the model indicates that the climate system's sensitivity to solar forcing is amplified by a factor of 5.1 ± 1.5 compared with the radiative sensitivity assumed in CMIP6 GCMs, suggesting that conventional models significantly underestimate the impact of solar variability on global temperature.

40.3. Results and comparison with paleoclimate data

Figures 40.1C and 40.1D present the simulations produced by the Empirical Climate Model (ECM), the latter incorporating pink noise to simulate internal climate system variability.

A qualitative comparison with major paleoclimate reconstructions—Figure 5.3, Figure 5.2D, and Figure 13.5 (cf. Esper et al., 2024a)[10]—reveals a striking similarity between the simulated sequence of warm and cold periods and those documented in proxy records. These include:

- *Roman Warm Period* (RWP) (~250 BC–400 AD): mild climate during the Roman Empire, conducive to agricultural expansion and demographic growth;
- *Dark Ages Cold Period* (~400–800 AD): post-Roman cooling, marked by solar minima and socioeconomic decline;
- *Medieval Warm Period* (MWP) (~800–1300 AD): moderate warming, associated with agricultural prosperity and urbanization;
- *Little Ice Age* (LIA) (~1300–1850 AD): global cooling, glacier expansion, and significant societal impacts; the coldest phase of the Holocene;
- *Modern Warm Period* (CWP) (1850–present): progressive warming, attributed to both natural and anthropogenic factors.

A quasi-millennial oscillation also emerges, with lower amplitude in the first millennium and stronger modulation in the second. This pattern aligns with the Bray–Hallstatt cycle (~2318 years), which influences transitions between climatic epochs and modulates major minima, such as those around 370 AD and 1530 AD.

---

[9] Scafetta, N., 2023c. Empirical assessment of the role of the Sun in climate change using balanced multi-proxy solar records. Geoscience Frontiers, 14(6), 101650.

[10] Esper, J., et al., 2024. The IPCC's reductive Common Era temperature history. Communications Earth & Environment, 5, 222.

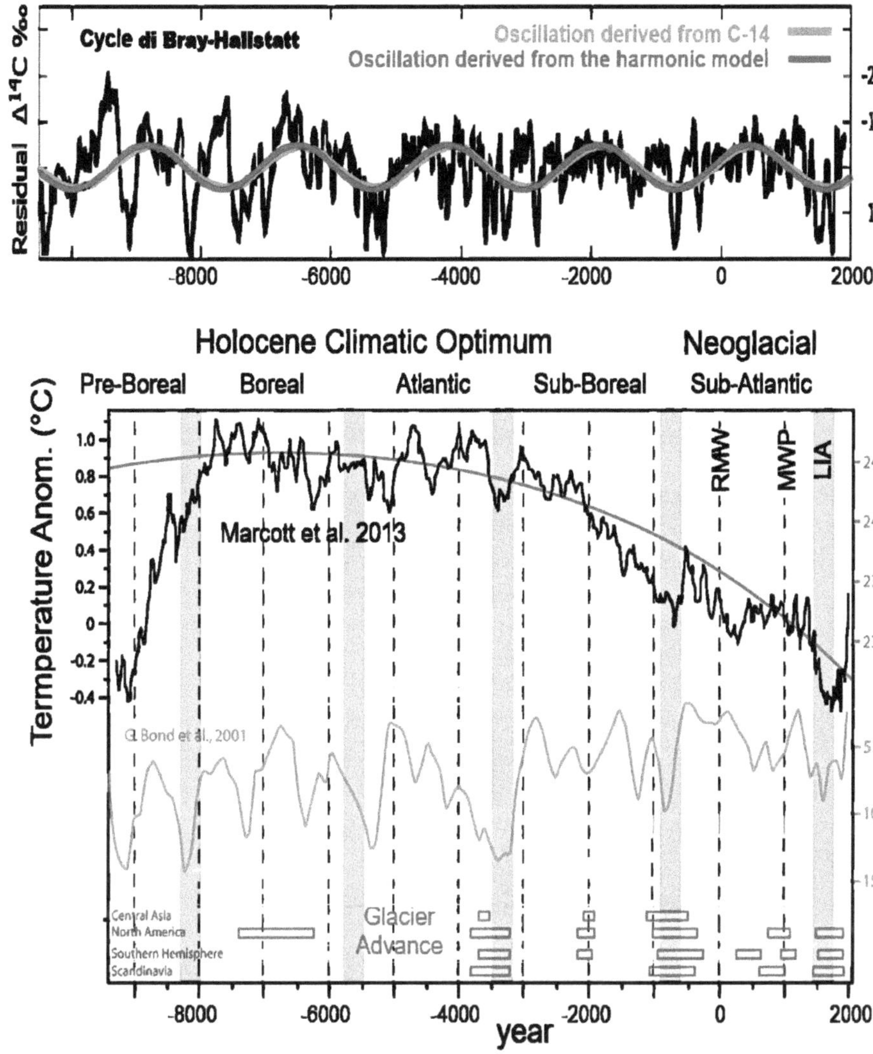

Figure 40.2. [Top] Comparison between the Bray–Hallstatt oscillation extracted from the $\Delta^{14}C$ (‰) record and the harmonic cycle derived from the orbital invariant inequality $(1, -3, 1, 1)$ discussed in Figure 37.6. (Adapted from Scafetta et al., 2016; Scafetta, 2020). [Bottom] Evidence of the Bray–Hallstatt cycle in climate series. Data from Bond et al. (2001); Mayewski et al. (2004); Marcott et al. (2013). (Recreated from Vinós, 2017).

The ability of the ECM to reproduce this historical sequence of climate phases using a compact set of physically interpretable parameters is currently unmatched by any General Circulation Model (GCM). This result further underscores the physical limitations of GCMs in simulating natural variability on secular and millennial timescales.

Holocene climate subdivisions and orbital dynamics. Figure 40.2 compares the Bray–Hallstatt cyclicity identified in the ^{14}C radiocarbon record with the same periodicity found in various Holocene climate series (Bond et al., 2001; Mayewski et al., 2004; Marcott et al., 2013).[11]

The planetary model based on the orbital invariant inequality (1, −3, 1, 1) successfully explains the emergence of the five main climatic subdivisions of the Holocene. Each phase exhibits a periodicity of approximately 2100–2500 years, offering a coherent framework for understanding long-term climate evolution and its influence on human societies:

- Preboreal (approx. 10500–8100 BC): the earliest phase of the Holocene, beginning just after the Younger Dryas; marked by rapid post-glacial warming and retreat of the continental ice sheets;
- Boreal (approx. 8100–5700 BC): climatic stabilization and forest expansion; favorable to Mesolithic cultures and early agriculture in the Near East;
- Atlantic (approx. 5700–3300 BC): Holocene Thermal Maximum; warm and humid conditions, widespread forest cover, and demographic growth; often considered the climatic optimum of the Holocene;
- Subboreal (approx. 3300–700 BC): gradual cooling and increased climatic variability; associated with the emergence and expansion of Bronze and Iron Age civilizations;
- Subatlantic (approx. 600 BC–present): the most climatically unstable phase, featuring alternating warm and cold periods, including the Roman Warm Period, the Little Ice Age, and the Modern Warm Period.

This sequence, modulated by the Bray–Hallstatt cycle, provides a valuable interpretive key for understanding millennial-scale climate variability and its connection to planetary orbital dynamics.

40.4. Scientific and epistemological implications

According to the ECM model, approximately 50% of the warming observed between the preindustrial period (1850–1900) and the 2011–2020 decade can be attributed to solar variability. This estimate is supported by multiple independent studies (Scafetta & West, 2006b, 2007; Scafetta, 2009, 2023c).[12]

This finding contrasts with the prevailing view that recent global warming is exclusively anthropogenic. Instead, the ECM model suggests a mixed contribution between natural and anthropogenic forcings.

---

[11] Bond, G., et al., 2001. Persistent solar influence on North Atlantic climate during the Holocene. Science, 294(5549), 2130–2136.

Mayewski, P.A., et al., 2004. Holocene climate variability. Quaternary Research, 62(3), 243–255.

Marcott, S.A., et al., 2013. A reconstruction of regional and global temperature for the past 11,300 years. Science, 339(6124), 1198–1201.

[12] Scafetta, N., West, B.J., 2006b. Phenomenological solar signature in 400 years of reconstructed Northern Hemisphere temperature record. Geophysical Research Letters, 33, L17718.

Scafetta, N., West, B.J., 2007. Phenomenological reconstructions of the solar signature in the Northern Hemisphere surface temperature records since 1600. Journal of Geophysical Research, 112, D24S03.

Scafetta, N., 2009. Empirical analysis of the solar contribution to global mean air surface temperature change. Journal of Atmospheric and Solar–Terrestrial Physics, 71, 17–18.

Scafetta, N., 2023c. Empirical assessment of the role of the Sun in climate change using balanced multi-proxy solar records. Geoscience Frontiers, 14(6), 101650.

Despite its structural simplicity, the model underscores the importance of adopting solar reconstructions that reflect significant secular variability. Furthermore, it proposes that Total Solar Irradiance (TSI) represents only one facet of the Sun's climatic influence. Other mechanisms may amplify the solar impact, including modulation of cloud cover and planetary albedo by galactic cosmic rays and interplanetary dust fluxes, both of which can be affected by the solar magnetic field.

These mechanisms, although not yet fully understood, could play a critical role in amplifying the climatic effects of solar variability.

40.5. Between empirical realism and the limits of GCM modeling

The Empirical Climate Model (ECM) presented in this chapter offers a coherent and observationally grounded reconstruction of Holocene climate variability, spanning from the Roman era to the present. Its strength lies in its ability to reproduce key paleoclimatic and historical features using a minimal set of physically interpretable parameters, without relying on the structural complexity, tuning procedures, or embedded assumptions characteristic of CMIP6 General Circulation Models (GCMs).

This approach challenges the prevailing notion that climate science is *"settled"* in its foundational assumptions. Rather, it suggests that the IPCC IPCC (2021) interpretation of global warming between 1850–1900 and 2010–2020 may be incomplete, particularly with respect to attribution. The marginalization of solar activity—as implied in Figure 6.3—appears increasingly difficult to sustain in light of empirical reconstructions that highlight its persistent and cyclical influence.

Instead of dismissing alternative models, the ECM encourages a broader epistemological shift: one that values empirical realism, embraces methodological pluralism, and remains open to revisiting core assumptions. Solar variability, planetary harmonics, and natural oscillations should not be treated as residual noise but as integral components of Earth's climate architecture.

In this light, the harmonic–astronomical approach does not seek to replace conventional models but to complement, enrich, and guide their development. It provides a framework capable of capturing long-term modulations and cyclical structures that GCMs currently fail to reproduce. A climate science grounded in observation, falsifiability, and openness to complexity is better positioned to advance theoretical understanding and to inform responsible policy.

FIGURE 40.3. **Evidence of periods warmer than today during the Holocene.** This figure presents visual documentation of Holocene intervals characterized by reduced glacier extent in the Central Alps, suggesting climatic phases warmer than today. [A–B] The Susten Pass (Switzerland) as it appears today (A) and as it likely looked during Roman times, approximately 2,000 years ago (B), with greener terrain and widespread forest cover (illustration based on a draft by Christoph Schlüchter). [C] Geologist Christian Schlüchter displays a tree trunk recovered near a Swiss glacier, part of a broader body of evidence indicating eight Holocene phases of glacier retreat in the Central Alps (Hormes et al., 2001). [D–F] Additional discoveries of subfossil tree trunks near retreating glaciers, reinforcing the interpretation of episodic warmth and ecological shifts throughout the Holocene.

Part 7

Toward a New Paradigm

CHAPTER 41

Science as a process, not a dogma

In recent decades, climate research has evolved from a primarily technical discipline into one of the central themes of global policy discourse. Although significant, this evolution has been accompanied by a tendency toward simplification in public narratives, which has often obscured the inherent complexity of Earth's climate system. The prevailing rhetoric has at times become polarized: in some contexts, "science" is presented as a source of definitive conclusions, while legitimate scientific skepticism is dismissed without distinction and occasionally conflated with denialism.

Yet scientific inquiry is not founded on immutable certainties. It is a dynamic process of hypothesis testing, empirical validation, and theoretical refinement. It relies on models that must be confronted with observations, and on interpretations that evolve in response to new data.

Earth's climate, as a nonlinear and dynamic system, is shaped by multifactorial interactions, feedback mechanisms, and diverse external influences. For this reason, it cannot be reduced to ideological or binary frameworks. Nevertheless, certain strands of political and media communication have tended to present climate science as a unified narrative, less open to critique and less reflective of the diversity of scientific perspectives.

This volume seeks to reaffirm the role of scientific inquiry at the heart of climate discourse—not as a rhetorical instrument, but as a methodological and epistemological foundation. It highlights how several widely accepted claims remain subject to ongoing debate, revision, and methodological caution. Scientific research is not static: it is iterative, open to alternative hypotheses, attentive to empirical detail, and committed to reassessing its own assumptions.

The power of detail: Kepler's eight arcminutes. To illustrate the importance of precision in scientific reasoning, consider Johannes Kepler's analysis of Tycho Brahe's observational data on the motion of Mars. Kepler identified a discrepancy of just eight arcminutes (approximately 0.13°) between the theoretical predictions based on circular orbits, as assumed by the prevailing models, and the actual measurements. While many of his contemporaries, including Galileo, regarded this deviation as negligible, Kepler insisted:

> *"If I had believed that this deviation of eight minutes could be ignored, I would have patched up my hypothesis accordingly. But since it cannot be ignored, those eight minutes have led to the reformation of astronomy."*

"*Science is the honest attempt to understand the universe through observation and doubt.*" (Richard Feynman)

"*No amount of experimentation can ever prove me right; a single experiment can prove me wrong.*" (Albert Einstein)

"*When a scholar goes in the wrong direction, his methods invariably bring him back to the right path.*" (Bertrand Russell)

FIGURE 41.1. Science is grounded in the scientific method and often requires meticulous attention to detail while maintaining a holistic view of the phenomenon under investigation. It is precisely in the details, in seemingly negligible nuances, that evidence may lie capable of challenging established theories or even triggering paradigm shifts. The strength of research lies in its ability to investigate deeply without losing coherence with the global picture.

This minor anomaly led Kepler to abandon the notion of circular orbits and to propose the ellipse as the true path of planetary motion. From this insight emerged Kepler's first law, which transformed celestial mechanics.

In climate science, however, the discrepancies between Global Climate Model (GCM) simulations and empirical data are not subtle deviations comparable to Kepler's eight arcminutes. As discussed in Part II of this volume, these divergences are substantial and observable across multiple timescales—from annual to millennial—extending throughout the Holocene to the present. Such differences warrant careful examination and cannot be lightly dismissed without rigorous analysis or treated as a vague manifestation of natural climate variability.

41.1. Consensus as a political tool

One of the themes explored in this book is the role of *"scientific consensus"* in climate communication. The frequently cited figure of "97% consensus," derived from studies such as Cook et al. (2013)[1], has become a prominent reference point, often invoked to reinforce policy positions and to marginalize dissenting views, even when those views are grounded in documented evidence and methodological critique.

This consensus figure has been subject to scrutiny, particularly regarding the criteria used to classify studies and the semantic breadth of statements such

[1]Cook, J., Nuccitelli, D., Green, S.A., Richardson, M., Winkler, B., et al., 2013. Quantifying the consensus on anthropogenic global warming in the scientific literature. Environmental Research Letters, 8, 024024.

as *"humans contribute to climate change."* Such formulations, while broadly accurate, encompass a wide range of interpretations and do not necessarily reflect agreement on the magnitude, dominant mechanisms, or projected impacts of climate change.

Historically, consensus has not always coincided with scientific validity. Many foundational advances have emerged from minority positions that challenged prevailing paradigms and introduced new interpretive frameworks. In contemporary discourse, however, consensus is often invoked to discourage debate and to limit epistemological diversity.

This tendency warrants reflection. If scientific consensus is used primarily to legitimize authority rather than to foster inquiry, critical thinking risks being conflated with opposition, and the dynamic nature of scientific discourse may be replaced by prescriptive messaging.

Yet climate science, like all scientific domains, remains characterized by uncertainty. Acknowledging this uncertainty is not a weakness but a strength; it reflects the openness of science to revision, refinement, and deeper understanding.

41.2. Climate sensitivity and overconfidence in CMIP GCMs

One of the most debated topics in contemporary climate science is the estimation of equilibrium climate sensitivity (ECS), defined as the expected increase in global mean surface temperature following a doubling of atmospheric CO_2. The CMIP6 models employed by the IPCC in its Sixth Assessment Report (AR6) have widened the uncertainty range relative to previous generations, with ECS values spanning from 1.8 °C to over 5.7 °C (see Figure 11.1).

As discussed in earlier chapters, these models rely on free parameters that are also calibrated ex post to reproduce observed climate variations, particularly during the 1970–2000 period. However, these General Circulation Models (GCMs) struggle to accurately simulate other historical intervals, such as:

- 1910–1940: greater warming than projected;
- 1940–1970: a cooling phase that is not adequately captured;
- 2000–2025: more moderate warming than the model outputs.

These discrepancies suggest that erroneously modeled strong positive feedbacks—especially those associated with water vapor and cloud cover, among the most uncertain components of climate modeling—may have contributed to an overestimation of the climate response to radiative forcing.

Empirical comparisons with satellite data (UAH-MSU v.6, Spencer et al., 2017)[2] and surface thermometer datasets (HadCRUT5) indicate warming trends significantly lower than those simulated by the CMIP6 GCMs (see Figures 11.2 and 11.3). Between 1980–1990 and 2012–2022, the observed warming was approximately 0.51±0.1 °C, with a total radiative forcing of about 1.8 W/m² (using the IPCC effective radiative forcing under SSP2-4.5). Depending on the empirical methodology adopted, the resulting average ECS value lies between

[2]Spencer, R.W., Christy, J.R., Braswell, W.D., 2017. UAH version 6 global satellite temperature products: Methodology and results. Asia-Pacific Journal of Atmospheric Sciences, 53(1), 121–130.

1.1±0.4 °C and 2.1±0.7 °C (cf. Figure 31.2; Lewis 2023; Scafetta 2023b,c, 2024).[3] Such estimates are significantly lower than those of most CMIP6 GCMs and than the IPCC central estimate of 2.5–4.0 °C (AR6, 2021).[4]

Comparisons with paleoclimate data further suggest that GCMs do not adequately reproduce past warm periods recurring on millennial scales, such as the Medieval Warm Period (MWP), the Roman Warm Period (RWP), and the Holocene Climatic Optimum. This limitation raises questions about attributing the warming observed since 1850–1900 exclusively to anthropogenic emissions. It is plausible that recent warming may partially reflect the ascending phase of a longer-term natural climate cycle, similar to those observed during previous Holocene warm episodes. However, such mechanisms are not adequately represented in current models.

41.3. Solar variability and the ACRIM–PMOD controversy

Among the less emphasized yet scientifically relevant aspects of climate uncertainty is the role of solar variability and, in particular, the reconstruction of Total Solar Irradiance (TSI) from satellite measurements. While the IPCC treats solar variability as a negligible factor in recent global warming, the issue remains the subject of ongoing investigation. Empirical evidence suggests that solar variability has been closely correlated with climate fluctuations throughout the Holocene.

Central to this debate is the challenge of estimating long-term solar activity and identifying the mechanisms—both radiative and particulate—that may influence Earth's climate. CMIP6 models tend to minimize the solar contribution by adopting TSI reconstructions that exhibit the lowest secular variability among those available in the scientific literature. In addition, these models exclude mechanisms such as cosmic rays and interplanetary dust, which several empirical studies suggest may influence cloud formation, likely the key component of climate dynamics.

The controversy becomes more pronounced when satellite-based TSI measurements since 1978 are examined. Two major composites have emerged (see Figure 24.3):

- the ACRIM composite (Willson & Hoyt), which indicates a significant increase in TSI between 1980 and 2000, based on NASA's experimental data without retroactive adjustments;
- the PMOD composite (Fröhlich & Lean), which suggests a stable or slightly declining TSI over the same period, applying post hoc corrections to datasets such as NIMBUS7/ERB to align with theoretical solar models.

---

[3]Lewis, N., 2023. Objectively combining climate sensitivity evidence. Climate Dynamics, 60, 3139–3165.
Scafetta, N., 2023b. CMIP6 GCM ensemble members versus global surface temperatures. Climate Dynamics, 60, 3091–3120.
Scafetta, N., 2023c. Empirical assessment of the role of the Sun in climate change using balanced multi-proxy solar records. Geoscience Frontiers, 14(6), 101650.
Scafetta, N., 2024. Impacts and risks of "realistic" global warming projections for the 21st century. Geoscience Frontiers, 15(2), 101774.

[4]IPCC, 2021. Climate Change 2021: The Physical Science Basis. Cambridge University Press.

The IPCC's adoption of the PMOD composite and its derived proxy models as solar forcing inputs in the GCMs does not reflect a unanimous technical consensus. Rather, it represents a methodological choice that emphasizes anthropogenic greenhouse gases as the primary drivers of recent warming. According to studies by Scafetta & West (2008) and Scafetta (2009),[5] the TSI increase reported by ACRIM could (indirectly) account for up to 50% of the warming observed between 1980 and 2000.

This divergence between ACRIM and PMOD raises important scientific questions. Prioritizing theoretical reconstruction over direct experimental observations in the absence of conclusive empirical justification may limit the scope of climate attribution studies. The exclusion of natural factors, not through rigorous refutation but through alignment with prevailing modeling frameworks, underscores the need for continued methodological pluralism in climate science.

41.4. An alternative framework: *"climate realism"*

In a scientific landscape shaped by models that frequently overestimate observed warming, characterized by wide uncertainty ranges and by the strategic use of consensus in public discourse, this volume proposes an alternative interpretive framework: *climate realism.*

Climate realism does not deny climate change, nor does it reject the role of human activity. Rather, it advocates a more cautious, empirically grounded, and context-sensitive interpretation of the available data. It emphasizes direct observation, the study of natural cycles, and historical climate variability, while acknowledging the inherent limitations of predictive models and simulation systems.

This approach is distinct from both alarmist and dismissive narratives, which are often shaped by ideological or political considerations. Climate realism promotes rigorous scientific debate, anchored in evidence and open to revising prevailing theories when warranted. It recognizes that the physical mechanisms underlying certain climate cycles—particularly those associated with solar variability—remain partially unresolved. Nevertheless, it favors empirical and semi-empirical models that more accurately reproduce observed climate dynamics, in contrast to complex numerical models such as GCMs, which are based on reductionist principles and often struggle to simulate documented climate variability with sufficient fidelity.

The holistic-empirical approach rests on several foundational principles:

- projections should be consistent with observed data and not retrofitted to support predetermined narratives;
- uncertainty margins must be acknowledged and communicated transparently, rather than minimized for rhetorical emphasis;
- climate policies should be proportionate to empirically assessed risks, rather than driven by extreme scenarios derived from models that are highly sensitive to anthropogenic forcings;

[5]Scafetta, N., West, B.J., 2008. Is climate sensitive to solar variability? Physics Today, 61(3), 50–51. Scafetta, N., 2009. Empirical analysis of the solar contribution to global mean air surface temperature change. Journal of Atmospheric and Solar–Terrestrial Physics, 71, 17–18.

- scientific discourse must remain open and pluralistic, free from political, ideological, or financial constraints.

This is not a call for inaction, but for informed and adaptive responses grounded in ecological realities, shaped by local contexts, and guided by empirical evidence.

Such an approach prioritizes:

- local adaptation strategies tailored to geographic and climatic specificities;
- energy resilience, understood as the diversification and robustness of supply systems;
- technological flexibility, favoring solutions with low economic and environmental impact.

It does not advocate automatic or uncritical adherence to broad decarbonization policies, particularly when such measures may entail unintended consequences that exceed the challenges they aim to address.

41.5. Semi-empirical climate modeling and policy implications

To address the discrepancies between CMIP6 climate-model projections and historical evidence, this volume presents a comprehensive body of scientific research that has contributed to the development of several semi-empirical approaches. These methods seek to model the climate on the basis of its observed dynamic characteristics, relying primarily on direct analysis of empirical data rather than on the theoretical and parametric framework of current GCMs. The emphasis is placed on consistency with observations rather than on adherence to abstract modeling schemes.

One straightforward strategy involves rescaling GCM outputs to better align with the observed warming. By comparing the temperature increase simulated by high-ECS models with actual measurements, a calibration factor can be derived and applied to future projections.

A more advanced approach identifies potential climate cycles and integrates them with anthropogenic and volcanic forcings recalibrated from GCM simulations. This allows for a more faithful reconstruction of observed climate dynamics. The plausibility of natural cycles requires supporting evidence, which may come from identifying similar frequencies in solar activity, planetary alignments, and lunar cycles.

Alternatively, energy-balance models can be employed, allowing the climate system to respond differently to each forcing. This methodology enables the use of forcings as proxies, making it possible to simulate effects directly or indirectly associated with the same variables even when the underlying physics is not yet fully understood.

Across these approaches, a consistent conclusion emerges: the ECS of the system is likely lower than that estimated by many GCMs, particularly those with ECS values above 3 °C. This perspective is supported by numerous independent studies, including:

- Lewis & Curry (2018),[6] who estimate ECS between 1.5 °C and 2.0 °C on the basis of updated radiative forcing and ocean heat uptake data;

[6]Lewis, N., Curry, J.A., 2018. The impact of recent forcing and ocean heat uptake data on estimates

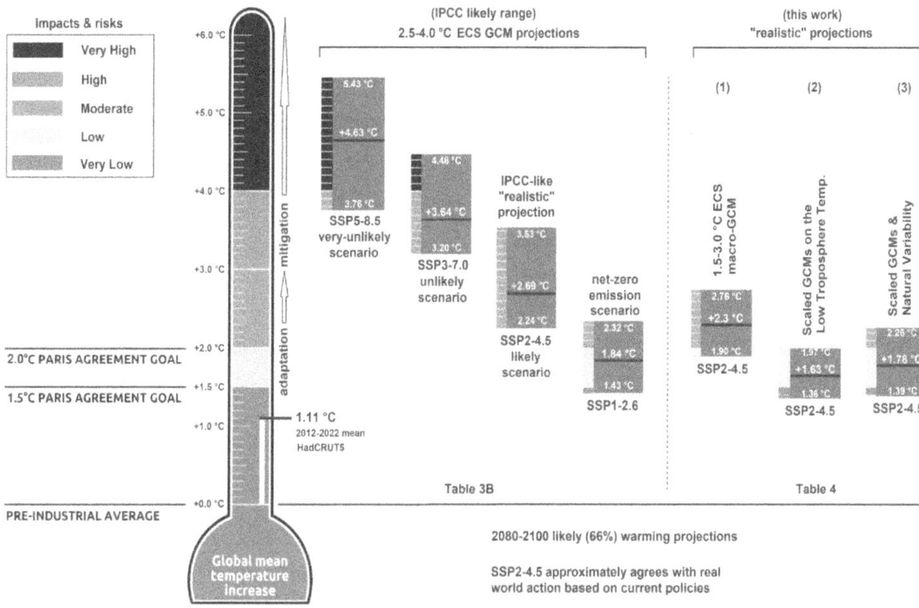

FIGURE 41.2. Summary and comparison of the impacts and risks associated with global warming over the period 2080–2100, through a comparative analysis between the projections of IPCC (2021), based on the four distinct socioeconomic development scenarios shown in Figure 6.4, and the semi-empirical climate simulations presented in Figures 29.1 and 30.6. (Adapted from Scafetta, 2024)

- Spencer (2021),[7] who reviews observational methodologies and finds ECS estimates frequently below 2 °C;
- Lewis (2023)[8] and Scafetta (2023b,c), who confirm ECS values likely below 3 °C, with consistent estimates in the range of 1 °C to 2 °C.

The systematic application of these corrective approaches to the CMIP6 GCM outputs underscores the need to reassess climate policies that rely on high-sensitivity projections.

Climate Thermometer and scenario comparison. Figure 41.2 summarizes these findings using the "*Climate Thermometer*" developed by Climate Action Tracker (2024), which compares the impacts and risks associated with global warming over the period 2080–2100. The figure includes projections from the IPCC CMIP6 models based on four socioeconomic development scenarios (Figure 6.4), alongside the semi-empirical simulations presented in Figures 29.1 and 30.6.

of climate sensitivity. Journal of Climate, 31(15), 6051–6071.

[7]Spencer, R.W., 2021. Lower estimates of climate sensitivity from observational data: A review. Remote Sensing, 13(8), 1496.

[8]Lewis, N., 2023. Objectively combining climate sensitivity evidence. Climate Dynamics, 60, 3139–3165.

Cost Category	SSP1-1.9 / SSP1-2.6	SSP2-4.5	Estimated Difference
Decarbonization and Energy	Massive and rapid transition	Gradual transition	Savings of trillions of dollars in infrastructure and energy system transformation
CO_2 Removal (CDR)	Required on a large scale	Limited or absent	Lower costs for carbon capture and storage technologies
Behavioral Changes	Widespread adoption of sustainable habits (consumption, mobility, diet)	Minor behavioral shifts	Reduced social and political costs
Impact on Economic Growth	Potential short-term slowdown due to aggressive mitigation	More linear and conventional development	Greater compatibility with existing economic trends
Adaptation Costs	Higher, due to strong mitigation efforts reducing long-term climate risks	Lower, but with less mitigation	Significant savings offset by reduced mitigation expenses

FIGURE 41.3. Qualitative estimate of the potential global savings under current SSP2-4.5 policies compared with *"Net-Zero"* scenarios (SSP1-1.9 or SSP1-2.6) (cf. Riahi et al., 2017; IPCC, 2023).

According to the CMIP6 GCM projections, only the SSP1-2.6 scenario, which imposes climate neutrality (*"Net-Zero"*) by 2075, keeps global surface warming near the +2 °C threshold by 2100. In agreement with this result, and in order to further increase the likelihood of meeting the Paris Climate Agreement (2016), the European Union has proposed the *"Green Deal,"* based on the more ambitious SSP1-1.9 scenario, which targets Net-Zero by 2050–2055 (Figure 2.7). However, achieving this scenario entails substantial costs, which may exceed the expected benefits in the early decades (Tol, 2023)[9] and remains speculative over longer time horizons.

In contrast, the semi-empirical simulations developed by Scafetta (2024)[10] suggest that the +2 °C threshold could be respected even under the more moderate SSP2-4.5 scenario, known as the *"Middle Road."* A low apparent climate sensitivity to radiative forcing may also arise from the heat capacity of the ocean,

[9]Tol, R.S.J., 2023. Costs and benefits of the Paris climate targets. Climate Change Economics, 14(3), 2340003.

[10]Scafetta, N., 2024. Impacts and risks of "realistic" global warming projections for the 21st century. Geoscience Frontiers, 15(2), 101774.

which is roughly three orders of magnitude larger than that of the atmosphere (Traverso, 2026).[11] This would imply climate impacts that are contained and largely manageable through affordable adaptation policies, without requiring radical decarbonization.

SSP2-4.5: a realistic and more balanced path. The SSP2-4.5 scenario represents an intermediate trajectory, both socioeconomically and in terms of emissions. It envisions a world of uneven growth, stable institutions without deep reforms, and gradual environmental investment. It does not aim for climate neutrality; mitigation is partial, and adaptation becomes progressively essential but remains feasible.

This scenario also appears to be the most consistent with current global environmental policies (Hausfather & Peters, 2020; Figure 2.7). If ECS is lower than the IPCC estimates—for example, between 1.5 °C and 2 °C, as suggested by the semi-empirical models—SSP2-4.5 could be sufficient to meet the Paris Climate Agreement targets. In that case, the savings relative to more ambitious scenarios (SSP1-1.9 and SSP1-2.6) would be substantial, both in terms of investment and of socioeconomic impact (Figure 41.3).

Economic costs of climate scenarios. The SSP trajectories differ significantly in the estimated economic costs associated with the implementation of their respective mitigation policies. Based on the available literature and indicative modeling exercises, the approximate cumulative costs (expressed in net present value, NPV, from 2025 to 2100) can be roughly summarized as follows:

- SSP1-1.9: 9–11 trillion USD;
- SSP1-2.6: 6–8 trillion USD;
- SSP2-4.5: 3–5 trillion USD.

These figures should be interpreted as illustrative estimates, subject to significant variation depending on assumptions about discount rates, technological progress, regional implementation strategies, and the inclusion or exclusion of adaptation costs. They do not represent official IPCC projections, but rather reflect estimates that can be deduced from independent economic analyses (Tol, 2023; Rao et al., 2017, etc.).

Under these assumptions, the SSP2-4.5 scenario appears to require less than half the investment of SSP1-1.9, while still offering a pathway that could nonetheless contain warming, particularly if the ECS is lower than the CMIP6 model average of 3 °C (cf. IPCC, 2022b).[12]

The European context. Europe presents notable peculiarities. Based on regional marginal cost analyses (Rao et al., 2017; Chen et al., 2023a),[13] the cost per ton of CO_2 avoided in Europe is among the highest globally, particularly in the industrial and transport sectors. Implementing SSP1-1.9 would entail a total

[11]Traverso, A., 2026. An engineering perspective on climate change: mitigation or adaptation? Energy Nexus, 21, 100646.

[12]IPCC, 2022b. Sixth Assessment Report (AR6), Working Group III: Mitigation of Climate Change – Chapter 3 and Technical Summary. 2022.

[13]Rao, S., Klimont, Z., Smith, S.J., Van Dingenen, R., Dentener, F., et al., 2017. Future air pollution in the Shared Socioeconomic Pathways. Global Environmental Change, 42, 346–358.
Chen, X., Deng, Y., Ou, Y., Cui, R.Y., Clarke, L.E., et al., 2023. Global cost and benefit analysis of climate mitigation pathways. Nature Climate Change, 13, 241–249.

mitigation cost for the EU (NPV 2025–2100) of 1.8–2.5 trillion USD, given its already advanced decarbonization and the additional technological constraints involved.

For SSP1-2.6, the cumulative costs range from 1.2 to 1.8 trillion USD. SSP2-4.5, which requires more limited and gradual efforts, is estimated at 0.8–1.2 trillion USD (cf. European Commission, 2024b; IPCC, 2022a).[14] These figures encompass investments in renewable energy, energy efficiency, transport electrification, industrial decarbonization, and infrastructure upgrades, but exclude the costs of climate impacts and adaptation.

In this context, the semi-empirical approach illustrated throughout this volume is not merely a methodological refinement. It represents a scientifically grounded alternative to polarized narratives in climate discourse.

Since the semi-empirical approach demonstrates that the warming expected by 2100 under SSP2-4.5 policies is compatible with the warming projected by the CMIP6 GCMs under SSP1-1.9 or SSP1-2.6, comparable social outcomes could be achieved while reducing global costs by up to 6 trillion USD, and EU costs by 1.0–1.3 trillion USD.

A scientific proposal with political consequences. This empirical reframing of climate attribution leads to a more nuanced interpretation:

- approximately 50% of the warming observed from 1850–1900 to 2010–2020 may be attributable to natural factors, potentially of solar or astronomical origin;
- approximately 30% may be linked to anthropogenic emissions;
- approximately 20% could result from local surface effects, such as urban heat islands and homogenization errors in the processing of meteorological records.

Such a distribution challenges the core assumption underlying Net-Zero policies, namely that the entirety of the observed warming since 1850–1900 is attributable to human activity, as asserted by the IPCC (2021) (cf. Figures 6.2 and 6.3). Consequently, the economic and social implications could result in avoidable global costs amounting to several trillion dollars. These costs could be significantly reduced through a more careful and balanced consideration of the empirical evidence.

Scientific implications and epistemic reorientation. From a scientific standpoint, the estimates above carry two important implications.

(1) The equilibrium climate sensitivity (ECS) in response to radiative forcing may be as low as $1.1 \pm 0.4\,°C$, primarily determined by the Planck response (1.16 °C), as discussed in Chapter 6. This implies that climate feedback mechanisms may offset one another, contributing to relative stability in Earth's climate system.

(2) The solar influence on millennial scales may involve a variability of 4–5 W/m^2; moreover, the Sun's role in climate change may operate primarily through non-radiative mechanisms, such as the modulation

[14]European Commission, 2024. Costs of climate mitigation in the EU: Background report for the 2040 climate target roadmap.

of the interplanetary magnetic field, rather than through irradiance alone.

41.6. Returning to science as method

Scientific progress rarely springs from consensus. It emerges from the courage to ask difficult questions and from the discipline to observe reality without preconception, even when doing so unsettles prevailing paradigms.

In today's climate-science discourse, complexity is too often flattened into simplified narratives. This volume has sought to re-anchor the conversation in the principles of scientific inquiry: methodological rigor, epistemological pluralism, and a sober recognition of uncertainty.

Climate change is shaped by both natural and anthropogenic factors. How we interpret these dynamics—and how we respond to them—determines whether our policies are guided by empirical evidence or shaped by ideological imperatives. Science is not a tribunal issuing final verdicts; it is a space for open-ended inquiry, a process of refinement, and an exercise in intellectual humility.

A return to science as method also demands a renewed commitment to epistemic transparency, which entails the explicit acknowledgment of theoretical boundaries, model assumptions, and interpretive caveats. In climate science, where multiple causal layers intersect and where long-term projections rely on incomplete data, intellectual honesty becomes a civic responsibility. Rather than concealing uncertainty for the sake of rhetorical clarity, scientists must cultivate a culture of didactic clarity, in which complexity is not erased but made legible. Only through such transparency can scientific discourse remain resilient against politicization and retain its role as a guide—rather than a weapon—in shaping informed and pluralistic responses to global change.

Some exchanges in the scholarly literature (e.g.: May & Crok, 2025; Pulles, 2025)[15] illustrate how easily scientific debate can fall short of its epistemic ideal. Critiques of prevailing climate interpretations may at times be selective or insufficiently developed, while the responses they elicit, though attempting counter-argument, often remain limited in scope and do not fully engage the deeper empirical and methodological complexities of the climate problem. The result is a discussion that advances less through cumulative clarification than through parallel lines of reasoning that only partially confront one another.

To embrace realism in climate science is to acknowledge the provisional nature of our knowledge. It means recognizing that models are heuristic frameworks—not oracles—and that the most resilient responses to change involve adaptive strategies, technological flexibility, and respect for diverse solutions. Above all, it means returning to science as a critical, iterative process rather than wielding it as an instrument of authority.

As the historian of science Thomas Kuhn reminded us: *"Scientific revolution never begins with consensus, but with the breaking of it."*

[15] May, A., Crok, M., 2025. Carbon dioxide and a warming climate are not problems. American Journal of Economics and Sociology, 84, 43–57.
Pulles, T., 2025. Climate Denialism. American Journal of Economics and Sociology, 84, 7–19.

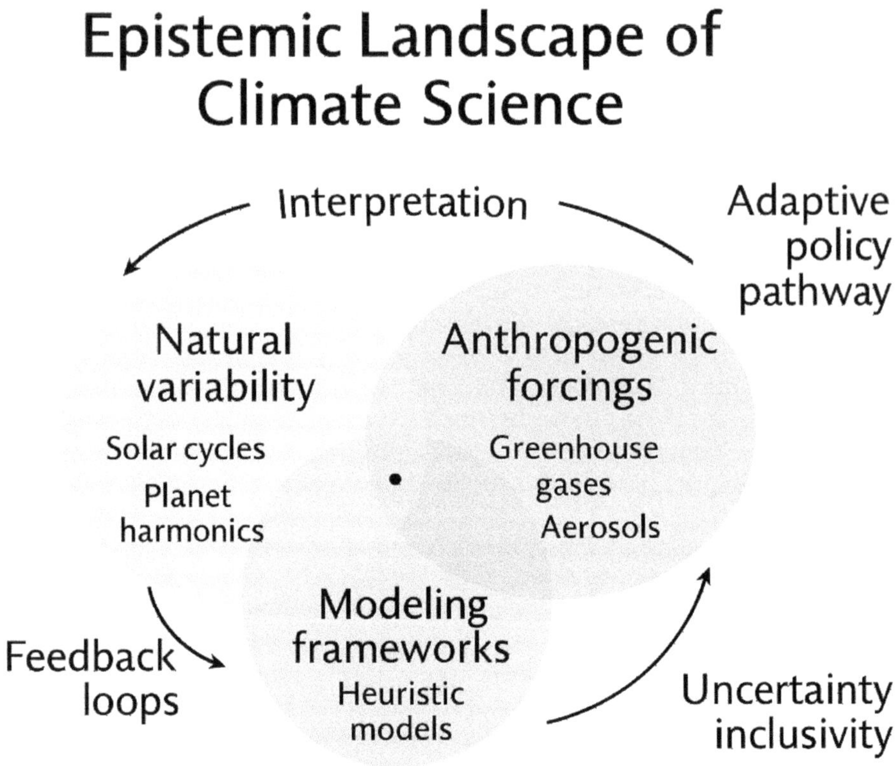

FIGURE 41.4. **Epistemic structure of Climate Science.** This conceptual diagram illustrates the dynamic interplay among three foundational domains in climate science: natural variability—such as solar cycles and planetary harmonics; anthropogenic forcings, including greenhouse gases and aerosols; and modeling frameworks, encompassing both heuristic and deterministic approaches. The overlapping regions represent zones of epistemic synthesis, where interpretive efforts, feedback mechanisms, and uncertainty management converge.Peripheral annotations highlight the broader implications of this integration by emphasizing the need for adaptive policy pathways, an inclusive treatment of uncertainty, and reflexive feedback between empirical observation and theoretical modeling. The diagram invites a pluralistic understanding of climate dynamics, in which no single domain holds epistemic primacy. It serves as a visual scaffold for interdisciplinary reasoning, bridging physical processes, human influence, and methodological abstraction.

CHAPTER 42

Toward the planetary hypothesis

Every great scientific revolution began when someone dared to look at the sky... and saw order where there had only been chaos.

42.1. Paradigm shifts in climate science

The history of scientific thought reveals that flawed theories can persist for centuries even in the presence of contradictory evidence, particularly when no conceptually robust and methodologically grounded alternative is available. A notable example is Aristotelian physics, which posited, among other principles, that a moving body requires a continuous force to maintain motion. Although questioned as early as the medieval period, Aristotelian science dominated intellectual discourse for nearly two millennia. Its internal inconsistencies became increasingly evident, but only with the rise of the experimental method—introduced by Galileo and developed by Kepler, Descartes, Newton, and others—was it ultimately replaced by classical physics, marking a decisive turning point in our understanding of nature.

In a similar way, the current paradigm in climate science—centered on the *Anthropogenic Global Warming Theory* (AGWT) and reliant on models that emphasize anthropogenic forcings while simplifying solar variability—exhibits notable epistemological and predictive limitations. As with all scientific transitions, critique alone is insufficient, and positions of denial that lack constructive alternatives tend to be unproductive. What is needed is a broader interpretive framework capable of addressing the inconsistencies of the current theory through systemic insight and empirical coherence.

This volume goes beyond a theoretical and methodological critique of the AGWT. It proposes an alternative conceptual framework grounded in the observation of natural cyclicity and in the empirical possibility of a synchronization among planetary dynamics, solar variability, and Earth's climate system. This integrative approach draws on planetary models, solar cycles, multi-scale oscillations, and hypotheses of large-scale solar modulation, offering the potential to advance a renewed paradigm in climate science. Such a paradigm could expand our understanding by reconnecting climate processes with their spatial and astronomical context, and invites a reconsideration of solar physics and of the architecture of the solar system itself.

In a similar context, Alfred Wegener's pioneering work *The Origin of Continents and Oceans* (1915) offers a valuable epistemological precedent. His theory of continental drift, long dismissed by mainstream geologists, was based on an interdisciplinary synthesis of geological, paleontological, climatological, and geophysical evidence. As Wegener remarked in the foreword to the fourth edition (1928), challenging the intellectual rigidity of his critics:

> *"Scientists still do not appear to understand sufficiently that all earth sciences must contribute evidence toward unveiling the state of our planet in earlier times, and that the truth of the matter can only be reached by combining all this evidence."*

This was not a neutral observation but a critique of fragmented and overly compartmentalized scientific thinking. A comparable episode occurred in 1931, when Albert Einstein faced coordinated criticism from conservative academics in the pamphlet *Hundert Autoren gegen Einstein* (*"One Hundred Authors Against Einstein"*). His famous reply—*"if I were wrong, one author would have been enough"*—highlighted the epistemological limitations of consensus-based resistance to paradigm shifts. Both Wegener and Einstein remind us that scientific revolutions are rarely embraced; they must confront established orthodoxy with methodological rigor and empirical depth.

This insight remains highly relevant. Just as Wegener recognized that no single discipline could fully explain the dynamics of Earth's crust, climate science must move beyond its current anthropocentric and reductionist boundaries. This shift must occur not through speculation, but in response to an expanding body of interdisciplinary scientific evidence.

The paradigm proposed in *The Frontier of Climate Science* calls for a synthesis of the Earth sciences with astronomical and solar sciences, recognizing that Earth's climate is not an isolated terrestrial phenomenon but an open system shaped by cosmic geometries, solar rhythms, and planetary interactions. Although still in the process of theoretical development, this proposal reflects the aspiration for an epistemological shift: a climate science that reclaims its astronomical dimension and conceives of climate as a system sensitive to astronomical harmonics—predictable, observable, and modelable on long timescales.

Astronomical origins of climate cycles. Figure 32.2A illustrates that Earth's climate has undergone substantial changes throughout its history, including phases significantly warmer than today, such as the Eemian interglacial (115–130 thousand years ago), when global temperatures were likely 2–3 °C higher and sea levels stood 6 to 9 meters above present levels.

Despite the complexity of climate dynamics, Figure 32.2B reveals that these variations follow a multiplicity of cycles—ranging from short-term (e.g., annual) to long-term oscillations spanning up to 150 million years—many of which appear to have astronomical origins. Among the most widely accepted are the multi-millennial Milankovitch cycles, based on Earth's orbital variations (eccentricity, axial tilt, and precession), and various solar cycles identified in paleoclimate archives throughout the Holocene.

These harmonic patterns suggest that Earth's climate is influenced not only by internal dynamics but also, and potentially predominantly, by external factors. These include the galactic motion of the solar system through the Milky Way's spiral arms and the gravitational and resonant forces exerted by the planets. The interaction of these cosmic dynamics may modulate climate variability on Earth.

Planetary synchronization and solar modulation. Recent research has converged on a key insight: solar activity exhibits cycles that correlate strongly with terrestrial climate cycles, and both align closely with major planetary periods. These include:

- rapid synodic interactions (0.3–0.33 years);
- the Earth–Jupiter synodic cycle ($\sim$1.1 years);
- the Schwabe cycle (10–12 years);
- the Hale cycle (20–24 years);
- the Gleissberg cycle (80–100 years);
- secular cycles ($\sim$103, 115, 130, 150 years);
- the Jose cycle (178–190 years);
- the Suess–de Vries cycle (200–240 years);
- the Eddy cycle (800–1100 years);
- the Bray–Hallstatt cycle (2100–2500 years).

The recurrence of these frequencies in planetary harmonics is unlikely to be coincidental, especially given the multiple lines of evidence from cosmogenic isotopes (^{14}C and ^{10}Be), dendrochronological records, and climate proxies in ice cores and ocean sediments (Scafetta, 2012c, 2014b; Scafetta et al., 2016; Scafetta & Bianchini, 2023).[1]

One of the key implications is that the solar cycles traditionally attributed to internal magnetohydrodynamic dynamo mechanisms may in fact arise, at least in part, from synchronization with planetary frequencies: an "*astronomical metronome*" formed by what Scafetta (2020)[2] has labeled "*invariant orbital inequalities.*" Conventional solar dynamo models, despite their sophistication, struggle to account for the regularity and persistence of many observed cycles. Similarly, the GCMs fail to reproduce major empirically observed climate oscillations.

The correspondence between solar activity frequencies and terrestrial climate cycles derived from planetary orbits and conjunctions suggests the need to reconsider the traditional view of the Sun as isolated from celestial dynamics. Earth, in turn, is influenced not only by solar variability but also by other astronomical effects, including soli-lunar long-range tidal perturbations with periodicities of approximately 9 years and longer, which appear in both paleoclimate and instrumental records.

42.2. Toward a cosmic climatology

Recognizing a planetary origin for these cycles implies that the Sun—and, by extension, Earth's climate—may respond to orbital geometries, gravitational tides, and large-scale heliospheric modulations. These may influence the cosmic ray flux, interplanetary dust dispersion, and the configuration of solar and interplanetary magnetic fields, producing indirect but potentially significant effects on climate variability.

---

[1]Scafetta, N., Bianchini, A., 2022. The planetary theory of solar activity variability: a review. Frontiers in Astronomy and Space Sciences, 9, 937930.
Scafetta, N., Bianchini, A., 2023. Overview of the spectral coherence between planetary resonances and solar and climate oscillations. Climate, 11(4), 77.
[2]Scafetta, N., 2020. Solar oscillations and the orbital invariant inequalities of the solar system. Solar Physics, 295(2), 33.

Part VI of this volume presents extensive empirical evidence in support of these hypotheses. Yet none of these physical processes are currently represented in the GCMs employed by the IPCC, which limits their ability to account for the natural components of climate dynamics.

In particular, Scafetta et al. (2016)[3] and Scafetta (2020) have demonstrated that the Bray–Hallstatt cycle (2318 years), which is clearly observed in cosmogenic and climate data, is consistent with a harmonic oscillation generated by the invariant orbital inequalities of Jupiter, Saturn, Uranus, and Neptune. Specifically, this inequality is represented by the orbital vector $(1, -3, 1, 1)$, combining the synodic cycles of Jupiter–Saturn $(1, -1, 0, 0)$, Saturn–Uranus $(0, 1, -1, 0)$, and Saturn–Neptune $(0, 1, 0, -1)$. These cycles resonate throughout the heliosphere.

The vectorial composition suggests that the Bray–Hallstatt cycle may result from a gravitational beat among major planetary harmonics, modulating the millennial *"pulse"* of solar activity. This oscillation also correlates with transitions between dynamically ordered and disordered phases in the Sun's motion around the solar system barycenter. Such a dynamic may influence the solar dynamo and the propagation of particle flows through the heliosphere.

Similarly, the 11-year solar cycle—known as the Schwabe cycle—appears reproducible through the combined tidal forces exerted by Venus, Earth, and Jupiter on the Sun. This interaction is described by the orbital vector $(3, -5, 2)$, which represents a harmonic beat between the third harmonic of the Venus–Earth synodic period and the second harmonic of the Earth–Jupiter synodic period. An alternative model incorporates the tidal cycles of the Jupiter–Saturn system (with a periodicity of 9.93 years) and Jupiter's orbital period (11.86 years). Thus, Rudolf Wolf (1859)[4] may have been correct: the Schwabe cycle could indeed be governed by the gravitational influences of Venus, Earth, Jupiter, and Saturn.

More broadly, the motion of the planets around the Sun generates a well-defined harmonic structure, characterized by distinct spectral clusters. These clusters exhibit notable coherence with all the major solar and climate cycles currently observed (Figures 30.1 and 37.5), offering a compelling interpretive framework for the origin and persistence of oscillations in the Sun–Earth system.

If further confirmed, the planetary hypothesis of solar and climate variability would imply that the cyclic structure of the Sun and of Earth's climate is at least partially predictable. Planetary cycles would act as cosmic metronomes, imposing a regular order on solar and climatic oscillations, which would simply fluctuate around periodic attractors determined by the complex geometry of planetary motion. Gravitational tides, often neglected in conventional solar models, could be amplified by internal resonance phenomena analogous to the vibration of a tuning fork, and could be capable of modulating even the Sun's luminosity.

Moreover, heliospheric oscillations triggered by seemingly weak planetary conjunctions (such as those between Uranus and Neptune, despite their distance

[3]Scafetta, N., Milani, F., Bianchini, A., Ortolani, S., 2016. On the astronomical origin of the Hallstatt oscillation found in radiocarbon and climate records throughout the Holocene. Earth-Science Reviews, 162, 24–43.

[4]Wolf, R., 1859. Extract of a letter from Prof. R. Wolf of Zurich to Mr. Carrington, dated Jan. 12, 1859. Monthly Notices of the Royal Astronomical Society, 19, 85.

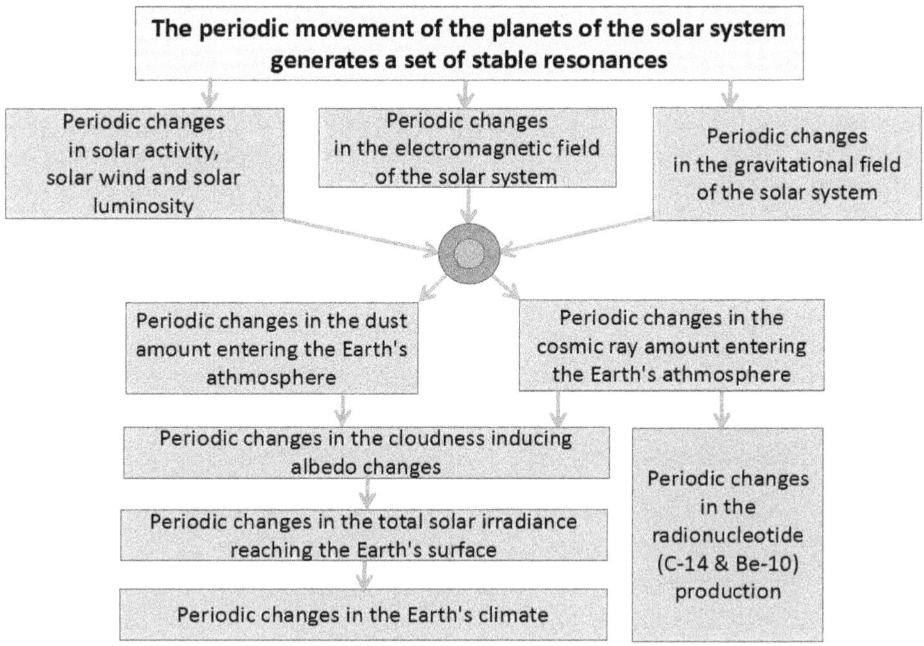

FIGURE 42.1. Schematic representation of the planetary model of solar activity changes and climate oscillations. (Adapted from Scafetta & Bianchini, 2023).

and modest mass) may nonetheless activate stable heliospheric synchronisms, generating significant solar and climatic effects on very long time scales.

An astronomical symphony. Figure 42.1 presents a schematic synthesis of the emerging paradigm explored throughout this volume: a planetary model of solar activity capable of explaining and anticipating climate oscillations across multiple temporal scales. It illustrates the principal cycles and the underlying theoretical mechanisms—planetary tides, gravitational resonances, orbital inequalities, and cosmic modulations—together with their potential effects on the heliosphere and, by extension, on Earth's climate system.

While future investigations through advanced simulations, long-term datasets, or space-based measurements may confirm, refine, or challenge this framework, the accumulated evidence already calls for a serious and open-minded reassessment of the planetary theory. The coherence of the observed harmonics, the recurrence of celestial cycles in paleoclimatic archives, and the limitations of conventional models in capturing long-term variability all point toward the need for a broader interpretive structure.

This hypothesis is not merely an intellectual provocation or a speculative detour. It represents a conceptual window onto a possible new paradigm for solar physics and climate science. It offers an alternative and scientifically stimulating interpretation of Earth's natural climate rhythms—one that moves beyond the constraints of anthropocentric modeling and reductive treatments of solar dynamics. It invites us to consider climate not as an isolated terrestrial phenomenon but as part of a larger cosmic choreography.

From this perspective, the interaction among the Sun, the planets, and Earth is not a peripheral curiosity but an astronomical symphony of remarkable complexity, governed by gravitational rhythms, orbital geometries, and resonant harmonics that remain only partially understood. It is a harmony once intuited by ancient thinkers such as Pythagoras, who envisioned the *"music of the spheres,"* and by Kepler, who sought mathematical order in planetary motion. Today, their intuition finds renewed expression in the language of science: spectral analysis, dynamical-systems theory, and heliophysical modeling.

With patience, rigor, and epistemological openness, we are beginning to decipher this score—not to romanticize the cosmos, but to understand it more fully.

42.3. Main critiques of the approach presented here and concise responses

The perspective developed throughout this volume departs in several important respects from the dominant framework adopted in mainstream climate assessments, particularly those centered on the IPCC reports and on the extensive use of GCMs. Consequently, the approach presented here has been met with a number of recurring critiques, both methodological and epistemological. Addressing these critiques systematically is essential for clarifying the scope, limits, and intent of the arguments developed in the preceding chapters, especially regarding the role of natural variability, solar–astronomical influences, and the interpretation of model-based projections.

A first and commonly raised criticism concerns the use of statistical correlations among astronomical cycles, solar variability, and climatic records. It is often argued that the identification of shared periodicities across different time series does not amount to a demonstration of causality, and that such correlations may be spurious, particularly when derived from relatively short and noisy climate records. As discussed in several chapters of this book, however, the empirical detection of stable and recurrent periodicities has historically constituted a legitimate starting point for scientific inquiry. Many well-established physical theories—such as those describing orbital forcing or long-term climate modulation—were initially guided by robust empirical regularities and only later supported by detailed mechanistic explanations. The key point is that the dominant climatic cycles identified in the empirical record coincide with independently known astronomical frequencies and recur across multiple, heterogeneous datasets, substantially reducing the likelihood of chance correlations.

A closely related criticism concerns the alleged risk of overfitting, particularly in the use of harmonic or semi-empirical models to reproduce past climate variability. According to this view, models constructed as sums of periodic components may achieve an excellent fit to historical data while lacking genuine predictive capability. In response, this book has consistently stressed that the principal frequencies employed are not selected freely to optimize statistical agreement, but are inferred from well-defined orbital, synodic, and solar cycles that can be calculated independently of the climate data. This greatly constrains the degrees of freedom involved. Moreover, several examples discussed in earlier chapters demonstrate that such models retain skill when applied outside their calibration intervals, including their ability to anticipate multidecadal fluctuations and periods of reduced warming. While no claim is made that these models offer complete or definitive forecasts, their predictive consistency challenges the notion that they are mere exercises in retrospective curve fitting.

Another major criticism targets the relative weighting assigned to natural versus anthropogenic forcings. It is often asserted that emphasizing solar and internal climate variability necessarily implies a downplaying of the role of greenhouse gases, particularly CO_2. As made explicit throughout the book, this is a mischaracterization. The framework presented here does not deny the influence of anthropogenic emissions on global temperature; rather, it questions the magnitude of the response inferred from high-sensitivity climate models. By anchoring sensitivity estimates more directly to observational constraints—especially satellite-based temperature records and ocean-heat-content data—the book argues for a lower effective climate sensitivity than that assumed in many GCMs. Within this framework, twentieth-century and early twenty-first-century warming emerges as the combined result of anthropogenic forcing and significant natural variability, rather than as the near-exclusive consequence of human emissions.

A further critique concerns the physical plausibility of solar–astronomical influences, particularly those involving planetary modulation of solar activity. Standard models of solar physics often regard planetary tidal or gravitational effects as negligible when compared with the energy scales of internal solar processes. As discussed in detail in the relevant chapters, this objection is based on a linear, classical-physics intuition applied to a system that is fundamentally nonlinear and non-classical. The argument advanced in this book is not that planetary forces directly drive solar output in an energetically dominant way, but that they may act as weak yet coherent pacemakers capable of synchronizing or modulating processes within the solar interior through resonance and feedback mechanisms that could amplify the signal, because the Sun is itself an energy generator. The widespread occurrence of synchronization in complex dynamical systems, together with the persistent empirical alignment between planetary periods and observed solar cycles, suggests that this possibility cannot be dismissed a priori.

Another frequently voiced criticism relates to the absence of explicit solar–astronomical forcings in mainstream climate models. It is sometimes claimed that, if such influences were climatically relevant, they would naturally appear in GCM simulations. This book has argued that such reasoning conflates model completeness with physical reality. GCMs, despite their sophistication, rely on incomplete representations of key processes such as cloud feedbacks, aerosol–cloud interactions, and internal ocean–atmosphere variability. They are also subject to structural tuning aimed at reproducing recent warming trends. In this context, the failure of the models to reproduce well-documented multidecadal climate oscillations suggests intrinsic limitations rather than the nonexistence of underlying forcings. The absence of a mechanism in a model therefore cannot be taken as conclusive evidence against its relevance in nature.

Finally, a more general epistemological criticism concerns the degree to which the perspective developed in this book aligns with the prevailing scientific consensus. As emphasized in Chapter 9, consensus reflects a sociological agreement at a given point in time rather than a definitive measure of scientific truth. Progress in climate science, as in other fields, depends on the continuous testing of assumptions against observations and on the willingness to revise dominant frameworks when persistent discrepancies emerge. The objective of this book has not been to replace one orthodoxy with another, but to contribute to a more pluralistic and empirically grounded discussion of climate change, in which uncertainties are acknowledged rather than obscured.

In summary, the critiques addressed here illuminate central tensions that permeate contemporary climate science: between empirical evidence and model-based inference, between explanatory completeness and observational realism, and between consensus-driven narratives and open scientific inquiry. The responses articulated throughout this volume do not claim to resolve these tensions definitively, but they reinforce the central thesis of the book: that a realistic understanding of climate change requires a balanced assessment of both natural variability and anthropogenic forcing, and a cautious approach to policies founded on models that remain subject to substantial and often underappreciated structural uncertainties.

42.4. Closing statement

The Frontier of Climate Science integrates climatology with solar physics, planetary dynamics, and historical epistemology. It has critically examined the limitations of prevailing models and theories while proposing an alternative interpretive framework grounded in empirical coherence and enriched by multidisciplinary scientific insight.

Rather than rejecting established knowledge, the research presented here seeks to broaden it by reintroducing natural variability, solar cyclicity, and planetary synchronization as legitimate components of climate dynamics. The evidence discussed across the chapters suggests that Earth's climate system may be more deeply embedded in cosmic rhythms than is currently acknowledged within dominant paradigms. This is not a call for rupture, but for refinement. Scientific progress depends not on consensus alone, but on the capacity to question, revise, and reframe understanding in the light of empirical evidence. The planetary hypothesis advanced here remains a work in progress, yet it outlines a promising direction for future research—one that may reconnect climate science with a broader spatial and temporal perspective.

This volume invites researchers, policymakers, and readers to approach climate not only as a terrestrial problem, but as a dynamic expression of Earth's position within the solar system. It calls for a restoration of science as method: open, adaptive, and grounded in the pursuit of understanding rather than closure.

In doing so, we may discover that Earth's climate is not merely a product of internal feedbacks and anthropogenic forcings, but a resonant expression of the solar system's deeper architecture.

Bibliography

Abreu, J.A., Beer, J., Ferriz-Mas, A., McCracken, K.G., Steinhilber, F., 2012. Is there a planetary influence on solar activity? *Astronomy and Astrophysics*, 548, A88.

Agnihotri, R., Dutta, K., 2003. Centennial scale variations in monsoonal rainfall: manifestations of solar variability. *Current Science*, 85, 459–463.

Agnoletti, M., Piras, F., Venturi, M., Santoro, A., 2022. Cultural values and forest dynamics: The Italian forests in the last 150 years. *Forest Ecology and Management*, 503, 119655.

Agol, E., Dorn, C., Grimm, S.L., Turbet, M., Ducrot, E., et al., 2021. Refining the transit-timing and photometric analysis of TRAPPIST-1. *The Planetary Science Journal*, 2, 1.

Agustí-Panareda, A., Barré, J., Balsamo, G., Basart, S., Bennouna, Y., et al., 2023. Technical note: The CAMS greenhouse gas reanalysis from 2003 to 2020. *Atmospheric Chemistry and Physics*, 23, 3829–3850.

Ahuir, J., Mathis, S., Amard, L., 2021. Dynamical Tide in Stellar Radiative Zones. *Astronomy and Astrophysics*, 651, A3.

al-Balkhi, Abu Ma'shar (IX secolo). *De Magnis Coniunctionibus*. Edited by K. Yamamoto & Ch. Burnett. Leiden: Brill, 2000, 2 vols. (Arabic and Latin text).

Alimonti, G., Mariani, L., Prodi, F., Ricci, R.A., 2022. A critical assessment of extreme events trends in times of global warming. *European Physical Journal Plus*, 137, Article 112.

Alimonti, G., Mariani, L., 2023. Is the number of global natural disasters increasing? *Environmental Hazards*, 22(4), 345–368.

Alimonti, G., Mariani, L., 2025. Quantifying the climate crisis: A data-driven framework using response indicators for evidence-based adaptation policies. *Environmental Hazards*, 1–30.

Alley, R.B., 2000. The Younger Dryas cold interval as viewed from central Greenland. *Quaternary Science Reviews*, 19, 213–226.

Anchukaitis, K.J., Smerdon, J.E., 2022. Progress and uncertainties in global and hemispheric temperature reconstructions of the Common Era. *Quaternary Science Reviews*, 293, 107537.

Antonioli, F., Ferranti, L., Stocchi, P., Deiana, G., Lo Presti, V., et al., 2018. Morphometry and elevation of the last interglacial tidal notches in tectonically stable coasts of the Mediterranean Sea. *Earth-Science Reviews*, 185, 600–623.

Ardalan, A.A., Hashemi-Farahani, H., 2007. A harmonic approach to global ocean tide analysis based on TOPEX/Poseidon satellite. *Marine Geophysical Research*, 28, 235–255.

Arent, D.J., Tol, R., Faust, E., Hella, J., Kumar, S., et al., 2014. Key economic sectors and services. In: *Climate Change 2014: Impacts, Adaptation, and Vulnerability*. IPCC WGII AR5.

Arlt, R., 2011. The sunspot observations by Samuel Heinrich Schwabe. *Astronomische Nachrichten*, 332(8), 805–814.

Arrhenius, S., 1896. On the Influence of Carbonic Acid in the Air upon the Temperature of the Ground. *Philosophical Magazine and Journal of Science*, Serie 5, 41, 237–276.

Arrhenius, S., 1906. "Die vermutliche Ursache der Klimaschwankungen". Meddelanden från K. Vetenskapsakademiens Nobelinstitut Band 1 No 2. "The Probable Cause of Climate Fluctuations – Svante Arrhenius, A Translation of his 1906 Amended View of "Global Warming", 2014. Available at https://friendsofscience.org/assets/documents/Arrhenius%201906,%20final.pdf

Arrhenius, S., 1908. *Worlds in the Making: The Evolution of the Universe* (H. Borns, Trans.). New York: Harper & Brothers. (Original work published in 1906).

Atwoli, L., Baqui, A.H., Benfield, T., Bosurgi, R., Godlee, F., et al., 2021. Call for Emergency Action to Limit Global Temperature Increases, Restore Biodiversity, and Protect Health. *The New England Journal of Medicine*, 385, 1134–1137.

Australian Institute of Marine Science, 2025. *Annual Summary Report of Coral Reef Condition 2024/25*.

AIMS. https://www.aims.gov.au/monitoring-great-barrier-reef/gbr-condition-summary-2024-25

Bahcall, J.N., Pinsonneault, M.H., Basu, S., 2001. Solar Models: Current Epoch and Time Dependences, Neutrinos, and Helioseismological Properties. *The Astrophysical Journal*, 555, 990–1012.

Bahcall, J.N., Serenelli, A.M., Basu, S., 2005. New Solar Opacities, Abundances, Helioseismology, and Neutrino Fluxes. *The Astrophysical Journal*, 621, L85–L88.

Baliunas, S.L., Soon, W.H., 1991. Are variations in the length of the activity cycle related to changes in brightness in solar-type stars? *The Astrophysical Journal*, 374, 463–470.

Bank, M.J., Scafetta, N., 2022. Scaling, mirror symmetries and musical consonances among the distances of the planets of the Solar System. *Frontiers in Astronomy and Space Sciences*, 8, 758184.

Bard, E., Raisbeck, G.M., Yiou, F., Jouzel, J., 2000. Solar irradiance during the last 1200 years based on cosmogenic nuclides. *Tellus Series B: Chemical and Physical Meteorology*, 52, 985–992.

Barker, A.J., Ogilvie, G.I., 2010. On internal wave breaking and tidal dissipation near the centre of a solar-type star. *Monthly Notices of the Royal Astronomical Society*, 404, 1849–1868.

Bates, J.R., 2016. Estimating climate sensitivity using two-zone energy balance models. *Earth and Space Science*, 3, 207–225.

Battistini, A., 2011. Il ciclo undecennale del Sole secondo Bendandi. *New Ice Age*. Available at: http://daltonsminima.altervista.org/?p=8669

Benestad, R.E., Schmidt, G.A., 2009. Solar trends and global warming. *Journal of Geophysical Research: Atmospheres*, 114, D14101.

Beer, J., Tobias, S.M., Weiss, N.O., 2018. On long-term modulation of the Sun's magnetic cycle. *Monthly Notices of the Royal Astronomical Society*, 473, 1596–1602.

Beck, J.G., 2000. A comparison of differential rotation measurements. *Solar Physics*, 191, 47–70.

Bendandi, R., 1931. *Un Principio Fondamentale dell'Universo*. Faenza: Osservatorio Bendandi.

Bereiter, B., Eggleston, S., Schmitt, J., Nehrbass-Ahles, C., Stocker, T.F., et al., 2015. Revision of the EPICA Dome-C CO_2 record from 800 to 600 kyr before present. *Geophysical Research Letters*, 42, 542–549.

Beck, E.-G., 2022. Reconstruction of atmospheric CO_2 background levels since 1826 from direct measurements near ground. *Science of Climate Change*, 2(2), 148–211.

Berg, M., Kuralbayeva, K., Venables, A.J., 2023. Climate change and economic development: Revisiting the evidence. *Oxford Economic Papers*, 75(1), 1–24.

Berner, R.A., 1994. GEOCARB II: A revised model of atmospheric CO_2 over Phanerozoic time. *American Journal of Science*, 294, 56–91.

Berner, R.A., Kothavala, Z., 2001. GEOCARB III: A revised model of atmospheric CO_2 over Phanerozoic time. *American Journal of Science*, 301, 182–204.

Bertolucci, S., Zioutas, K., Hofmann, S., Maroudas, M., 2017. The Sun and its planets as detectors for invisible matter. *Physics of the Dark Universe*, 17, 13.

Bigg, E.K., 1967. Influence of the planet Mercury on sunspots. *The Astronomical Journal*, 72, 463–466.

Birol, F., Bignalet-Cazalet, F., Cancet, M., Daguze, J.-A., Fkaier, W., et al., 2025. Understanding uncertainties in the satellite altimeter measurement of coastal sea level: insights from a round-robin analysis. *Ocean Science*, 21, 133–150.

Bjørk, A.A., Kjær, K.H., Korsgaard, N.J., Khan, S.A., Kjeldsen, K.K., et al., 2012. An aerial view of 80 years of climate-related glacier fluctuations in southeast Greenland. *Nature Geoscience*, 5, 427–432.

Black, D.E., Peterson, L.C., Overpeck, J.T., Kaplan, A., Evans, M.N., Kashgarian, M., 1999. Eight centuries of North Atlantic Ocean atmosphere variability. *Science*, 286, 1709–1713.

Bollinger, C.J., 1952. A 44.77 year Jupiter-Earth-Venus configuration Sun-Tide period in solar-climate cycles. *Proceedings of the Oklahoma Academy of Science*, 307–311.

Bond, G., Kromer, B., Beer, J., Muscheler, R., Evans, M.N., et al., 2001. Persistent solar influence on North Atlantic climate during the Holocene. *Science*, 294(5549), 2130–2136.

Boretti, A., 2025. Subsidence: critical factor in local sea level rise assessments. *Ocean & Coastal Management*, 245, 107646.

Bova, S., Rosenthal, Y., Liu, Z., Yan, M., 2021. Seasonal origin of the thermal maxima at the Holocene and the last interglacial. *Nature*, 589, 548–553.

Box, G.E.P., 1976. Science and Statistics. *Journal of the American Statistical Association*, 71(356), 791–799.

Bragato, P.L., 2015. Italian seismicity and Vesuvius' eruptions synchronize on a quasi 60 year oscillation. *Earth and Space Science*, 2(10), 453–467.

Brakenridge, G.R., 2025. Late Quaternary supernovae in Earth history. *Monthly Notices of the Royal Astronomical Society*, 539(4), 3201–3219.

Bray, J.R., 1968. Glaciation and solar activity since the fifth century BC and the solar cycle. *Nature*,

220, 672–674.
Bray, J.R., 1971. Solar–climate relationships in the post-Pleistocene. *Science*, 171(3972), 1242–1243.
Briffa, K.R., Schweingruber, F.H., Jones, P.D., Osborn, T.J., Harris, I.C., Shiyatov, S.G., Vaganov, E.A., 1998. Trees tell of past climates: but are they speaking less clearly today? *Philosophical Transactions of the Royal Society B: Biological Sciences*, 353(1365), 65–73.
Brohan, P., Kennedy, J.J., Harris, I., Tett, S.F.B., Jones, P.D., 2006. Uncertainty estimates in regional and global observed temperature changes: A new dataset from 1850. *Journal of Geophysical Research: Atmospheres*, 111, D12106.
Brown, E.W., 1900. A possible explanation of the sunspot period. *Monthly Notices of the Royal Astronomical Society*, 60, 599–606.
Brussato, G., 2021. *Energia verde? Prepariamoci a scavare. I costi ambientali e sociali delle energie rinnovabili.* Montaonda.
Büntgen, U., Kirdyanov, A.V., Krusic, P.J., Shishov, V.V., Esper, J., 2021. Arctic aerosols and the 'divergence problem' in dendroclimatology. *Dendrochronologia*, 67, 125837.
Burgess, M.G., Ritchie, J., Shapland, J., Pielke, R., 2021. IPCC baseline scenarios have over-projected CO_2 emissions and economic growth. *Environmental Research Letters*, 16(1), 014016.
Burke, M., Hsiang, S.M., Miguel, E., 2015. Global non-linear effect of temperature on economic production. *Nature*, 527(7577), 235–239.
Buzzanga, B., Shirzaei, M., Wdowinski, S., 2023. Localized uplift, widespread subsidence, and implications for sea level rise in the New York City metropolitan area. *Science Advances*, 9(45), eadi8259.
Cai, M., Lu, J., 2023. Research progress on the divergence problem of tree ring growth-climate relationship. *Journal of Sichuan Forestry Science and Technology*, 44(6), 7–13.
Callebaut, D.K., de Jager, C., Duhau, S., 2012. The influence of planetary attractions on the solar tachocline. *Journal of Atmospheric and Solar-Terrestrial Physics*, 80, 73–78.
Callebaut, D.K., de Jager, C., Duhau, S., 2013. Reply to "The influence of planetary attractions on the solar tachocline" by N. Scafetta, O. Humlum, J.E. Solheim, K. Stordahl. *Journal of Atmospheric and Solar–Terrestrial Physics*, 102, 372–373.
Callendar, G.S., 1938. The artificial production of carbon dioxide and its influence on temperature. *Quarterly Journal of the Royal Meteorological Society*, 64, 223–240.
Camuffo, D., Bertolin, C., Barriendos, M., Dominguez-Castro, F., Cocheo, C., et al., 2010. 500-year temperature reconstruction in the Mediterranean Basin using documentary data and instrumental observations. *Climatic Change*, 101, 169–199.
Capron, E., Govin, A., Stone, E.J., Masson-Delmotte, V., Mulitza, S., Otto-Bliesner, B., Rasmussen, T.L., 2017. Temporal and spatial structure of multi-millennial temperature changes at high latitudes during the Last Interglacial. *Quaternary Science Reviews*, 103, 116–133.
Capron, E., Rasmussen, S.O., Popp, T.J., Erhardt, T., Fischer, H., et al., 2021. The anatomy of past abrupt warmings recorded in Greenland ice. *Nature Communications*, 12, 2106.
Cardoso, R.D.S., Dorigon, L.P., Teixeira, D.C.F., Amorim, M.C.d.C.T., 2017. Assessment of urban heat islands in small- and mid-sized cities in Brazil. *Climate*, 5(1), 14.
Cartapanis, O., Jonkers, L., Moffa-Sanchez, P., Jaccard, S.L., de Vernal, A., 2022. Complex spatio-temporal structure of the Holocene Thermal Maximum. *Nature Communications*, 13, 5662.
Cauquoin, A., Raisbeck, G.M., Jouzel, J., Bard, E., and ASTER Team, 2014. No evidence for planetary influence on solar activity 330,000 years ago. *Astronomy and Astrophysics*, 561, A132.
Cazenave, A., Moreira, L., 2022. Contemporary sea-level changes from global to local scales: a review. *Proceedings of the Royal Society A: Mathematical, Physical and Engineering Sciences*, 478(2261), 20220049.
CEA-OMB, Council of Economic Advisers & Office of Management and Budget. *Methodologies and Considerations for Integrating the Physical and Transition Risks of Climate Change into Macroeconomic Forecasting for the President's Budget*. White Paper, March 13, 2023. Available at: https://bidenwhitehouse.archives.gov/wp-content/uploads/2023/03/CEA-OMB-White-Paper.pdf
Chand, S.S., Tory, K.J., Walsh, K.J.E., Camargo, S.J., Kuleshov, Y., Ye, H., 2022. Declining tropical cyclone frequency under global warming. *Nature Climate Change*, 12, 655–660.
Chand, S.S., Tory, K.J., Walsh, K.J.E., Camargo, S.J., Kuleshov, Y., Ye, H., 2023. Reply to: Limitations of reanalyses for detecting tropical cyclone trends. *Nature Climate Change*, 13, 381–383.
Chambers, D.P., Merrifield, M.A., Nerem, R.S., 2012. Is there a 60-year oscillation in global mean sea level? *Geophysical Research Letters*, 39, L18607.

Chapman, G.A., Cookson, A.M., Preminger, D.G., 2013. Modeling Total Solar Irradiance with San Fernando Observatory ground-based photometry. *Solar Physics*, 283, 295–305.

Chang, S., Yu, Z., 1981. Historical records of meteorite falls in China and their time series analysis. *Memoirs of the National Institute of Polar Research* (Special Issue), 20, 276–284.

Charbonneau, P., 2022. External Forcing of the Solar Dynamo. *Frontiers in Astronomy and Space Sciences*, 9, 853676.

Charney, J.G., Arakawa, A., Baker, D.J., Bolin, B., Dickinson, R.E., et al., 1979. *Carbon dioxide and climate: A scientific assessment*. Washington, DC: The National Academies Press. Available at: https://doi.org/10.17226/12181

Charvátová, I., 1988. The solar motion and the variability of solar activity. *Advances in Space Research*, 8(7), 147–150.

Charvátová-Jakubcová, I., Střeštík, J., Křivský, L., 1988. The periodicity of aurorae in the years 1001–1900. *Studia Geophysica et Geodaetica*, 32(1), 70–77.

Charvátová, I., 1989. On the relation between solar motion and the long-term variability of solar activity. *Studia Geophysica et Geodaetica*, 33(3), 230–241.

Charvátová, I., 2009. Long-term predictive models based on solar motion. *Pattern Recognition in Physics*, 1(1), 1–12.

Chatzistergos, T., Krivova, N.A., Yeo, K.L., 2023. Long-term changes in solar activity and irradiance. *Journal of Atmospheric and Solar-Terrestrial Physics*, 252, 106150.

Chatzistergos, T., 2024. A discussion of implausible total solar-irradiance variations since 1700. *Solar Physics*, 299, 21.

Chen, C., Park, T., Wang, X., Piao, S., Xu, B., et al., 2019. China and India lead in greening of the world through land-use management. *Nature Sustainability*, 2, 122–129.

Chen, X., Deng, Y., Ou, Y., Cui, R.Y., Clarke, L.E., et al., 2023a. Global cost and benefit analysis of climate mitigation pathways. *Nature Climate Change*, 13, 241–249.

Chen, Y., van der Werf, G.R., Randerson, J.T., Rogers, B.M., Mu, M., Li, F., Giglio, L., 2023b. Multi-decadal trends and variability in burned area from the fifth version of the Global Fire Emissions Database (GFED5). *Earth System Science Data*, 15(12), 5227–5247.

Christiansen, B., Ljungqvist, F.C., 2012. The extra-tropical Northern Hemisphere temperature in the last two millennia: Reconstructions of low-frequency variability. *Climate of the Past*, 8, 765–786.

Christy, J.R., Spencer, R.W., Braswell, W.D., Junod, R., 2018. Examination of space-based bulk atmospheric temperatures used in climate research. *International Journal of Remote Sensing*, 39, 3580–3607.

Christy, J.R., Curry, J.A., Koonin, S.E., McKitrick, R., Spencer, R.W., 2025. *A critical review of impacts of greenhouse gas emissions on the U.S. climate*. U.S. Department of Energy, Washington, DC. Available at: https://www.energy.gov/topics/climate

Chylek, P., Klett, J.D., Lesins, G., Dubey, M.K., Hengartner, N., 2014. The Atlantic Multidecadal Oscillation as a dominant factor of oceanic influence on climate. *Geophysical Research Letters*, 41, 1689–1697.

Climate Action Tracker, 2024. *The CAT Thermometer, CAT Emissions Gap*. Available at: https://climateactiontracker.org

Cionco, R.G., Kudryavtsev, S.M., Soon, W.W.H., 2021. Possible Origin of Some Periodicities Detected in Solar-Terrestrial Studies: Earth's Orbital Movements. Earth and Space Science, 8, e2021EA001805.

Cisneros, M., Cacho, I., Frigola, J., Canals, M., Masqué, P., et al., 2016. Sea surface temperature variability in the central-western Mediterranean Sea during the last 2700 years: A multi-proxy and multi-record approach. *Climate of the Past*, 12(4), 849–869.

Coddington, O., Lean, J.L., Pilewskie, P., Snow, M., Lindholm, D., 2016. A solar irradiance climate data record. *Bulletin of the American Meteorological Society*, 97, 1265.

Coe, D., Fabinski, W., Wiegleb, G., 2021. The Impact of CO_2, H_2O and Other "Greenhouse Gases" on Equilibrium Earth Temperatures. *International Journal of Atmospheric and Oceanic Sciences*, 5(2), 29–40.

Cohler, J., 2025. The father of lies hijacking climate science: global mean surface temperature does not exist. *Journal of American Physicians and Surgeons*, 30(4), 122–126.

Connolly, R., Connolly, M., Soon, W., Legates, D.R., Cionco, R.G., Velasco Herrera, V.M., 2019. Northern hemisphere snow-cover trends (1967–2018): A comparison between climate models and observations. *Geosciences*, 9(3), 135.

Connolly, R., Connolly, M., Carter, R.M., Soon, W., 2020. How much human-caused global warming should we expect with business-as-usual climate policies? *Energies*, 13(6), 1365.

Connolly, R., Soon, W., Connolly, M., Baliunas, S., Berglund, J., Butler, C. J., Cionco, R. G., Elias, A. G., Fedorov, V. M., Harde, H., Henry, G. W., Hoyt, D. V., Humlum, O., Legates, D. R., Lüning, S., Scafetta, N., Solheim, J.-E., Szarka, L., van Loon, H., Velasco Herrera, V. M., Willson, R. C., Yan, H., Zhang, W. (2021). How much has the Sun influenced Northern Hemisphere temperature trends? An ongoing debate. *Research in Astronomy and Astrophysics*, 21(6), 131.

Connolly, R., Soon, W., Connolly, M., Baliunas, S., Berglund, J., Butler, C. J., Cionco, R. G., Elias, A. G., Fedorov, V. M., Harde, H., Henry, G. W., Hoyt, D. V., Humlum, O., Legates, D. R., Scafetta, N., Solheim, J.-E., Szarka, L., Velasco Herrera, V. M., Yan, H., Zhang, W., 2023. Challenges in the Detection and Attribution of Northern Hemisphere Surface Temperature Trends Since 1850. *Research in Astronomy and Astrophysics*, 23(10), 105015.

Connolly, R., Soon, W., Connolly, M., Cionco, R.G., Elias, A. G., Henry, G. W., Scafetta, N., Velasco Herrera, V.M., 2024. Multiple new or updated satellite total solar irradiance (TSI) composites (1978–2023). *The Astrophysical Journal*, 975(1), 102.

Cook, J., Nuccitelli, D., Green, S.A., Richardson, M., Winkler, B., et al., 2013. Quantifying the consensus on anthropogenic global warming in the scientific literature. *Environmental Research Letters*, 8, 024024.

Copernicus Marine Service, 2023. Global Ocean acidification – mean sea water pH time series and trend from Multi-Observations Reprocessing. LSCE. Available at https://data.marine.copernicus.eu/product/GLOBAL_OMI_HEALTH_carbon_ph_area_averaged/description

Cramer, B.S., Miller, K.G., Barrett, P.J., Wright, J.D., 2009. Late Cretaceous–Neogene trends in deep ocean temperature and continental ice volume. *Science*, 327, 1187–1191.

Crawford, J., 1991. Climate bifurcations and nonlinear transitions. *Climatic Dynamics*, 6(3), 145–158.

Creel, R.C., Austermann, J., Kopp, R.E., Khan, N.S., Albrecht, T., Kingslake, J., 2024. Global mean sea level likely higher than present during the Holocene. *Nature Communications*, 15, 10731.

Crok, M., May, A. (Eds.), 2023. The Frozen Climate Views of the IPCC. Clintel Foundation. Available at: https://clintel.org

Cronin, T.W., Dutta, I., 2023. How well do we understand the Planck feedback? *Journal of Advances in Modeling Earth Systems*, 15(7), e2023MS003729.

Crowley, T.J., 2000. Causes of climate change over the past 1000 years. *Science*, 289, 270–277.

Cruz-Silva, E., Harrison, S.P., Prentice, I.C., Marinova, E., Bartlein, P.J., et al., 2023. Pollen-based reconstructions of Holocene climate trends in the eastern Mediterranean region. *Climate of the Past*, 19, 2093–2108.

Cuffey, K.M., Clow, G.D., 1997. Temperature, accumulation, and ice sheet elevation in central Greenland through the last deglacial transition. *Journal of Geophysical Research: Oceans*, 102(C12), 26383–26396.

Curry, J., 2023. Climate Uncertainty and Risk: Rethinking Our Response. Anthem Press.

Czymzik, M., Muscheler, R., Brauer, A., 2016. Solar modulation of flood frequency in central Europe during spring and summer on interannual to multi-centennial timescales. *Climate of the Past*, 12, 799–805.

D'Aleo, J.S., 2016. A critical look at surface temperature records. In: Easterbrook, D.J. (Ed.) *Evidence-Based Climate Science*, 2nd Edition. Elsevier, pp. 11–48.

Dalmonech, D., Marano, G., Collalti, A., et al., 2022. Forest management as a strategy for climate change mitigation: insights from southern European plantations. *Science of the Total Environment*, 838, 159361.

Dansgaard, W., Johnsen, S.J., Møller, J., Langway, C.C. Jr., 1969. One thousand centuries of climatic record from Camp Century on the Greenland ice sheet. *Science*, 166(3903), 377–380.

Dansgaard, W., Johnsen, S.J., Clausen, H.B., Dahl-Jensen, D., Gundestrup, N., Hammer, C.U., Oeschger, H., 1984. North Atlantic climatic oscillations revealed by deep Greenland ice cores. In: Hansen, J.E., Takahashi, T. (Eds.) *Climate Processes and Climate Sensitivity*.

Datseris, G., Stevens, B., 2021. Earth's albedo and its symmetry. *AGU Advances*, 2, e2021AV000440.

Davis, J.C., Bohling, G., 2001. The search for patterns in ice-core temperature curves. In: Gerhard, L.C., Harrison, W.E., Hanson, B.M. (Eds.) *Geological Perspectives of Global Climate Change*, pp. 213–229.

Davis, W.J., 2017. The relationship between atmospheric carbon dioxide concentration and global temperature for the last 425 million years. *Climate*, 5(4), 76.

Davis, L.L.B., Thompson, D.W.J., Kennedy, J.J., Kent, E.C., 2019. The importance of unresolved biases in twentieth-century sea surface temperature observations. *Bulletin of the American Meteorological*

Society, 100, 621–629.
Davis, W.J., 2023. Mass extinctions and their relationship with atmospheric carbon dioxide concentration: Implications for Earth's future. *Earth's Future*, 11, e2022EF003336.
Dayaratna, K., McKitrick, R., Kreutzer, D., 2017. Empirical climate sensitivity and the social cost of carbon. *Climate Change Economics*, 8(2), 1750006.
Dayaratna, K., McKitrick, R., Kreutzer, D., 2023. Updated estimates of climate sensitivity and implications for policy. *Heritage Foundation Technical Report*.
de Jager, C., Versteegh, G.J.M., 2005. Do planetary motions drive solar variability? *Solar Physics*, 229, 175–179.
de Laat, A.T.J., Maurellis, A.N., 2006. Evidence for influence of anthropogenic surface processes on lower tropospheric and surface temperature trends. *International Journal of Climatology*, 26(7), 897–913.
Dell, M., Jones, B.F., Olken, B.A., 2012. Temperature shocks and economic growth: Evidence from the last half century. *American Economic Journal: Macroeconomics*, 4(3), 66–95.
de Menocal, P.B., Tierney, J.E., 2012. Green Sahara: African humid periods paced by Earth's orbital changes. *Nature Education Knowledge*, 3(10), 12.
Deming, D., 2006. U.S. Senate Hearing on Climate Change and the Media (Dec 6, 2006). Available at: https://www.epw.senate.gov/ public/index.cfm/hearings?Id=BFE4D91D-802A-23AD-4306-B4121BF7EC ED&Statement_id=361256C4-11DC-4E5D-8D1D-9FEDF082D081
Denton, G.H., Karlén, W., 1973. Holocene climatic variations—their pattern and possible cause. *Quaternary Research*, 3, 155–205.
de Vries, H., 1958. Variations in concentration of radiocarbon with time and location on Earth. *Proceedings of the Royal Netherlands Academy of Arts and Sciences B*, 61, 94–102.
de Wit, T.D., Kopp, G., Fröhlich, C., Schöll, M., 2017. Methodology to create a new total solar irradiance record: making a composite out of multiple data records. *Geophysical Research Letters*, 44(3), 1196–1203.
Dewitte, S., Clerbaux, N., 2017. Measurement of the Earth radiation budget at the top of the atmosphere — a review. *Remote Sensing*, 9(11), 1143.
Diamond, J., 2005. *Collapse: How Societies Choose to Fail or Succeed*. Penguin.
Dietz, S., Rising, J., Stoerk, T., Wagner, G., 2021. Economic impacts of tipping points in the climate system. *Proceedings of the National Academy of Sciences*, 118(34), e2103081118.
Döhne, T., Horwath, M., Thomas, M., 2024. Ocean mass change from GRACE/GRACE-FO and satellite altimetry: updated estimates and uncertainty assessment. *EGU General Assembly 2024*, Abstract EGU24-288.
Domesday Book (1086). A Complete Translation. (1992). London: Penguin Books/Alecto Historical Editions.
Dong, Y., Wu, N., Li, F., Zhang, D., Zhang, Y., et al., 2022. The Holocene temperature conundrum answered by mollusk records from East Asia. *Nature Communications*, 13, 5153.
Dong, B., Sutton, R.T., Wilcox, L.J., 2023. Decadal trends in surface solar radiation and cloud cover over the North Atlantic sector during the last four decades: drivers and physical processes. *Climate Dynamics*, 60, 2533–2546.
Duft, D., Nachbar, M., Eritt, M., Vrtala, A., Tomsche, L., et al., 2019. Unravelling the microphysics of polar mesospheric cloud formation. *Atmospheric Chemistry and Physics*, 19(4), 2871–2879.
Easterbrook, D.J., 2019. *Solar Magnetic Cause of Climate Changes and Origin of the Ice Ages*. Independently Published.
Eddy, J.A., 1976. The Maunder Minimum. *Science*, 192, 1189–1202.
Eddy, J.A., 1977. Climate and the changing Sun. *Climatic Change*, 1, 173–190.
Egorova, T., Schmutz, W., Rozanov, E., Shapiro, A.I., Usoskin, I., et al., 2018. Revised historical solar irradiance forcing. *Astronomy and Astrophysics*, 615, A85.
Eichler, A., Olivier, S., Henderson, K., Laube, A., Beer, J., Papina, T., Gäggeler, H. W., Schwikowski, M., 2009. Temperature response in the Altai region lags solar forcing. *Geophysical Research Letters*, 36, L01808.
Emanuel, K., 2023. Limitations of reanalyses for detecting tropical cyclone trends. *Nature Climate Change*, 13, 379–380.
England, M.R., Polvani, L.M., Screen, J., Chan, A.C., 2025. Minimal Arctic seaice loss in the last 20 years, consistent with internal climate variability. *Geophysical Research Letters*, 52, e2025GL116175.
Enzi, S., Bertolin, C., Diodato, N., 2014. Snowfall time-series reconstruction in Italy over the last 300 years. *The Holocene*, 24(3), 346–356.

EPA, 2023. *Technical Support Document: Social Cost of Greenhouse Gases*. U.S. Environmental Protection Agency.

Erb, M.P., McKay, N.P., Steiger, N., Dee, S., Hancock, C., Ivanovic, R.F., Gregoire, L.J., Valdes, P., 2022. Reconstructing Holocene temperatures in time and space using paleoclimate data assimilation. *Climate of the Past*, 18(12), 2599–2629.

Esper, J., St. George, S., Anchukaitis, K., D'Arrigo, R., Ljungqvist, F.C., et al., 2018. Large-scale, millennial-length temperature reconstructions from tree rings. *Dendrochronologia*, 50, 81–90.

Esper, J., Smerdon, J.E., Anchukaitis, K.J., Cook, E.R., D'Arrigo, R., et al., 2024a. The IPCC's reductive Common Era temperature history. *Communications Earth & Environment*, 5, 222.

Esper, J., Torbenson, M., Büntgen, U., 2024b. 2023 summer warmth unparalleled over the past 2,000 years. *Nature*, 627, 769–773.

Essex, C., McKitrick, R., Andresen, B., 2007. Does a global temperature exist? *Journal of Non-Equilibrium Thermodynamics*, 32(1), 1–27.

Etminan, M., Myhre, G., Highwood, E.J., Shine, K.P., 2016. Radiative forcing of carbon dioxide, methane, and nitrous oxide: A significant revision of the methane radiative forcing. *Geophysical Research Letters*, 43, 12614–12623.

Etz, D.V., 2000. Conjunctions of Jupiter and Saturn. *Journal of the Royal Astronomical Society of Canada*, 94, 174–178.

European Union, 2024. *European Green Deal*. Available at: https://commission.europa.eu/strategy-and-policy/priorities-2019-2024/european-green-deal_en

EDGARD: Emission Database for Global Atmospheric Research. European Commission, 2025. *GHG emissions of all world countries*. Available at: https://edgar.jrc.ec.europa.eu/report_2025

European Commission, 2024b. *Costs of climate mitigation in the EU: Background report for the 2040 climate target roadmap*.

Evin, G., Ribes, A., Corre, L., 2024. Assessing CMIP6 uncertainties at global warming levels. *Climate Dynamics*, 62, 8057–8072.

Eyring, V., Bony, S., Meehl, G.A., Senior, C.A., Stevens, B., et al., 2016. Overview of the coupled model intercomparison project phase 6 (CMIP6) experimental design and organization. *Geoscientific Model Development*, 9(5), 1937–1958.

Fagan, B., 2008. *The Great Warming: Climate Change and the Rise and Fall of Civilizations*. Bloomsbury Press.

Fagan, B., 2019. *The Little Ice Age: How Climate Made History 1300–1850*. Basic Books.

Fairbridge, R.W., 1961. Eustatic changes in sea level. *Physics and Chemistry of the Earth*, 4, 99–185.

Fairbridge, R.W., Shirley, J.H., 1987. Prolonged minima and the 179-yr cycle of the solar inertial motion. *Solar Physics*, 110, 191.

Faust, J.C., Fabian, K., Milzerc, G., Giraudeau, J. Kniesa, J., 2016. Norwegian fjord sediments reveal NAO related winter temperature and precipitation changes of the past 2800 years. *Earth and Planetary Science Letters*, 435, 84–93.

Filippelli, G., Beal, L., Rajaram, H., AghaKouchak, A., Balikhin, M.A., et al., 2021. Geoscientists, who have documented the rapid and accelerating climate crisis for decades, are now pleading for immediate collective action. *Geophysical Research Letters*, 48, e2021GL096644.

Fiorentin, P., Cavazzani, S., Bertolo, A., Ortolani, S., Binotto, R., Saviane, I., 2025. SQM ageing and atmospheric conditions: how do they affect the long-term trend of night sky brightness measurements? *Sensors*, 25(2), 516.

Firestone, R.B., West, A., Kennett, J.P., Becker, L., Bunch, T.E., et al., 2007. Evidence for an extraterrestrial impact 12,900 years ago that contributed to the megafaunal extinctions and the Younger Dryas cooling. *Proceedings of the National Academy of Sciences*, 104(41), 16016–16021.

Fleitmann, D., Matter, A., 2009. The speleothem record of climate variability in Southern Arabia. *Comptes Rendus Geoscience*, 341 (8–9), 633–642.

Foote, E., 1856. Circumstances affecting the heat of the Sun's rays. *The American Journal of Science and Arts*, 22(66), 382–383.

Forster, P., Gregory, J.M., 2006. The climate sensitivity and its components diagnosed from Earth Radiation Budget data. *Journal of Climate*, 19, 39–52.

Foster, G.L., Lear, C.H., Rae, J.W.B., 2012. The evolution of pCO_2 over the past 65 Myr. *Earth and Planetary Science Letters*, 341–344, 243–254.

Forte, E., Azzaro, M., Cannone, N., Longhi, A., Santin, I, Guglielmin, M., 2025. A warming pulse in the Antarctic continent changed the landscape during the Middle Ages. *Communications Earth & Environment*, 6, 281.

Fourier, J., 1827. Mémoire sur les températures du globe terrestre et des espaces planétaires. *Mémoires de l'Académie Royale des Sciences de l'Institut de France*, 6, 569–604.

Friis-Christensen, E., Lassen, K., 1991. Length of the solar cycle: An indicator of solar activity closely associated with climate. *Science*, 254, 698–700.

Fröhlich, C., Lean, J., 1998. The Sun's total irradiance: Cycles, trends and related climate change uncertainties since 1976. *Geophysical Research Letters*, 25 (23), 4245–4392.

Fröhlich, C., 2006. Solar irradiance variability since 1978. *Space Science Reviews*, 125, 53–65.

Fröhlich, C., 2009. Evidence of a long-term trend in total solar irradiance. *Astronomy & Astrophysics*, 501(3), L27–L30.

Fröhlich, C., 2012. Total solar irradiance observations. *Surveys in Geophysics*, 33(3–4), 453–473.

Fyfe, J., Gillett, N., Zwiers, F., 2013. Overestimated global warming over the past 20 years. *Nature Climate Change*, 3, 767–769.

Gallavotti, G., 1981. Perturbation theory for classical Hamiltonian systems. *Springer Lecture Notes in Physics*, 93, 32–93.

Ge, Q.S., Liu, H.L., Ma, X., Zheng, J.Y., Hao, Z.X., 2017. Characteristics of temperature change in China over the last 2000 years. *Advances in Atmospheric Sciences*, 34, 941–951.

Geddes, A.B., King-Hele, D.G., 1983. Equations for mirror symmetries among the distances of the planets. *Quarterly Journal of the Royal Astronomical Society*, 24, 10–13.

Gelaro, R., McCarty, W., Suárez, M. J., et al., 2017. The Modern-Era Retrospective Analysis for Research and Applications, Version 2 (MERRA-2). *Journal of Climate*, 30, 5419–5454.

Gerhart, L.M., Ward, J.K., 2010. Plant responses to low [CO_2] of the past. *New Phytologist*, 188(3), 674–695.

Gervais, F., 2016. Anthropogenic CO_2 warming challenged by 60-year cycle. *Earth-Science Reviews*, 155, 129–135.

Global Coal Plant Tracker: 2024. Available at: https://globalenergymonitor.org /projects/global-coal-plant-tracker/tracker/

Ghil, M., Allen, M.R., Dettinger, M.D., Ide, K., Kondrashov, D., Mann, M. E., ... & Yiou, P., 2002. Advanced spectral methods for climatic time series. *Reviews of Geophysics*, 40(1), 3.1–3.41.

Gibbs, J. (Dir.), Moore, M. (Exec. Prod.), 2019. *Planet of the Humans* [Film]. Rumble Media.

Gillon, M., Triaud, A.H.M.J., Demory, B.-O., Jehin, E., Agol, E., et al., 2017. Seven temperate terrestrial planets around the nearby ultracool dwarf star TRAPPIST-1. *Nature*, 542(7642), 456–460.

Golaz, J.-C., Caldwell, P.M., Van Roekel, L.P., Petersen, M.R., Tang, Q., et al., 2019. The DOE E3SM coupled model version 1: Overview and evaluation at standard resolution. *Journal of Advances in Modeling Earth Systems*, 11, 2089–2129.

Gold, T., Soter, S., 1969. Atmospheric tides and the resonant rotation of Venus. *Icarus*, 11(3), 356–366.

Goodman, J., Dickson, E.S., 1998. Dynamical tide in solar-type binaries. *The Astrophysical Journal*, 507, 938–944.

Gore, A., 2006. *An Inconvenient Truth* [Film documentary]. Paramount Classics. Available at: https://participant.com/campaign/inconvenient-truth/

Grossman, E.L., Joachimski, M.M., 2022. Oxygen isotope stratigraphy of the Phanerozoic. *Annual Review of Earth and Planetary Sciences*, 50, 1–35.

Gray, S.T., Graumlich, L.J., Betancourt, J.L., Pederson, G.T., 2004. A tree-ring based reconstruction of the Atlantic Multidecadal Oscillation since 1567 A.D. *Geophysical Research Letters*, 31, L12205.

Groves, D.I., Santosh, M., Yang, C.-X., 2023. Net zero climate remediations and potential terminal depletion of global critical metal resources. *Geosystems and Geoenvironment*, 2(1), 100136.

Guidoboni, E., Navarra, A., Boschi, E., 2011. *The Spiral of Climate: Civilizations of the Mediterranean and Climate Change in History*. Bononia University Press.

Gussmann, E.A., 1967. Saturation in Fraunhofer lines. *The Observatory*, 87, 233–236.

Hafner, A., Schwörer, C., 2018. Vertical mobility around the high-alpine Schnidejoch Pass: Indications of Neolithic and Roman pastoralism in the Swiss Alps from paleoecological and archaeological sources. *Quaternary International*, 495, 101–113.

Halevy, I., Bachan, A., 2017. The geologic history of seawater pH. *Science*, 355(6329), 1069–1071.

Hall, C., 2023 (December 1). *Carbon dioxide: Movie night*. Watts Up With That? Available at: https://wattsupwiththat.com/2023/12/01/carbon-dioxide-movie-night-the-global-picture/

Hansen, J., Sato, M., Ruedy, R., Kharecha, P., Lacis, A., et al., 2007. Climate simulations for 1880–2003 with GISS modelE. *Climate Dynamics*, 29, 661–696.

Harde, H., 2014. Advanced two-layer climate model for the assessment of global warming by CO_2. *Open Journal of Atmospheric and Climate Change*, 1(3), 1–51.

Harriss, B., 1971. The Monte Moro Pass and the Col D'Hérens. *Alpine Journal*, 76, 127–132.
Harris, I., Osborn, T.J., Jones, P., Lister, D., 2020. Version 4 of the CRU TS monthly high-resolution gridded multivariate climate dataset. *Scientific Data*, 7, 109.
Hartlep, T., Mansour, N.N., 2005. Acoustic wave propagation in the Sun. *Center for Turbulence Research Annual Research Briefs*, Stanford University, 357–365.
Hausfather, Z., Peters, G.P., 2020. Emissions — the 'business as usual' story is misleading. *Nature*, 577(7792), 618–620.
Hausfather, Z., Marvel, K., Schmidt, G.A., Nielsen-Gammon, J.W., Zelinka, M., 2022. Climate simulations: recognize the 'hot model' problem. *Nature*, 605(7908), 26–29.
Haynes, L.L., Hönisch, B., 2020. The seawater carbon inventory at the Paleocene–Eocene Thermal Maximum. *Proceedings of the National Academy of Sciences*, 117(39), 24088–24095.
Helbling, M., Meierrieks, D., 2023. Global warming and urbanization. *Journal of Population Economics*, 36, 1187–1223.
Hellmuth, F., Carlsen, T., Daloz, A.S., David, R.O., Che, H., Storelvmo, T., 2025. Evaluation of biases in mid-to-high-latitude surface snowfall and cloud phase in ERA5 and CMIP6 using satellite observations. *Atmospheric Chemistry and Physics*, 25, 1353–1383.
Hersbach, H., Bell, B., Berrisford, P., et al., 2020. The ERA5 global reanalysis. *Quarterly Journal of the Royal Meteorological Society*, 146, 1999–2049.
Hille, K. (25 April 2016). Carbon dioxide fertilization greening Earth, study finds. NASA. Retrieved December 27, 2020.
Hodnebrog, Ø., Myhre, G., Samset, B.H., Sager, P.L., Smith, C.J., et al., 2024. Global aerosol emission reductions drive positive trend in Earth's energy imbalance. Nature *Communications Earth & Environment*, 5, 317.
Hormes, A., Müller, B. U., Schlüchter, C., 2001. The Alps with little ice: Evidence for eight Holocene phases of reduced glacier extent in the Central Swiss Alps. *The Holocene*, 11(3), 255–265.
Horrox, R. (Ed.), 1994. *The Black Death*. Manchester University Press.
Hoyt, D.V., Schatten, K.H., 1993. A discussion of plausible solar irradiance variations, 1700–1992. *Journal of Geophysical Research: Space Physics*, 98(A11), 18895–18906.
Howe, R., Christensen-Dalsgaard, J., Hill, F., Komm, R.W., Larsen, R.M., Schou, J., Thompson, M.J., Toomre, J., 2000. Dynamic variations at the base of the solar convection zone. *Science*, 287(5462), 2456–2460.
Hoyt, D.V., Schatten, K.H., 1997. *The Role of the Sun in Climate Change*. Oxford University Press.
Hu, Y., McKitrick, R., 2015. Insurance payouts and asset exposure in climate-sensitive regions. *Climate Economics Review*, 9(1), 33–49.
Hudson, R., 2023. The problem of tropospheric temperature trend measurements for global climate models: complex versus direct empiricism. *New Technology and the Humanities*, 3(2), 117–123.
Humlum, O., Stordahl, K., Solheim, J.-E., 2013. The phase relation between atmospheric carbon dioxide and global temperature. *Global and Planetary Change*, 100, 51–69.
Hung, C.-C., 2007. Apparent relations between solar activity and solar tides caused by the planets. *NASA Technical Memorandum*, TM-2007-214817.
International Energy Agency, 2024. *World Energy Outlook 2024*. Available at: https://www.iea.org/reports/world-energy-outlook-2024
International Energy Agency, 2025. *CO_2 Emissions: Global Energy Review 2025*. Available at: https://www.iea.org/reports/global-energy-review-2025
Ilin, E., Poppenhäger, K., Chebly, J., Ilić, N., Alvarado-Gómez, J.D., 2024. Planetary perturbers: Flaring star–planet interactions in Kepler and TESS. *Monthly Notices of the Royal Astronomical Society*, 527(2), 3395–3417.
Imbrie, J., Boyle, E.A., Clemens, S.C., Duffy, A.C., Howard, W.R., et al., 1992. On the structure and origin of major glaciation cycles. *Paleoceanography*, 7, 701–738.
IPCC, 1990. *Climate Change: The IPCC Scientific Assessment*. Cambridge University Press.
IPCC, 1995. *Climate Change 1995: The Science of Climate Change*. Cambridge University Press.
IPCC, 2001. *Climate Change 2001: The Scientific Basis*. Cambridge University Press.
IPCC, 2007. *Climate Change 2007: The Physical Science Basis*. Cambridge University Press.
IPCC, 2013. *Climate Change 2013: The Physical Science Basis*. Cambridge University Press.
IPCC, 2018. *Global Warming of 1.5°C*. IPCC Special Report.
IPCC, 2021. *Climate Change 2021: The Physical Science Basis*. Cambridge University Press.
IPCC, 2022. *Climate Change 2022a: Impacts, Adaptation and Vulnerability*. Cambridge University Press.
IPCC, 2022b. *Sixth Assessment Report (AR6), Working Group III: Mitigation of Climate Change* – Chapter

3 and Technical Summary. 2022.
IPCC, 2023. *Climate Change 2023: Synthesis Report*. IPCC, Geneva, Switzerland.
Jenkins, S., Smith, C., Allen, M., Grainger, R., 2023. Tonga eruption increases chance of temporary surface temperature anomaly above 1.5 °C. *Nature Climate Change*, 13(2), 127–129.
Jevrejeva, S., Moore, J.C., Grinsted, A., Woodworth, P.L., 2008. Recent global sea level acceleration started over 200 years ago? *Geophysical Research Letters*, 35, L08715.
Jiang, J., Chatterjee, P., Choudhuri, A.R., 2007. Solar activity forecast with a dynamo model. *Monthly Notices of the Royal Astronomical Society*, 381, 1527–1542.
Jones, P.D., Groisman, P.Ya., Coughlan, M., Plummer, N., Wang, W.-C., Karl, T.R., 1990. Assessment of urbanization effects in time series of surface air temperature over land. *Nature*, 347(6289), 169–172.
Jones, T.R., Cuffey, K.M., Roberts, W.H.G., Markle, B.R., Steig, E.J., et al., 2023. Seasonal temperatures in West Antarctica during the Holocene. *Nature*, 613, 292–297.
Jose, P.D., 1965. Sun's motion and sunspots. *The Astronomical Journal*, 70, 193.
Jouzel, J., Masson-Delmotte, V., Cattani, O., Dreyfus, G., Falourd, S., et al., 2007. Orbital and millennial Antarctic climate variability over the past 800,000 years. *Science*, 317, 793–796.
Judge, P.G., Egeland, R., Henry, G.W., 2020. Sun-like stars shed light on solar climate forcing. *The Astrophysical Journal*, 891(1), 96.
Juckett, D.A., 2000. Solar activity cycles, north–south asymmetries, and differential rotation associated with the solar inertial motion. *Solar Physics*, 191, 201–226.
Karl, T.R., Hassol, S.J., Miller, C.D., Murray, W.L., 2006. *Temperature Trends in the Lower Atmosphere: Steps for Understanding and Reconciling Differences*. U.S. Climate Change Science Program and the Subcommittee on Global Change Research. Washington, DC, USA.
Katata, G., Connolly, R., O'Neill, P., 2023. Evidence of urban blending in homogenized temperature records in Japan and the United States. *Journal of Applied Meteorology and Climatology*, 62(8), 1095–1114.
Kaufmann, R.K., Kauppi, H., Mann, M.L., Stock, J.H., 2011. Reconciling anthropogenic climate change with observed temperature 1998–2008. *Proceedings of the National Academy of Sciences*, 108(29), 11790–11793.
Kaufman, D., McKay, N., Routson, C., Erb, M., Dätwyler, C., et al., 2020. Holocene global mean surface temperature: a multi-method reconstruction. *Scientific Data*, 7, 201.
Kaufman, D.S., Broadman, E., 2023. Revisiting the Holocene global temperature conundrum. *Nature*, 614, 425–435.
Keeling, C.D., 1960. The Concentration and Isotopic Abundances of Carbon Dioxide in the Atmosphere. *Tellus*, 12(2), 200–203.
Keeling, C.D., Whorf, T.P., 2000. The 1800-year oceanic tidal cycle: A possible cause of rapid climate change. *Proceedings of the National Academy of Sciences*, 97(8), 3814–3819.
Kennedy, J.J., Rayner, N.A., Smith, R.O., Parker, D.E., Saunby, M., 2011. Reassessing biases and other uncertainties in sea surface temperature observations measured in situ since 1850: 2. Biases and homogenization. *Journal of Geophysical Research: Atmospheres*, 116, D14104.
Kennedy, J.J., 2014. A review of uncertainty in situ measurements and data sets of sea surface temperature. *Reviews of Geophysics*, 52, 1–32.
Kennedy, J.J., Rayner, N.A., Atkinson, C.P., Killick, R.E., 2019. An ensemble dataset of sea-surface temperature change from 1850. *Journal of Geophysical Research: Atmospheres*, 124, 7719–7763.
Kepler, J., 1596. *Mysterium Cosmographicum* (The Cosmographic Mystery).
Kepler, J., 1601. On the more certain fundamentals of astrology. In: Brackenridge, J.B., Rossi, M.A. (Eds.), *Proceedings of the American Philosophical Society*, 123(2), 85–116 (1979).
Kepler, J., 1606. *De Stella Nova in Pede Serpentarii* (On the new star in Ophiuchus's foot).
Kepler, J., 1609. *Astronomia Nova* (New Astronomy). Trans. Donahue, W.H., Cambridge University Press.
Kepler, J., 1619. *Harmonices Mundi* (The Harmony of the World). Trans. Field, J., American Philosophical Society, 1997.
Kerr, R.A., 2001. A variable sun paces millennial climate. *Science*, 294(5546), 1431–1433.
Kim, D., Christy, J.R., 2022. Detecting impacts of surface development near weather stations since 1895 in the San Joaquin Valley of California. *Theoretical and Applied Climatology*, 149, 1223–1238.
Kirkby, J., 2007. Cosmic rays and climate. *Surveys in Geophysics*, 28(5–6), 333–375.
Kirkby, J., Curtius, J., Almeida, J., Dunne, E., Duplissy, J., et al., 2011. Role of sulphuric acid, ammonia and galactic cosmic rays in atmospheric aerosol nucleation. *Nature*, 476, 429–433.

Kitchen, D.E., 2014. *Global Climate Change: Turning Knowledge into Action*. Pearson.
Klimenko, V.V., Matskovsky, V.V., Dahlmann, D., 2014. Multi-archive temperature reconstruction of the Russian Arctic for the past two millennia. *Geography, Environment, Sustainability*, 4(12), 84–95.
Klyashtorin, L.B., 2001. Climate change and long-term fluctuations of commercial catches: the possibility of forecasting. *FAO Fisheries Technical Paper*, No. 410, Rome, FAO, 86p. Available at: https://open knowledge.fao.org/items/6c2a808e-ed70-4a5d-a574-1f68745be7c3.
Klyashtorin, L.B., Lyubushin, A.A., 2007. *Cyclic Climate Changes and Fish Productivity*. Moscow, VNIRO Publishing. Available at: http://alexeylyubushin.narod.ru/Climate_Changes_and_Fish_Productivity.pdf.
Klyashtorin, L.B., Borisov, V., Lyubushin, A.A., 2009. Cyclic changes of climate and major commercial stocks of the Barents Sea. *Marine Biology Research*, 5(1), 4–17.
Knight, J., Kennedy, J.J., Folland, C., Harris, G., Jones, G.S., et al., 2009. Do global temperature trends over the last decade falsify climate predictions? *Bulletin of the American Meteorological Society*, 90(8), S22–S23.
Knudsen, M., Seidenkrantz, M.S., Jacobsen, B., Kuijpers, A., 2011. Tracking the Atlantic Multidecadal Oscillation over 8,000 years. *Nature Communications*, 2, 178.
Knutson, T.R., Camargo, S.J., Chan, J.C.L., Emanuel, K., Ho, C.H., et al., 2020. Tropical cyclones and climate change assessment: Part II. Projected response to anthropogenic warming. *Bulletin of the American Meteorological Society*, 101(3), E303–E322.
Knutti, R., Hegerl, G.C., 2008. The equilibrium sensitivity of the Earths temperature to radiation changes. *Nature Geoscience*, 1, 735–743.
Knutti, R., Rugenstein, M.A., Hegerl, G.C., 2017. Beyond equilibrium climate sensitivity. *Nature Geoscience*, 10, 727–736.
Kobayashi, S., Ota, Y., Harada, Y., et al., 2015. The JRA-55 Reanalysis: General Specifications and Basic Characteristics. *Journal of the Meteorological Society of Japan*, 93(1), 5–48.
Komitov, B., 2009. The "Sun-climate" relationship. II. The "cosmogenic" beryllium and the middle latitude aurora. *Bulgarian Astronomical Journal*, 12, 85–93.
Komm, R., Howe, R., Durney, B.R., Hill, F., 2003. Temporal variation of angular momentum in the solar convection zone. *The Astrophysical Journal*, 586(1), 650–662.
Koonin, S.E., 2024. *Unsettled* (Updated and Expanded Edition)*: What climate science tells us, what it doesn't, and why it matters*. BenBella Books.
Kopp, G., Lawrence, G., 2005. The Total Irradiance Monitor (TIM): Instrument design. *Solar Physics*, 230, 91–109.
Kopp, G., Heuerman, K., Lawrence, G., 2005. The Total Irradiance Monitor (TIM): Instrument calibration. *Solar Physics*, 230, 111–127.
Kopp, G., Lean, J.L., 2011. A new, lower value of total solar irradiance: Evidence and climate significance. *Geophysical Research Letters*, 38, L01706.
Kopp, G., Krivova, N., Wu, C.J., Lean, J., 2016. The impact of the revised sunspot record on solar irradiance reconstructions. *Solar Physics*, 291, 2951–2965.
Kotov, V.A., Haneychuk, V.I., 2020. Oscillations of solar photosphere: 45 years of observations. *Astronomische Nachrichten*, 341, 595–599.
Kopp, G., 2025. Solar Irradiance Measurements. Living Reviews in *Solar Physics*, 22, 1.
Kotz, M., Levermann, A., Wenz, L., 2024. The economic commitment of climate change. *Nature*, 628, 551–557.
Koutsoyiannis, D., 2024. Stochastic assessment of temperature–CO_2 causal relationship in climate from the Phanerozoic through modern times. *Mathematical Biosciences and Engineering*, 21(7), 6560–6602.
Krivova, N. A., Solanki, S. K., Fligge, M., 2003. Reconstruction of solar total and spectral irradiance variations in cycle 23. *Astronomy & Astrophysics*, 399, L1–L4.
Krivova, N.A., Vieira, L.E.A., Solanki, S.K., 2010. Reconstruction of solar spectral irradiance since the Maunder minimum. *Journal of Geophysical Research: Space Physics*, 115, A12112.
Křivský, L., Pejml, K., 1988. Solar activity aurorae and climate in Central Europe in the last 1000 years. *Bulletin of the Astronomical Institute of the Czechoslovak Academy of Sciences* No 75. http://www.ngdc.noaa.gov/stp/aeronomy/aurorae.html
Kuramoto, Y., 1984. *Chemical Oscillations, Waves, and Turbulence*. Springer, Berlin.
Kutschera, W., Patzelt, G., Schaefer, J.M., Schlüchter, C., Steier, P., Wild, E.M., 2020. The movements of Alpine glaciers throughout the last 10,000 years as sensitive proxies of temperature and climate changes. *EPJ Web of Conferences*, 232, 02002.

Lamb, H.H., 1965. The early medieval warm epoch and its sequel. *Palaeogeography, Palaeoclimatology, Palaeoecology*, 1, 13–37.

Lamb, H.H., 1972. *Climate: Present, Past and Future*. Vol. 1. Methuen.

Lamb, H.H., 1995. *Climate, History and the Modern World* (2nd ed.). London: Routledge. ISBN: 978-0415127341.

Landscheidt, T., 1988. Solar rotation, impulses of the torque in the sun's motion, and climatic variation. *Climatic Change*, 12(3), 265–295.

Landscheidt, T., 1999. Extrema in sunspot cycle linked to Sun's motion. *Solar Physics*, 189(2), 413–424.

Lanza, A.F., 2018. Star–Planet Interactions. In Proceedings of the 18th Cambridge Workshop on Cool Stars, Stellar Systems, and the Sun (Cool Stars 18).

Lasher, G.E., Axford, Y., 2019. Medieval warmth confirmed at the Norse Eastern Settlement in Greenland. *Geology*, 47(3), 267–270.

Laskar, J., Robutel, P., Joutel, F., Gastineau, M., Correia, A.C.M., Levrard, B., 2004. A long-term numerical solution for the insolation quantities of the Earth. *Astronomy & Astrophysics*, 428, 261–285.

Laskar, J., Fienga, A., Gastineau, M., Manche, H., 2011. La2010: a new orbital solution for the long-term motion of the Earth. *Astronomy and Astrophysics*, 532, A89.

Lassen, K., Friis-Christensen, E., 2000. Reply to "Solar cycle lengths and climate: A reference revisited". *Journal of Geophysical Research: Space Physics*, 105(A12), 27493–27495.

Lawrimore, J.H., Menne, M.J., Gleason, B.E., Williams, C.N., Wuertz, D.B., et al., 2011. An overview of the Global Historical Climatology Network monthly mean temperature data set, version 3. *Journal of Geophysical Research: Atmospheres*, 116, D19121.

Lean, J., Beer, J., Bradley, R., 1995. Reconstruction of solar irradiance since 1610: implications for climate change. *Geophysical Research Letters*, 22, 3195–3198.

Lean, J., 2000. Evolution of the Sun's spectral irradiance since the Maunder Minimum. *Geophysical Research Letters*, 27, 2425–2428.

Lean, J., Wang, Y.M., Sheeley, N.R., 2005. SORCE contributions to new understanding of global change and solar variability. *Solar Physics*, 230, 27–53.

Legates, D.R., Soon, W., Briggs, W.M., Monckton, C., 2015. Climate consensus and 'misinformation': A rejoinder to agnotology, scientific consensus, and the teaching and learning of climate change. *Science & Education*, 24, 299–318.

Le Mouël, J. L., Lopes, F., Courtillot, V., 2020. Characteristic time scales of decadal to centennial changes in global surface temperatures over the past 150 years. Earth and Space Science, 7(4), e2019EA000671.

Le Roy Ladurie, E., 1967. *Histoire du Climat depuis l'an mil*. Flammarion, Paris (trad. B. Bray, Doubleday & Co., 1971).

Lewis, N., Curry, J.A., 2018. The impact of recent forcing and ocean heat uptake data on estimates of climate sensitivity. *Journal of Climate*, 31(15), 6051–6071.

Lewis, N., 2023. Objectively combining climate sensitivity evidence. *Climate Dynamics*, 60, 3139–3165.

Lewis, N., 2025. Comment on "Opinion: Can uncertainty in climate sensitivity be narrowed further?" by Sherwood and Forest (2024). *Atmospheric Chemistry and Physics*, 25, 8821–8829.

Li, Y., Wu, D., Wang, T., Chen, L., Zhang, C., Guo, S., 2023. Late Holocene temperature and precipitation variations in an alpine region of the northeastern Tibetan Plateau. *Palaeogeography, Palaeoclimatology, Palaeoecology*, 615, 111442.

Li, Z., Liu, T., Huang, Y., Peng, J., Ling, Y., 2022. Evaluation of the CMIP6 precipitation simulations over global land. *Earth's Future*, 10, e2021EF002500.

Lin, Y.C., Fan, C.Y., Damon, P.E., Wallick, E.J., 1975. Long-term modulation of cosmic-ray intensity and solar activity cycle. *14th International Cosmic Ray Conference*, 3, 995–999.

Lindsey, R., 2003. *Under a Variable Sun*. NASA Earth Observatory Available at: https://earthobservatory.nasa.gov/features/VariableSun/variable4.php

Lindzen, R.S., Chou, M.-D., Hou, A.Y., 2001. Does the Earth have an adaptive infrared iris? *Bulletin of the American Meteorological Society*, 82, 417–432.

Lindzen, R.S., Choi, Y.-S., 2011. On the observational determination of climate sensitivity and its implications. *Asia-Pacific Journal of Atmospheric Sciences*, 47, 377–390.

Liu, Z., Zhu, J., Rosenthal, Y., Timm, O.E., 2014. The Holocene temperature conundrum. *Proceedings of the National Academy of Sciences*, 111(34), E3501–E3505.

Ljungqvist, F.C., 2010. A new reconstruction of temperature variability in the extra-tropical northern hemisphere during the last two millennia. *Geografiska Annaler: Series A, Physical Geography*, 92,

339–351.
Lockwood, M., Stamper, R., Wild, M., 1999. A doubling of the Sun's coronal magnetic field during the past 100 years. *Nature*, 399, 437–439.
Loeb, N.G., Ham, S.-H., Allan, R.P., Thorsen, T.J., Meyssignac, B., Kato, S., Johnson, G.C., Lyman, J.M., 2024. Observational Assessment of Changes in Earth's Energy Imbalance Since 2000. *Surveys in Geophysics*, 45(6), 1757–1783.
Loehle, C., Scafetta, N., 2011. Climate change attribution using empirical decomposition of climatic data. *The Open Atmospheric Science Journal*, 5, 74–86.
Lomborg, B., 2020. *False Alarm: How Climate Change Panic Costs Us Trillions, Hurts the Poor, and Fails to Fix the Planet*. Basic Books.
Lomborg, B., 2023 (29 novembre). 'Net Zero' fails the cost-benefit test. *The Wall Street Journal*. Available at: https://totallyengagedamericans.org/2023/12/net-zero-fails-the-cost-benefit-test/
Lovett, E.O., 1895. The Great Inequality of Jupiter and Saturn. *The Astronomical Journal*, 15, 113–127.
Lüdecke, H.J., Müller-Plath, G., Lüning, S., 2024. Central-European sunshine hours, relationship with the Atlantic Multidecadal Oscillation, and forecast. *Scientific Reports*, 14, 25152.
Luger, R., Sestovic, M., Kruse, E., 2017. A Seven-Planet Resonant Chain in TRAPPIST-1. *Nature Astronomy*, 1, 0129.
Lüning, S., 2022. *Climate reconstructions of the 'Medieval Warm Period' 1000–1200 AD*. Available at: http://t1p.de/mwp
Luterbacher, J., Xoplaki, E., Dietrich, D., Jones, P.D., Davies, T.D., et al., 2002. Extending North Atlantic Oscillation reconstructions back to 1500. *Atmospheric Science Letters*, 2(1–4), 114–124.
Luterbacher, J., Werner, J.P., Smerdon, J.E., Fernández-Donado, L., González-Rouco, F.J., et al., 2016. European summer temperatures since Roman times. *Environmental Research Letters*, 11, 024001.
Lutsko, N.J., Luongo, M.T., Wall, C.J., Myers, T.A., 2022. Correlation between cloud adjustments and cloud feedbacks in CMIP6. *Journal of Geophysical Research: Atmospheres*, 127, e2022JD037486.
Lynas, M., Houlton, B.Z., Perry, S., 2021. Greater than 99% consensus on human-caused climate change in the peer-reviewed scientific literature. *Environmental Research Letters*, 16, 114005.
Ma, N., Zhang, Y., Yang, Y., 2025. Recent decline in global ocean evaporation due to wind stilling. *Geophysical Research Letters*, 52, e2024GL114256.
Ma, P.-L., Harrop, B.E., Larson, V.E., Neale, R.B., Gettelman, A., et al., 2022. Better calibration of cloud parameterizations and subgrid effects increases the fidelity of the E3SM Atmosphere Model version 1. *Geoscientific Model Development*, 15, 2881–2916.
Macario-Rojas, A., Smith, K.L., Roberts, P.C.E., 2018. Solar activity simulation and forecast with a flux-transport dynamo. *Monthly Notices of the Royal Astronomical Society*, 479, 3791–3803.
Maibach, E., Perkins, D., Francis, Z., Myers, T., Engblom, A., et al., 2016. *A 2016 national survey of American Meteorological Society member views on climate change*. George Mason University, Center for Climate Change Communication.
Mahaney, W.C., Allen, C.C.R., Pentlavalli, P., Kulakova, A., Young, J.M., et al., 2017. Biostratigraphic evidence relating to the age-old question of Hannibal's invasion of Italy, I: history and geological reconstruction. Archaeometry, 59, 164–178.
Mallet, M., Voldoire, A., Solmon, F., Nabat, P., Drugé, T., Roehrig, R., 2024. Impact of biomass burning aerosols (BBA) on the tropical African climate in an ocean–atmosphere–aerosol coupled climate model. *Atmospheric Chemistry and Physics*, 24, 12509–12535.
Manabe, S., Strickler, R.F., 1964. Thermal Equilibrium of the Atmosphere with a Convective Adjustment. *Journal of the Atmospheric Sciences*, 21(4), 361–385.
Manabe, S., Wetherald, R.T., 1967. Thermal Equilibrium of the Atmosphere with a Given Distribution of Relative Humidity. *Journal of the Atmospheric Sciences*, 24(3), 241–259.
Manabe, S., Wetherald, R.T., 1975. The effects of doubling the CO_2 concentration on the climate of a general circulation model. *Journal of the Atmospheric Sciences*, 32, 3–15.
Mann, M.E., Bradley, R.S., Hughes, M.K., 1999. Northern hemisphere temperatures during the past millennium: inferences, uncertainties, and limitations. *Geophysical Research Letters*, 26, 759–762.
Mann, M., Zhang, Z., Hughes, M., Bradley, R., Miller, S., et al., 2008. Proxy-based reconstructions of hemispheric and global surface temperature variations over the past two millennia. *Proceedings of the National Academy of Sciences*, 105(36), 13252–13257.
Mann, M.E., Zhang, Z., Rutherford, S., Bradley, R.S., Malcolm, K., et al., 2009. Global signatures and dynamical origins of the Little Ice Age and Medieval Climate Anomaly. *Science*, 326(5957), 1256–1260.
Mantua, N.J., Hare, S.R., Zhang, Y., Wallace, J.M., Francis, R.C., 1997. A Pacific interdecadal climate

oscillation with impacts on salmon production. *Bulletin of the American Meteorological Society*, 78, 1069–1080.

Marano, G., Dalmonech, D., Collalti, A., et al., 2022. Forest management intensity and carbon sequestration potential under climate change scenarios in temperate Europe. *Agricultural and Forest Meteorology*, 319, 109203.

Marchitelli, V., Harabaglia, P., Troise, C., De Natale, G., 2020. On the correlation between solar activity and large earthquakes worldwide. *Scientific Reports*, 10, 11495.

Marcott, S.A., Shakun, J.D., Clark, P.U., Mix, A.C., 2013. A reconstruction of regional and global temperature for the past 11,300 years. *Science*, 339(6124), 1198–1201.

Margaritelli, G., Cacho, I., Català, A., Barra, M., Bellucci, L. G., Lubritto, C., Rettori, R., Lirer, F., 2020. Persistent warm Mediterranean surface waters during the Roman period. *Scientific Reports*, 10, 10430.

Matskovsky, V.V., Helama, S., 2014. Testing long-term summer temperature reconstruction based on maximum density chronologies obtained by reanalysis of tree-ring data sets from northernmost Sweden and Finland. *Climate of the Past*, 10(4), 1473–1487.

Matsumoto, H., Svensmark, H., Enghoff, M.B., 2022. Effects of Forbush decreases on clouds determined from PATMOS-x. *Journal of Atmospheric and Solar–Terrestrial Physics*, 230, 105845.

Matthes, K., Funke, B., Andersson, M.E., Barnard, L., Beer, J., et al., 2017. Solar forcing for CMIP6 (v3.2). *Geoscientific Model Development*, 10, 2247–2302.

Mauritsen, T., Stevens, B., 2015. Missing iris effect as a possible cause of muted hydrological change and high climate sensitivity in models. *Nature Geoscience*, 8(5), 346–351.

Mauritsen, T., Bader, J., Becker, T., Behrens, J., Bittner, M., et al., 2019. Developments in the MPI-M Earth System Model version 1.2 and its response to increasing CO_2. *Journal of Advances in Modeling Earth Systems*, 11, 998–1038.

Mauritsen, T., Roeckner, E., 2020. Tuning the MPI-ESM1.2 global climate model to improve the match with instrumental record warming. *Journal of Advances in Modeling Earth Systems*, 12, e2019MS002037.

May, A., Crok, M., 2025. Carbon dioxide and a warming climate are not problems. *American Journal of Economics and Sociology*, 84, 43–57.

Mayewski, P.A., Rohling, E.E., Stager, J.C., Karlén, W., Maasch, K.A., Meeker, L.D., Meyerson, E.A., Gasse, F., van Kreveld, S., Holmgren, K., Lee-Thorp, J., Rosqvist, G., Rack, F., Staubwasser, M., Schneider, R.R., Steig, E.J., 2004. Holocene climate variability. *Quaternary Research*, 62(3), 243–255.

Mazzarella, A., Scafetta, N., 2012. Evidences for a quasi 60-year North Atlantic Oscillation since 1700 and its meaning for global climate change. *Theoretical and Applied Climatology*, 107, 599–609.

Mazzarella, A., Giuliacci, A., Scafetta, N., 2013. Quantifying the Multivariate ENSO Index (MEI) coupling to CO_2 concentration and to the length of day variations. *Theoretical and Applied Climatology*, 111, 601–607.

McCracken, K.G., Beer, J., Steinhilber, F., Abreu, J., 2013. A phenomenological study of the cosmic ray variations over the past 9400 years. *Solar Physics*, 286, 609.

McCracken, K.G., Beer, J., Steinhilber, F., 2014. Evidence for planetary forcing of the cosmic ray intensity and solar activity. *Solar Physics*, 289, 3207–3229.

McIntyre, S., McKitrick, R., 2003. Corrections to the Mann et al. (1998) proxy database and Northern Hemispheric average temperature series. *Energy & Environment*, 14(6), 751–771.

McKitrick, R.R., Michaels, P.J., 2007. Quantifying the influence of anthropogenic surface processes and inhomogeneities on gridded global climate data. *Journal of Geophysical Research*, 112, D24S09.

McKitrick, R., 2013. Encompassing Tests of Socioeconomic Signals in Surface Climate Data. *Climatic Change*, 120(1–2), 95–107.

McKitrick, R., Christy, J., 2018. A test of the tropical 200–300-hPa warming rate in climate models. Earth and Space Science, 5, 529–536.

McKitrick, R., 2019. Climategate: 10 years after. Available at: https://www.ross mckitrick.com/uploads/4/8/0/8/4808045/climategate.10yearsafter.pdf

McKitrick, R., Christy, J., 2020. Pervasive warming bias in CMIP6 tropospheric layers. Earth and Space Science, 7, e2020EA001281.

McKitrick, R., 2025. Extended crop yield meta-analysis data do not support upward SCC revision. *Scientific Reports*, 15, 5575.

McNutt, M., 2015. The beyond-two-degree inferno. *Science*, 349, 7.

McSweeney, R., Hausfather, Z., 2018. Q&A: How do climate models work? *Carbon Brief*. Available at: https://www.carbonbrief.org/qa-how-do-climate-models-work/#tune

Mears, C.A., Wentz, F.J., 2016. Sensitivity of satellite-derived tropospheric temperature trends to the diurnal cycle adjustment. *Journal of Climate*, 29(10), 3629–3646.

Medhaug, I., Stolpe, M.B., Fischer, E.M., Knutti, R., 2017. Reconciling controversies about the global warming hiatus. *Nature*, 545(7652), 41–47.

Meinshausen, M., Nicholls, Z.R.J., Lewis, J., Gidden, M.J., Vogel, E., et al., 2020. The shared socio-economic pathway (SSP) greenhouse gas concentrations and their extensions to 2500. *Geoscientific Model Development*, 13(8), 3571–3605.

Melo, M., Lapin, M., Pecho, J., 2022. Climate in the Past and Present in the Slovak Landscapes—The Central European Context. In: Lehotský, M., Boltižiar, M. (eds.) *Landscapes and Landforms of Slovakia*, Springer.

Menne, M.J., Williams, C.N., Gleason, B.E., Rennie, J.J., Lawrimore, J.H., 2018. The Global Historical Climatology Network monthly temperature dataset, version 4. *Journal of Climate*, 31(24), 9835–9854.

Migliario, E., 2022. *Mobilità sui valichi alpini in età imperiale romana. In Atti del convegno "Alpi e transiti"*. Museo Tridentino di Scienze Naturali, Trento, 15–16 ottobre 2022.

Mignot, J., Hourdin, F., Deshayes, J., Boucher, O., Gastineau, G., Musat, I., et al., 2021. The tuning strategy of IPSL-CM6A-LR. *Journal of Advances in Modeling Earth Systems*, 13, e2020MS002340.

Mironova, I.A., Aplin, K.L., Arnold, F., Bazilevskaya, G.A., Beer, J., et al., 2015. Energetic particle influence on the Earth's atmosphere. *Space Science Reviews*, 194(1–4), 1–96.

Mishnaevsky, L. Jr., 2022. Root causes and mechanisms of failure of wind turbine blades: overview. *Materials*, 15, 2959.

Mitalas, R., Sills, K.R., 1992. On the photon diffusion time scale for the Sun. *The Astrophysical Journal*, 401, 759–760.

Mitchell, D.M., Lo, Y.T.E., Seviour, W.J.M., Haimberger, L., Polvani, L.M., 2020. The vertical profile of recent tropical temperature trends. *Environmental Research Letters*, 15(10), 104004.

Miyahara, H., Kusano, K., Kataoka, R., Shima, S.-i., Touber, E., 2023. Response of high-altitude clouds to galactic cosmic ray cycles in tropical regions. *Frontiers in Earth Science*, 11, 1157753.

Mohaddes, K., Ng, S., Raissi, M., 2023. *Climate Change and Macroeconomic Forecasting*. IMF Working Paper No. WP/23/45. International Monetary Fund.

Moberg, A., Sonechkin, D.M., Holmgren, K., Datsenko, N.M., Karlén, W., 2005. Highly variable Northern Hemisphere temperatures reconstructed from proxy data. *Nature*, 433, 613–617.

Mohajerani, A., Bakaric, J., Jeffrey-Bailey, T., 2017. The urban heat island effect, its causes, and mitigation, with reference to the thermal properties of asphalt concrete. *Journal of Environmental Management*, 197, 522–538.

Molchanov, A.M., 1968. The resonant structure of the solar system: The law of planetary distances. *Icarus*, 8, 203–215.

Molchanov, A.M., 1969. The reality of resonances in the solar system. *Icarus*, 11, 104–110.

Moldwin, M., 2008. *An Introduction to Space Weather*. Cambridge University Press.

Möller, F., 1963. Influence of changes in CO_2 concentration on the radiation balance of Earth's surface and on the climate. *Journal of Geophysical Research*, 68, 3877–3886.

Monckton, C., Soon, W.-H., Legates, D.R., Briggs, W.M., 2015. Why models run hot: results from an irreducibly simple climate model. *Science Bulletin*, 60(1), 122–135.

Monnin, E., Steig, E.J., Siegenthaler, U., Kawamura, K., Schwander, J., et al., 2004. Evidence for substantial accumulation rate variability in Antarctica during the Holocene, through synchronization of CO_2 in the Taylor Dome, Dome C and DML ice cores. *Earth and Planetary Science Letters*, 224, 45–54.

Montford, A.W., 2010. *The hockey stick illusion: Climategate and the corruption of science*. London: Stacey International.

Montford, A., 2013. Consensus? What consensus? GWPF Note 5. Available at https://www.thegwpf.org/content/uploads/2013/09/Montford-Consensus.pdf

Moore, S., Smith, J., Taylor, L., 2017. Reassessing the EPA's social cost of carbon. *Regulation*, 40(3), 14–20.

Moons, M., Morbidelli, A., 1995. Secular resonances in mean motion commensurabilities: the 4/1, 3/1, 5/2, and 7/3 cases. *Icarus*, 114, 33–50.

Moons, M., Morbidelli, A., Migliorini, F., 1998. Dynamical structure of the 2/1 commensurability with Jupiter and the origin of the resonant asteroids. *Icarus*, 135, 458–468.

Morice, C.P., Kennedy, J.J., Rayner, N.A., Jones, P.D., 2012. Quantifying uncertainties in global and regional temperature change using an ensemble of observational estimates: the HadCRUT4

dataset. *Journal of Geophysical Research: Atmospheres*, 117, D08101.
Morice, C.P., Kennedy, J.J., Rayner, N.A., Winn, J.P., Hogan, E., et al., 2021. An updated assessment of near-surface temperature change from 1850: the HadCRUT5 data set. *Journal of Geophysical Research: Atmospheres*, 126, e2019JD032361.
Mörner, N.-A., Tooley, M., Possnert, G., 2004. New perspectives for the future of the Maldives. *Global and Planetary Change*, 40(1–2), 177–182.
Mörner, N.-A., Scafetta, N., Solheim, J.-E., 2015. The January 7 giant solar flare, the simultaneous triple planetary conjunction and additional records. In: Mörner, N.-A. (Ed.) *Planetary Influence on the Sun and the Earth*. Nova Science Publishers.
Murata, A., Mori, T., Kato, H., Hu, H.-M., Shen, C.-C., et al., 2025. Holocene Temperature Trend Inferred From Oxygen and Carbonate Clumped Isotope Profiles. *Island Arc*, 34, e70002.
Murray-Wallace, C.V., Woodroffe, C.D., 2014. *Quaternary Sea-Level Changes: A Global Perspective*. Cambridge University Press.
Myhre, G., 1998. New estimates of radiative forcing due to well-mixed greenhouse gases. *Geophysical Research Letters*, 25(14), 2715–2718.
Nance, R.D., Murphy, J.B., Santosh, M., Pisarevsky, S.A., Stern, R.J., 2014. The supercontinent cycle. *Geoscience Frontiers*, 5, 395–411.
Nataf, H.-C., 2022. Tidally synchronized solar dynamo: a rebuttal. *Solar Physics*, 297, 107.
Neff, U., Burns, S.J., Mangini, A., Mudelsee, M., Fleitmann, D., Matter, A., 2001. Strong coherence between solar variability and the monsoon in Oman between 9 and 6 kyr ago. *Nature*, 411, 290–293.
Newton, I., 1704. *Manuscript letter containing prophetic calculations based on the Book of Daniel*. Newton Archive, Hebrew University of Jerusalem.
Newell, R.G., Prest, B.C., Sexton, S., 2021. The social cost of carbon: Advances in long-term probabilistic modeling. *Brookings Papers on Economic Activity*, spring 2021, 1–65.
Ney, E.P., 1959. Cosmic Radiation and the Weather. *Nature*, 183, 451–452.
Nikolov, N., Zeller, K., 2017. New Insights on the Physical Nature of the Atmospheric Greenhouse Effect Deduced from an Empirical Planetary Temperature Model. *Environment Pollution and Climate Change*, 1(2), 1000112.
Nikolov, N., Zeller, K.F., 2024. Roles of Earth's Albedo Variations and Top-of-the-Atmosphere Energy Imbalance in Recent Warming: New Insights from Satellite and Surface Observations. *Geomatics*, 4, 311–341.
Nissan, H., Goddard, L., Coughlan de Perez, E., Furlow, J., Baethgen, W., Thomson, M.C., Mason, S.J., 2019. On the use and misuse of climate change projections in international development. *WIREs Climate Change*, 10(6), e579.
NOAA (Amministrazione Nazionale Oceanica e Atmosferica), 2023. Bacino degli uragani del Nord Atlantico (1851–2022): confronto tra HURDAT originale e rivisto. Available at https://www.aoml.noaa.gov/hrd/hurdat/comparison_table.html
Nordhaus, W.D., 1993. Rolling the 'DICE': An optimal transition path for controlling greenhouse gases. *Resource and Energy Economics*, 15(1), 27–50.
Nordhaus, W.D., 2018. Projections and uncertainties about climate change in an era of minimal climate policies. *American Economic Journal: Economic Policy*, 10(3), 333–360.
Ogurtsov, M.G., Nagovitsyn, Y.A., Kocharov, G.E., Jungner, H., 2002. Long-period cycles of the Sun's activity recorded in direct solar data and proxies. *Solar Physics*, 211, 371.
Okal, E., Anderson, D.L., 1975. On the planetary theory of sunspots. *Nature*, 253, 511.
ÓhAiseadha, C., Quinn, G., Connolly, R., Connolly, M., Soon, W., 2020. Energy and climate policy — an evaluation of global climate change expenditure 2011–2018. *Energies*, 13(18), 4839.
Ollier, C.D., 2013. Global warming and climate change: Science and politics. *Quaestiones Geographicae*, 32(1), 63–76.
Olmsted, D., 1856. *On the recent secular period of the aurora Borealis*. Smithsonian Contributions to Knowledge, 8, 1–52. Washington, D.C.: Smithsonian Institution.
O'Neill, P., Connolly, R., Connolly, M., Soon, W., Chimani, B., et al., 2022. Evaluation of the homogenization adjustments applied to European temperature records in the Global Historical Climatology Network dataset. *Atmosphere*, 13(2), 285.
O'Neill, B.C., 2023. Global development trends and climate resilience. *Journal of Climate Policy*, 23(2), 145–162.
Ortolani, S., Saviane, I., Cavazzani, S. 2019. Cloud coverage and sky brightness at La Silla. *Zenodo*, doi:10.5281/zenodo.3248992.
Osborn, T.J., Jones, P.D., Lister, D.H., Morice, C.P., Simpson, I.R., et al., 2021. Land surface air

temperature variations across the globe updated to 2019: the CRUTEM5 dataset. *Journal of Geophysical Research: Atmospheres*, 126, e2019JD032352.

PAGES-2K Consortium, 2019. Consistent multi-decadal variability in global temperature reconstructions and simulations over the Common Era. *Nature Geoscience*, 12(8), 643–649.

Palmer, T., Stevens, B., 2019. The scientific challenge of understanding and estimating climate change. *Proceedings of the National Academy of Sciences*, 116(49), 24390–24395.

Paris Climate Agreement, 2016. Available at: https://unfccc.int/process-and-meetings/the-paris-agreement

Parker, D.E., Legg, T.P., Folland, C.K., 1992. A new daily Central England Temperature Series, 1772-1991. *International Journal of Climatology*, 12, 317–342.

Parker, D.E., 2006. A Demonstration That Large-Scale Warming Is Not Urban. *Journal of Climate*, 19(12), 2882–2895.

Parsons, T.E., Wu, P.-C., Wei, M., D'Hondt, S., 2023. The weight of New York City: possible contributions to subsidence from anthropogenic sources. *Earth's Future*, 11(5), e2022EF003465.

Patterson, R.T., Prokoph, A., Chang, A., 2004. Late Holocene sedimentary response to solar and cosmic ray activity influenced climate variability in the NE Pacific. *Sedimentary Geology*, 172, 67–84.

Penza, V., Berrilli, F., Bertello, L., Cantoresi, M., Criscuoli, S., Giobbi, P., 2022. Total solar irradiance during the last five centuries. *The Astrophysical Journal*, 937(2), 84.

Peterson, T.C., Gallo, K.P., Lawrimore, J., Owen, T.W., Huang, A., McKittrick, D., 1999. Global rural temperature trends. *Geophysical Research Letters*, 26(3), 329–332.

Peterson, T.C., Connolley, W.M., Fleck, J., 2008. The Myth of the 1970s Global Cooling Scientific Consensus. *Bulletin of the American Meteorological Society*, 89(9), 1325–1338.

Petit, J.R., Jouzel, J., Raynaud, D., Barkov, N.I., Barnola, J.-M., et al., 1999. Climate and atmospheric history of the past 420,000 years from the Vostok ice core. *Nature*, 399, 429–436.

Petit, J.R., et al., 2001. Vostok Ice Core Data for 420,000 Years. *IGBP PAGES/World Data Center for Paleoclimatology Data Contribution* Series #2001-076. NOAA/NGDC Paleoclimatology Program, Boulder CO, USA.

Peyrou-Lauga, R., 2017. Using real Earth albedo and Earth IR flux for spacecraft thermal analysis. Proceedings of the 47th International Conference on Environmental Systems (ICES), Charleston, South Carolina, USA.

Philipona, R., Mears, C., Fujiwara, M., Jeannet, P., Thorne, P., et al., 2018. Radiosondes show that after decades of cooling, the lower stratosphere is now warming. *Journal of Geophysical Research: Atmospheres*, 123(23), 12,509–12,522.

Piao, S., Wang, X., Park, T., Chen, C., Lian, X., et al., 2019. Characteristics, drivers and feedbacks of global greening. *Nature Reviews Earth & Environment*, 1(1), 14–27.

Pielke Jr., R., 2018. Normalized disaster losses in the United States: 1900–2017. *Environmental Hazards*, 17(2), 123–136.

Pielke Jr., R., 2019. Tracking progress on the economic costs of disasters under the indicators of the sustainable development goals. *Environmental Hazards*, 18(1), 1–6.

Pielke Jr., R., 2020. Tracking progress on disaster risk reduction. *Nature Sustainability*, 3(9), 706–709.

Pielke, R., Burgess, M.G., Ritchie, J., 2022. Plausible 2005–2050 emissions scenarios project between 2°C and 3°C of warming by 2100. *Environmental Research Letters*, 17, 024027.

Pierrehumbert, R.T., 2010. *Principles of Planetary Climate*. Cambridge University Press.

Pinho, O., Orgaz, M., 2000. The urban heat island in a small city in coastal Portugal. *International Journal of Biometeorology*, 44, 198–203.

Pitron, G., 2020. *The Rare Metals War: The dark side of clean energy and digital technologies*. Scribe Publications.

Plass, G.N., 1956. The carbon dioxide theory of climatic change. *Tellus*, 8(2), 140–154.

Plavcová, E., Lhotka, O., Beranová, R., Dubrovský, M., Skalák, P., 2025. Precipitation variability in CMIP6 climate models across the North Atlantic–European region. *Climate Dynamics*, 63, 98.

Pletser, V., 2017. Generalized Titius–Bode relations for the exoplanets of main sequence stars. *Astrophysics and Space Science*, 362(11), 203.

Prasad, S., Vos, H., Negendank, J.F.W., Waldmann, N., Goldstein, S.L., Stein, M., 2004, Evidence from Lake Lisan of solar influence on decadal- to centennial-scale climate variability during marine oxygen isotope stage 2. *Geology*, 32, 581.

Pulles, T., 2025. Climate Denialism. *American Journal of Economics and Sociology*, 84, 7–19.

Rae, J.W.B., Burke, A., Robinson, L.F., Adkins, J.F., Chen, T., et al., 2021. CO_2 over the past 66 million years from marine archives. *Annual Review of Earth and Planetary Sciences*, 49, 599–631.

Rahmstorf, S., 2008. Anthropogenic climate change: revisiting the facts. In: Zedillo, E. (Ed.), *Global Warming: Looking Beyond Kyoto*. Brookings Institution Press, Ch. 3, pp. 34–53.

Rampino, M.R., Caldeira, K., 2020. A 32-million year cycle detected in sea-level fluctuations over the last 545 Myr. *Geoscience Frontiers*, 11(6), 2061–2065.

Rao, S., Klimont, Z., Smith, S.J., Van Dingenen, R., Dentener, F., et al., 2017. Future air pollution in the Shared Socioeconomic Pathways. *Global Environmental Change*, 42, 346–358.

Rasool, S.I., Schneider, S.H., 1971. Atmospheric carbon dioxide and aerosols: effects of large increases on global climate. *Science*, 173, 138–141.

Reimer, P.J., Baillie, M.G.L., Bard, E., 2004. IntCal04 Terrestrial Radiocarbon Age Calibration, 0–26 Cal Kyr BP. *Radiocarbon*, 46, 1029–1058.

Rempel, M., 2020. On the contribution of quiet Sun magnetism to solar irradiance variations: Constraints from small-scale dynamo simulations. *The Astrophysical Journal*, 894(2), 140.

Rezaei, E.E., Webber, H., Asseng, S., Boote, K., Durand, J.L., et al., 2023. Climate change impacts on crop yields. *Nature Reviews Earth & Environment*, 4: 831–846.

Riahi, K., van Vuuren, D.P., Kriegler, E., Edmonds, J., O'Neill, B.C., et al., 2017. The Shared Socioeconomic Pathways and their energy, land use, and greenhouse gas emissions implications: An overview. *Global Environmental Change*, 42, 153–168.

Ripple, W. J., Wolf, C., Galetti, M., et al., 2017. World Scientists' Warning to Humanity: A Second Notice. *BioScience*, 67(12), 1026–1028.

Ritchie, J., Dowlatabadi, H., 2017. Why do climate change scenarios return to coal? *Energy*, 140, 1276–1291.

Robinson, T.D., Catling, D.C., 2014. Common 0.1 bar tropopause in thick atmospheres set by pressure-dependent infrared transparency. *Nature Geoscience*, 7(1), 12–15.

Rodrigo-Gámiz, M., Martínez-Ruiz, F., Rampen, S.W., Schouten, S. Sinninghe Damsté, J.S., 2014. Sea surface Temperature variations in the western Mediterranean Sea over the last 20 kyr: A dual-organic proxy (UK´37 and LDI) approach. *Paleoceanography*, 29, 87–98.

Rojas, J., Duprat, J., Engrand, C., Dartois, E., Delauche, L., et al., 2021. The micrometeorite flux at Dome C (Antarctica), monitoring the accretion of extraterrestrial dust on Earth. *Earth and Planetary Science Letters*, 560, 116794.

Royer, D.L., Berner, R.A., Beerling, D.J., 2001. Phanerozoic atmospheric CO_2 change: evaluating geochemical and paleobiological approaches. *Earth-Science Reviews*, 54(4), 349–392.

Royer, D.L., Berner, R.A., Montañez, I.P., Tabor, N.J., Beerling, D.J., 2004. CO_2 as a primary driver of Phanerozoic climate. *Nature*, 446, 530–532.

Rugenstein, M., Zelinka, M.D., Karnauskas, K.B., Ceppi, P., Andrews, T., 2023. Patterns of surface warming matter for climate sensitivity. *Eos*, 104.

Rugenstein, M., Hakuba, M., 2023. Connecting hemispheric asymmetries of planetary albedo and surface temperature. *Geophysical Research Letters*, 50(6), e2022GL101802.

Ruiz-Lara, T., Gallart, C., Bernard, E.J., Cassisi, S., 2020. The recurrent impact of the Sagittarius dwarf on the Milky Way star formation history. *Nature Astronomy*, 4(10), 965–973.

Salvador, R., 2013. A mathematical model of the sunspot cycle for the past 1000 yr. *Pattern Recognition in Physics*, 1, 117–122.

Santer, B., Fyfe, J., Pallotta, G., Flato, G.M., Meehl, G.A., et al., 2017. Causes of differences in model and satellite tropospheric warming rates. *Nature Geoscience*, 10, 478–485.

Santer, B.D., Po-Chedley, S., Zhao, L., Zou, C.-Z., Fu, Q., et al., 2023. Exceptional stratospheric contribution to human fingerprints on atmospheric temperature. *Proceedings of the National Academy of Sciences*, 120(20), e2300758120.

Scafetta, N., West, B.J., 2005. Estimated solar contribution to the global surface warming using the ACRIM TSI satellite composite. *Geophysical Research Letters*, 32, L18713.

Scafetta, N., West, B.J., 2006a. Phenomenological solar contribution to the 1900–2000 global surface warming. *Geophysical Research Letters*, 33, L05708.

Scafetta, N., West, B.J., 2006b. Phenomenological solar signature in 400 years of reconstructed Northern Hemisphere temperature record. *Geophysical Research Letters*, 33, L17718.

Scafetta, N., West, B.J., 2007. Phenomenological reconstructions of the solar signature in the Northern Hemisphere surface temperature records since 1600. *Journal of Geophysical Research*, 112, D24S03.

Scafetta, N., West, B.J., 2008. Is climate sensitive to solar variability? *Physics Today*, 61(3), 50–51.

Scafetta, N., 2009. Empirical analysis of the solar contribution to global mean air surface temperature change. *Journal of Atmospheric and Solar–Terrestrial Physics*, 71, 17–18.

Scafetta, N., Willson, R.C., 2009. ACRIM-gap and TSI trend issue resolved using a surface magnetic flux TSI proxy model. *Geophysical Research Letters*, 36, L05701.
Scafetta, N., 2010. Empirical evidence for a celestial origin of the climate oscillations and its implications. *Journal of Atmospheric and Solar–Terrestrial Physics*, 72, 951–970.
Scafetta, N., 2012a. Testing an astronomically based decadal-scale empirical harmonic climate model versus the IPCC (2007) general circulation climate models. *Journal of Atmospheric and Solar–Terrestrial Physics*, 80, 124–137.
Scafetta, N., 2012b. Multi-scale harmonic model for solar and climate cyclical variation throughout the Holocene based on Jupiter–Saturn tidal frequencies plus the 11-year solar dynamo cycle. *Journal of Atmospheric and Solar–Terrestrial Physics*, 80, 296–311.
Scafetta, N., 2012c. A shared frequency set between the historical mid-latitude aurora records and the global surface temperature. *Journal of Atmospheric and Solar–Terrestrial Physics*, 74, 145–163.
Scafetta, N., 2012d. Does the Sun work as a nuclear fusion amplifier of planetary tidal forcing? *Journal of Atmospheric and Solar–Terrestrial Physics*, 81–82, 27–40.
Scafetta, N., 2013a. Discussion on climate oscillations: CMIP5 general circulation models versus a semi-empirical harmonic model based on astronomical cycles. *Earth-Science Reviews*, 126, 321–357.
Scafetta, N., 2013b. Discussion on common errors in analyzing sea level accelerations, solar trends and global warming. *Pattern Recognition in Physics*, 1, 37–57.
Scafetta, N., Humlum, O., Solheim, J.-E., Stordahl, K., 2013. Comment on "The influence of planetary attractions on the solar tachocline" by Callebaut, de Jager and Duhau. *Journal of Atmospheric and Solar-Terrestrial Physics*, 102, 368–371.
Scafetta, N., Willson, R.C., 2013a. Planetary harmonics in the historical Hungarian aurora record (1523–1960). *Planetary and Space Science*, 78, 38–44.
Scafetta, N., Willson, R.C., 2013b. Empirical evidences for a planetary modulation of total solar irradiance and the TSI signature of the 1.09-year Earth–Jupiter conjunction cycle. *Astrophysics and Space Science*, 348, 25–39.
Scafetta, N., 2014a. Multi-scale dynamical analysis (MSDA) of sea level records versus PDO, AMO, and NAO indexes. *Climate Dynamics*, 43(1–2), 175–192.
Scafetta, N., 2014b. Discussion on the spectral coherence between planetary, solar and climate oscillations: a reply to some critiques. *Astrophysics and Space Science*, 354, 275–299.
Scafetta, N., 2014c. The complex planetary synchronization structure of the solar system. *Pattern Recognition in Physics*, 2, 1–19.
Scafetta, N., Willson, R.C., 2014. ACRIM total solar irradiance satellite composite validation versus TSI proxy models. *Astrophysics and Space Science*, 350, 421–442.
Scafetta, N., Mazzarella, A., 2015a. The Arctic and Antarctic sea-ice area index records versus measured and modeled temperature data. *Advances in Meteorology*, 481834, 1–8.
Scafetta, N., Mazzarella, A. 2015b. Spectral coherence between climate oscillations and the $M \geq 7$ earthquake historical worldwide record. *Natural Hazards*, 76, 1807–1829.
Scafetta, N., Milani, F., Bianchini, A., Ortolani, S., 2016. On the astronomical origin of the Hallstatt oscillation found in radiocarbon and climate records throughout the Holocene. *Earth-Science Reviews*, 162, 24–43.
Scafetta, N., 2016. High resolution coherence analysis between planetary and climate oscillations. *Advances in Space Research*, 57, 2121–2135.
Scafetta, N., 2018. Reply on Comment on "High resolution coherence analysis between planetary and climate oscillations" by S. Holm. *Advances in Space Research*, 62, 334–342.
Scafetta, N., Ouyang, S., 2019. Detection of UHI bias in China climate network using Tmin and Tmax surface temperature divergence. *Global and Planetary Change*, 181, 102989.
Scafetta, N., Willson, R.C., Lee, J.N., Wu, D.L., 2019a. Modeling quiet solar luminosity variability from TSI satellite measurements and proxy models during 1980–2018. *Remote Sensing*, 11(21), 2569.
Scafetta, N., Milani, F., Bianchini, A., 2019b. Multiscale analysis of the instantaneous eccentricity oscillations of the planets of the solar system from 13,000 BC to 17,000 AD. *Astronomy Letters*, 45, 778–789.
Scafetta, N., 2020. Solar oscillations and the orbital invariant inequalities of the solar system. *Solar Physics*, 295(2), 33.
Scafetta, N., Milani, F., Bianchini, A., 2020. A 60-year cycle in the meteorite fall frequency suggests a possible interplanetary dust forcing of the Earth's climate driven by planetary oscillations. *Geophysical Research Letters*, 47(18), e2020GL089954.

Scafetta, N., 2021a. Detection of non-climatic biases in land surface temperature records by comparing climatic data and their model simulations. *Climate Dynamics*, 56(9–10), 2959–2982.
Scafetta, N., 2021b. Reconstruction of the interannual to millennial scale patterns of the global surface temperature. *Atmosphere*, 12, 147.
Scafetta, N., 2021c. Testing the CMIP6 GCM simulations versus surface temperature records from 1980–1990 to 2011–2021: High ECS is not supported. *Climate*, 9, 161.
Scafetta, N., 2022. Advanced testing of low, medium, and high ECS CMIP6 GCM simulations versus ERA5-T2m. *Geophysical Research Letters*, 49(6), e2022GL097716.
Scafetta, N., Bianchini, A., 2022. The planetary theory of solar activity variability: a review. *Frontiers in Astronomy and Space Sciences*, 9, 937930.
Scafetta, N., 2023a. CMIP6 GCM validation based on ECS and TCR ranking for 21st-century temperature projections and risk assessment. *Atmosphere*, 14, 345.
Scafetta, N., 2023b. CMIP6 GCM ensemble members versus global surface temperatures. *Climate Dynamics*, 60, 3091–3120.
Scafetta, N., 2023c. Empirical assessment of the role of the Sun in climate change using balanced multi-proxy solar records. *Geoscience Frontiers*, 14(6), 101650.
Scafetta, N., 2023d. Reply to "Comment on 'Advanced Testing of Low, Medium, and High ECS CMIP6 GCM Simulations Versus ERA5-T2m' by Schmidt et al. (2023). *Geophysical Research Letters*, 50, e2023GL104960.
Scafetta, N., 2023e. Comment on "Tidally Synchronized Solar Dynamo: A Rebuttal" by Nataf (Solar Phys. 297, 107, 2022). *Solar Physics*, 298, 24.
Scafetta, N., Bianchini, A., 2023. Overview of the spectral coherence between planetary resonances and solar and climate oscillations. *Climate*, 11(4), 77.
Scafetta, N., 2024. Impacts and risks of "realistic" global warming projections for the 21st century. *Geoscience Frontiers*, 15(2), 101774.
Scafetta, N., Mörner, N.-A., 2025. The giant solar flare event of January 7, 2014 in light of the planetary theory of solar variability. *Science of Climate Change*, 5.2, 161–165. Also in: Mörner, N.-A. (Ed.) *Planetary Influence on the Sun and the Earth*. Nova Science Publishers (2015).
Scafetta, N., 2026. Detection, attribution, and modeling of climate change: Key open issues. *Gondwana Research*, 152, 92–128.
Scafetta, N., Milani, F., 2025. Spectral Structure of the Solar Inertial Motion from 12999 BC to 16998 AD. *Publications of the Astronomical Society of the Pacific*, 137, 054402.
Scafetta, N., Bianchini, A., 2026. Planetary modulation of solar and climate oscillations. *Habitable Planet*, 2(1), 46–62.
Scardigli, F., 2003. A quantum-like description of the planetary systems. *Foundations of Physics Letters*, 16(4), 393–398.
Schelling, T.C., 1992. Some economics of global warming. *American Economic Review*, 82(1), 1–14.
Schildknecht, D., 2020. Saturation of the infrared absorption by carbon dioxide in the atmosphere. *International Journal of Modern Physics B*, 34(30), 2050321.
Schilliger, L., Wild, M., Folini, D., Hakuba, M., 2024. An investigation on causes of the detected surface solar radiation brightening in Europe using satellite data. *Journal of Geophysical Research: Atmospheres*, 129(6), e2024JD041101.
Schlesinger, M., Ramankutty, N., 1994. An oscillation in the global climate system of period 65–70 years. *Nature*, 367, 723–726.
Schmidt, G.A., Jones, G.S., Kennedy, J.J., 2023. Comment on "Advanced Testing of Low, Medium, and High ECS CMIP6 GCM Simulations Versus ERA5-T2m" by N. Scafetta (2022). *Geophysical Research Letters*, 50, e2022GL102530.
Schmutz, W.K., 2021. Changes in the total solar irradiance and climatic effects. *Journal of Space Weather and Space Climate*, 11, 40.
Schoeberl, M.R., Wang, Y., Taha, G., Zawada, D.J., Ueyama, R., Dessler, A., 2024. Evolution of the climate forcing during the two years after the Hunga Tonga–Hunga Ha'apai eruption. *Journal of Geophysical Research: Atmospheres*, 129, e2024JD041296.
Schönwiese, C.-D., 1995. *Klimaänderungen: Daten, Analysen, Prognosen* (2nd ed.). Springer, Berlin.
Schötz, C., 2025. Spatial correlation in economic analysis of climate change. *Nature*, 644, E27–E30.
Scotese, C.R., Song, H., Mills, B.J.W., van der Meer, D.G., 2021. Phanerozoic paleotemperatures: The earth's changing climate. *Earth-Science Reviews*, 215, 103503.
Semenov, V. A., Aldonina, T. A., Li, F., Keenlyside, N. S., Wang, L., 2024. Arctic sea ice variations in the first half of the 20th century: A new reconstruction based on hydrometeorological data.

Advances in Atmospheric Sciences, 41, 1483–1495.
Shakun, J., Clark, P., He, F., Marcott, S.A., Mix, A.C., Liu, Z., et al., 2012. Global warming preceded by increasing carbon dioxide concentrations during the last deglaciation. *Nature*, 484, 49–54.
Shapiro, A. I., Schmutz, W., Rozanov, E., Schoell, M., Haberreiter, M., Shapiro, A.V., Nyeki, S., 2011. A new approach to the long-term reconstruction of the solar irradiance leads to large historical solar forcing. *Astronomy & Astrophysics*, 529, A67.
Shaviv, N.J., 2002. Cosmic ray diffusion from the galactic spiral arms, iron meteorites, and a possible climatic connection. *Physical Review Letters*, 89(5), 051102.
Shaviv, N.J., 2003. The spiral structure of the Milky Way, cosmic rays, and ice age epochs on Earth. *New Astronomy*, 8, 39–77.
Shaviv, N.J., Veizer, J., 2003. Celestial driver of Phanerozoic climate? *GSA Today*, 13(7), 4–10.
Shaviv, N.J., 2008. Using the oceans as a calorimeter to quantify the solar radiative forcing. *Journal of Geophysical Research: Space Physics*, 113, A11101.
Shaviv, N.J., Prokoph, A., Veizer, J., 2014. Is the Solar System's Galactic Motion Imprinted in the Phanerozoic Climate? *Scientific Reports*, 4, 6150.
Shaviv, N.J., Svensmark, H., Veizer, J., 2023. The Phanerozoic climate. *Annals of the New York Academy of Sciences*, 1519, 7–19.
Shellenberger, M., 2020. *Apocalypse Never: Why Environmental Alarmism Hurts Us All*. HarperCollins.
Shen, X., Liu, Y., Wang, H., 2023. Stable climate states and bifurcation thresholds. *Earth System Dynamics*, 14(1), 89–104.
Sherwood, S., Webb, M., Annan, J., Armour, K., Forster, P., et al., 2020. An assessment of Earth's climate sensitivity using multiple lines of evidence. *Reviews of Geophysics*, 58, e2019RG000678.
Sherwood, S.C., Forest, C.E., 2024. Opinion: Can uncertainty in climate sensitivity be narrowed further? *Atmospheric Chemistry and Physics*, 24, 2679–2688.
Shkolnik, E., Walker, G.A.H., Bohlender, D.A., 2003. Evidence for planet-induced chromospheric activity on HD 179949. *The Astrophysical Journal*, 556, 296–301.
Shkolnik, E., Walker, G.A.H., Bohlender, D.A., Gu, P.G., Kürster, M., 2005. Hot Jupiters and hot spots: the short- and long-term chromospheric activity on stars with giant planets. *The Astrophysical Journal*, 622, 1075–1090.
Sibony, Y., Helled, R., Feldmann, R., 2022. The rotation of planet-hosting stars. *Monthly Notices of the Royal Astronomical Society*, 513(2), 2057–2075.
Simpson, I.R., McKinnon, K.A., Kennedy, D., Lawrence, D.M., Lehner, F., Seager, R., 2024. Observed humidity trends in dry regions contradict climate models. *Proceedings of the National Academy of Sciences*, 121(1), e2302480120.
Smythe, C.M., Eddy, J.A., 1977. Planetary tides during the Maunder Minimum. *Nature*, 266, 434–435.
Solanki, S.K., Fligge, M., 1998. Solar irradiance since 1874 revisited. *Geophysical Research Letters*, 25(3), 341–344.
Song, H., Scotese, C.R., Mills, B.J.W., van der Meer, D.G., 2019. Phanerozoic climate change reflected in global marine oxygen-isotope data. *Gondwana Research*, 67, 172–187.
Soon, W., Posmentier, E., Baliunas, S., 2000. Climate hypersensitivity to solar forcing? *Annales Geophysicae*, 18(5), 583–588.
Soon, W., Baliunas, S., 2003. Proxy climatic and environmental changes of the past 1000 years. *Climate Research*, 23(2), 89–110.
Soon, W., Baliunas, S., Idso, C., Idso, S., Legates, D.R., 2003a. Reconstructing climatic and environmental changes of the past 1000 years: A reappraisal. *Energy and Environment*, 14(2–3), 233–296.
Soon, W., Legates, D.R., 2013. Solar irradiance modulation of equator-to-pole (Arctic) temperature gradients: Empirical evidence for climate variation on multi-decadal timescales. *Journal of Atmospheric and Solar-Terrestrial Physics*, 93, 45–56.
Soon, W., Connolly, R., Connolly, M., 2015. Re-evaluating the role of solar variability on Northern Hemisphere temperature trends since the 19th century. *Earth-Science Reviews*, 150, 409–452.
Soon, W., Connolly, R., Connolly, M., Akasofu, S.-I., Baliunas, S., Berglund, J., Bianchini, A., Briggs, W.M., Butler, C.J., Cionco, R.G., Crok, M., Elias, A. G., Fedorov, V.M., Gervais, F., Harde, H., Henry, G.W., Hoyt, D.V., Humlum, O., Legates, D.R., Lupo, A.R., Maruyama, S., Moore, P., Ogurtsov, M., ÓhAiseadha, C., Oliveira, M.J., Park, S.-S., Qiu, S., Quinn, G., Scafetta, N., Solheim, J.-E., Steele, J., Szarka, L., Tanaka, H.L., Taylor, M.K., Vahrenholt, F., Velasco Herrera, V.M., Zhang, W., 2023. The detection and attribution of Northern Hemisphere land surface warming (1850–2018) in terms of human and natural factors: Challenges of inadequate data. *Climate*, 11(9), 179.

Soon, W., and CERES team (2024, March 10). The challenges of trying to reconstruct changes in Total Solar Irradiance (TSI) since 1700: A response to Chatzistergos (2024)'s remake of Hoyt & Schatten (1993). CERES-Science. Available at: https://www.ceres-science.com/post/response-to-chatzistergos-2024

Spencer, R.W., 2012. USHCN Surface Temperatures, 1973-2012: Dramatic Warming Adjustments, Noisy Trends. Available at: https://www.drroyspencer.com/2012/04/ushcn-surface-temperatures-1973-2012-dramatic-warming-adjustments-noisy-trends/

Spencer, R.W., Christy, J.R., Braswell, W.D., 2017. UAH version 6 global satellite temperature products: Methodology and results. *Asia-Pacific Journal of Atmospheric Sciences*, 53(1), 121–130.

Spencer, R.W., 2021. Lower estimates of climate sensitivity from observational data: A review. *Remote Sensing*, 13(8), 1496.

Spencer, R.W., 2023. Epic fail in America's Heartland: climate models greatly overestimate Corn Belt warming. Available at: https://www.drroyspencer.com/2023/06/epic-fail-in-americas-heartland-climate-models-greatly-overestimate-corn-belt-warming/

Spencer, R.W., Christy, J.R., 2023. Effective climate sensitivity distributions from a 1D model of global ocean and land temperature trends, 1970–2021. *Theoretical and Applied Climatology*, 155, 299–308.

Spencer, R.W., Christy, J.R., Braswell, W.D., 2025. Urban heat island effects in U.S. summer surface temperature data, 1895–2023. *Journal of Applied Meteorology and Climatology*, 64(7), 717–728.

Squintu, A.A., van der Schrier, G., Štěpánek, P., Zahradníček, P., Tank, A.K., 2020. Comparison of homogenization methods for daily temperature series against an observation-based benchmark dataset. *Theoretical and Applied Climatology*, 140, 285–301.

Stefani, F., Giesecke, A., Weber, N., Weier, T., 2016. Synchronized helicity oscillations: A link between planetary tides and the solar cycle? *Solar Physics*, 291, 2197–2212.

Stefani, F., Giesecke, A., Weier, T., 2019. A model of a tidally synchronized solar dynamo. *Solar Physics*, 294, 60.

Stefani, F., 2021. Solar and anthropogenic influences on climate: Regression analysis and tentative predictions. *Climate*, 9, 163.

Stefani, F., Horstmann, G.M., Klevs, M., Mamatsashvili, G., Weier, T., 2024. Rieger, Schwabe, Suess-de Vries: the sunny beats of resonance. *Solar Physics*, 299, 51.

Stephens, G.L., O'Brien, D., Webster, P., Pilewski, P., Kato, S., Li, J.-L., 2015. The albedo of Earth. *Reviews of Geophysics*, 53(1), 141–163.

Steinhilber, F., Beer, J., Fröhlich, C., 2009. Total solar irradiance during the Holocene. *Geophysical Research Letters*, 36, L19704.

Steinhilber, F., Abreu, J.A., Beer, J., Brunner, I., Christl, M., et al., 2012. 9,400 years of cosmic radiation and solar activity from ice cores and tree rings. *Proceedings of the National Academy of Sciences*, 109(16), 5967–5971.

Stewart, I.D., Oke, T.R., 2012. Local climate zones for urban temperature studies. *Bulletin of the American Meteorological Society*, 93(12), 1879–1900.

Stix, M., 2003. On the time scale of energy transport in the Sun. *Solar Physics*, 212, 3–6.

Strogatz, S.H., 2009. Exploring complex networks. *Nature*, 410, 268.

Suess, H.E., 1965. Secular variations of the cosmic-ray-produced carbon-14 in the atmosphere and their interpretations. *Journal of Geophysical Research*, 70, 5937–5952.

Svensmark, H., Friis-Christensen, E., 1997. Variation of cosmic ray flux and global cloud coverage — a missing link in solar-climate relationships. *Journal of Atmospheric and Solar-Terrestrial Physics*, 59(11), 1225–1232.

Svensmark, H., 2007. Cosmoclimatology: a new theory emerges. *Astronomy & Geophysics*, 48(1), 1.18–1.24.

Svensmark, H., Friis-Christensen, E., 2007. Reply to Lockwood and Fröhlich – The persistent role of the Sun in climate forcing. *Danish National Space Center Scientific Report* 3/2007. ISBN: 978-87-91694-14-1

Svensmark, J., Enghoff, M.B., Shaviv, N.J., Svensmark, H., 2016. The response of clouds and aerosols to cosmic ray decreases. *Journal of Geophysical Research: Space Physics*, 121(9), 8152–8181.

Svensmark, H., 2019. *Force Majeure: The Sun's Role in Climate Change. The Global Warming Policy Foundation*. ISBN 978-0993119095. Available at: gwpf.org

Svensmark, H., 2022. Supernova rates and burial of organic matter. *Geophysical Research Letters*, 49(1), e2021GL096376.

Tang, Y., Poddar, S., Kay, M., Rougieux, F.E., 2026. Understanding and reducing the risk of

extreme photovoltaic degradation. *IEEE Journal of Photovoltaics*, 16(1), 150–159.
Tattersall, R., 2013. The Hum: log-normal distribution and planetary-solar resonance. *Pattern Recognition in Physics*, 1, 185–198.
Thoung, T., 2017. Why we should abandon the social cost of carbon. *Energy Policy*, 105, 512–520.
Thunberg, G., Ernman, M., Ernman, B., Thunberg, S., 2020. *Our house is on fire: Scenes of a family and a planet in crisis*. Penguin Books.
Tinsley, B.A., 2008. The global atmospheric electric circuit and its effects on cloud microphysics. *Pattern Recognition in Physics*, 71(6), 066801.
Tinsley, B.A., Zhou, L., 2014. Effects of global circuit downward current flow on atmospheric dynamics: modeling of relevant cloud microphysics. *Proceedings of the International Conference on Atmospheric Electricity* (ICAE 2014), Paper 282.
Titius, J.D., 1766. Anmerkungen über die Abstände der Planeten. In C. Bonnet, *Betrachtungen über die Natur* (Übers. J. D. Titius). Leipzig: Weidmanns Erben und Reich.
Tobias, S.M., 2002. The solar dynamo. *Philosophical Transactions of the Royal Society A: Mathematical, Physical and Engineering Sciences*, 360(1801), 2741–2756.
Tol, R.S.J., 2023. Costs and benefits of the Paris climate targets. *Climate Change Economics*, 14(3), 2340003.
Toohey, M., Sigl, M., 2017. Volcanic stratospheric sulfur injections and aerosol optical depth from 500 BCE to 1900 CE. *Earth System Science Data*, 9, 809–831.
Traverso, A., 2026. An engineering perspective on climate change: mitigation or adaptation? *Energy Nexus*, 21, 100646.
Tung, K.K., Camp, C.D., 2008. Solar cycle warming at the Earth's surface in NCEP and ERA-40 data: A linear discriminant analysis. *Journal of Geophysical Research: Atmospheres*, 113, D05114.
Tyndall, J., 1861. The Bakerian Lecture: On the Absorption and Radiation of Heat by Gases and Vapours, and on the Physical Connexion of Radiation, Absorption, and Conduction. *Philosophical Transactions of the Royal Society of London*, 151, 1–36.
Undorf, S., Pulkkinen, K., Wikman-Svahn, P., Bender, F.A.-M., 2022. How do value-judgments enter model-based assessments of climate sensitivity? *Climatic Change*, 174, 19.
Van der Meer, D.G., Stap, L., Scotese, C.R., Mills, B.J.W., Sluijs, A., van Hinsbergen, D.J.J., 2025. Phanerozoic orbital-scale glacio-eustatic variability. *Earth and Planetary Science Letters*, 667, 119526.
van Wijngaarden, W. A., Happer, W. H., 2020. Dependence of Earth's thermal radiation on five most abundant greenhouse gases. Available at: https://doi.org/10.48550/arXiv.2006.03098.
van Wijngaarden, W.A., Happer, W., 2023. Atmosphere and greenhouse gas primer. Available at: https://doi.org/10.48550/arXiv.2303.00808.
van Wijngaarden, W.A., Happer, W., 2025. Radiation transport in clouds. *Science of Climate Change*, 5(1), 1–12.
Vecchi, G., Knutson, T., 2011. Estimating annual numbers of Atlantic hurricanes missing from the HURDAT database (1878–1965) using ship track density. *Journal of Climate*, 24(6), 1736–1746.
Veizer, J., Godderis, Y., Francois, L.M., 2000. Evidence for decoupling of atmospheric CO_2 and global climate during the Phanerozoic eon. *Nature*, 408, 698–701.
Veretenenko, S.V., Ogurtsov, M.G., 2018. 60-year cycle in the Earth's climate and dynamics of correlation links between solar activity and circulation of the lower atmosphere. *Geomagnetism and Aeronomy*, 58, 973–981.
Vieira, L.E.A., Solanki, S.K., Krivova, N.A., Usoskin, I., 2011. Evolution of the solar irradiance during the Holocene. *Astronomy & Astrophysics*, 531, A6.
Vinós, J., 2017. *Nature Unbound III—Holocene Climate Variability* (Part B). Available at: https://judithcurry.com/2017/05/28/nature-unbound-iii-holocene-climate-variability-part-b/
von Storch, H., Zorita, E., Jones, J.M., Dimitriev, Y., González-Rouco, F., Tett, S.F.B., 2004. Reconstructing past climate from noisy data. *Science*, 306(5696), 679–682.
Voortman, H. G., De Vos, R., 2025. A global perspective on local sea level changes. *Journal of Marine Science and Engineering*, 13(9), 1641.
Vos, H., Brüchmann, C., Lücke, A., Negendank, J.F.W., Schleser, G.H., Zolitschka, B., 2004. Phase stability of the solar Schwabe cycle in Lake Holzmaar, Germany, and GISP2, Greenland, between 10,000 and 9,000 cal. BP. In: Fischer, H., Kumke, T., Lohmann, G., Flöser, G., Miller, H., von Storch, H., et al. (Eds.), *The Climate in Historical Times: Towards a Synthesis of Holocene Proxy Data and Climate Models*. Springer, Berlin, pp. 293–317.
Wanner, H., Mercolli, L., Grosjean, M., Ritz, S.P., 2015. Holocene climate variability and change: a

data-based review. *Journal of the Geological Society*, 172(2), 254–263.

Webster, P.J., Holland, G.J., Curry, J.A., Chang, H.-R., 2005. Changes in tropical cyclone number, duration, and intensity in a warming environment. *Science*, 309(5742), 1844–1846.

Wagner, G., Beer, J., Masarik, J., Muscheler, R., Kubik, P.W., Mende, W., et al., 2001. Presence of the solar de Vries cycle (205 years) during the last ice age. *Geophysical Research Letters*, 28(15), 303–306.

Wang, Y.-M., Lean, J.L., Sheeley, N.R. Jr., 2005. Modeling the Sun's magnetic field and irradiance since 1713. The Astrophysical Journal. 625, 522–538.

Watts, A., 2022. *Corrupted climate stations: The official U.S. temperature record remains fatally flawed*. The Heartland Institute. Available at: heartland.org(accessed April 10, 2025).

Weiss, N.O., Tobias, S.M., 2016. Supermodulation of the Sun's magnetic activity: the effects of symmetry changes. *Monthly Notices of the Royal Astronomical Society*, 456, 2654–2661.

WCRP Coupled Model Intercomparison Project (CMIP), 2024. Available at: https://wcrp-cmip.org/.

Westerhold, T., Marwan, N., Drury, A.J., Liebrand, D., Agnini, C., et al., 2020. An astronomically dated record of Earth's climate and its predictability over the last 66 million years. *Science*, 369(6509), 1383–1387.

Westhoff, J., Sinnl, G., Svensson, A., Freitag, J., Kjær, H.A., et al., 2022. Melt in the Greenland EastGRIP ice core reveals Holocene warm events. *Climate of the Past*, 18, 1011–1034.

Willson, R.C., Mordvinov, A.V., 2003. Secular total solar irradiance trend during solar cycles 21–23. *Geophysical Research Letters*, 30(5), 1199.

White, S., Pfister, C., Mauelshagen, F. (Eds.), 2018. *The Palgrave Handbook of Climate History*. Palgrave Macmillan. Available at: springer.com.

Wickham, C., Rohde, R., Muller, R.A., Wurtele, J., Curry, J., et al., 2013. Influence of Urban Heating on the Global Temperature Land Average using Rural Sites Identified from MODIS Classifications. *Geoinformatics & Geostatistics: An Overview*, 1(2), doi:10.4172/2327-4581.1000104.

Wilson, I.R.G., Carter, B.D., Waite, I.A., 2008. Does a spin–orbit coupling between the Sun and the Jovian planets govern the solar cycle? *Publications of the Astronomical Society of Australia*, 25, 85–93.

Wilson, I.R.G., 2013. The Venus–Earth–Jupiter spin-orbit coupling model. *Pattern Recognition in Physics*, 1, 147–158.

Wolf, R., 1859. Extract of a letter from Prof. R. Wolf of Zurich to Mr. Carrington, dated Jan. 12, 1859. *Monthly Notices of the Royal Astronomical Society*, 19, 85.

Wolff, C.L., Mayr, H.G., 2004. The Sun's reversing flows and heat spike as caused by g-modes. *Astrophysical Journal Letters*, 606(2), L163–L166.

Wolff, C.L., Patrone, P.N., 2010. A new way that planets can affect the Sun. *Solar Physics*, 266, 227–246.

Wood, K.D., 1972. Physical Sciences: Sunspots and Planets. *Nature*, 240, 91–93.

Woods, T.N., Leibacher, J.W., 2023. The Solar Radiation and Climate Experiment (SORCE) mission: Final calibrations and data products. *Solar Physics*, 298, 25.

Wu, C.-J., Krivova, N.A., Solanki, S.K., Usoskin, I.G., 2018. Solar total and spectral irradiance reconstruction over the last 9000 years. *Astronomy and Astrophysics*, 620, A120.

Wyatt, M.G., Curry, J.A., 2014. Role of Eurasian Arctic shelf sea ice in a secularly varying hemispheric climate signal during the 20th century. *Climate Dynamics*, 42, 2763–2782.

Xiao, Z., Zhao, L., Zhou, L., Huo, W., Mironova, I.A., Miyahara, H., 2024. Editorial: Impact of solar activities on weather and climate. *Frontiers in Earth Science*, 11, 1338416.

Yamaguchi, Y.T., Yokoyama, Y., Miyahara, H., Sho, K., Nakatsuka, T., 2010. Synchronized Northern Hemisphere climate change and solar magnetic cycles during the Maunder Minimum. *Proceedings of the National Academy of Sciences*, 107, 20697–20702.

Yang, B., Wang, J., Bräuning, A., Dong, Z., Esper, J., 2009. Late Holocene climatic and environmental changes in arid central Asia. *Quaternary International*, 194(1–2), 68–78.

Yeo, K.L., Krivova, N.A., Solanki, S.K., Glassmeier, K.H., 2014. Reconstruction of total and spectral solar irradiance from 1974 to 2013 based on KPVT, SoHO/MDI and SDO/HMI observations. *Astronomy and Astrophysics*, 570, A85.

Yeo, K.L., Solanki, S.K., Krivova, N.A., Rempel, M., Anusha, L.S., et al., 2020. The dimmest state of the Sun. *Geophysical Research Letters*, 47(19), e2020GL090243.

Yndestad, H., 2022. Lunar-driven Control of Climate and Barents Sea Ecosystems. *Science of Climate Change*, 2(1), 74–77.

Yu, Z., Chang, S., Kumazawa, M., Furumoto, M., Yamamoto, A., 1983. Presence of periodicity in meteorite falls. *National Institute of Polar Research Memoirs*, Special Issue, 30, 362–366.

Zachos, J., Pagani, M., Sloan, L., Thomas, E., Billups, K., 2001. Trends, rhythms, and aberrations in global climate 65 Ma to present. *Science*, 292, 686–693.

Zachos, J.C., Dickens, G.R., Zeebe, R.E., 2008. An early Cenozoic perspective on greenhouse warming and carbon-cycle dynamics. *Nature*, 451, 279–283.

Zanchettin, D., Timmreck, C., Khodri, M., Schmidt, A., Toohey, M., et al., 2022. Effects of forcing differences and initial conditions on inter-model agreement in the VolMIP volc-pinatubo-full experiment. *Geoscientific Model Development*, 15, 2265–2292.

Zander, P.D., Böhl, D., Sirocko, F., Auderset, A., Haug, G.H., Martínez-García, A., 2024. Reconstruction of warm-season temperatures in central Europe during the past 60,000 years from lacustrine branched glycerol dialkyl glycerol tetraethers (brGDGTs). *Climate of the Past*, 20, 841–864.

Zhang, Q., Soon, W.H., Baliunas, S.L., Lockwood, G.W., Skiff, B.A., Radick, R.R., 1994. A method of determining possible brightness variations of the Sun in past centuries from observations of solar-type stars. *The Astrophysical Journal*, 427, L111–L114.

Zhang, X., Evans, J.P., Burrell, A. L., 2024a. Less than 4% of dryland areas are projected to desertify despite increased aridity under climate change. *Communications Earth & Environment*, 5(1), 1–9.

Zhang, J., Xue, W., Megner, L., Plane, J.M.C., Rapp, M., 2024b. Dominant role of charged meteoric smoke particles in mesospheric ice nucleation. EGUsphere: https://doi.org/10.5194/egusphere-2024-1259

Zhang, B., Song, S., Wang, H., Guo, T., Ding, Y., 2025a. Evaluation of the performance of CMIP6 models in simulating extreme precipitation and its projected changes in global climate regions. *Natural Hazards*, 121, 1737–1763.

Zhang, L., Lü, T., Xue, L., Mo, W., Wang, C., Zhao, X., Hu, D., 2025b. An Overview of the Holocene High Sea Level Around the South China Sea: Age, Height, and Mechanisms. *Atmosphere*, 16(8), 993.

Zhao, C., Liu, B., Piao, S., Wang, X., Lobell, D.B., et al., 2017. Temperature increase reduces global yields of major crops in four independent estimates. *Proceedings of the National Academy of Sciences*, 114(35), 9326–9331.

Zhao, Z., Deng, Y., Zhang, J., 2018. Temperature and economic performance in developing countries. *World Development*, 105, 314–326.

Zhu, Z., Piao, S., Myneni, R.B., Huang, M., Zeng, Z., et al., 2016. Greening of the Earth and its drivers. *Nature Climate Change*, 6(8), 791–795.

Ziskin, S., Shaviv, N.J., 2012. Quantifying the role of solar radiative forcing over the 20th century. *Advances in Space Research*, 50(6), 762–776.

Zou, C.-Z., Xu, H., Hao, X., Liu, Q., 2023. Mid-tropospheric layer temperature record derived from satellite microwave sounder observations with backward merging approach. *Journal of Geophysical Research: Atmospheres*, 128, e2022JD037472.

Published Books by the Author (English and Italian Editions)

West, B.J., Scafetta, N., 2010. *Disrupted Networks: from Physics to Climate Change*. World Scientific Publishing Company. ISBN: 9789814304306.

Scafetta, N., 2010. *Fractal and Diffusion Entropy Analysis of Time Series: Theory, concepts, applications and computer codes for studying fractal noises and Lévy walk signals*. VDM Verlag Dr. Müller, ISBN: 9783639257953.

Scafetta, N., 2010. *Climate Change and Its Causes, A Discussion About Some Key Issues*. Science & Public Policy Institute. Available at: https://web.archive.org/web/20100306143828/http://scienceandpublicpolicy.org/originals/climate_change_causes.html

Battaglia, F., Crescenti, U., Giaccio, M., Mariani, L., Miccadei, E., Scafetta, N., 2019. *Clima, basta catastrofismi. Riflessioni scientifiche su passato e futuro*. 21mo Secolo. ISBN: 9788887731712.

Scafetta, N., 2020. *Napoli: la Città del Sole e di Partenope: Il ruolo dell'astronomia, della mitologia e di Pitagora nella pianificazione urbana di Neapolis*. Borè-Youcanprint. ISBN: 9788831657570.

Fortelli, A., Mazzarella, A., Scafetta, N., Toti, M., 2021. *La neve a Napoli: Immagini ed emozioni*. Aracne. ISBN: 9788825540321.

Aletta, N., Scafetta, N., 2021. *Gaeta: l'Abbazìa del Tirreno*. Reprint by Borè-Youcanprint. ISBN: 9791220343237.

Aletta, N., Scafetta, N., 2022. *Gaeta: guida storico-artistico-archeologica*. Reprint by Borè-Youcanprint. ISBN: 9791221415391.

Scafetta, N., 2024. *Naples: the City of the Sun and Parthenope: The role of astronomy, mythology, and Pythagoras in the urban planning of Neapolis*. FedOA – Federico II University Press, Naples; Cavoliniana: monographs of the Society of Naturalists in Naples. ISBN: 9791222757278.

Battaglia, F., Berkhout, G., Scafetta, N., 2025. *Clima, lasciamo parlare i dati: Una revisione critica dell'impatto delle emissioni di gas-serra sul clima*. 21mo Secolo. ISBN: 9788887731903.

Scafetta, N., 2026. *The Frontier of Climate Science: Solar variability, natural cycles and model uncertainty*. Complexity Insight Press. (This Volume).

Chapters in edited volumes:

Vahrenholt, F., Lüning, S., 2013. *The Neglected Sun: How the Sun Precludes Climate Catastrophe*. Stacey International. ISBN: 9781909022249

Mörner, N.-A., (editor), 2015. *Planetary Influence on the Sun and the Earth, and a Modern Book-Burning*. Nova Science Pub Inc. ISBN: 9781634828376.

Marohasy, J., (editor), 2017. *Climate Change: The Facts 2017*. Institute of Public Affairs. ISBN: 978-0909536800.

Ceradelli, G., Mastrangelo, L., 2020. *Cambiamento climatico, Covid-19 e finanze*. Key Editore. ISBN: 978-8827906941.

Prestininzi, A., (editor), 2022. *Dialoghi sul clima. Tra emergenza e conoscenza*. Rubbettino. ISBN: 9788849871746.

Crok, M., May, A., (editors), 2023. *The Frozen Climate Views of the IPCC: An Analysis of AR6*. Andy May Petrophysicist. ISBN: 9798890748614.

West, B.J., Bianco, S., (editors), 2023. *Paolo Grigolini and 50 Years of Statistical Physics*. Cambridge Scholars Publishing. ISBN: 9781527502222.

Porro, N. (editor), 2024. *La grande bugia verde. Gli scienziati smontano, con dati reali, i dogmi dell'allarmismo climatico*. Liberilibri. ISBN: 9791280447449.

Beisner, E.C., Legates, D.R., (Editors), 2024. *Climate and Energy: The Case for Realism*. Regnery. ISBN: 9781684512676.

Acknowledgments

I sincerely thank Prof. M. Santosh,[5] Prof. Alberto Prestininzi,[6] and Prof. Judith Curry[7] for their kind contributions to this volume. Their forewords offer unique and thoughtful views from different areas of Earth Science, adding real depth to the work. I am truly grateful for their support and encouraged by their strong dedication to open and honest discussion in climate science.

---

[5]Prof. M. Santosh is a globally renowned geoscientist specializing in petrology, geochemistry, geochronology, metallogeny, and supercontinent tectonics. He currently serves as Professor at the China University of Geosciences in Beijing and as Emeritus Professor at Kochi University in Japan. He previously worked as Senior Scientist at the Centre for Earth Science Studies (India) for 18 years and was Full Professor at Kochi University from 2000 to 2012. He is the Founding Editor of the journals *"Gondwana Research"* and *"Habitable Planet,"* and the Founding Secretary General of the *"International Association for Gondwana Research"* (IAGR). He also serves as Editorial Advisor for *"Geoscience Frontiers,"* Advisor for *"Geosystems and Geoenvironment,"* and Associate Editor for *"Ore Geology Reviews"* and *"Geological Journal."* Prof. Santosh has authored over 1600 peer-reviewed research papers (Scopus), edited numerous volumes and special issues, and co-authored the influential book *"Continents and Supercontinents"* (Oxford University Press, 2004). He has received several prestigious awards, including the National Mineral Award, Outstanding Geologist Award, Thomson Reuters Research Front Award, and Global Talent Award. He has also been recognized as a Highly Cited Researcher by Clarivate for eight consecutive years. According to the 2025 Stanford–Elsevier ranking of the *"World's Top 2% Scientists,"* Prof. Santosh is ranked #1 in the subfield of Geology.

[6]Prof. Alberto Prestininzi is a renowned Italian geologist and (retired) Full Professor of Applied Geology at Sapienza University of Rome and teaches *Risk Analysis* at the University eCampus, Civil Engineering. He is internationally recognized for his expertise in hydrogeological risk, slope instability, and geotechnical hazards. He founded and directed the Research center on Prediction, Prevention and Control of Geological Risks (*"Centro di Ricerca Previsione, Prevenzione e Controllo dei Rischi Geologici"* — CERI) of Sapienza University of Rome. He has been Chair of the National Commission for Major Risks of the Italian Civil Protection Department and Secretary of the Italian Section of the *"International Association for Engineering Geology"* (IAEG), and also founded and served as editor-in-chief of the *"Italian Journal of Engineering Geology and Environment."* In 2023, he was appointed coordinator of the scientific committee overseeing the technical update of the *"Strait of Messina Bridge"* project, contributing his expertise in geological stability and seismic risk. His work combines scientific rigor with public outreach, and he has played a key role in shaping Italy's approach to geological risk management.

[7]Prof. Judith Curry is an American climatologist and former Chair of the School of Earth and Atmospheric Sciences at Georgia Tech. She is known for her research on atmospheric dynamics, climate modeling, and hurricane variability. Curry has contributed extensively to the scientific debate on climate uncertainty and risk assessment, and she is Co-founder and President of the climate consulting firm *"Climate Forecast Applications Network"* (CFAN, https://www.cfanclimate.net/). Her work emphasizes the complexity of climate systems and the importance of open scientific dialogue. Curry is one of the five contributing authors of the US Department of Energy (DOE) report *"A Critical Review of Impacts of Greenhouse Gas Emissions on the U.S. Climate"* (https://www.energy.gov/topics/climate; July 29, 2025). The report has renewed debate over rescinding the Endangerment Finding and its vehicle emission standards, potentially marking a pivotal shift in U.S. climate regulation.

I wish to express my sincere gratitude to the *International Association for Gondwana Research* (IAGR, https://iagrglobal.org/), to its Secretary General, Prof. M. Santosh, and to the designated members of the IAGR Scientific Committee for endorsing *The Frontier of Climate Science* and for authorizing the use of the official IAGR logo in this publication.

I also wish to express my sincere gratitude to the *"Centro di Ricerca Previsione, Prevenzione e Controllo dei Rischi Geologici"* (*"Centre for Research on Prediction, Prevention, and Mitigation of Geological Risks"*) — CERI, Sapienza University of Rome (https://ceri.web.uniroma1.it/) — and to its Director, Prof. Gabriele Scarascia Mugnozza, for endorsing *The Frontier of Climate Science* and for authorizing the use of the official CERI logo in this publication.

The recognition of the book by IAGR and CERI as a meaningful contribution to geoethical and interdisciplinary research is deeply appreciated.

Finally, I am also grateful to Prof. Antonio Bianchini[8] and Prof. Sergio Ortolani[9] for their careful reading of the manuscript and for their valuable scientific insights, which greatly contributed to enhancing its clarity and style.

---

[8]Prof. Antonio Bianchini is a (retired) Associate Professor of Astronomy at the University of Padua and is currently affiliated with the National Institute for Astrophysics (INAF) – Astronomical Observatory of Padua. His research has primarily focused on interacting binary stars and cataclysmic variables. In particular, he investigated how magnetic activity cycles in the solar-type components of cataclysmic variables shed light on both their variability and long-term evolution. In recent years, he has also extended his scientific interests to climatology. He has authored over 190 publications in stellar astrophysics, observational astronomy, and climate-related studies (NASA/ADS).

[9]Prof. Sergio Ortolani is a (retired) Full Professor of Astronomy at the University of Padua. His research has focused on stellar populations, galactic structure, and photometric surveys. With over 500 publications indexed in NASA/ADS, he has made significant contributions to the study of the Milky Way and the dynamics of globular clusters, often in collaboration with major international observatories and space missions. He is also recognized as an expert in astronomical site selection and atmospheric science.

The author

Nicola Scafetta earned his Physics degree from the University of Pisa and, in 2001, a Ph.D. in Physics at the University of North Texas (USA), specializing in Statistical Physics and Complex Systems. From 1999 to 2014, he conducted research and taught in several U.S. institutions — including the University of North Texas, Duke University, UNC Chapel Hill, UNC Greensboro, and Elon University — offering courses in Physics and Astronomy.

He was co-investigator of NASA-JPL's ACRIM team, analyzing satellite measurements of Total Solar Irradiance (TSI) and the Sun's role in Earth's climate.

Since 2014, he has been Associate Professor of Oceanography, Meteorology, and Climatology at the University of Naples Federico II, and in 2018 he was appointed Laboratory Teaching and Research Supervisor (RaDOR) of the Federico II Meteorological Observatory.

His research spans climatology, solar physics, meteorology, oceanography, astronomy, archaeology, and environmental medicine, with a focus on statistical methodologies for complex signals and astronomical models of climate variability. He has also worked on predictive modeling and environmental risk assessment of phenomena such as earthquakes, floods, and air pollution.

Over the course of his career, he has authored more than 130 peer-reviewed scientific papers and several books, and he has delivered over 200 scientific presentations at academic conferences as well as to the general public. Since 2017, he has been ranked among the prestigious *"World's Top 2% Scientists,"* based on data from Stanford University and Elsevier.

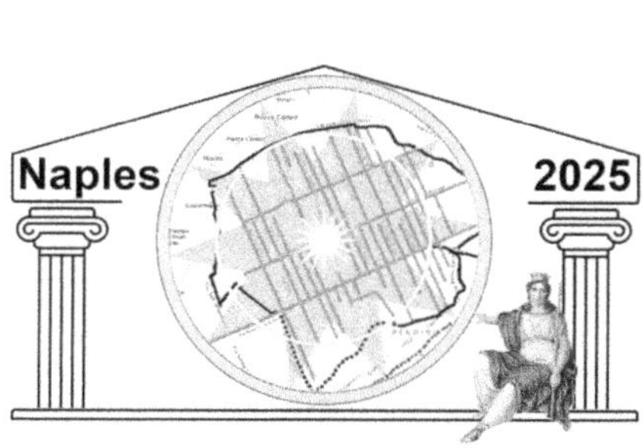

2500 years from the founding of Neapolis

Books by the Author

Bruce J. West & Nicola Scafetta
"Disrupted Networks: From Physics To Climate Change".

The book argues that modern society relies on complex networks for its stability and that understanding these systems requires a new, cross-disciplinary science of networks. Written in non-mathematical language, it reviews how science has evolved in the twenty-first century, shifting from the twentieth-century model of experiment + theory (data and knowledge) to a three-tiered structure of experiment + computation + theory (data, information, and knowledge). This new paradigm affects fields ranging from psychophysics to climate science.

ISBN : 9789814304306

Nicola Scafetta
"Fractal and Diffusion Entropy Analysis of Time Series: Theory, concepts, applications and computer codes for studying fractal noises and Lévy walk signals".

The book examines how scale invariance appears in many complex systems and why accurately determining the scaling exponents of a time series is essential for understanding the underlying physical processes. Traditional methods for estimating these exponents all rely on the variance, but this creates ambiguity because complex phenomena fall into two distinct statistical classes: fractal Brownian motion and Lévy flights/walks. To address these limitations, the book introduces alternative fractal analysis techniques, focusing on a complementary entropy-based method called Diffusion Entropy Analysis (DEA). Using synthetic data as well as solar, geophysical, sociological, physiological, and biological time series, the author compares these methods and highlights the conceptual problems inherent in variance-based approaches.

ISBN : 9783639257953

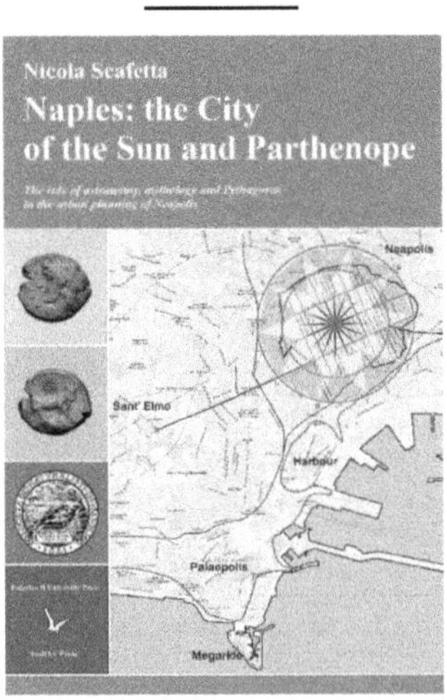

Nicola Scafetta
"Naples – the City of the Sun and Parthenope: The role of astronomy, mythology, and Pythagoras in the urban planning of Neapolis".

The book uncovers the deepest symbolic meaning of ancient Naples through an archaeoastronomical analysis of the religious and philosophical ideas that guided the foundation of Neapolis in the 6th–5th centuries BC. It argues that the city's street plan and its precise geographical–astronomical orientation at solstices and equinoxes were intentionally designed to evoke the cults of Apollo, the Greek Sun god, and Parthenope, the protective deity linked to the Cumaean Sibyl. Drawing on Pythagorean cosmology, the layout of Neapolis appears to follow the harmonic proportions of the golden ratio and the symbolic number ten, representing the ordered structure of the cosmos. The city was therefore conceived as a microcosm—a sacred "city-temple" with the divine Sun at its centre and Parthenope as its oracular guardian. The essay shows how the decumani and cardines, as well as Naples' cultural traditions, churches, cathedral, symbols, chants, sweets, mosaics, and relics, still preserve the memory of this ancient solar and Apollonian heritage embedded in the original Greek city.

ISBN : 9791222757278

www.ingramcontent.com/pod-product-compliance
Ingram Content Group UK Ltd.
Pitfield, Milton Keynes, MK11 3LW, UK
UKHW022033041026
15591UKWH00012B/660

9 791298 617605